Microorganisms for Sustainability

Volume 49

Series Editor

Naveen Kumar Arora, Environmental Microbiology, School for Environmental Science, Babasaheb Bhimrao Ambedkar University, Lucknow, Uttar Pradesh, India

This multidisciplinary book series covers the wide aspects of microbes and unravels the role of microbes towards achieving a sustainable world. It focuses on various microbial technologies related to sustenance of ecosystems and achieving targets of Sustainable Development Goals (SDGs). Series brings together content on microbe-based technologies for replacing harmful chemicals in agriculture, green alternatives to fossil fuels, use of microorganisms for reclamation of wastelands & stress affected regions, bioremediation of contaminated habitats, biodegradation purposes etc.

Volumes in 'Microorganisms for Sustainability' offer readers an in-depth understanding on a range of themes with respect to the role and use of microbes for various agro-ecological and industrial purposes including enzymes for pollutant degradation, extremophilic microbes, microbial biotechnology for agriculture crops, effluent treatment, role of microbes in food products and applications in bioenergy.

The series brings forth chapters and volumes from all parts of the world, at one place for its readers, thereby consolidating the international and latest trends about the crucial role of microbes in sustaining the ecosystems.

This series is a comprehensive and updated collection of thematically diverse volumes providing insights to leading scientists, young researchers, educators and decision-makers in government, the private sector, and non-governmental organizations in their efforts to achieve the global goals.

The book series is a single- blinded peer reviewed compendium focused on bringing up contemporary themes. It operates on editorial peer review system. Each volume is reviewed by editorial board members and 1-2 external independent reviewers. Each chapter is reviewed by the volume editors and 2-3 external reviewers.

The series accepts proposals and suggestions for future volumes. However, the final decision of acceptance is up to the series editor.

Kubilay Kurtulus Bastas • Ajay Kumar •
U. Sivakumar
Editors

Microbial Biocontrol: Molecular Perspective in Plant Disease Management

Springer

Editors
Kubilay Kurtulus Bastas
Department of Plant Protection
Selcuk University
Konya, Türkiye

Ajay Kumar
Amity Institute of Biotechnology
Amity University
Noida, India

U. Sivakumar
Department of Agricultural Microbiology
Tamil Nadu Agricultural University
Coimbatore, India

ISSN 2512-1901 ISSN 2512-1898 (electronic)
Microorganisms for Sustainability
ISBN 978-981-99-3946-6 ISBN 978-981-99-3947-3 (eBook)
https://doi.org/10.1007/978-981-99-3947-3

This Springer imprint is published by the registered company Springer Nature Singapore Pte Ltd.
The registered company address is: 152 Beach Road, #21-01/04 Gateway East, Singapore 189721, Singapore

Contents

Editors and Contributors

About the Editors

Kubilay Kurtulus Bastas is currently a full professor of plant pathology at Selcuk University, Turkey. Dr. Bastas is a plant protection graduate from Faculty of Agriculture, Ankara University. He earned his doctoral degrees in plant pathology-bacteriology from Selcuk University and earned a postdoctoral degree in molecular phytopathology from Warwick University, UK. During his 26 years as a faculty member, Dr. Bastas has received several awards and fellowships from national (TUBITAK, TMMOB, and YOK) and international (British Council and ERASMUS) foundations.

Ajay Kumar is Assistant Professor of Agricultural biotechnology/microbiology in Amity Institute of Biotechnology, Amity University, Noida-201313, India, and his research area includes microbiome, postharvest management, microbial biocontrol, plant–microbe interactions, cyanobacterial biology and microbial endophytes. Dr. Kumar completed his doctoral research from the Department of Botany, Banaras Hindu University, Varanasi, India, and Postdoc from the Agricultural Research Organization, Volcani Center, Rishon Leziyon, Israel.

Sivakumar Uthandi is Professor of Agricultural Microbiology in Tamil Nadu Agricultural University (TNAU), Coimbatore, India, received doctorate degrees from TNAU and gained postdoctoral experience at the University of Florida, USA. He has been visiting Associate Professor at the University of Florida, USA; and also visited Wageningen University, The Netherlands; BIC, Russian Academy of Sciences, Novosibirsk, Russia; and the University of Warwick, UK; AVRDC, Taiwan; and the University of Jaffna, Sri Lanka. He is now serving as Visiting Adjunct Professor at the University of Tokyo, Japan.

Contributors

Dharmaraj Amalamol ICAR-Sugarcane Breeding Institute, Indian Council of Agricultural Research, Coimbatore, India

R. Anandham Department of Agricultural Microbiology, Tamil Nadu Agricultural University (TNAU), Coimbatore, Tamil Nadu, India

Nivetha Andichamy Agricultural College and Research Institute, Killikulam, Tuticurin, India

P. Arul Jose Department of Agricultural Microbiology, TNAU, Madurai, Tamil Nadu, India

N. M. R. Ashwin ICAR-Sugarcane Breeding Institute, Indian Council of Agricultural Research, Coimbatore, India

Yesim Aysan Department of Plant Protection, Faculty of Agriculture, Cukurova University, Adana, Turkey

Kubilay Kurtulus Bastas Department of Plant Protection, Faculty of Agriculture, Selcuk University, Konya, Turkey

Ömür Baysal Molecular Microbiology Unit, Department of Molecular Biology and Genetics, Faculty of Science, Muğla Stkı Koçman University, Muğla, Turkey

Haris Butt Department of Plant Protection, Faculty of Agriculture, Selcuk University, Konya, Turkey

Zhicong Dai Institute of Environment and Ecology, Academy of Environmental Health and Ecological Security, Jiangsu University, Zhenjiang, P.R. China
School of the Environment and Safety Engineering, Jiangsu University, Zhenjiang, P.R. China

Daolin Du Institute of Environment and Ecology, Academy of Environmental Health and Ecological Security, Jiangsu University, Zhenjiang, P.R. China
School of the Environment and Safety Engineering, Jiangsu University, Zhenjiang, P.R. China

N. O. Gopal Department of Agricultural Microbiology, Tamil Nadu Agricultural University (TNAU), Coimbatore, Tamil Nadu, India

Sabarinathan Kuttalingam Gopalasubramanian Agricultural College and Research Institute, Killikulam, Tuticurin, India

S. Harish Department of Plant Pathology, Tamil Nadu Agricultural University, Coimbatore, India

Sumer Horuz Department of Plant Protection, Faculty of Agriculture, Erciyes University, Kayseri, Turkey

Sridharan Jegadeesan Faculty of Life Sciences, School of Plant Sciences and Food Security, Tel Aviv University, Tel Aviv, Israel

I. Johnson Department of Plant Pathology, TNAU, Coimbatore, Tamil Nadu, India

J. Beslin Joshi Department of Plant Biotechnology, Centre for Plant Molecular Biology and Biotechnology, Tamil Nadu Agricultural University, Coimbatore, India

R. Kannan Department of Plant Pathology, Tamil Nadu Agricultural University, Coimbatore, Tamil Nadu, India

Pervin Kinay-Teksur Department of Plant Protection, Faculty of Agriculture, Ege University, Izmir, Turkiye

R. Krishnamoorthy Department of Agricultural Microbiology, TNAU, Madurai, Tamil Nadu, India
Department of Crop Management, Vanavarayar Institute of Agriculture, Pollachi, Tamil Nadu, India

Ajay Kumar Amity Institute of Biotechnology, Amity University, Noida, India

Mouriya Suraj Kumar ICAR-Sugarcane Breeding Institute, Indian Council of Agricultural Research, Coimbatore, India

Kana Valiyaveettil Lakshana ICAR-Sugarcane Breeding Institute, Indian Council of Agricultural Research, Coimbatore, India

Palaniyandi Malathi ICAR-Sugarcane Breeding Institute, Indian Council of Agricultural Research, Coimbatore, India

A. Manikandan Department of Agricultural Microbiology, Tamil Nadu Agricultural University (TNAU), Coimbatore, Tamil Nadu, India
Institute of Ecology and Earth Sciences, University of Tartu, Tartu, Estonia

Bharani Manoharan Institute of Environment and Ecology, Academy of Environmental Health and Ecological Security, Jiangsu University, Zhenjiang, P.R. China
School of the Environment and Safety Engineering, Jiangsu University, Zhenjiang, P.R. China

S. T. Mehetre Nuclear Agriculture and Biotechnology Division, Bhabha Atomic Research Centre, Mumbai, India

Mustafa Mirik Department of Plant Protection, Agricultural Engineering Faculty, Tekirdag Namik Kemal University, Tekirdag, Turkey

Manisha Mishra University Department of Botany, T.M. Bhagalpur University, Bhagalpur, India

Geethanjali Muthuramalingam Biocatalysts Laboratory, Department of Agricultural Microbiology, Tamil Nadu Agricultural University, Coimbatore, India

M. Mythili Department of Agricultural Microbiology, Tamil Nadu Agricultural University, Coimbatore, Tamil Nadu, India

Shobana Narayanasamy Biocatalysts Laboratory, Department of Agricultural Microbiology, Tamil Nadu Agricultural University, Coimbatore, India

Senthil Natesan Department of Plant Molecular Biology and Bioinformatics, Centre for Plant Molecular Biology and Biotechnology, Tamil Nadu Agricultural University, Coimbatore, India

Cansu Oksel Department of Plant Protection, Agricultural Engineering Faculty, Tekirdag Namik Kemal University, Tekirdag, Turkey

R. Oviya Department of Plant Pathology, Agricultural College and Research Institute, Tamil Nadu Agricultural University, Madurai, Tamil Nadu, India

S. Parthasarathy Department of Plant Pathology, Amrita School of Agricultural Sciences, Amrita Vishwa Vidyapeetham, Coimbatore, India

Shanshan Qi Institute of Environment and Ecology, Academy of Environmental Health and Ecological Security, Jiangsu University, Zhenjiang, P.R. China
School of the Environment and Safety Engineering, Jiangsu University, Zhenjiang, P.R. China

R. Raghu Department of Agricultural Microbiology, Tamil Nadu Agricultural University (TNAU), Coimbatore, Tamil Nadu, India
Department of Plant Biotechnology, TNAU, Coimbatore, India

T. Raguchander Department of Plant Pathology, Tamil Nadu Agricultural University, Coimbatore, India

Pavithra Raj Agricultural College and Research Institute, Killikulam, Tuticurin, India

L. Rajendran Department of Plant Pathology, Tamil Nadu Agricultural University, Coimbatore, India

Monisha Rajkumar Biocatalysts Laboratory, Department of Agricultural Microbiology, Tamil Nadu Agricultural University, Coimbatore, India

A. Ramalakshmi Department of Agricultural Microbiology, Tamil Nadu Agricultural University, Coimbatore, Tamil Nadu, India

V. Ramamoorthy Department of Plant Pathology, Agricultural College and Research Institute, Tamil Nadu Agricultural University, Thanjavur, Tamil Nadu, India

Kannan Rengasamy Department of Plant Pathology, Tamil Nadu Agricultural University, Coimbatore, India

Uma Sankareswari Rengasamy Agricultural College and Research Institute, Madurai, India

V. Sendhilvel Department of Plant Pathology, Tamil Nadu Agricultural University, Coimbatore, India

M. Senthilkumar Agricultural College and Research Institute, TNAU, Eachangkottai, Tamil Nadu, India

Haripriya Shanmugam Centre for Agricultural Nanotechnology, Tamil Nadu Agricultural University, Coimbatore, India

U. Sivakumar Department of Agricultural Microbiology, Tamil Nadu Agricultural University, Coimbatore, Tamil Nadu, India

G. Sobanbabu Department of Plant Pathology, Agricultural College and Research Institute, Tamil Nadu Agricultural University, Madurai, Tamil Nadu, India

Chitra Sudalaimani Biocatalysts Laboratory, Department of Agricultural Microbiology, Tamil Nadu Agricultural University, Coimbatore, India

Amalraj Ramesh Sundar ICAR-Sugarcane Breeding Institute, Indian Council of Agricultural Research, Coimbatore, India

M. Theradimani Agricultural College and Research Institute, Tamil Nadu Agricultural University, Killikulam, Tamil Nadu, India

Sivakumar Uthandi Biocatalysts Laboratory, Department of Agricultural Microbiology, Tamil Nadu Agricultural University, Coimbatore, India
Department of Agricultural Microbiology, Tamil Nadu Agricultural University, Coimbatore, India

Ramamoorthy Vellaisamy Department of Plant Pathology, Agricultural College and Research Institute, Tamil Nadu Agricultural University, Eachangkottai, Thanjavur, Tamil Nadu, India

Rasappa Viswanathan ICAR-Sugarcane Breeding Institute, Indian Council of Agricultural Research, Coimbatore, India

Chapter 1
Microbial Metabolites: A Potential Weapon Against Phytopathogens

Shobana Narayanasamy, Monisha Rajkumar, Geethanjali Muthuramalingam, Chitra Sudalaimani, and Sivakumar Uthandi

Abstract Microorganisms and plants are constantly interacting with each other. While many of them have the potential to cause infection and disease, others may interact mutually, enabling plants to absorb vital nutrients and increase their disease resistance. Plants must control the expression of hundreds of genes during these interactions, which eventually activates many hormonal signaling pathways and alters the concentrations of several metabolites. The characterization of metabolites responsible for such reactions has been made possible by metabolomics, and this information has offered crucial inputation for enhancing existing plant protection strategies for yield enhancement. A wide variety of bioactive small molecules produced by bacteria, fungi, and actinobacteria have an excellent potential for application in agriculture and plant protection tactics. Antibiotics, growth hormones, pigments, nitrogen-containing compounds, fatty acid derivatives, terpenes, volatile metabolites, and aromatics compounds are the well-known metabolites produced by microorganisms that are not necessary for their growth and development. However, they have demonstrated an enormous potential to enhance plant health. Microbial metabolites have a diverse range of functions, including defending plants against infections, pests, and herbivores; regulating plant growth; responding to environmental challenges; acting as signaling cue; and regulating interactions between organisms. Likewise, metabolites also have a potential role in many of the aforementioned activities, either directly or indirectly, via altering the plant metabolism. This chapter discusses the importance of metabolomics, tools and techniques, advanced metabolomics analyses, sources of metabolites, and their potential to ward off pathogens.

Keywords Secondary metabolites · Plant defense · Microbes · Phytopathogens · Metabolomics · Disease management · *Bacillus* · *Pseudomonas* · *Trichoderma*

S. Narayanasamy · M. Rajkumar · G. Muthuramalingam · C. Sudalaimani · S. Uthandi (✉)
Biocatalysts Laboratory, Department of Agricultural Microbiology, Tamil Nadu Agricultural University, Coimbatore, India
e-mail: usiva@tnau.ac.in

K. K. Bastas et al. (eds.), *Microbial Biocontrol: Molecular Perspective in Plant Disease Management*, Microorganisms for Sustainability 49,
https://doi.org/10.1007/978-981-99-3947-3_1

1.1 Introduction

Plants have evolved intricate defensive systems to combat biotic and abiotic stressors, as natural systems present myriad competing forces (Ballhorn et al. 2011; Singh et al. 2018). Plants are confronted with a wide array of microorganisms throughout their lifecycle, and their interactions with these microorganisms could be either beneficial or harmful, resulting in mutualism or deleterious interactions, respectively (Rodriguez et al. 2019). It can adjust their innate immune system depending on the strategy prompted by microbes to respond to distinct beneficial and harmful microorganisms and exhibit suitable responses (Pieterse et al. 2014). Plants have an advanced immune system that may manifest spontaneously or in response to a microbial attack. When a microorganism overcomes these protective barriers, it forms a persistent interface with the plant, resulting in either a beneficial partnership or a disease. Mutualistic associations may trigger immune responses against other microorganisms. Additionally, harmful microbial interactions induce the immune system, but they do so against themselves. Despite the presence of pathogens in the surrounding soil, water, and air, significant crop loss due to disease is unusual. This represents the plant's defensive systems and innate biological control strategies for pest and disease control (Nishad et al. 2020). Further, plants have also adapted to defend themselves against further microbe attacks if they become infected.

Plant diseases produced by biological causes result in massive losses in crop output and quality. Plant diseases may be prevented and controlled using a variety of methods. The majority of plant producers rely on chemical fertilizers or insecticides despite using effective agricultural management methods (Pal and Gardener 2006; Kumari et al. 2022). Regardless of the fact that chemical agents have greatly enhanced crop yields for many years, their possible limitations and excessive usage result in major impairment to the soil ecology and increasing environmental pollution. As a result, much effort has been put into developing environmentally benign methods of controlling plant diseases while at the same time lowering associated health concerns. Application of live microorganisms that can prevent the growth of plant pathogens is one way to achieve biological control of phyto-pathogens (Barupal et al. 2019; Heimpel and Mills 2017; Kumar 2022). Microbial biological control agents (MBCAs) are naturally occurring, highly inhibitory, and may thrive on artificial conditions and suppress plant diseases. Use of such antag-onistic biological control agents (BCAs) once or repeatedly in high concentrations during crop growing seasons to improve plant protection is referred to as "augmen-tative biological control" (van Lenteren et al. 2018). There is rapidly growing ecological concern in the application of microbe-based products as substitutes to, or in concert with, chemicals for suppressing the prevalence and impact of a wide range of plant pathogens (Jayaprakashvel and Mathivanan 2011; Mathivanan and Srinivasan 1997). Not only an intact microbe is exploited as biological control agent, but also their derivatives, such as microbial metabolites, can be utilized (Glare et al. 2012; Barupal et al. 2020; Singh et al. 2020).

In addition to being effectors of ecological competition and coexistence, secondary metabolites (SMs) produced by microorganisms, such as antibiotics, toxins, alkaloids, and others, have also been proven to be pheromones, immune modulators, enzyme inhibitors, antagonists, pesticides, and growth promoters. It plays a pivotal role in plant health and fitness. Interestingly, several of the microbial metabolites employed in plant disease management strategies are quite important. Antimicrobial metabolite production is frequently seen during the microbial interaction between pathogens and antagonistic bacteria. One or more antibiotics have been demonstrated to play a significant role in disease suppression of several biocontrol systems, and this has received attention from various researchers (Handelsman and Stabb 1996; Mathivanan et al. 2008; Yamaguchi 1996; Singh et al. 2017). For example, at least 20 different metabolite groups with antifungal action were capable of being produced by *Pseudomonas* spp. Similarly, viscosinamide is an antifungal compound produced by several bacterial species, including *Pseudomonas* and *Burkholderia*. Furthermore, it has been proven that several sets of metabolites have analogous biological properties. For instance, two distinct antibiotics such as mycosubtilin and zwittermicin A have similar ranges of efficacy against oomycetous infections (Leclere et al. 2005). A diverse group of organic, low-molecular-weight chemicals produced by microbes are referred to as antibiotics. Antibiotics are fatal to the development or metabolic processes of other microbes at low dosages (Handelsman and Stabb 1996). In comparison to their rivals, the chemical fungicides, the antimicrobial secondary metabolites of biological control agents, are distinctive and desirable (Kim and Hwang 2007). The majority of the in vitro studies on these metabolites have indicated the possibility of several types of metabolites, both chemically and functionally. Furthermore, the secondary metabolites produced by the bacteria and fungi are extensively studied against a wide array of phytopathogens. With this context, the present chapter discusses the recent and anticipated uses of microbial secondary metabolites in plant growth and defense.

1.2 Metabolomics: A Way Forward in System Biology

In recent years, the sciences of genomes, transcriptomics, proteomics, phenomics, and metabolomics, which collectively make up the term "Omics," have undergone enormous progress in plant–microbe interactions. Among omics techniques, metabolomics is the most complicated and has received more attention in agricultural science, paving the way for precise metabolite profiling in microorganisms, plants, and mammals (Heyman and Dubery 2015; Kumar et al. 2017; Wuolikainen et al. 2016). Metabolomics has become a significant keystone in systems biology techniques due to a contemporary renaissance of research in metabolism and a growing understanding of the physiological implications that can be derived by characterizing the whole small-molecule complement of a biological system (Ray 2010; Tugizimana et al. 2018). Besides, it also provides a complete picture of cellular metabolites, small organic molecules that partake in many cellular activities and thus represent a cell's intrinsic physiological state. Therefore, metabolic profiling, or metabolomics, may be thought of as a quantitative evaluation of the multiparametric physiological adaptation of the biological system to genetic or environmental stressors (Nicholson et al. 2002; Worley and Powers 2013). Based on this description, metabolomics is the most excellent option for "functionally" exploring metabolism since it provides the most precise details and is a molecular-level integration of all prior biological information of a cell, including genomic, transcriptomic, and proteomic levels (Beisken et al. 2015; Worley and Powers 2013). These small molecules, or metabolites (molecular weight ranges up to 1500 Da), are the byproducts of gene expression and, under specific physiological conditions, determine the phenotype of a cell or tissue at the biochemical level. Thus, metabolite profile trends can offer a holistic fingerprint of the physiological state and a better understanding of particular biochemical processes in a microbe or plant. Furthermore, an integrated strategy of incorporating insights from transcriptomics, proteomics, and metabolomics will allow researchers to identify and prioritize genes and differential metabolites to enhance plant defense against a wide range of phytopathogens.

1.2.1 Analytical Tools and Techniques in Metabolomics Studies

Modern metabolomics systems have witnessed significant advancements in the collection of metabolome data employing two essential methods, notably nuclear magnetic resonance (NMR) and mass spectrometry (MS). The NMR method of metabolite identification uses the magnetic characteristics of atom nuclei in a magnetic field. The NMR is a quasi-technique frequently used to identify metabolites with lower molecular weight (50 kDa) for various applications, including metabolite profiling, fingerprinting, and identifying metabolic flux present in biological samples (Emwas and Roy 2019). However, a significant obstacle that prevents this approach from being widely used is poor sensitivity. Contrary to NMR, MS had better sensitivity enabling researchers to get extensive metabolome data coverage. This prompted the researchers to discover new metabolic biomarkers and substances that might assist in modeling metabolic pathways and networks. Issaq et al. (2009) reported that recent advancements in ionization techniques, including air pressure chemical ionization (APCI), electrospray ionization (ESI), and Matrix assisted laser desorption/ionization-Time of Flight (MALDI-TOF), have increased the accuracy of MS. Typically, chromatography methods, including capillary electrophoresis, gas chromatography, liquid chromatography, Fourier transform ion cyclotron resonance, and field asymmetric waveforms ion mobility spectrometry (FAIMS), are used with MS to increase throughput. Modern metabolomics techniques may offer massive metabolite surveys that encompass targeted and untargeted metabolites (Lei et al. 2011). National Institute of Standards and Technology (NIST), METLIN, and the Golm Metabolome Database (GMD) can be used to detect metabolites (Johnson and Lange 2015). Furthermore, the detected metabolite data are subjected to statistical analysis such as principal component analysis (PCA), correlation map, partial least squares (PLS), heatmap, reconstructing metabolic pathways, and so on, with the help of online tools and software, including Cytoscape, MetaboAnalyst, and so on (Tsugawa et al. 2015; Xie et al. 2015).

1.3 Types of Secondary Microbial Metabolites

Microbes produce a broad range of bioactive metabolites during idiophase of their growth that are not necessary for growth and development but are instead used in defense against pathogens, interaction modulation for mutual benefits, environmental adaptation, and triggering signaling molecules to interact among taxa. These metabolites also showed more significant potential in plant growth and establishment (Bérdy 2005; Ruiz et al. 2010). The peptide pathway, non-ribosomal polypeptide synthase pathway, polyketide synthase pathway, shikimate pathway, and β-lactam synthetic pathway are the seven vital metabolic pathways involved in the biosynthesis of bioactive secondary metabolites.

Based on their function, metabolites are categorized into several groups, including antibiotics, chelating compounds, pigments, and growth regulators (Ullah et al. 2020). They help to establish a symbiotic and associative relationship between the plant roots and microorganisms (Cooper 2007). Under the stressed conditions, a subset of microorganisms known as plant growth-promoting rhizobacteria (PGPR) form symbiotic relationships with plant roots and regulate their secondary metabolites (Ryffel et al. 2016; Etalo et al. 2018; Vimal and Singh 2019). For example, *Pseudomonas aeruginosa* and *Bacillus subtilis* primed in *Pisum sativum* exhibited the maximum production of gallic acid and phenols over control (Jain et al. 2015). The results indicated that root exudates are the primary factor in developing the communities of beneficial microorganisms near the roots, which in turn protect plants against a wide range of insects, pathogenic organisms, and other environmental stresses.

Beneficial microorganisms produce a wide array of secondary metabolites that are noteworthy. Glycosides, pyrroles, pyrrolidones, lactones, terpenes, alkaloids, polyoxins, phenazines, siderophores, blasticidins, strobilurins, indole types, benzene derivatives, oligopeptides, azole compounds, fatty acids, and terpenoids are some of the chemical group of metabolites produced by microorganisms that are used in crop protection (Chu et al. 2010; Hammami et al. 2009; Shen 2003; Song et al. 2015; Dickschat et al. 2010; Vimal et al. 2018). Lipopeptides produced by *Bacillus* sp. are widely employed as an antimicrobial compound against plant pathogens (Kavitha et al. 2005). *B. subtilis* produced Bac 14B, and putidacin produced by *Pseudomonas putida*, is effective against pathogens (Hammami et al. 2009; Parret et al. 2003). Siderophores (desferrioxamine, catechol, bacillibactin, pyoverdines, etc.) are iron-chelating compounds produced by several PGPR organisms that showed inhibitory action against a wide range of plant diseases (Chu et al. 2010; Hider and Kong 2010). In addition, iron-binding siderophores are also reported to inhibit mycelial growth (Idris et al. 2007; Jayaprakashvel 2008).

Various secondary metabolites with antibacterial properties that are either easily soluble in water or soluble in other organic solvents have been identified. Phenazine-1-carboxide, a water-soluble metabolite purified from *Pseudomonas aeruginosa* MML2212, showed potential antagonistic activity against *Rhizoctonia solani* (Shanmugaiah et al. 2010). Volatile metabolites have gained greater attention and are more commonly used in plant disease management strategies (Garbeva et al. 2014; Tyc et al. 2015). Six potent antifungal organic volatile chemicals, including cyclohexanol, benzothiazole, dimethyl trisulfide, *n*-decanal, and nonanal, have been produced by bacteria that inhibit the sclerotial germination and mycelial development of *Sclerotinia sclerotiorum* (Fernando et al. 2005). A total of 21 volatile compounds were synthesized by *Bacillus subtilis* PPCB001, 3-hydroxy-2-butanone (acetoin) being the most common ketone type of volatile compound that suppressed *Penicillium digitatum*, *Penicillium crustosum*, and *Penicillium italicum*, which causes post-harvest diseases in citrus crops (Arrebola et al. 2010). *Trichoderma longibrachiatum* EF5, an endophytic fungi producng volatile compounds, including longifolene, 1-butanol 2-methyl, cedrene, cuprenene, and caryophyllene, inhibits the growth of *Macrophomina phaseolina* and *Sclerotium rolfsii* (Sridharan et al. 2020). However, the stability of metabolites to resist diverse variables in natural context is crucial since most microbial metabolites exhibit significant antioxidant activities under laboratory conditions. The environmental factors (light, pH, and temperature) might either favorably or adversely influence the microbial metabolites' stability.

1.3.1 Source of Secondary Metabolites

Microorganisms produce a wide range of chemical classes of photoprotective metabolites. The synthesis of metabolites by bacteria is not selective. Numerous reports on several well-known metabolites come from diverse microorganisms. The list of microbial metabolites widely used in plant protection is provided in Table 1.1. Based on their biological origin, several studies conducted over a period of decades have firmly concluded that they may be readily classed as bacterial, fungal, and actinobacterial metabolites.

1.3.1.1 Metabolites of Bacteria

Bacteria belonging to the genera *Bacillus*, *Enterobacter*, *Agrobacterium*, *Erwinia*, *Lysobacter*, *Pseudomonas*, *Burkholderia*, and *Serratia* are widely used as biocontrol agents against many pest and plant diseases. Primarily, antibiotics produced by *Bacillus* and *Pseudomonas* are widely used to control plant pathogens. Fengycin, iturin A, mycosubtilin, bacillomycins, saltavalin, kanosamine, and zwittermicin A

Table 1.1 Examples of microbial metabolites involved in plant defense

Metabolite	Target disease	Source of organism	References
Agrocin 84	Crown gall disease *Agrobacterium tumefaciens*	*Agrobacterium radiobacter* K84	Ryder and Jones (1991)
Fengycin and Iturin A	*Rhizoctonia solani* *Gaeumannomyces graminis* var. *tritici*	*Bacillus subtilis* Z-14 *Bacillus amyloliquefaciens*	Yu et al. (2002) Zohora et al. (2016) Xiao et al. (2021)
Zwittermicin A Kanosamine	*Phytophthora medicaginis*	*Bacillus cereus*	Silo-Suh et al. (1998)
Phenazine-1-carboxylic acid	*Gaeumannomyces graminis* var. *tritici* *Botrytis cinerea* *Acidovorax citrulli*	*Pseudomonas fluorescens* *Pseudomonas chlororaphis* *Pseudomonas aeruginosa* *Pseudomonas aureofaciens*	Simionato et al. (2017) Liu et al. (2021) Pierson and Pierson (1996)
Mycosubtilin	*Pythium aphanidermatum*	*Bacillus subtilis* BBG100	Leclere et al. (2005)
Bacillomycin	*Fusarium oxysporum* *Alternaria alternata*, *Rhizoctonia solani*, and *Phytophthora capsici*	*Bacillus amyloliquefaciens* FZB42 *Bacillus vallismortis* ZZ185	Koumoutsi et al. (2004) Zhao et al. (2010)
Wuyiencin	*Botrytis cinerea*	*Streptomyces hygroscopicus* var.	Zhong et al., 2004
Pyrrolnitrin	*Aphanomyces cochlioides*, *Rhizoctonia solani*, *Pyrenophora tritici-repentis*, and *Fusarium* sp.	*Pseudomonas cepacia* *P. fluorescens*	Pawar et al. (2019)
Volatile organic compound (VOC)	Soil-borne fungal pathogens	*Pseudomonas* spp., *Serratia* spp., *Stenotrophomonas* spp., *Bacillus* sp., etc.	Dwivedi and Johri (2003), Kai et al. (2007), Kanchiswamy et al. (2015)
Hydrogen cyanide	*Rhizoctonia solani* *Meloidogyne javanica*	*P. fluorescens* *P. aeruginosa*	Jayaprakashvel and Mathivanan (2011)
2,4 Diacetylphloroglucinol (DAPG)	*G. graminis* var. *tritici*., *Pectobacterium carotovorum*, *Staphylococcus aureus*	*P. fluorescens*	Julian et al. (2020), Raaijmakers and Mazzola (2012)
Trichothecenes	Soil-borne fungal pathogens	*Trichoderma* sp.	Malmierca et al. (2012)
Strobilurin A and B	Fungal pathogens in cereal, fruit, and vegetables	*Strobilurus tenacellus*	Anke et al. (1977)

(continued)

Table 1.1 (continued)

Metabolite	Target disease	Source of organism	References
1-pentanol, 1-hexanol, glucopyranose, longifolene, caryophyllene, 1-butanol-2-methyl, cedrene, and cuprenene	*R. solani Macrophomina phaseolina*	*Trichoderma longibrachiatum* EF5	Sridharan et al. (2020), Sridharan et al. (2021), Sornakili et al. (2020)
Fusapyrone deoxyfusapyrone	*Ascochyta rabiei*, *Alternaria alternate*, *Aspergillus* sp., *B. cinerea*, *Cladosporium cucumerinum*, and *Penicillium verrucosum*	*Fusarium* sp.	Altomare et al. (2000)
Polyoxin B and D	Fungal pathogens in fruits, vegetables, and ornamentals	*Streptomyces cacaoi* var. *asensis*	Barka et al. (2016), Devi et al. (2022)
Reveromycins A and B	*Rhizopus stolonifer*, *B. cinerea*, *Sclerotinia sclerotiorum*, and *Mucor hiemalis*	*Streptomyces* sp. 3–10	Lyu et al. (2017)
Viridin	*Colletotrichum* sp., *Botrytis allii*, *Penicillium expansum*, *Aspergillus niger*, and *Fusarium caeruleum*	*Trichoderma virens Trichoderma koningii*	Vinale and Sivasithamparam (2020)
Kasugamycin, oxytetracycline, and streptomycin	*Erwinia amylovora*	*Streptomyces* sp.	Slack et al. (2021)
Avermectins	Nematodes and insect pests	*Streptomyces avermitilis*	Khalil and Darwesh (2019)
Blasticidin S	Rice blast	*Streptomyces griseochromogenes*	Stokowa-Sołtys and Jeżowska-Bojczuk (2013)

(Kloepper et al. 2004; Koumoutsi et al. 2004; Smith et al. 1993; Zhao et al. 2010) produced by *Bacillus* sp. are most widely used for controlling plant pathogens. Azole compounds produced by *Bacillus licheniformis* MML2501 were well characterized for their potential antagonistic activity against *M. phaseolina* and other soil-borne fungal pathogens (Prashanth 2007). Among various BCAs, *Pseudomonas* sp. is the prolific producer of metabolites like phenazines (Chin-A-Woeng et al. 2003), phenolics (Keel et al. 1992), pyrrole-type compounds (Pfender et al. 1993), polyketides (Kraus and Loper 1995), peptides (Soerensen et al. 2001), 2,4-diacetylphloroglucinol (Dwivedi and Johri 2003), hydrogen cyanide (Siddiqui 2005), siderophores (Meyer et al. 2002), oomycin A, pyocyanin (PCN), anthranilate, pyrrolnitrin (PRN), pyoluteorin, ammonia, gluconic acid, and viscosinamide (Chin-A-Woeng et al. 2003; Sandra et al. 2001; Zhao et al. 2010). Bacteriocins such as thuricin, thuricin 7, thuricin S, thuricin CD 19, thuricin 439A, thuricin 439B, bacthuricin F4, tochicin, kurstakin 18, and entomocin are the precursors of antibiotics (Bais et al. 2004; Sansinenea and Ortiz 2011). Furthermore, *Lysobacter* sp., *Pantoea agglomerans*, *Burkholderia cepacia*, and *Agrobacterium radiobacter* have been demonstrated to produce antimicrobial compounds, including xanthobaccin A, pyrrolnitrin, agrocin 84, pseudane, and herbicolin (Islam et al. 2005; Sandra et al. 2001; Zhao et al. 2010).

1.3.1.2 Metabolites of Fungi

Several fungal genera, namely, *Ampelomyces*, *Aspergillus*, *Gliocladium*, *Coniothyrium*, *Laetisaria*, *Penicillium*, *Sporodesmium*, *Phlebiopsis*, *Tilletiopsis*, *Talaromyces*, *Trichothecium*, *Trichoderma*, and *Fusarium* are renowned for producing bioactive secondary metabolites that may be used to combat plant diseases (Ramesh and Mathivanan 2009). Metabolites such as 1-hexanol, diethyl trisulfide, 1-pentanol, myristoyl pantothenate, *D*-alanine, and bisabolol produced by *Trichoderma longibrachiatum* EF5 showed potential activity against *M. phaseolina.* Similarly, a novel *Trichoderma asperellum* GDFS1009 produced several antimicrobial SMs, including alkanes and polyketides (Wu et al. 2017), enabling researchers to pursue the investigation into the genetic analysis of SMs and their widespread application in plant disease management. Selvaraj et al. (2020) reported that the metabolites produced by *Glomus intraradices* reduced the feeding capability of *Spodoptera litura* in blackgram. Viridin from *Trichoderma viride* and koninginins from *Trichoderma koningii* are two examples of antibiotic compounds made by various *Trichoderma* spp. Other compounds produced by these species include trichodermol, pyrones, mannitol, peptides, 2-hydroxymalonate acid, polyketides, peptaibols, cytosporone, hydroxy-lactones, and azaphilones (Vinale and Sivasithamparam 2020). Secondary metabolite production has been associated with various stages of morphological distinction and active development in numerous fungi (Chiang et al. 2009). Besides, antimicrobial metabolites such as *p*-disubstitutes aromatic compounds, deoxyfusapyrone, and fusapyrone from non-pathogenic *Fusarium* sp. exhibited antagonistic activity against *Ascochyta rabiei*, *Aspergillus* sp., *Botrytis cinerea*, and *Penicillium verrucosum* (Altomare et al. 2000; Babalola 2010; Mathivanan and Murugesan 1999). It has been observed that several fungal taxa, including *Acremonium obclavatum*, *Aphanocladium album*, *Myrothecium verrucaria*, *Penicillium* sp., *Coniothyrium minitans*, and *Verticillium chlamydosporium*, produce antimicrobial compounds that are effective against wide range of phytopathogens (Liu and Li 2004). However, their significance in plant defense has not been fully exploited.

1.3.1.3 Metabolites of Actinobacteria

Actinobacteria are the soil or plant microbiome that has a potential role in plant growth and development. The potential of actinobacteria in biocontrol has been broadly studied (Silva et al. 2022) as they possess secondary metabolite production, including a range of antibacterial, nematicidal, antifungal, and herbicidal compounds and anti-infection agents against several phytopathogenic fungi in nature, when it comes to the control of plant diseases (Chen et al. 2018). Tetracyclines, neomycin, chloramphenicol, vancomycin, erythromycin, cephalosporin, rifamycin, and kanamycin were few of the antibacterial antibiotics produced by actinobacteria (Prabavathy et al. 2008). Phenazine compound extracted from *Nocardiopsis* OPC-15 is used in plant protection (Tsujibo et al. 1988). Additionally, actinobacteria have produced many insecticidal metabolites, including avermectins, milbemycin, and tetranactin (Isono 1990; Misato 1983; Tanaka and Omura 1993). Actinobacteria have been shown to synthesize secondary metabolites with herbicidal action, including bialaphos, anisomycins, herbimycins, herbicidines, phthoxazolin, hydantocidin, homoalanosin, and cornexistin, and many of them were also used as weedicides (Copping 1996; Mio et al. 1991; Nakajima et al. 1991; Shen 1997; Stephen and Lydon 1987; Zhang et al. 1987). The investigations on the actinobacterial metabolites used as microbial herbicides, including the metabolites, have been reviewed by Li et al. (2003). Besides, blasticidin S, streptomycin, natamycin, validamycin, polyoxins, kasugamycin, oxytetracycline, and mildiomycin are some secondary metabolites produced by actinobacteria. Several SMs produced by various species act as fungicides and bactericides (Aggarwal et al. 2016). *Streptomyces* are the promising source of secondary metabolites among diverse actinobacteria group that can colonize rhizosphere and plant tissues efficiently (Olanrewaju and Babalola 2019). Production of secondary metabolites that hinder growth, such as toxins, antibiotics, biosurfactants, volatiles, and others, can suppress or kill microbial rivals through interference competition (Vurukonda et al. 2018). *Actinoplanes* sp. HBDN08-produced antifungal compound 5-hydroxyl-5-methyl-2-hexenoic acid effectively, which inhibits the growth of pathogens such as *Cladosporium cucumerinum* and *Botrytis cinerea* (Zhang et al. 2010).

1.4 Microbial Metabolites as Signaling Cues

One of the most exciting phenolic signaling molecules is 2,4-diacetylphloroglucinol (DAPG), which apparently exhibits phytotoxic, anthelminthic, and antibacterial characteristics when employed at a significant level. However, it increases root exudation at low concentrations, thereby regulating the auxin-dependent signaling pathway and improving lateral root formation. Co-inoculation of wheat plants with DAPG-producing *P. fluorescens* and *Azospirillum* Sp245-Rif significantly enhanced the root diameter and lateral roots. Plant growth regulation and nutritional balance have been linked to signaling molecules produced by PGPR. According to Ortiz-Castro et al. (2011), cyclodipeptide is an auxin-like signaling cue that promotes lateral root growth. Plants respond specifically to several signaling chemicals produced by bacteria (Schikora et al. 2016). The compound *N*-acyl-homoserine lactone (AHL), which has an acyl side chain with a length spanning from C6 to C18, is a primary cue for this task (Chernin 2011; Ortíz-Castro et al. 2008; Schenk et al. 2014). For example, priming of 3-oxo-C8-homoserine lactone (HSL) from a microbial origin altered the regulation of proteins involved in plant protection, phytohormone activation, and cytoskeleton remodeling in *Arabidopsis thaliana* (Miao et al. 2012). Likewise, 3-oxo-C10-HSL application in *Vigna radiata* significantly upregulated the genes responsible for lateral root development (Bai et al. 2012). Further, when the bacterial pathogen *Pseudomonas syringae* and the fungal pathogen *Golovinomyces orontii* infect *Arabidopsis thaliana*, AHL molecules trigger systemic resistance (Schikora et al. 2011). *Dyella* sp., *Pseudomonas* sp., *Burkholderia* sp., *Paenibacillus* sp., and *Janthinobacterium* sp. (Schulz-Bohm et al. 2015) are among the rhizosphere organisms that emit volatiles during microbial interactions in the rhizosphere and have a strong long-distance impact on nearby microbes and in nutrient-depleted soil. Many species of *Pseudomonas* and *Bacillus* were used as biocontrol agents against plant pathogens, and these species produce VOCs with antibacterial activity against several microbial communities (Raza et al. 2016). For example, a study examined that VOCs produced by *Bacillus* spp., including benzaldehyde, 1,2-benzisothiazol-3(2H)-one, and 1,3-butadiene, had strong inhibitory activity against *Ralstonia solanacearum*, the causal agent of bacterial wilt disease (Schulz-Bohm et al. 2017). A range of filamentous fungi produce the most conspicuous fungal VOC, 1-octen-3-ol, sometimes known as the mushroom smell, which serve as a development signal among fungi (Miyamoto et al. 2014; Schulz-Bohm et al. 2017). One of the most prevalent microbial volatiles, dimethyl disulfide, can drastically reduce the expression of the genes involved in ergosterol biosynthesis in various fungi, which in turn affects the pathogenicity and proliferation of those organisms (Tyagi et al. 2020). Fungal VOCs can have inhibitory effects and drive antagonistic interactions among fungi. Plant-pathogenic fungi like *Fusarium oxysporum*, *Rhizoctonia solani*, *Sclerotium rolfsii*, *Sclerotinia sclerotiorum*, and *Alternaria brassicicola* are strongly exaggerated by the VOCs released by *Trichoderma* species (Amin et al. 2010). For instance, it has been discovered that yeast employs chemical ammonia to facilitate intercolony signaling (Palková et al. 1997). Furthermore, Spraker et al. (2014) reported a two-way volatile interaction between two economically critical soil-borne pathogens of peanut, *Aspergillus flavus* and *Ralstonia solanacearum*, a fungus and bacterium, respectively. *A. flavus* volatiles and *R. solanacearum* reduced production of the virulence factor extracellular polysaccharide (EPS). *Aspergillus flavus* and *Ralstonia solanacearum*, two commercially significant soil-borne pathogens of peanuts, were found to interact in a two-way volatile manner, reducing the synthesis of virulence factor. By producing metabolites like aliphatic organic acids and aromatic nitroamino compounds (Benzyl glycinate), *T. longibrachiatum* EF5 had a significant impact on thwarting soil-borne phytopathogens like *Magnaporthe grisea*,

Rhizoctonia solani, and *Macrophomina phaseolina* (Sornakili et al. 2020; Sridharan et al. 2020).

Besides, physical interaction between plants and mycorrhizal fungi and the synthesis of bioactive compounds like volatiles and metabolites appear to be key factors in how a mycorrhizal network develops (Bonfante and Anca 2009). A mycorrhiza helper bacterium, *Streptomyces* strain AcH 505 producing auxofuran, inhibits the growth of plant-pathogenic fungi while promoting the mycelial growth of ectomycorrhizal fungi and the development of ectomycorrhizas between *Amanita muscaria* and Spruce. This suggests that the strain produces the molecules that both hinder and enhance fungal growth (Schrey et al. 2005).

1.5 Microbial Metabolites as Plant Growth Regulators

Metabolites produced by microbes belonging to distinct group of secondary metabolites affect diverse aspects of plant growth and development and offer protection against abiotic and biotic stress at lower concentrations (Davies 2013; Vimal and Singh 2019). Auxins (glucosinolate), gibberellins (GAs), salicylic acid (SA), cytokinins, abscisic acid (ABA), brassinosteroids, and jasmonate are some of them (polyhydroxy steroids). Auxins aid in apical dominance and cell growth, whereas cytokinins stimulate cell division and development; in addition, they also promote the synthesis of proteins and RNA and forestall senescence (Piotrowska and Czerpak 2009; Vimal et al. 2018; Vimal et al. 2017). Salicylic acids take part in photosynthesis, transpiration, and ion transport, aiding in plant defense against infections. Brassinosteroids are steroidal phytohormones employed in vascular immunity, stomatal regulation, and reproduction. Plant-produced ABA aids in leaf abscission signaling and offers resistance against diseases, whereas jasmonic acid protects plants against pest attack. PGPR are diverse bacteria that either directly or indirectly influence plant development by producing phytohormones or 1-aminocyclopropane 1-carboxylic acid (ACC) deaminase, nitrogen fixation, iron chelation, phosphate solubilization, and induced systemic resistance (ISR). ACC deaminase is an enzyme produced by PGPR organisms that reduces ethylene-induced stress (Pandey and Gupta 2019; Vimal et al. 2019). PGPR make an association with plants as symbiotic or as free-living rhizobacteria and are categorized based on root proximity as bacteria living in the soil near the roots (rhizosphere), colonizing the root surface (rhizoplane), and residing in root tissues (endophytes) inhabiting the space between cortical cells or inside cells in specialized structures (nodules) (Vimal 2018). Among this association, symbiotic intracellular PGPR are localized in specialized structures of plant cells and produce nodules, whereas free-living extracellular PGPR do not produce nodules but promote plant growth (Vessey 2003; Vimal 2018).

1.6 Microbial Metabolites as a Weapon Against Pathogens and Disease

Plant defense mechanisms are triggered in response to different metabolites generated by microorganisms that act as signaling cues under stressful environments (Singh 2013; Vimal et al. 2017). The peptidoglycans, pattern recognition receptor proteins, lipopolysaccharides, pathogen-associated molecular patterns, and microbe-associated molecular patterns are among them (Boller and Felix 2009; Rivas and Thomas 2005). After being recognized, plants emit signaling molecules, including ethylene, salicylic acid, and jasmonate, to activate the defensive mechanism (Yi et al. 2014). Besides, phenolic compounds produced by microorganisms would facilitate shielding the plants from pathogenic organisms (Chomel et al. 2016). Pea plants inoculated with *P. aeruginosa* and *B. subtilis* had enhanced phenols and exhibited no symptoms over control plants (Jain et al. 2015). Further, the treated pea plant shoots showed increased salicylic acid accumulation, a critical signal of systemic acquired resistance in plants (Jain et al. 2015). Plants innate benzoxazinoid metabolites provide immunity in addition to their ability to combat pathogenic fungi and aphids (Ahmad et al. 2011). Besides, it has been discovered that microbial metabolites of fungal, bacterial, and actinobacterial antagonists can inhibit the growth of several phytopathogens. Plant defense against pathogens relies mainly on synthesizing antimicrobial secondary metabolites (SMs), siderophores, lytic enzymes, and microbial cyanide (Keswani et al. 2020). The chemical 2,4-dihydroxy-7-methoxy-2H-1,4-benzoxazin-3(4H)-one is present in several crops, including maize and other cereals. Maize inoculated with different *Azospirillum* strains resulted in the modification of benzoxazinoid levels (Chamam et al. 2013; Couillerot et al. 2013). The glutathione S-transferase enzyme present in nematodes aids in the infection of the plant host. However, the presence of *Bacillus simplex* in soybean roots altered piperine alkaloids, preventing nematode penetration and cyst development (Babu et al. 2012; Kang et al. 2018). Some of the metabolites of microbial origins, such as acetoin (bacterial VOCs), DAPG, nod factors, polyamine amide (PAA), pyoverdine, and AHLs, have a dual role in plant nutrition and growth while also increasing plant resilience and tolerance to biotic and abiotic challenges (Rosier et al. 2018). The significance of microbial metabolites in plant growth, signaling, and plant defense is depicted in Fig. 1.1.

1.6.1 Key Metabolic Events in Plant Defense

According to Conrath (2009), the modification of cell walls, expression of defense genes, alteration of primary metabolism, and production of secondary metabolites are all associated with the priming potential of PGPR. Induction of ISR in *Arabidopsis* plants by *P. fluorescens* showed differential regulation of 50 metabolites. Sugars and amino acids were the predominant metabolites that stood out

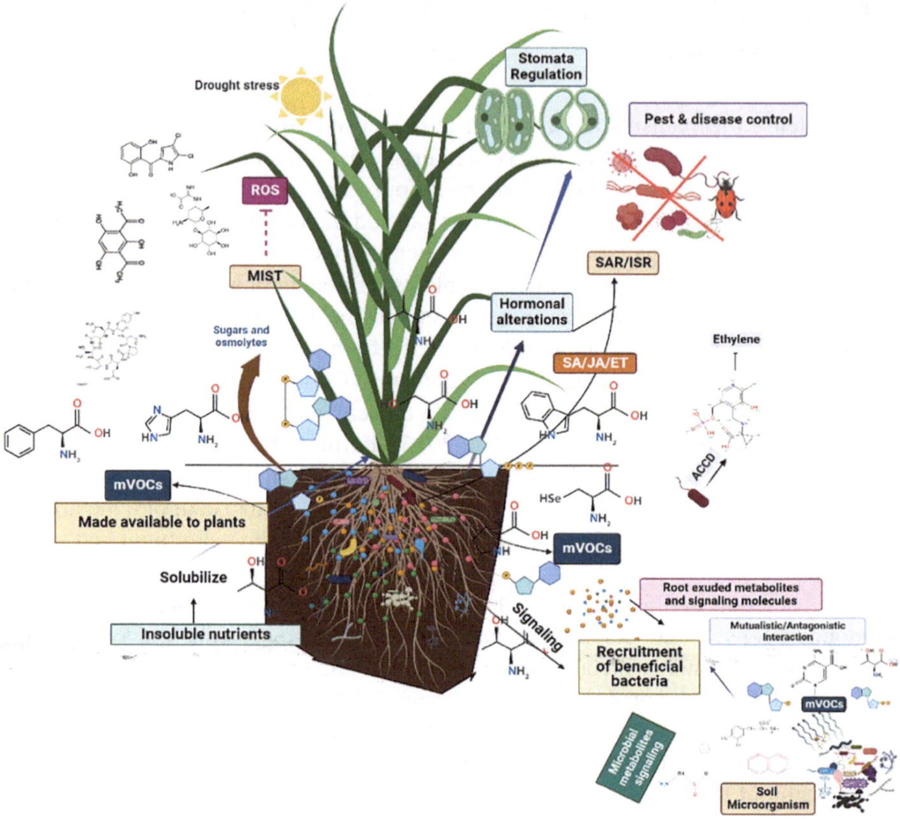

Fig. 1.1 Schematic representation of microbial metabolites in plant growth, signaling, stress resistance, and plant defense. (Image created using biorender.com)

among them (van de Mortel et al. 2012). Studies on ISR/PGPR priming frequently use molecular methods rather than metabolomics ones. As a result, little is known about metabolome alterations during ISR/PGPR priming and their importance. Despite the use of various stimuli, the metabolic activities in priming, in addition to chemical elicitation, are more comparable (Balmer et al. 2015; Mhlongo et al. 2016; Pastor et al. 2014). As a result, metabolic investigations using various molecules might be employed to describe how both primary and secondary metabolites contribute to plant priming (Djami-Tchatchou et al. 2017). Primary metabolism mainly acts as a source of energy for the onset of plant priming and for synthesizing and activating phytoanticipins, phytoalexins, and phytohormones during plant defense. Besides, PGPR inoculation significantly affects tricarboxylic acid (TCA) and sugar conjugation during plant defense (Gamir et al. 2014). Furthermore, citrate, malate, and 2-oxalate are the TCA intermediates found to be over-accumulated in defense priming (Mhlongo et al. 2018).

The presence of secondary metabolites changes in response to various environmental stimuli and is crucial for plant defense and environmental adaptability (Dörnenburg 2004). PGPR can activate secondary metabolism by the production of a wide array of chemical substances. Numerous studies have also demonstrated that phenolic compounds, terpenoids and alkaloids in plants, are substantially altered by rhizobacterial and mycorrhizal root colonization (Ramos-Solano et al. 2015; Toussaint et al. 2007). Ramos-Solano et al. (2015) demonstrated that the application of nine PGPR strains on berry plants significantly modified the secondary metabolites, such as flavonoids and phenolics, which postponed post-harvest fungal development on berries. Additionally, plants associated with PGPR demonstrated substantial production of secondary metabolites, including coumarins and flavonoids (van de Mortel et al. 2012). Additionally, *Pseudomonas putida* KT2440 primed in maize roots prompted the plants to develop systemic resistance by altering their metabolism. The jasmonic acid and abscisic acid-dependent pathways were used for their primary responses, and the phospholipids were identified as the key metabolites in the KT2440 interaction.

Moreover, microbial substances like lipopeptides (LPs) and AHLs could also prime plants by altering secondary metabolites (Han et al. 2016; Schenk et al. 2014). The lipoxygenase enzyme (LOX)-regulated pathway was triggered by *Bacillus*, as reported by Blée (2002). Fengycin treatments activated the metabolism of the phenylpropanoid pathway in potato tuber cells (Ongena et al. 2005). In addition, AHLs also triggered callose deposition and phenolic, SA, and oxylipin accumulation in several plants (Schenk et al. 2014; Schikora et al. 2016).

1.6.2 Bacterial Metabolites Against Pathogens

Many species of *Pseudomonas* and *Bacillus* produce metabolites and volatile compounds that have a potential antagonistic role against various disease-causing plant pathogens and were widely employed as biological control agents (BCAs) (Palmieri et al. 2022). *Pseudomonas aureofaciens* and *Pseudomonas fluorescens* synthesize a metabolite pyrrolnitrin (PRN); it is found actively against various fungi belonging to basidiomycota, deuteromycota, and ascomycota. *Burkholderia cepacia* releases pyrrolnitrin (PRN), which is toxic to *R. solani*, and *Pseudomonas fluorescens* BL915 suppresses *R. solani* in cotton. Besides, pyrrolnitrin can be widely used as fungicide in plant protection operations (Ligon et al. 2000). Pathma et al. (2011) reported that *P. fluorescens* produces phenazine-1-carboxylic acid, which inhibits the fungal pathogens like *Rhizoctonia solani*, *Pythium* sp., *Polyporus* sp., *Macrophomina phaseolina*, and other bacterial pathogens. *P. aeruginosa* produces blue-colored compound pyocyanin (PCN) (1-hydroxy-5-methyl-phenazine), which is highly toxic to various species of fungi and bacteria. The antifungal activity of PCN has been broadly documented against *Sitophilus oryzae*, *R. solani*, *Pythium*, and *Fusarium oxysporum*. Pyocyanin is more stable and effective than phenazine-1-carboxylic acid (El-Fouly et al. 2015). Siderophores are the chelating compounds produced by several rhizosphere bacteria that inhibit many phytopathogens. BCAs produce siderophore pyoverdine and pyochelin, which have antimicrobial activity (Haas and Défago 2005). Siderophore-biocontrol potential of *Rhizobium meliloti* has been documented against the causative agent of groundnut charcoal rot by *Macrophomina phaseolina* (Arora et al. 2001). Mercado-Blanco et al. (2004) reported siderophore production by rhizospheric *Pseudomonas* sp. suppressed the *Verticillium* wilt. Siderophores recorded increased inhibition of mycelial growth of the pathogen in the iron-deficient environment (Jayaprakashvel 2008). Idris et al. (2007) have reported that the growth of *Pythium ultimum* can be strongly inhibited by many rhizobacteria by producing metabolites and siderophores. Wuyiencin produced by *Streptomyces hygroscopicus var. wuyiensis* showed broad-spectrum activity against various diseases like leaf mold, powdery mildew, and gray mold and effectively controlled fungal and bacterial phytopathogens (Cui et al. 2010). Rajer et al. (2017) reported some species of *Bacillus* to produce four volatile compounds, nonanal, acetophenone, benzaldehyde, and benzothiazole, which can act against bacterial ring rot-causing *Clavibacter michiganensis* spp. *Sepedonicus*, and benzaldehyde, 1,2-benzisothiazol-3(2H)-one, and 1,3-butadiene, when tested against *Ralstonia solanacearum*, the causal organism of bacterial wilt disease (Tahir et al. 2017).

1.6.3 Fungal Metabolites Against Pathogens

Trichoderma viride and *Trichoderma harzianum* produce volatile compounds, inhibiting *Aspergillus flavus* and *Fusarium moniliforme*. The volatile organic compounds (VOCs) produced by *Trichoderma aureoviride* and *T. viride* inhibited protein synthesis and mycelial growth in *Serpula lacrymans* (Humphris et al. 2002). A strain of *T. harzianum* produces T39 butenolide (Vinale et al. 2006). Vinale et al. (2006) reported that this compound showed antagonism toward the growth of *Gaeumannomyces graminis var. Tritici.* Mutawila et al. (2016) reported that the filtrates of fungal culture contain secondary metabolites like 6-pentyl-α-pyrone (6PP), which can protect pruning wounds of grapevines from trunk disease pathogens. Fungi are the prolific producers of secondary metabolites comprised of peptaibols, pyrones, volatile, polyketide, non-ribosomal peptides, and non-volatile terpenes (Crutcher et al. 2013), and they are found as a fundamental element to the health and prosperity of every terrestrial ecosystem (Bills and Gloer 2016). Apart from the fact that secondary metabolites produced by fungi help promote plant growth and vitality, they can potentially induce plant-induced systemic resistance (ISR) to protect plants from infectious pathogens and diseases (O'Brien and Wright 2011).

1.6.4 Actinobacterial Metabolites Against Pathogens

Synthesis of growth-inhibitory metabolites such as biosurfactants, antibiotics, volatiles, toxins, and others can suppress or kill microbial rivals through interference competition (Vurukonda et al. 2018). The *Streptomyces* strain TKA-5 produces secondary metabolites that exhibited antagonistic activity against *Phytophthora capsici*, phytophthora blight of bell pepper, and *Alternaria brassicicola*, black leaf spot of spoon cabbage (Ko et al. 2010). These antimicrobial substances impair spore germination and cause morphological distortions to include tortuosity, shrinkage, and collapse (Chen et al. 2018), and inhibit protein biosynthesis and neurotransmission of pathogenic microorganisms (Aggarwal et al. 2016). They eventually contribute to developing a disease-suppressive soil (Viaene et al. 2016). Disease suppression in soil appeared to be due to the concerted activities of multiple microbial genera working together at specific sites or operating at different stages of the infection process of the pathogen (Gómez Expósito et al. 2017).

1.7 Conclusion and Future Perspectives

Biological control of plant diseases through microbial metabolites produced by rival microorganisms is a desirable alternative to chemically synthesized pesticides. In the proximity of plant roots, microorganisms produce a broad spectrum of secondary metabolites. While some are employed to establish mutualistic relationships with plant roots, others act as pathogen-suppressing antagonists. In nutrient-deficient soil, for example, microorganisms produce certain chelator chemicals that aid in the absorption of nutrients by plants. Plant environmental factors also influence the production of metabolites by bacteria. Some metabolites are signaling molecules that control plant metabolism to increase stress tolerance. Microorganisms release a range of metabolites in such circumstances, enhancing the plant's hormonal system, defense system, and plant growth-associated attributes. The major challenge of metabolomics, particularly in plant defense research, is the detection of unknown molecules. There are no specialized metabolite databases for plants or plant diseases like those used in yeast (Jewison et al. 2012) and human (Wishart et al. 2018) investigations. The intricate secondary metabolism, which plays a significant role in plant defense and microbial infection, and the complexity of the interactions between plant hosts and their pathogens do not make cataloguing metabolites much easier. Therefore, the most effective method of building a strong and trustworthy plant–pathogen metabolomic database may involve sharing that knowledge across organizations and making the spectrum data on secondary metabolites public. Microbial metabolites influence plant health, and applying or spraying microbial extracts to plant surfaces can trigger induced systemic resistance (ISR) and aid in plant disease recovery. Moreover, chemical priming with metabolites is a recognized agricultural technique for intervening in various plant systems to modify the salicylic acid and jasmonic acid signaling results in ISR, since the phenotype associated with the expression of many genes is driven by metabolites, which are the eventual outcome. Next-generation sequencing and other "omics" technologies have provided new insights into microbial and chemical ecology that the suppressive soils contain a valuable source of biocontrol agents, which do not represent when the cultures are tested in vitro. In addition, these microbial metabolites bring up new opportunities for entrepreneurs and manufacturers by serving as the critical compounds for synthesizing plant protection chemicals.

References

Aggarwal N, Thind S, Sharma S (2016) Role of secondary metabolites of actinomycetes in crop protection. In: Plant growth promoting actinobacteria. Springer, pp 99–121

Ahmad S, Veyrat N, Gordon-Weeks R, Zhang Y, Martin J, Smart L, Glauser G, Erb M, Flors V, Frey M, Ton J (2011) Benzoxazinoid metabolites regulate innate immunity against aphids and fungi in maize. Plant Physiol 157(1):317–327. https://doi.org/10.1104/pp.111.180224

Altomare C, Perrone G, Zonno MC, Evidente A, Pingue R, Fanti F, Polonelli L (2000) Biological characterization of fusapyrone and deoxyfusapyrone, two bioactive secondary metabolites of *Fusarium semitectum*. J Nat Prod 63:1131–1135

Amin F, Razdan V, Mohidin F, Bhat K, Sheikh P (2010) Effect of volatile metabolites of *Trichoderma* species against seven fungal plant pathogens in-vitro. J Phytology 2:34–37

Anke T, Oberwinkler F, Steglich W, Schramm G (1977) The strobilurins-new antifungal antibiotics from the basidiomycete *Strobilurus tenacellus*. J Antibiot 30:806–810

Arora N, Kang SC, Maheshwari DK (2001) Isolation of siderophore-producing strains of *Rhizobium meliloti* and their biocontrol potential against *Macrophomina phaseolina* that causes charcoal rot of groundnut. Curr Sci 81(6):673–677

Arrebola E, Sivakumar D, Korsten L (2010) Effect of volatile compounds produced by *Bacillus* strains on postharvest decay in citrus. Biol Control 53:122–128

Babalola OO (2010) Improved mycoherbicidal activity of *Fusarium arthrosporioides*. Afr J Microbiol Res 4:1659–1662

Babu RO, Moorkoth D, Azeez S, Eapen SJ (2012) Virtual screening and *in vitro* assay of potential drug like inhibitors from spices against glutathione-S-transferase of *Meloidogyne incognita*. Bioinformation 8:319

Bai X, Todd CD, Desikan R, Yang Y, Hu X (2012) N-3-oxodecanoyl-L-homoserine-lactone activates auxin induced adventitious root formation via hydrogen peroxide- and nitric oxide-dependent cyclic GMP signaling in mung bean. Plant Physiol 158:725–736

Bais HP, Fall R, Vivanco JM (2004) Biocontrol of *Bacillus subtilis* against infection of *Arabidopsis* roots by *Pseudomonas syringae* is facilitated by biofilm formation and surfactin production. Plant Physiol 134(1):307–319. https://doi.org/10.1104/pp.103.028712

Ballhorn DJ, Kautz S, Jensen M, Schmitt I, Heil M, Hegeman AD (2011) Genetic and environmental interactions determine plant defences against herbivores. J Ecol 99(1):313–326. https://doi.org/10.1111/j.1365-2745.2010.01747.x

Balmer A, Pastor V, Gamir J, Flors V, Mauch-Mani B (2015) The 'prime-ome': towards a holistic approach to priming. Trends Plant Sci 20(7):443–452. https://doi.org/10.1016/j.tplants.2015.04.002

Barka EA, Vatsa P, Sanchez L, Gaveau-Vaillant N, Jacquard C, Klenk H-P, Clément C, Ouhdouch Y, van Wezel GP (2016) Taxonomy, physiology, and natural products of actinobacteria. Microbiol Mol Biol Rev 80(1):1–43

Barupal T, Meena M, Sharma K (2019) Inhibitory effects of leaf extract of *Lawsonia inermis* on *Curvularia lunata* and characterization of novel inhibitory compounds by GC-MS analysis. Biotechnol Rep 23:e00335. https://doi.org/10.1016/j.btre.2019

Barupal T, Meena M, Sharma K (2020) A study on preventive effects of *Lawsonia inermis* L. bioformulations against leaf spot disease of maize. Biocatal Agric. Biotechnol 23:101473. https://doi.org/10.1016/j.bcab.2019.101473

Beisken S, Eiden M, Salek RM (2015) Getting the right answers: understanding metabolomics challenges. Exp Rev Mol Diagnost 15(1):97–109. https://doi.org/10.1586/14737159.2015.974562

Bérdy J (2005) Bioactive microbial metabolites. J Antibiot 58(1):1–26. https://doi.org/10.1038/ja.2005.1

Bills GF, Gloer JB (2016) Biologically active secondary metabolites from the fungi. Microbiol Spec 4(6). https://doi.org/10.1128/microbiolspec.FUNK-0009-2016

Blée E (2002) Impact of phyto-oxylipins in plant defense. Trends Plant Sci 7(7):315–322. https://doi.org/10.1016/s1360-1385(02)02290-2

Boller T, Felix G (2009) A renaissance of elicitors: perception of microbe-associated molecular patterns and danger signals by pattern-recognition receptors. Annu Rev Plant Biol Contr 60(379e406)

Bonfante P, Anca I-A (2009) Plants, mycorrhizal fungi, and bacteria: a network of interactions. Ann Rev Microbiol 63(1):363–383

Chamam A, Sanguin H, Bellvert F, Meiffren G, Comte G, Wisniewski-Dye F et al (2013) Plant secondary metabolite profiling evidences strain-dependent effect in the *Azospirillum-Oryza sativa* association. Phytochem Rev 87:65–77

Chen Y, Zhou D, Qi D, Gao Z, Xie J, Luo Y (2018) Growth promotion and disease suppression ability of a *Streptomyces* sp. CB-75 from banana rhizosphere soil. Front Micobiol 8:2704

Chernin LS (2011) Quorum-sensing signals as mediators of PGPRs' beneficial traits. In: Bacteria in agrobiology: plant nutrient management. Springer, Berlin, Heidelberg, pp 209–236

Chiang YM, Lee KH, Sanchez JF, Keller NP, Wang CC (2009) Unlocking fungal cryptic natural products. Nat Prod Commun 4(11):1505–1510

Chin-A-Woeng TF, Bloemberg GV, Lugtenberg BJ (2003) Phenazines and their role in biocontrol by *Pseudomonas* bacteria. New Phytol 157(3):503–523

Chomel M, Guittonny-Larchevêque M, Fernandez C, Gallet C, DesRochers A, Paré D, Jackson BG, Baldy V (2016) Plant secondary metabolites: a key driver of litter decomposition and soil nutrient cycling. J Ecol 104(6):1527–1541. https://doi.org/10.1111/1365-2745.12644

Chu BC, Garcia-Herrero A, Johanson TH, Krewulak KD, Lau CK, Peacock RS, Slavinskaya Z, Vogel HJ (2010) Siderophore uptake in bacteria and the battle for iron with the host; a bird's eye view. Biometals 23(4):601–611

Conrath U (2009) Chapter 9 Priming of induced plant defense responses. In: Advances in botanical research, vol 51. Academic Press, pp 361–395. https://doi.org/10.1016/S0065-2296(09)51009-9

Cooper JE (2007) Early interactions between legumes and rhizobia: disclosing complexity in a molecular dialogue. J Appl Microbiol 103(5):1355–1365. https://doi.org/10.1111/j.1365-2672.2007.03366.x

Copping LG (1996) Crop protection agents from nature: natural products and analogues, vol 501. The Royal Society of Chemistry, Cambridge

Couillerot O, Ramírez-Trujillo A, Walker V, von Felten A, Jansa J, Maurhofer M et al (2013) Comparison of prominent *Azospirillum* strains in *Azospirillum-pseudomonas- glomus* consortia for promotion of maize growth. Appl Microbiol Biotechnol 97:4639–4649

Crutcher FK, Parich A, Schuhmacher R, Mukherjee PK, Zeilinger S, Kenerley CM (2013) A putative terpene cyclase, *vir4*, is responsible for the biosynthesis of volatile terpene compounds in the biocontrol fungus *Trichoderma virens*. Fungal Genet Biol 56:67–77

Cui Z, Zhang K, She G, Lin Y, Sun L, Yea C (2010) Antimicrobial activity investigation on wuyiencin fractions of different polarity. Molecules 15:3179–3185

Davies PJ (2013) Plant hormones: physiology. Springer Science & Business Media, Biochemistry and molecular biology

Devi AP, Jesumaharaja GL, Balasundaram K, Sahana N, Battacharya PM, Roy A, Bandyopadhyay S, Khalko S (2022) Chapter 19 - *Streptomyces* sp.: a feasible biocontrol agent for sustainable management of crop diseases. In: Malik JA (ed) Microbes and microbial biotechnology for green remediation. Elsevier, pp 377–388. https://doi.org/10.1016/B978-0-323-90452-0.00025-6

Dickschat JS, Wickel S, Bolten CJ, Nawrath T, Schulz S, Wittmann C (2010) Pyrazine biosynthesis in *Corynebacterium glutamicum*. Eur J Org Chem 14:2687–2695

Djami-Tchatchou AT, Sanan-Mishra N, Ntushelo K, Dubery IA (2017) Functional roles of microRNAs in agronomically important plants—potential as targets for crop improvement and protection. Front Plant Sci 8. https://doi.org/10.3389/fpls.2017.00378

Dörnenburg H (2004) Evaluation of immobilisation effects on metabolic activities and productivity in plant cell processes. Process Biochem 39:1369–1375. https://doi.org/10.1016/S0032-9592(03)00262-0

Dwivedi D, Johri BN (2003) Antifungals from fluorescent pseudomonads: biosynthesis and regulation. Curr Sci 12:1693–1703

El-Fouly MZ, Sharaf AM, Shahin AAM, El-Bialy HA, Omara AMA (2015) Biosynthesis of pyocyanin pigment by *Pseudomonas aeruginosa*. J Rad Res App Sci 8(1):36–48. https://doi.org/10.1016/j.jrras.2014.10.007

Emwas AH, Roy R (2019) NMR spectroscopy for metabolomics research. Metabolites 9(7):123. https://doi.org/10.3390/metabo9070123

Etalo DW, Jeon JS, Raaijmakers JM (2018) Modulation of plant chemistry by beneficial root microbiota. Nat Prod Rep 35:398–409

Fernando WG, Ramarathnam R, Krishnamoorthy AS, Savchuk SC (2005) Identification and use of potential bacterial organic antifungal volatiles in biocontrol. Soil Biol Biochem 37:955–964

Gamir J, Sánchez-Bel P, Flors V (2014) Molecular and physiological stages of priming: how plants prepare for environmental challenges. Plant Cell Rep 33(12):1935–1949. https://doi.org/10.1007/s00299-014-1665-9

Garbeva P, Hordijk C, Gerards S, De-Boer W (2014) Volatiles produced by the mycophagous soil bacterium *Collimonas*. FEMS Microbiol Ecol 87:639–649

Glare T, Caradus J, Gelernter W, Jackson T, Keyhani N, Köhl J, Marrone P, Morin L, Stewart A (2012) Have biopesticides come of age? Trends Biotechnol 30(5):250–258. https://doi.org/10.1016/j.tibtech.2012.01.003

Gómez Expósito R, de Bruijn I, Postma J, Raaijmakers JM (2017) Current insights into the role of rhizosphere bacteria in disease suppressive soils. Front Microbiol 8. https://doi.org/10.3389/fmicb.2017.02529

Haas D, Défago G (2005) Biological control of soil-borne pathogens by fluorescent pseudomonads. Nat Rev Microbiol 4:307–319

Hammami I, Rhouma A, Jaouadi B, Rebai A, Nesme X (2009) Optimization and biochemical characterization of a bacteriocin from a newly isolated *Bacillus subtilis* strain 14B for biocontrol of *Agrobacterium* spp. strains. Lett Appl Microbiol 48(2):253–260

Han S, Li D, Trost E, Mayer KF, Vlot AC, Heller W, Schmid M, Hartmann A, Rothballer M (2016) Systemic responses of barley to the 3-hydroxy-decanoyl-homoserine lactone producing plant beneficial endophyte *Acidovorax radicis* N35. Front Plant Sci 7:1868. https://doi.org/10.3389/fpls.2016.01868

Handelsman J, Stabb E (1996) Biocontrol of soil-borne plant pathogens. Plant Cell 8:1855–1869

Heimpel G, Mills N (2017) Biological control: ecology and applications. Am Entomol 64 (1), spring, E2. https://doi.org/10.1017/9781139029117

Heyman H, Dubery I (2015) The potential of mass spectrometry imaging in plant metabolomics: a review. Phytochem Rev 15. https://doi.org/10.1007/s11101-015-9416-2

Hider RC, Kong X (2010) Chemistry and biology of siderophores. Nat Prod Rep 27:637–657

Humphris S, Bruce A, Buultjens E, Wheatley R (2002) The effects of volatile microbial secondary metabolites on protein synthesis in *Serpula lacrymans*. FEMS Microbiol Lett 210:215–219. https://doi.org/10.1016/S0378-1097(02)00604-3

Idris A, Labuschagne N, Korsten L (2007) Screening rhizobacteria for biological control of *Fusarium* root and crown rot of sorghum in Ethiopia. Biol Control 40:97–106

Islam M, Hashidoko Y, Deora A, Ito T, Tahara S (2005) Suppression of damping-off disease in host plants by the rhizoplane bacterium *Lysobacter* sp. strain SB-K88 is linked to plant colonization and antibiosis against soil-borne peronosporomycetes. Appl Environ Microbiol 71:3786–3796

Isono K (1990) Antibiotics as non-pollution agricultural pesticides. Comments Agr Food Chem 2: 123–142

Issaq HJ, Van QN, Waybright TJ, Muschik GM, Veenstra TD (2009) Analytical and statistical approaches to metabolomics research. J Sep Sci 32(13):2183–2199

Jain A, Singh A, Singh S, Singh HB (2015) Phenols enhancement effect of microbial consortium in pea plants restrains *Sclerotinia sclerotiorum*. Biol Control 89:23–32

Jayaprakashvel M (2008) Development of a synergistically performing bacterial consortium for sheath blight suppression in rice. University of Madras, Madras

Jayaprakashvel M, Mathivanan N (2011) Management of plant diseases by microbial metabolites. In: Maheshwari DK (ed) Bacteria in agrobiology: plant nutrient management. Springer, Berlin Heidelberg, Berlin, Heidelberg, pp 237–265. https://doi.org/10.1007/978-3-642-21061-7_10

Jewison T, Knox C, Neveu V, Djoumbou Y, Guo AC, Lee J, Liu P, Mandal R, Krishnamurthy R, Sinelnikov I, Wilson M (2012) YMDB: the yeast metabolome database. Nucleic Acids Res 40 (D1):D815–D820

Johnson SR, Lange BM (2015) Open-access metabolomics databases for natural product research: present capabilities and future potential. Front Bioeng Biotechnol 3:22. https://doi.org/10.3389/fbioe.2015.00022

Julian WT, Vasilchenko AV, Shpindyuk DD, Poshvina DV, Vasilchenko AS (2020) Bacterial-derived plant protection metabolite 2,4-Diacetylphloroglucinol: effects on bacterial cells at inhibitory and subinhibitory concentrations. Biomol Ther 11(1):13. https://doi.org/10.3390/biom11010013

Kai M, Effmert U, Berg G, Piechulla B (2007) Volatiles of bacterial antagonists inhibit mycelial growth of the plant pathogen *Rhizoctonia solani*. Arch Microbiol 187:351–360

Kanchiswamy CN, Malnoy M, Maffei ME (2015) Chemical diversity of microbial volatiles and their potential for plant growth and productivity. Front Plant Sci 6:151. https://doi.org/10.3389/fpls.2015.00151

Kang W, Zhu X, Wang Y, Chen L, Duan Y (2018) Transcriptomic and metabolomic analyses reveal that bacteria promote plant defense during infection of soybean cyst nematode in soybean. BMC Plant Biol 18:86

Kavitha S, Senthilkumar S, Gnanamanickam S, Inayathullah M, Jayakumar J (2005) Isolation and partial characterization of antifungal protein from *Bacillus polymyxa* strain VLB16. Process Biochem 40:3236–3243

Keel C, Schnider U, Maurhofer M, Voisard C, Laville J, Burger U, Wirthner PJ, Haas D, Défago G (1992) Suppression of root diseases by *Pseudomonas fluorescens* CHA0: importance of the bacterial secondary metabolite 2, 4-diacetylphloroglucinol. Mol Plant-Microbe Interac 5(1): 4–13

Keswani C, Singh SP, Cueto L, García-Estrada C, Mezaache-Aichour S, Glare TR, Borriss R, Singh SP, Blázquez MA, Sansinenea E (2020) Auxins of microbial origin and their use in agriculture. Appl Microbiol Biotechnol 104(20):8549–8565. https://doi.org/10.1007/s00253-020-10890-8

Khalil MS, Darwesh DM (2019) Avermectins: the promising solution to control plant parasitic nematodes. J Plant Sci Phytopathol 3:81–85

Kim B, Hwang B (2007) Microbial fungicides in the control of plant diseases. J Phytopathol 155: 641–653

Kloepper JW, Ryu C-M, Zhang S (2004) Induced systemic resistance and promotion of plant growth by *Bacillus* spp. Phytopathology 94(11):1259–1266

Ko WH, Tsou YJ, Lin MJ, Chern LL (2010) Activity and characterization of secondary metabolites produced by a new microorganism for control of plant diseases. New Biotechnol 27(4): 397–402. https://doi.org/10.1016/j.nbt.2010.05.014

Koumoutsi A, Chen X-H, Henne A, Liesegang H, Hitzeroth G, Franke P, Vater J, Borriss R (2004) Structural and functional characterization of gene clusters directing nonribosomal synthesis of bioactive cyclic lipopeptides in *Bacillus amyloliquefaciens* strain FZB42. J Bacteriol 186(4): 1084–1096

Kraus J, Loper JE (1995) Characterization of a genomic region required for production of the antibiotic pyoluteorin by the biological control agent *Pseudomonas fluorescens* Pf-5. Appl Environ Microbiol 61(3):849–854

Kumar A (2022) Microbial biocontrol: sustainable agriculture and phytopathogen management, vol 1. Springer Nature Chem, Cham, Switzerland, pp 1–369. https://doi.org/10.1007/978-3-030-87512-1

Kumar R, Bohra A, Pandey AK, Pandey MK, Kumar A (2017) Metabolomics for plant improvement: status and prospects. Front Plant Sci 8:1302. https://doi.org/10.3389/fpls.2017.01302

Kumari M, Qureshi KA, Jaremko M, White JF, Singh SK, Sharma VK, Singh KK, Santoyo G, Puopolo G, Kumar A (2022) Deciphering the role of endophytic microbiome in postharvest diseases management of fruits: opportunity areas in commercial up-scale production. Front Plant Sci 13:1026575. https://doi.org/10.3389/fpls.2022.1026575

Leclere V, Bechet A, Adam JS, Guez B, Wathelet M, Ongena P, Thonart F, Gancel M, Chollet I, P J (2005) Mycosubtilin overproduction by *Bacillus subtilis* BBG100 enhances the organism's antagonistic and biocontrol activities. Appl Environ Microbiol 71:4577–4584

Lei Z, Huhman DV, Sumner LW (2011) Mass spectrometry strategies in metabolomics. J Biol Chem 286(29):25435–25442. https://doi.org/10.1074/jbc.R111.238691

Li Y, Sun Z, Zhuang X, Xu L, Chen S, Li M (2003) Research progress on microbial herbicides. Crop Prot 22:247–252

Ligon JM, Hill DS, Hammer PE, Torkewitz NR, Hofmann D, Kempf H-J, Pée K-HV (2000) Natural products with antifungal activity from *Pseudomonas* biocontrol bacteria. Pest Manag Sci 56(8):688–695. https://doi.org/10.1002/1526-4998

Liu X, Li S (2004) Fungal secondary metabolites in biological control of crop pests. In: An Z (ed) Handbook of industrial mycology. CRC, New York, pp 723–747

Liu Y, Zhou Y, Qiao J, Yu W, Pan X, Zhang T, Liu Y, Lu S-E (2021, 2012) Phenazine-1-carboxylic acid produced by *Pseudomonas chlororaphis* YL-1 is effective against Acidovorax citrulli. Microorganisms 9(10)

Lyu A, Liu H, Che H, Yang L, Zhang J, Wu M, Chen W, Li G (2017) Reveromycins A and B from *Streptomyces* sp. 3–10: antifungal activity against plant pathogenic fungi *in vitro* and in a strawberry food model system. Front Microbiol 8:550. https://doi.org/10.3389/fmicb.2017.00550

Malmierca MG, Cardoza RE, Alexander NJ, McCormick SP, Hermosa R, Monte E, Gutiérrez S (2012) Involvement of *Trichoderma* trichothecenes in the biocontrol activity and induction of plant defense-related genes. Appl Environ Microbiol 78(14):4856–4868. https://doi.org/10.1128/aem.00385-12

Mathivanan N, Murugesan K (1999) Isolation and purification of an antifungal metabolite from *Fusarium chlamydosporum*, a mycoparasite to *Puccinia arachidis*, the rust pathogen of groundnut. Indian J Exp Biol 37:98–101

Mathivanan N, Srinivasan K, Chelliah S (1997) Evaluation of *Trichoderma viride* and carbendazim and their integration for the management of root diseases in cotton. Indian J Microbiol 37:107–108

Mathivanan N, Prabavathy VR, Vijayanandraj VR (2008) The effect of fungal secondary metabolites on bacterial and fungal pathogens. In: Karlovsky P (ed) Secondary metabolites in soil ecology. Springer, Berlin, pp 129–140

Mercado-Blanco J, Rodrıguez-Jurado D, Hervás A, Jiménez-Dıaz RM (2004) Suppression of *Verticillium* wilt in olive planting stocks by root-associated fluorescent *Pseudomonas* spp. Biol Control 30(2):474–486

Meyer JM, Geoffroy VA, Baysse C, Cornelis P, Barelmann I, Taraz K, Budzikiewicz H (2002) Siderophore-mediated iron uptake in fluorescent *Pseudomonas*: characterization of the pyoverdine-receptor binding site of three cross-reacting pyoverdines. Arch Biochem Biophys 397(2):179–183. https://doi.org/10.1006/abbi.2001.2667

Mhlongo MI, Steenkamp PA, Piater LA, Madala NE, Dubery IA (2016) Profiling of altered metabolomic states in *Nicotiana tabacum* cells induced by priming agents. Front. Plant Sci 7

Mhlongo MI, Piater LA, Madala NE, Labuschagne N, Dubery IA (2018) The chemistry of plant–microbe interactions in the rhizosphere and the potential for metabolomics to reveal signaling related to defense priming and induced systemic resistance. Front Plant Sci 9. https://doi.org/10.3389/fpls.2018.00112

Miao C, Liu F, Zhao Q, Jia Z, Song S (2012) A proteomic analysis of *Arabidopsis thaliana* seedling responses to 3-oxo-octanoyl-homoserine lactone, a bacterial quorum-sensing signal. Biochem Biophys Res Commun 427:293–298

Mio S, Sano H, Shindou M, Honma T, Sugai S (1991) Synthesis and herbicidal activity of deoxy derivatives of (+) hydantocidin. Agric Biol Chem 55:1105–1109

Misato T (1983) Recent status and future aspects of agricultural antibiotics. In: Miyamoto J, Kaerney PC (eds) Pesticide chemistry: human welfare and the environment. Pergamon, Oxford, pp 241–246

Miyamoto K, Murakami T, Kakumyan P, Keller NP, Matsui K (2014) Formation of 1-octen-3-ol from aspergillus flavus conidia is accelerated after disruption of cells independently of Ppo oxygenases, and is not a main cause of inhibition of germination. PeerJ 2:e395. https://doi.org/10.7717/peerj.395

Mutawila C, Vinale F, Halleen F, Lorito M, Mostert L (2016) Isolation, production and in vitro effects of the major secondary metabolite produced by *Trichoderma* species used for the control of grapevine trunk diseases. Plant Pathol 65(1):104–113

Nakajima M, Itoi K, Takamatsu Y, Sato S, Furukawa Y (1991) Cornexistin: a new fungal metabolite with herbicidal activity. J Antibiot (Tokyo) 44:1065–1072

Nicholson JK, Connelly J, Lindon JC, Holmes E (2002) Metabonomics: a platform for studying drug toxicity and gene function. Nat Rev Drug Disc 1(2):153–161. https://doi.org/10.1038/nrd728

Nishad R, Ahmed T, Rahman VJ, Kareem A (2020) Modulation of plant defense system in response to microbial interactions. Front Microbiol 11:1298. https://doi.org/10.3389/fmicb.2020.01298

O'Brien J, Wright G (2011) An ecological perspective of microbial secondary metabolism. Cur Opin Biotechnol 22:552–558. https://doi.org/10.1016/j.copbio.2011.03.010

Olanrewaju OS, Babalola OO (2019) *Streptomyces*: implications and interactions in plant growth promotion. App Microbiol Biotechnol 103(3):1179–1188

Ongena M, Jacques P, Touré Y, Destain J, Jabrane A, Thonart P (2005) Involvement of fengycin-type lipopeptides in the multifaceted biocontrol potential of *Bacillus subtilis*. App Microbiol Biotechnol 69:29–38. https://doi.org/10.1007/s00253-005-1940-3

Ortíz-Castro R, Martínez-Trujillo M, López-Bucio J (2008) N acyl-L-homoserine lactones: a class of bacterial quorum sensing signals alter post-embryonic root development in *Arabidopsis thaliana*. Plant Cell Environ 31:1497–1509

Ortiz-Castro R, Díaz-Pérez C, Martínez-Trujillo M, Rosa E, Campos-García J, López-Bucio J (2011) Transkingdom signalling based on bacterial cyclodipeptides with auxin activity in plants. PNAS 108:7253–7258

Pal KK, Gardener BM (2006) Biological control of plant pathogens. The Plant Health Instructor:1–25

Palková Z, Janderová B, Jí G, Zikánová B, Pospíŝek M, Forstová J (1997) Ammonia mediates communication between yeast colonies. Nature 390(6659):532–536

Palmieri D, Ianiri G, Del Grosso C, Barone G, De Curtis F, Castoria R, Lima G (2022) Advances and perspectives in the use of biocontrol agents against fungal plant diseases. Horticulturae 8(7): 577

Pandey S, Gupta S (2019) ACC deaminase producing bacteria with multifarious plant growth promoting traits alleviates salinity stress in French bean (*Phaseolus vulgaris*) plants. Front Microbiol 10:1506

Parret AH, Schoofs G, Proost P, De Mot R (2003) Plant lectin-like bacteriocin from a rhizosphere-colonizing pseudomonas isolate. J Bacteriol 185(3):897–908

Pastor V, Balmer A, Gamir J, Flors V, Mauch-Mani B (2014) Preparing to fight back: generation and storage of priming compounds. Front Plant Sci 5:295. https://doi.org/10.3389/fpls.2014.00295

Pathma J, Kennedy RK, Sakthivel N (2011) Mechanisms of fluorescent pseudomonads that mediate biological control of phytopathogens and plant growth promotion of crop plants. In: Maheshwari DK (ed) Bacteria in agrobiology: plant growth responses, vol 4. Springer, Berlin Heidelberg, pp 77–105

Pawar S, Chaudhari A, Prabha R, Shukla R (2019) Microbial pyrrolnitrin: natural metabolite with immense practical utility. Biomol Ther 9(9):443. https://doi.org/10.3390/biom9090443

Pfender WF, Kraus J, Loper JE (1993) A genomic region from *Pseudomonas fluorescens* Pf-5 required for pyrrolnitrin production and inhibition of *Pyrenophora tritici-repentis* in wheat straw. Phytopathology 83(11):1223–1228

Pierson LS, Pierson EA (1996) Phenazine antibiotic production in *Pseudomonas aureofaciens:* role in rhizosphere ecology and pathogen suppression. FEMS Microbiol Lett 136:101–108

Pieterse CM, Zamioudis C, Berendsen RL, Weller DM, Van Wees SC, Bakker PA (2014) Induced systemic resistance by beneficial microbes. Annu Rev Phytopathol 52:347–375. https://doi.org/10.1146/annurev-phyto-082712-102340

Piotrowska A, Czerpak R (2009) Cellular response of light/ dark-grown green alga *Chlorella vulgaris Beijerinck* (Chlorophyceae) to exogenous adenine-and phenylurea-type cytokinins. Acta Physiol Plant 31:573–585

Prabavathy VR, Vajayanandraj VR, Malarvizhi K, Mathivanan N, Mohan N, Murugesan K (2008) Role of actinomycetes and their metabolites in crop protection. In: Khachatourian GC, Arora DK, Rajendran TP, Srivastava AK (eds) Agriculturally important microorganisms. Academic World International, Bhopal, pp 243–255

Prashanth S (2007) Biological control of *Macrophomina* root rot and plant growth promotion in groundnut by *Bacillus licheniformis* MML2501, an azole compound producing rhizobacterium. University of Madras, Madras

Raaijmakers JM, Mazzola M (2012) Diversity and natural functions of antibiotics produced by beneficial and plant pathogenic bacteria. Annu Rev Phytopathol 50:403–424

Rajer FU, Wu H, Xie Y, Xie S, Raza W, Tahir HAS, Gao X (2017) Volatile organic compounds produced by a soil-isolate, *Bacillus subtilis* FA26 induce adverse ultra-structural changes to the cells of *Clavibacter michiganensis* ssp. Sepedonicus, the causal agent of bacterial ring rot of potato. Microbiology 163:523–530

Ramesh S, Mathivanan N (2009) Screening of marine actinomycetes isolated from the bay of Bengal, India for antimicrobial activity and industrial enzymes. W J Microbiol Biotechnol 25(12):2103–2111

Ramos-Solano B, Algar E, Gutierrez-Mañero FJ, Bonilla A, Lucas JA, García-Seco D (2015) Bacterial bioeffectors delay postharvest fungal growth and modify total phenolics, flavonoids and anthocyanins in blackberries. LWT Food Sci Technol 61:437–443. https://doi.org/10.1016/j.lwt.2014.11.051

Ray LB (2010) Metabolism is not boring. Science 330(6009):1337–1337. https://doi.org/10.1126/science.330.6009.1337

Raza W, Ling N, Liu D, Wei Z, Huang Q, Shen Q (2016) Volatile organic compounds produced by Pseudomonas fluorescens WR-1 restrict the growth and virulence traits of *Ralstonia solanacearum*. Microbiol Res 192:103–113. https://doi.org/10.1016/j.micres.2016.05.014

Rivas S, Thomas CM (2005) Molecular interactions between tomato and the leaf mold pathogen *Cladosporium fulvum*. Annu Rev Phytopathol 43:395–436

Rodriguez PA, Rothballer M, Chowdhury SP, Nussbaumer T, Gutjahr C, Falter-Braun P (2019) Systems biology of plant-microbiome interactions. Mol Plant 12(6):804–821. https://doi.org/10.1016/j.molp.2019.05.006

Rosier A, Medeiros FH, Bais HP (2018) Defining plant growth promoting rhizobacteria molecular and biochemical networks in beneficial plant-microbe interactions. Plant Soil 428:35–55

Ruiz B, Chávez A, Forero A, García-Huante Y, Romero A, Sánchez M, Rocha D, Sánchez B, Rodríguez-Sanoja R, Sánchez S, Langley E (2010) Production of microbial secondary metabolites: regulation by the carbon source. Crit Rev Microbiol 36(2):146–167. https://doi.org/10.3109/10408410903489576

Ryder M, Jones D (1991) Biological control of crown gall using using *Agrobacterium* strains K84 and K1026. Funct Plant Biol 18:571–579. https://doi.org/10.1071/PP9910571

Ryffel F, Helfrich EJ, Kiefer P, Peyriga L, Portais JC, Piel J et al (2016) Metabolic footprint of epiphytic bacteria on *Arabidopsis thaliana* leaves. ISME J 10:632

Sandra A, Wright CH, Zumoff LS, Steven VB (2001) *Pantoea agglomerans* strain EH318 produces two antibiotics that inhibit *Erwinia amylovora in vitro*. Appl Environ Microbiol 67:282–292

Sansinenea E, Ortiz A (2011) Secondary metabolites of soil bacillus spp. Biotechnol Lett 33(8): 1523–1538

Schenk ST, Hernández-Reyes C, Samans B, Stein E, Neumann C, Schikora M et al (2014) N-acyl-homoserine lactone primes plants for cell wall reinforcement and induces resistance to bacterial pathogens via the salicylic acid/oxylipin pathway. Plant Cell 26:2708–2723

Schikora A, Schenk ST, Stein E, Molitor A, Zuccaro A, Kogel KH (2011) N-acyl-homoserine lactone confers resistance toward biotrophic and hemibiotrophic pathogens via altered activation of *AtMPK6*. Plant Physiol 157(1407–1418)

Schikora A, Schenk ST, Hartmann A (2016) Beneficial effects of bacteria-plant communication based on quorum sensing molecules of the N-acyl homoserine lactone group. Plant Mol Biol 90: 605–612

Schrey SD, Schellhammer M, Ecke M, Hampp R, Tarkka MT (2005) Mycorrhiza helper bacterium *Streptomyces* AcH 505 induces differential gene expression in the ectomycorrhizal fungus *Amanita muscaria*. New Phytol 168(1):205–216. https://doi.org/10.1111/j.1469-8137.2005.01518.x

Schulz-Bohm K, Zweers H, de Boer W, Garbeva P (2015) A fragrant neighborhood: volatile mediated bacterial interactions in soil. Front Microbiol 6. https://doi.org/10.3389/fmicb.2015.01212

Schulz-Bohm K, Martín-Sánchez L, Garbeva P (2017) Microbial volatiles: small molecules with an important role in intra-and inter-kingdom interactions. Front Microbiol 8:2484. https://doi.org/10.3389/fmicb.2017.02484

Selvaraj A, Thangavel K, Uthandi S (2020) Arbuscular mycorrhizal fungi (*Glomus intraradices*) and diazotrophic bacterium (*Rhizobium* BMBS) primed defense in blackgram against herbivorous insect (*Spodoptera litura*) infestation. Microbiol Res 231:126355. https://doi.org/10.1016/j.micres.2019.126355

Shanmugaiah V, Mathivanan N, Varghese B (2010) Purification, crystal structure and antimicrobial activity of phenazine-1-carboxamide produced by a growth-promoting biocontrol bacterium, *Pseudomonas aeruginosa* MML2212. J App Microbiol 108:703

Shen YC (1997) Recent progress on the research and development in agricultural antibiotic. Plant Prot Technol Ext 17:35–37

Shen B (2003) Polyketide biosynthesis beyond the type I, II and III polyketide synthase paradigms. Cur Opin Chem Biol 7(2):285–295

Siddiqui ZA (2005) PGPR: prospective biocontrol agents of plant pathogens. In: PGPR: biocontrol and biofertilization. Springer, pp 111–142

Silo-Suh LA, Stabb EV, Raffel SJ, Handelsman J (1998) Target range of zwittermicin A, an aminopolyol antibiotic from Bacillus cereus. Cur Microbiol 37(1):6–11

Silva GC, Kitano IT, Ribeiro IAF, Lacava PT (2022) The potential use of actinomycetes as microbial inoculants and biopesticides in agriculture. Front Soil Sci 2. https://doi.org/10.3389/fsoil.2022.833181

Simionato AS, Navarro MOP, de Jesus MLA, Barazetti AR, da Silva CS, Simões GC, Balbi-Peña MI, de Mello JCP, Panagio LA, de Almeida RSC, Andrade G, de Oliveira AG (2017) The effect of phenazine-1-carboxylic acid on mycelial growth of *Botrytis cinerea* produced by *Pseudomonas aeruginosa* LV strain. Front Microbiol 8. https://doi.org/10.3389/fmicb.2017.01102

Singh JS (2013) Plant growth promoting rhizobacteria. Resonance 18:275–281

Singh VK, Singh AK, Kumar A (2017) Disease management of tomato through PGPB: current trends and future perspective. 3 Biotech 7(4):1–10

Singh VK, Singh AK, Singh PP, Kumar A (2018) Interaction of plant growth promoting bacteria with tomato under abiotic stress: a review. Agric Ecosyst Environ 15(267):129–140

Singh M, Srivastava M, Kumar A, Singh AK, Pandey KD (2020) Endophytic bacteria in plant disease management. In: Kumar A, Singh KV (eds) Microbial endophytes: prospects for sustainable agriculture. Woodhead Publication, Cambridge USA, pp 61–89. https://doi.org/10.1016/B978-0-12-818734-0.00004-8

Slack SM, Walters KJ, Outwater CA, Sundin GW (2021) Effect of kasugamycin, oxytetracycline, and streptomycin on in-orchard population dynamics of *Erwinia amylovora* on apple flower stigmas. Plant Dis 105(6):1843–1850. https://doi.org/10.1094/pdis-07-20-1469-re

Smith K, Havey M, Handelsman J (1993) Suppression of cottony leak of cucumber with *Bacillus cereus* strain UW85. Plant Dis 77(2):139–142

Soerensen D, Nielsen TH, Christophersen C, Soerensen J, Gajhede M (2001) Cyclic lipoundecapeptide amphisin from *Pseudomonas* sp. strain DSS73. Acta Crystallogr Sect C Cryst Struct Commun 57(9):1123–1124

Song C, Schmidt R, de Jager V, Krzyzanowska D, Jongedijk E, Cankar K, Beekwilder J, van Veen A, de Boer W, van Veen JA (2015) Exploring the genomic traits of fungus-feeding bacterial genus Collimonas. BMC Genomics 16(1):1–17

Sornakili A, Thankappan S, Sridharan A, Nithya P, Uthandi S (2020) Antagonistic fungal endophytes and their metabolite-mediated interactions against phytopathogens in rice. Physiol Mol Plant Pathol 112:101525

Spraker JE, Jewell K, Roze LV, Scherf J, Ndagano D, Beaudry R, Linz JE, Allen C, Keller NP (2014) A volatile relationship: profiling an inter-kingdom dialogue between two plant pathogens, *Ralstonia solanacearum* and *Aspergillus flavus*. J Chem Ecol 40(5):502–513

Sridharan AP, Thankappan S, Karthikeyan G, Uthandi S (2020) Comprehensive profiling of the VOCs of *Trichoderma longibrachiatum* EF5 while interacting with *Sclerotium rolfsii* and *Macrophomina phaseolina*. Microbiol Res 236:126436. https://doi.org/10.1016/j.micres.2020.126436

Sridharan AP, Sugitha T, Karthikeyan G, Nakkeeran S, Sivakumar U (2021) Metabolites of *Trichoderma longibrachiatum* EF5 inhibits soil borne pathogen, *Macrophomina phaseolina* by triggering amino sugar metabolism. Microb Pathog 150:104714. https://doi.org/10.1016/j.micpath.2020.104714

Stephen OD, Lydon J (1987) Herbicides from natural compounds. Weed Technol 1:122–128

Stokowa-Sołtys K, Jeżowska-Bojczuk M (2013) A rice fungicide blasticidin S efficiently binds cu (II) ions and prevents DNA from metal-induced damage. J Inorg Biochem 127:73–78. https://doi.org/10.1016/j.jinorgbio.2013.06.014

Tahir HAS, Gu Q, Wu H, Raza W, Safdar A, Huang Z et al (2017) Effect of volatile compounds produced by *Ralstonia solanacearum* on plant growth promoting and systemic resistance inducing potential of *Bacillus* volatiles. BMC Plant Biol 17:133

Tanaka Y, Omura S (1993) Agroactive compounds of microbial origin. Annu Rev Microbiol 47: 57–87

Toussaint J-P, Smith F, Smith S (2007) Arbuscular mycorrhizal fungi can induce the production of phytochemicals in sweet basil irrespective of phosphorus nutrition. Mycorrhiza 17:291–297. https://doi.org/10.1007/s00572-006-0104-3

Tsugawa H, Cajka T, Kind T, Ma Y, Higgins B, Ikeda K, Kanazawa M, VanderGheynst J, Fiehn O, Arita M (2015) MS-DIAL: data-independent MS/MS deconvolution for comprehensive metabolome analysis. Nat Methods 12(6):523–526. https://doi.org/10.1038/nmeth.3393

Tsujibo H, Sato T, Inui M, Yamamoto H, Inamori Y (1988) Intracellular accumulation of phenazine antibiotics produced by an alkalophilic actinomycete. I. Taxonomy, isolation and identification of the phenazine antibiotics. Agric Biol Chem 52:301–306

Tugizimana F, Mhlongo MI, Piater LA, Dubery IA (2018) Metabolomics in plant priming research: the way forward? Int J Mol Sci 19(6). https://doi.org/10.3390/ijms19061759

Tyagi S, Lee K-J, Shukla P, Chae J-C (2020) Dimethyl disulfide exerts antifungal activity against sclerotinia minor by damaging its membrane and induces systemic resistance in host plants. Sci Rep 10(1):6547. https://doi.org/10.1038/s41598-020-63382-0

Tyc O, Zweers H, de-Boer W, Garbeva P (2015) Volatiles in inter-specific bacterial interactions. Front Microbiol 6:1412

Ullah A, Bano A, Janjua HT (2020) Chapter 3 - Microbial secondary metabolites and defense of plant stress. In: Singh JS, Vimal SR (eds) Microbial services in restoration ecology. Elsevier, pp. 37–46. doi:https://doi.org/10.1016/B978-0-12-819978-7.00003-8

van de Mortel JE, de Vos RC, Dekkers E, Pineda A, Guillod L, Bouwmeester K, van Loon JJ, Dicke M, Raaijmakers JM (2012) Metabolic and transcriptomic changes induced in *Arabidopsis* by the rhizobacterium *Pseudomonas fluorescens* SS101. Plant Physiol 160(4):2173–2188. https://doi.org/10.1104/pp.112.207324

van Lenteren JC, Bolckmans K, Köhl J, Ravensberg WJ, Urbaneja A (2018) Biological control using invertebrates and microorganisms: plenty of new opportunities. Biol Control 63(1):39–59. https://doi.org/10.1007/s10526-017-9801-4

Vessey JK (2003) Plant growth promoting rhizobacteria as biofertilizers. Plant Soil 255:571–586

Viaene T, Langendries S, Beirinckx S, Maes M, Goormachtig S (2016) *Streptomyces* as a plant's best friend? FEMS Microbiol Ecol 92(8). https://doi.org/10.1093/femsec/fiw119

Vimal SR (2018) Effect of plant growth promoting rhizobacteria (PGPR) and farmyard manure (FYM) amendment on growth parameters and antioxidant level in paddy (*Oryza Sativa* L.) crop under soil salinity Ambedkar (central). University Lucknow, Uttar Pradesh

Vimal SR, Singh JS (2019) Salt tolerant PGPR and FYM application in saline soil paddy agriculture sustainability. Climate Change Environ Sustain 7:61–71

Vimal SR, Singh JS, Arora NK, Singh S (2017) Soil-plant-microbe interactions in stressed agriculture management: a review. Pedosphere 27:177–192

Vimal SR, Gupta J, Singh JS (2018) Effect of salt tolerant *Bacillus* sp. and *Pseudomonas* sp. on wheat (*Triticum aestivum* L.) growth promotion under soil salinity: a comparative study. Microbiol Res 9:26–32

Vimal SR, Patel VK, Singh JS (2019) Plant growth promoting *Curtobacterium albidum* strain SRV4: an agriculturally important microbe to alleviate salinity stress in paddy plants. Ecol Indic 105:553–562

Vinale F, Sivasithamparam K (2020) Beneficial effects of *Trichoderma* secondary metabolites on crops. Phytother Res 34(11):2835–2842. https://doi.org/10.1002/ptr.6728

Vinale F, Marra R, Scala F, Ghisalberti EL, Lorito M, Sivasithamparam K (2006) Major secondary metabolites produced by two commercial *Trichoderma* strains active against different phytopathogens. Lett Appl Microbiol 43(2):143–148. https://doi.org/10.1111/j.1472-765X.2006.01939.x

Vurukonda S, Giovanardi D, Stefani E (2018) Plant growth promoting and biocontrol activity of *Streptomyces* spp. as endophytes. Int J Mol Sci 19(4):952. https://doi.org/10.3390/ijms19040952

Wishart DS, Feunang YD, Marcu A, Guo AC, Liang K, Vázquez-Fresno R, Sajed T, Johnson D, Li C, Karu N, Sayeeda Z (2018) HMDB 4.0: the human metabolome database for 2018. Nucleic Acids Res 46(D1):D608–D617

Worley B, Powers R (2013) Multivariate analysis in metabolomics. Curr Metabol 1(1):92–107. https://doi.org/10.2174/2213235x11301010092

Wu Q, Sun R, Ni M, Yu J, Li Y, Yu C, Dou K, Ren J, Chen J (2017) Identification of a novel fungus, *Trichoderma asperellum* GDFS1009, and comprehensive evaluation of its biocontrol efficacy. PLoS One 12(6):e0179957. https://doi.org/10.1371/journal.pone.0179957

Wuolikainen A, Jonsson P, Ahnlund M, Antti H, Marklund SL, Moritz T, Forsgren L, Andersen PM, Trupp M (2016) Multi-platform mass spectrometry analysis of the CSF and plasma metabolomes of rigorously matched amyotrophic lateral sclerosis, Parkinson's disease and control subjects. Mol BioSyst 12(4):1287–1298. https://doi.org/10.1039/c5mb00711a

Xiao J, Guo X, Qiao X, Zhang X, Chen X, Zhang D (2021) Activity of fengycin and iturin A isolated from *Bacillus subtilis* Z-14 on *Gaeumannomyces graminis* var. *Tritici* and soil microbial diversity. Front Microbiol 12. https://doi.org/10.3389/fmicb.2021.682437

Xie L-J, Chen Q-F, Chen M-X, Yu L-J, Huang L, Chen L, Wang F-Z, Xia F-N, Zhu T-R, Wu J-X (2015) Unsaturation of very-long-chain ceramides protects plant from hypoxia-induced damages by modulating ethylene signaling in *Arabidopsis*. PLoS Genet 11(3):e1005143

Yamaguchi I (1996) Pesticides of microbial origin and applications of molecular biology. In: Copping LG (ed) Crop protection agents from nature: natural products and analogues. The Royal Society of Chemistry, Cambridge, pp 27–49

Yi SY, Shirasu K, Moon JS, Lee SG, Kwon SY (2014) The activated SA and JA signaling pathways have an influence on flg22-triggered oxidative burst and callose deposition. PLoS One 9:e88951

Yu GY, Sinclair JB, Hartman GL, Bertagnolli BL (2002) Production of iturin A by *Bacillus amyloliquefaciens* suppressing *Rhizoctonia solani*. Soil Biol Biochem 34:955–963. https://doi.org/10.1016/S0038-0717(02)00027-5

Zhang J, Banko G, Wolfe S, Demain AL (1987) Methionine induction of ACV synthetase in *Cephalosporium acremonium*. J Ind Microbiol 2:251–255

Zhang J, Wang XJ, Yan YJ, Jiang L, Wang JD, Li BJ, Xiang WS (2010) Isolation and identification of 5-hydroxyl-5-methyl-2-hexenoic acid from *Actinoplanes* sp. HBDN08 with antifungal activity. Bioresour Technol 101:8383–8388

Zhao Z, Wang Q, Wang K, Brian K, Liu C, Gu Y (2010) Study of the antifungal activity of *Bacillus vallismortis* ZZ185 *in vitro* and identification of its antifungal components. Bioresour Technol 101(1):292–297

Zohora US, Ano T, Rahman MS (2016) Biocontrol of *Rhizoctonia solani* K1 by iturin A producer *Bacillus subtilis* RB14 seed treatment in tomato plants. Adv Microbiol 6(06):424

Chapter 2
Role of Cerato-Platanins in Inducing PAMP-Triggered Immunity in Plants

N. M. R. Ashwin, Dharmaraj Amalamol, Kana Valiyaveettil Lakshana, M Remya, Amalraj Ramesh Sundar, Palaniyandi Malathi, and Rasappa Viswanathan

Abstract Cerato-platanins (CPs) are a group of small, cysteine-rich, non-catalytic proteins secreted abundantly by most of the filamentous fungi. They were characterised either as microbe-associated molecular pattern (MAMP) or pathogen-associated molecular pattern (PAMP) based on their presence in the non-pathogenic or pathogenic fungi, respectively. CPs possess diverse biophysical and biochemical properties, viz. chitin-binding, cellulose loosening, surface active, polarity altering, hydrophobin-like or expansin-like activities that altogether aid the biocontrol agent *Trichoderma* spp., the mycoparasitism and host defence-triggering attributes. On the contrary, CPs also act as virulence determinant in some fungal pathogens. Meanwhile, priming/pre-treatment of CPs induces an array of defence responses in host plants, thereby activating PAMP-triggered immunity (PTI), which is often endowed with broad spectrum of resistance. Because of these paradoxical biological properties, CPs are characterised as phytotoxins, effectors, PAMPs, elicitors, biosurfactant and bioemulsifier in different microbial pathosystems. Hence, the cerato-platanin proteins (CPPs) possess immense potential to be employed in plant disease management strategies and other biotechnological (industrial) applications.

Keywords Cerato-platanins (CPs) · Filamentous fungi · Plant immune system · Plant protection · Plant–pathogen interactions

N. M. R. Ashwin · D. Amalamol · K. V. Lakshana · M. Remya · A. Ramesh Sundar (✉) · P. Malathi · R. Viswanathan
ICAR-Sugarcane Breeding Institute, Indian Council of Agricultural Research, Coimbatore, India

K. K. Bastas et al. (eds.), *Microbial Biocontrol: Molecular Perspective in Plant Disease Management*, Microorganisms for Sustainability 49,
https://doi.org/10.1007/978-981-99-3947-3_2

2.1 Introduction

In the microbial world, micro-organisms acquire energy from external environment through various modes, ranging from photo-assimilation to parasitism for growth and survival. Based on the nutritional relationship with external biota, the interaction may be symbiotic, antagonistic or synergistic in nature, which leads mutualism, commensalism and parasitism kind of interactions (Moënne-Loccoz et al. 2015; Drew et al. 2021). At the molecular level, these microbes and all other living organisms manoeuvre mainly through few specialised biomolecules like proteins, metabolites and polysaccharides, which will interact among themselves and with the external environment to accomplish their biological processes (Braga et al. 2016; Ashwin et al. 2017a, b; Kumar 2022). Similarly, during microbial biocontrol, antagonism or antibiosis sort of interactions occur when the microbial agents or their derived (molecular) products like proteins, metabolites are employed to control a range of pathogens and pests from invading the plants (Köhl et al. 2019; Kumari et al. 2022). These microbial agents either directly interact with the pathogen or indirectly elicit host defence to act against the pathogen.

All living entities right from a single-celled micro-organism to a well-developed, complex multicellular plants and animals do continuously evolve to get adapted to their dynamic habitat perturbations and co-evolve to compete with their dependant and associative biota for growth and survival (Mendes et al. 2013; Pathak et al. 2022). For instance, in a phytopathosystem, both the pathogen and the plant continuously evolve to overcome the attack by host defence mechanism and to circumvent pathogenicity determinants, respectively (Lyu et al. 2021). At the molecular level, this dynamic co-evolution engenders insertion/deletion mutations on potential proteins, which mediate the perception, interaction, signalling, etc. between the host and the pathogen (Jones and Dangl 2006; Ashwin et al. 2020a, b). Sometimes, a new set of proteins may also get evolved (acquired) by means of horizontal or vertical gene transfer from closely related species to tackle the evolutionary crisis against the aggressor (Gluck-Thaler et al. 2020). Over the years, these alterations facilitated both the host and the pathogen to develop protective structural barriers, and secrete toxic proteins and metabolites to act against the attacker. Thus, plants have developed the ability to perceive the presence of pathogen using their conserved molecular signatures called pathogen-associated molecular patterns (PAMPs) to trigger the first line of active defence called PAMP-triggered immunity (PTI) (Bohm et al. 2014). Pathogen on its counterpart secrete a range of small cysteine-rich proteins called effectors on host's intercellular spaces and cytoplasm to suppress the immunity triggered by PTI (Sundar et al. 2018). The second line of active defence prompted by plants against these effectors is called as effector-triggered immunity (ETI) and the counter action against a plant by a pathogen is called as effector-triggered susceptibility (ETS) (Lapin and Van den Ackerveken 2013; Wang and Wang 2018). On the host side, the defence mechanisms are triggered by the perception of PAMPs and effectors with a range of receptors like pattern-recognition receptors (PRRs) and R genes, respectively (Macho and Zipfel 2014). The specificity and the abundance of these receptors against the multitude effectors decide the qualitative and quantitative resistance level of the host against particular diseases. Conventionally, the term elicitor is referred to PAMP molecules most often, with some exceptions to certain effectors, which were also reported to induce host defence upon priming before the real-time interaction of pathogens (Miwa and Okazaki 2017; Ashwin et al. 2018).

Cerato-platanins (CPs) or cerato-platanin proteins (CPPs) are a highly conserved, small secreted, cysteine-rich, non-catalytic proteins produced abundantly by all types of filamentous fungi with diverse nutritional lifestyles (Gaderer et al. 2014). The protein was first reported by Pazzagli et al. (1999) in *Ceratocystis platani.* Thereafter, many research groups reported similar kind of proteins in many filamentous fungi and described them as Cerato-platanin and categorised such proteins as Cerato-platanin protein family (CPPs) (Chen et al. 2013). Evolutionarily, CPPs were subjected to many events of lateral gene transfer and gene duplication events, which led to evolve new homologues in most of the filamentous dikaryotic fungi (Yu and Li 2014; Gao et al. 2020). CPPs secreted by the multicellular filamentous fungi often form a protective external layer around the hyphal cell wall structures and self-assembles to form a thin protein layer on the air–water interface or on the surface of aqueous solutions (Baccelli 2015). Biological function-wise, this non-catalytic protein has been reported to act as phytotoxin, elicitor and effector in different pathosystems, as it is present in both beneficial and harmful filamentous fungi (Pazzagli et al. 2014; Baccelli 2015). Besides, CPPs are also reported to possess the properties of hydrophobins and expansins during host–pathogen interactions (Martellini et al. 2013; Baccelli et al. 2014a). Biochemical properties of CPs include self-assembling, hydrophobicity, biosurfactant and bioemulsification (Pitocchi et al. 2020; Gao et al. 2020). However, these biological and biochemical properties differ among their homologues and orthologues.

Plant disease management mediated by biological control agents (BCA) is a safe, environment friendly and sustainable practice in modern-day agriculture. *Trichoderma* spp. is one of the predominant BCAs, the potential of which is being continuously harnessed worldwide for effective plant disease management. Deciphering the molecular dialogue between *Trichoderma* and their hosts has the potential to offer new avenues in the biocontrol of plant pathogens. In this context, CPPs are rightly classified as PAMPs or microbe-associated molecular patterns (MAMPs), as documented very well in *Trichoderma* spp. CPs of *Trichoderma* spp. has been reported to confer a major contribution in the overall efficacy of microbial biocontrol by both mycoparasitism and triggering host resistance (Gomes et al. 2015; Cheng et al. 2018; Singh et al. 2017; Singh et al. 2020). Hence, this chapter is focussed on the diverse properties of CPPs, how it triggers host defence and their application potential in terms of agriculture and industrial perspectives.

2.2 Significance of Secretory Systems in Plant–Pathogen Interactions

Secretory systems play a critical role in driving growth, development, sustenance, biogenesis, associations and interactions of both plants and microbes by acting as an arterial network transporting a range of molecules, viz. ion, solutes, food, proteins/ enzymes and metabolites, to an external environment (Lucke et al. 2020). During plant–pathogen interactions, plants secrete antimicrobial molecules including phytoalexins and the receptor molecules to be transported to the apoplastic spaces and cell wall, respectively, through the well-developed intercellular membranous structures like endoplasmic reticulum, Golgi apparatus and vesicles (Wang and Dong 2011; Zhang et al. 2019). For instance, proteins that are targeted to extracellular space or cell membrane are often tagged with signal peptides for guiding them to the destination during transport. Similar mechanisms are also present in fungi as well. Unlike bacterial secretion systems where at least five different secretion mechanisms have been well characterised, fungal secretion mechanism, especially in the long filamentous fungal hyphae secretion system, remains largely unexplored to date (Amaral et al. 2015; Wang et al. 2020). Fungal secretion system is highly polarised as it is mostly active at the hyphal tips and wherever do interactions take place. The compartmentalised porous septa of the hyphae allow the transport of nutrients, organelles, carbohydrate-active enzymes, effectors and polysaccharides to the extracellular spaces for nutrient assimilation, self-protection, trafficking and symbiotic/ antagonistic interactions (Mergaert 2018). CPs, the small secretory proteins of filamentous fungi, also play a versatile role like self-protection, symbiotic as well as antagonistic interactions in biocontrol agents like *Trichoderma* spp. and phytopathogens like *Ceratocystis* spp., respectively.

2.3 Phylogenetic Distribution of CPs Among Filamentous Fungi

CP proteins are delimited to filamentous fungi belonging to the Ascomycete and Basidiomycete phyla. The group represents evolutionarily conserved protein homologues with diverse functions from defence elicitation to virulence factors (Gaderer et al. 2014). Genomic distribution of CP-encoding genes reveals several gene homologues within a species (Yu and Li 2014). For example, Barsottini et al. (2013) reported 12 genes (*MpCP1–MpCP12*) coding for CP from *Moniliophthora perniciosa*, which are differentially expressed during fungal growth and development, whereas, in genus like *Trichoderma*, CP is represented by 3 genes, *epl1*, *epl2* and *epl3*, among which *epl1* is the only predominant form. In an attempt to decipher the phylogenetic distribution of CP proteins, 153 protein sequences of major CP homologues from different fungal species were used to develop a phylogram using iQTree 1.6.12 (Fig. 2.1). The entire tree can be divided into ten different clades

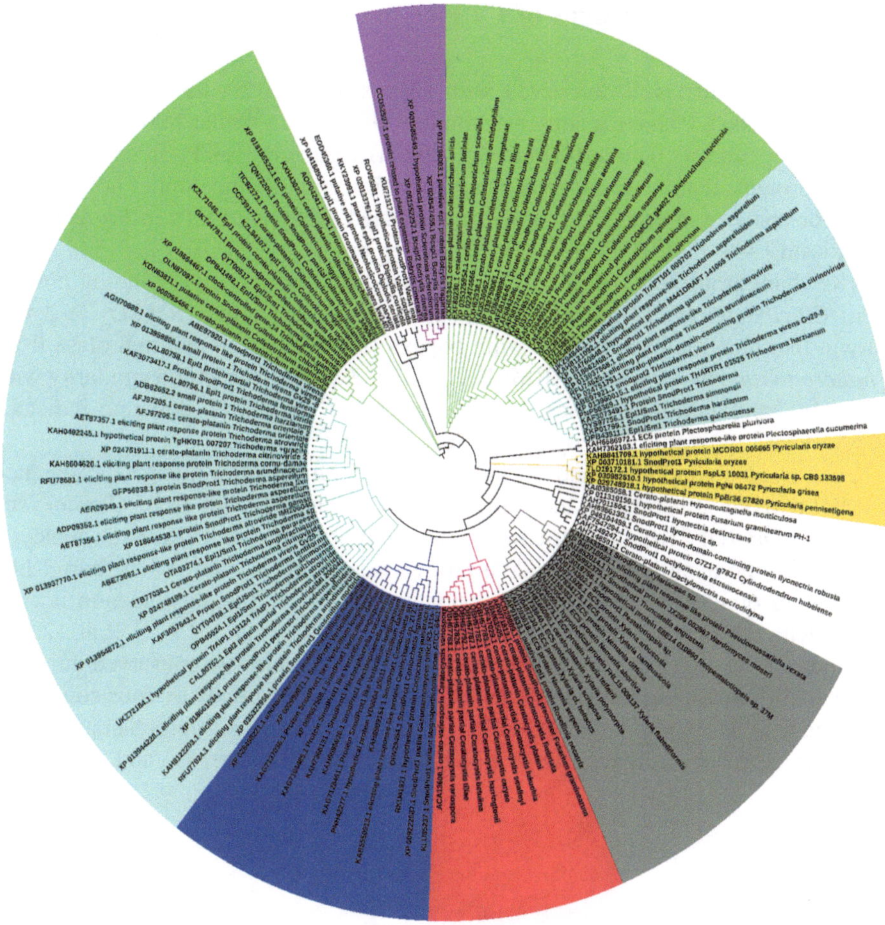

Fig. 2.1 Phylogenetic distribution of major CPPs among the filamentous fungi; maximum-likelihood phylogram developed using iQTree (http://www.iqtree.org/) (1000 bootstrap replicates) and visualised by iTOL (https://itol.embl.de/). Clades are given different shades

wherein *Trichoderma* and *Colletotrichum* are placed in two clades each. Intriguingly, CPPs of the most common BCA – *Trichoderma* spp. – are distributed separately as two different clades (magenta shade) and also in one of the *Colletotrichum* clades (green shade), most of which are demonstrated to elicit host defence responses.

2.4 Structural Features of CPPs

CPPs are generally well conserved with approximately 120–140 amino acids or 11.5–14.7 KDa. CPs have a conserved double ψβ-barrel globular fold composed of two α-helices and six β-strands (Luti et al. 2020a). Structurally, CPs are related to the type of proteins that are engaged in the recognition and modification of polysaccharide. For instance, a search for structurally related proteins of CPs in Protein Data Bank using CO-FACTOR (https://zhanggroup.org/COFACTOR/) and TM-Align (https://zhanggroup.org/TM-align/) servers matched with the sugar-binding protein from *Clavibacter michiganensis* (Root mean square deviation (RMSD) = 2.40), barwin-like plant-defence protein from papaya (RMSD = 2.2) and hydrolase from *Phanerodontia chrysosporium* (RMSD 2.1) (Fig. 2.2). This structural similarity with expansins and endoglucanases, especially in the ψβ-barrel domain, indicates polysaccharide-binding activity of CPs (Gaderer et al. 2014).

Further, computational blind-docking analysis for the nuclear magnetic resonance (NMR)-derived structure of the fungal elicitor cerato-platanin from *Ceratocystis platani* (PDB ID: 2KQA) with chitin polymer (PubChemID: 21252321) depicts a model with the ligand placed inside a groove located on one side of the barrel-like domain surrounded by hydrophobic residues (Fig. 2.3). This phenomenon is consistent with the previous observations of Luti et al. (2017) and Baroni et al. (2021) that most of the oligosaccharide-binding amino acid residues are conserved throughout the CP family (Fig. 2.4). Presence of two di-sulphide bonds by four conserved cysteine residues (Cys^{21-59} and $Cyst^{62-117}$), nine glycine, two tryptophan as well as signature sequences like CSDG/CSNG (Baccelli et al. 2015), and geometry,

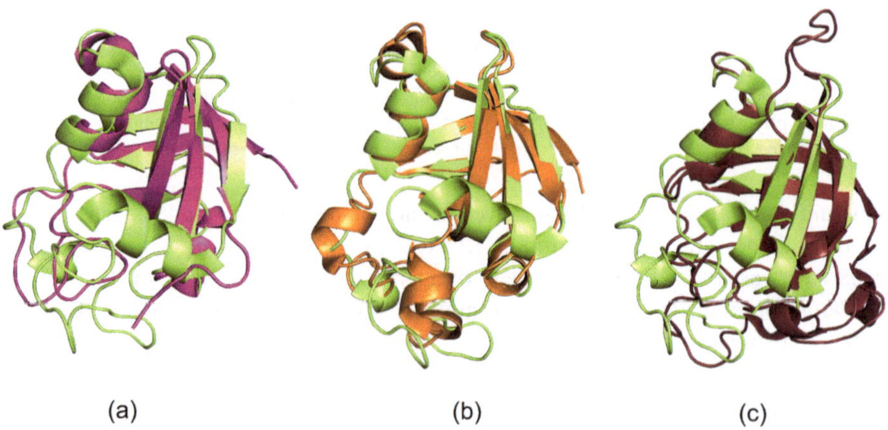

Fig. 2.2 Structural resemblances of CPs with other classes of proteins. Superimposed images of cerato-platanin protein (green; PDB ID: 2KQA) with (**a**) *Clavibacter michiganensis* expansin (PDB ID: 4JCW), (**b**) barwin-like protein from papaya (PDB ID: 4JP7) and (**c**) PcCel45A-endoglucanase from *Phanerodontia chrysosporium* (PDB ID: 3X2H)

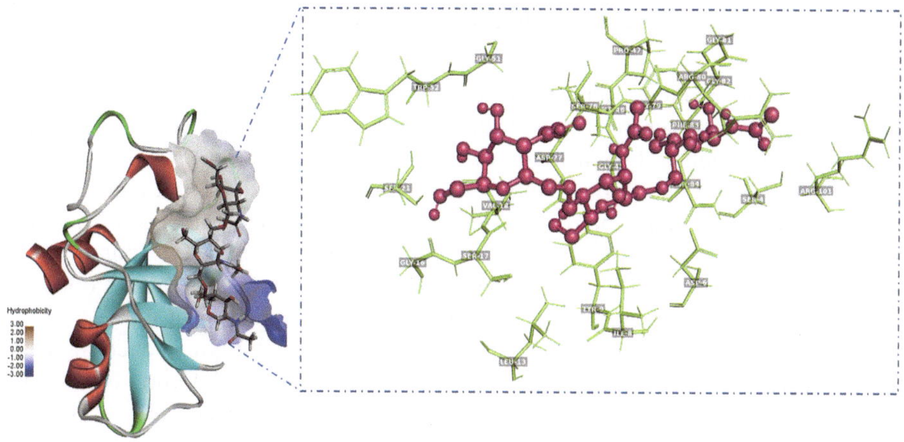

Fig. 2.3 Docked model of CP with chitin polysaccharide. The polysaccharide-binding groove located on one side of the ψβ-barrel and the chitin-interacting residues of the CP protein. Docking simulation was performed with AutoDock Vina v.1.2.0 (https://vina.scripps.edu/) and visualised via BIOVIA Discovery Studio Visualizer (https://www.3ds.com/products-services/biovia/products/molecular-modeling-simulation/biovia-discovery-studio/visualization/)

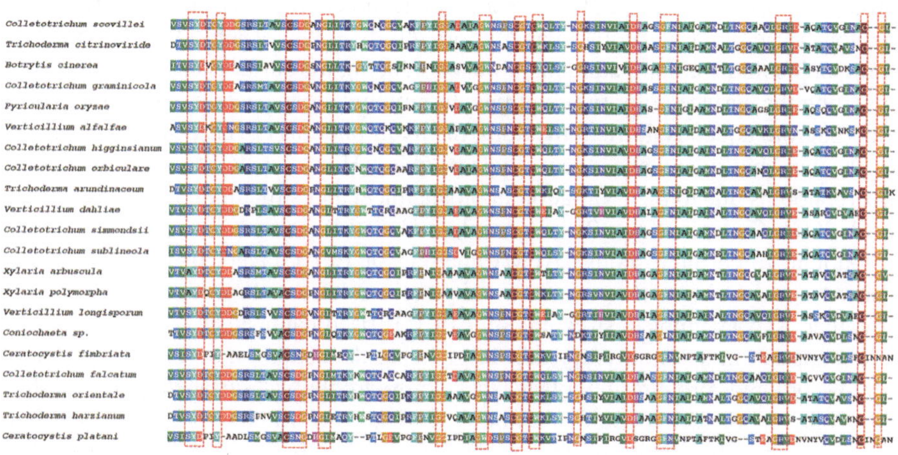

Fig. 2.4 Conservation of amino acid sequences of major CP homologues as aligned with Clustal Omega. Boxes indicate conserved residues/signatures

frequency, and non-covalent Interaction (GFN) further reiterates the conserved structure of CP proteins.

2.5 Physico-Chemical Properties of CPPs

CP family proteins are cysteine-rich, secreted and somewhat hydrophobic proteins involved in fungal growth and pathogenicity (Chen et al. 2013). They are non-catalytic proteins, demonstrated to be highly stable even in extreme temperatures (>90 °C) and pH (Pazzagli et al. 2014; Ashwin et al. 2017a, b). CPPs possess signal peptides that guide them to get secreted to the extracellular spaces. Though it gets secreted abundantly into the culture media, it was also found to be localised (loosely attached) in the fungal cell wall in some investigations (Gaderer et al. 2014; Pazzagli et al. 2014). Owing to its high abundance in the secretome, many CPPs have been found to be human allergens and phytotoxins, rendering them easy recognition by the immune system and classified as PAMP. Being localised on the cell wall, CPPs might protect the fungal cell wall polysaccharides from plant pathogenesis-related (PR) proteins, since CPP mutants were observed to be more vulnerable to chitinase and β-1,3-glucanase enzymatic treatments (Quarantin et al. 2016; Zhang et al. 2017a).

EPL1 (eliciting plant-response like protein) that belongs to CPPs from *Trichoderma atroviride* was reported to readily self-assemble and form a protein biofilm at air–water interfaces, which are the properties of hydrophobin class of proteins (Frischmann et al. 2013; Landeta-Salgado et al. 2021). Being secreted as a monomeric protein, hydrophobins tend to self-assemble at air–water interfaces to form an insoluble amphipathic membrane, reducing the surface tension and thereby inverting the surface polarity properties. In biological perspective, when the hyphae emerge from aqueous nutrient medium, the hydrophobins cover the aerial structures as a thin biofilm make them hydrophobic. However, CPPs do vary from hydrophobins due to their variation in biochemical and structural properties (Sbrana et al. 2007; Gaderer et al. 2014; Pazzagli et al. 2014).

Atomic force microscopy (AFM) imaging of EPL1 layers from *T. atroviride*, carried upon drying of large amount of protein solution, showed a meshwork-like irregular, self-assembled structure of EPL1 (Frischmann et al. 2013). However, when applied in solution (very thin layer being deposited on surface), it formed a highly ordered protein monolayer at the hydrophobic surface/liquid interface (Bonazza et al. 2015). Self-assembling potential of EPL1 from *Colletotrichum falcatum* was also demonstrated by Ashwin et al. (2017a, b), wherein a drop of protein solution incubated for few hours at room temperature gets self-assembled by forming irregularly crinkled globular crystal structures.

Sbrana et al. (2007) proposed a hypothesis for the dynamic aggregation mechanism of CP, wherein the process of formation of annular ring-shaped oligomers that ultimately resulted in large macrofibrillar assemblies on surface was explained. Later, Pazzagli et al. (2009) elaborated the process of aggregation of CPs from *Ceratocystis platani* via a nucleated growth mechanism of self-assembled protein layers. This nucleation model of aggregation process ended up with formation of ordered aggregates through the soluble prefibrillar structures. But their assembly was found to be affected by its interaction with a hydrophobic surface (Carresi et al.

2006). Similarly, CP protein aggregation in *C. platani* and *Ceratocystis populicola* mimicked the behaviour of amyloid-like protein assembly (Comparini et al. 2009). Self-assembly mechanism of CP from a marine fungus, *Paradendryphiella salina*, also showed the formation of ovoidal–fibrillar structures together with many small oligomers as a result of the hierarchical aggregation mechanism. The aggregation process was reported to be mediated by the polysaccharide content of seaweed that was used as a carbon source (Landeta-Salgado et al. 2021).

2.6 Hydrophobins vs CPs vs Expansins

The structural and/or biochemical properties of CPs in many ways resemble the properties of hydrophobins and expansins. CPs are speculated to be hydrophobin-like proteins owing to their ability to self-assemble and also due to the presence of conserved cysteine residues (Chen et al. 2013). For instance, similar to CPs, hydrophobins also get self-assembled at air–water interfaces, possess surface-active properties and alter the polarity of the solution or any aqueous surroundings (Frischmann et al. 2013; Bonazza et al. 2015). The hydrophilic and hydrophobic residues at the surface of hydrophobins easily invert the polarity of the interface between the fungus and the environment (Linder et al. 2005). Hydrophobins are also cysteine-rich, filamentous, fungal-specific proteins that secrete and self-assemble around the hyphal structure to aid as hydrophobic self-protective interface for adherence (Gaderer et al. 2014). Even though the hydrophobins do not possess sequence or structural similarity with CPPs, the surface-altering properties of both proteins are similar as stated above.

Unlike hydrophobins, self-assembly of EPL1 was found to be reversible, as these protein layers are easily dissolved by agitating or stirring the solution. Hence, CPPs are more soluble than hydrophobins (Linder et al. 2005). This is because 60% of the amino acid residues of CP homologues are non-polar, when analysed using EMBOSS PepStats (https://www.ebi.ac.uk/Tools/seqstats/emboss_pepstats/). The solvent accessible surface area (SASA) and hydrophobicity of the CP protein 2KQA, estimated using BioVia Discovery Studio Visualizer, depicts hydrophilic and hydrophobic patches and more solvent accessible area in its protein topology (Fig. 2.5a, b). This is in agreement with the observation made by Gaderer et al. (2014) regarding good solubility of EPL1 in aqueous solutions. Besides that, the surface architecture of CPs is at par with hydrophobins and can be called amphipathic.

Furthermore, hydropathy plots reveal that CPPs have less hydrophobic areas than hydrophobins (Seidl et al. 2006). When a mixture of equimolar solutions of EPL1 protein and hydrophobin was analysed under AFM, it showcased a similar topographical feature to pure EPL1/hydrophobin, but the hybrid protein layer so formed unprecedentedly exhibited new properties. The mixed protein layer did not invert, rather increased the hydrophobicity of surface like EPL1 and it was water insoluble like hydrophobins (Bonazza et al. 2015). The new properties might be due to slightly

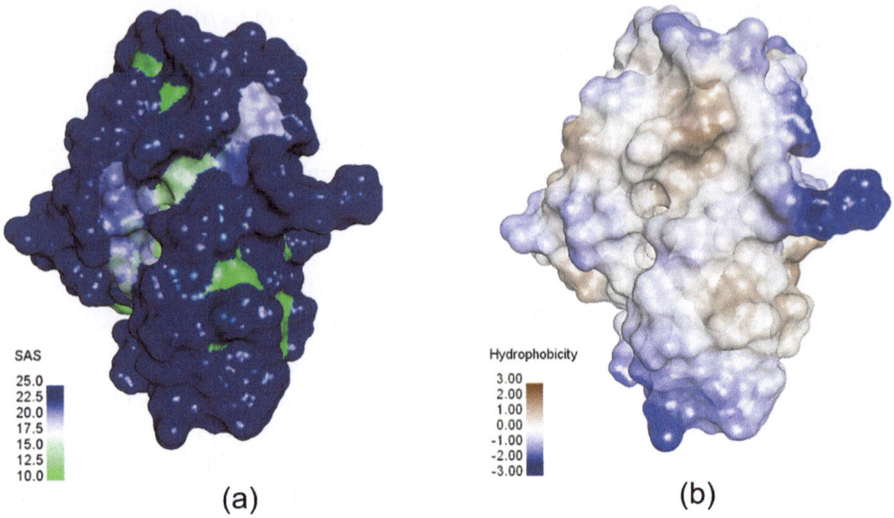

Fig. 2.5 Cerato-platanin protein surface attributes: Solvent-accessible surface area (**a**) and hydrophobicity (**b**) as displayed by BioVia Discovery Studio Visualizer

altered orientation of hydrophilic/hydrophobic areas of the proteins, which would aid in studying combined action of CPPs and hydrophobins present on fungal cell walls.

On the contrary to hydrophobins, expansins and CPs share structural similarity, specifically in their tertiary structure. Luti et al. (2017) defined CP as a novel monodomain expansin due to the presence of conserved Aspartate-D^{77} in the expansin-like domain. Similar to expansins, CPs also possess double ψβ-barrel structure with four conserved cysteine residues (de Oliveira et al. 2011). Evolutionary conservation of amino acids of a protein reflects its tendency to mutate, but at the same time it often retains structural integrity and functional characteristics of the macromolecule (Ashkenazy et al. 2016). The roles of microbial expansins are digestion of cell wall and cellulose fibril-loosening activities during interaction with plants (Cosgrove 2017). The chitin-binding and cellulose-loosening ability of CPPs, which closely resemble the properties of expansin proteins (de Oliveira et al. 2011; Baccelli et al. 2014a), suggest that they could weaken fungal or plant cell wall, and possibly involve in cell wall remodelling to promote hyphal growth or its progression in plants (Baccelli et al. 2015; Eranthodi et al. 2022). Luti et al. (2020a) reported that the expansin-like activity of CPs depends on the net charge for binding with cellulose and the carboxyl group of D77 residue is crucial for PAMP and expansin-like activities of CPs. The synergistic activities of expansin-like activities of CPPs along with other hydrolytic enzymes are proposed to aid the fungal pathogen and BCA to invade the host plant tissues easily (Baccelli 2015; Quarantin et al. 2019; Rovera et al. 2021).

2.7 Biological Significance of CPPs: As Elicitor, Effector and Biosurfactants

CPPs are one of the most abundantly secreted proteins of filamentous fungus during both in vitro culturing and in planta colonisation stages. Over the years many research groups conducted a series of investigations to explore the biological significance of CPPs of filamentous fungus, during which the diverse and contrasting functional roles of CPPs in different pathosystems and their potential for industrial applications have been reported. In order to present an overall glimpse of the research findings on CPPs reported to date, the information is compiled and illustrated in Table 2.1.

Although most of the gene-disruption and silencing studies of CPPs indicated that it is dispensable for growth and survival, there have been reports that CPPs are involved in virulence, act as an effector and possess the ability to trigger cell death in plants (Frias et al. 2011; Zhang et al. 2017b; Wang et al. 2018; Yang et al. 2018). Nevertheless, an equivalent or even more number of reports highlighted their involvement in inducing defence and enhancing disease resistance upon priming (pre-treatment) in other pathosystems and hence, reported to act as an elicitor or PAMP (Djonovic et al. 2006, 2007; Vargas et al. 2008; Yang et al. 2009; Wang et al. 2013; Baccelli et al. 2014a, b; Gomes et al. 2015, 2017; Ashwin et al. 2017a, b, b; Hong et al. 2017; Yu et al. 2018). Besides inducing defence-related genes/products or triggering PTI, CPPs are reported to suppress the expression level of pathogenicity genes during biocontrol, reduce the surface hydrophobicity and modify surfactant properties that could facilitate nutritional versatility around the hydrosphere of fungal hyphae.

CPPs have high affinity to chitin binding and interact with hydrophobic surfaces like wax cuticle of host leaves, which induces the (partial) unfolding of CP to make it biologically active to trigger PTI (Martellini et al. 2013). Though not all CPPs are glycosylated to regulate dimerisation of CP monomer, very few CPs were reported to do so. Studies on dimerisation of CPPs from *Trichoderma* spp. provided a preliminary indication of self-assembling potential of the protein (Seidl et al. 2006; Vargas et al. 2008). EPL1s of *Trichoderma* spp. and *C. falcatum* were able to forge a stable dimer, which was apparent even after treatment with various protein-denaturing agents (Seidl et al. 2006; Ashwin et al. 2017a, b). However, EPL1 of *C. falcatum* was not reported to be glycosylated. Mass spectrometry (MS) analysis of EPL1 of *Trichoderma* spp. revealed that the dimeric form had a double-oxidised tryptophan residue. Furthermore, the presence of glycosylation site negatively impacted the aggregation ability of CPPs, which also altered its recognition by the host (Vargas et al. 2008). Aggregated CP was found to be more effective in eliciting defence responses (PTI) in plants (Pazzagli et al. 2009), but on the contrary, Sm1 and EPL1 showcased reduced potential to induce defence response upon dimerisation (Vargas et al. 2008).

Table 2.1 List of cerato-platanin family proteins (CPPs) characterised and their biological significance under different interactions

S. No.	CPPs	Protein description	Organism/source	Host/substrate	Biological significance	References
1.	Cerato-platanin (CP)	*Ceratocystis platani* protein	*Ceratocystis platani*	Tobacco, *Platanus acerifolia*	Elicitor, phytotoxin, polysaccharide recognition, involved in hyphal growth and conidiogenesis; suppresses *Ceratocystis platani* colonisation after priming; CP localises in the cell wall of conidia and hyphae structures	Pazzagli et al. (1999, 2006, 2009, 2014); Boddi et al. (2004); Fontana et al. (2008); de Oliveira et al. (2011); Baccelli et al. (2012)
2.	CP, Pop1	*Ceratocystis platani* protein, Cerato-populin	*C. platani*, *Ceratocystis populicola*, recombinant protein (*Pichia pastoris*)	Plane tree, poplar tree	CP self-assembles in solution; it tends to form early annular ring-shaped oligomers that act as fundamental bricks of aggregation process and microfibrillar assembly; acts as elicitor; induces mitogen-activated protein kinase (MAPK) phosphorylation; reactive oxygen species (ROS), nitric oxide (NO) generation and upregulation of defence-related genes; affinity to chitin binding; interaction with hydrophobic surface like wax cuticle in leaves induces the (partial) unfolding of CP to make it biologically active to trigger PTI	Carresi et al. (2006); Sbrana et al. (2007); Comparini et al. (2009); Lombardi et al. (2013); Martellini et al. (2013); Baroni et al. (2021)

3.	CP, Sm1	*Ceratocystis platani* protein; small protein (recombinant/ mutant)	Recombinant protein – *P. pastoris/Escherichia coli*	*Arabidopsis thaliana*, maize	Elicitor, PAMP, induces a series of primary defence responses; expansin-like activity dependant on net charge for binding with cellulose; eliciting activity not dependent on cellulose-loosening activity; carboxyl group of D77 is crucial for PAMP and expansin-like activities	Buensanteai et al. (2010); Luti et al. (2016, 2017, 2020a, b); Cosgrove (2017)
4.	SnodProt1	*Stagonospora nodorum* protein	*Stagonospora nodorum*	Wheat	Elicitor, virulence factor	Hall et al. (1999)
5.	Sp1	SnodProt1	*Leptosphaeria maculans*	*Brassica napus*	Elicitor; induces hypersensitive response (HR); not required for pathogenicity	Wilson et al. (2002)
6.	Msp1	*Magnaporthe oryzae* SnodProt1	*Magnaporthe oryzae*	Rice	Required for virulence; no phytotoxicity at sub-lethal concentration; induces PTI via MAPK, transcription factors (TFs), hormone-signalling proteins and PR proteins similar to flg22 after secretion into apoplastic space in rice	Jeong et al. (2007); Meng et al. (2018), (2019); Wang et al. (2016)
7.	MgSM1, MoSM1	*Magnaporthe (grisea) oryzae* small protein	*Magnaporthe oryzae*	*Arabidopsis*, rice	Elicitor; induced HR; ectopic expression of EPL1 confers broad spectrum of disease resistance against bacterial, viral and fungal diseases via both salicylic acid (SA)- and jasmonic acid	Yang et al. (2009); Baccelli et al. (2014a, b); Baccelli et al. (2015); Hong et al. (2017)

(continued)

Table 2.1 (continued)

S. No.	CPPs	Protein description	Organism/source	Host/substrate	Biological significance	References
					(JA)-mediated defence pathways; possess expansin-like cellulose-loosening activity	
8.	MpCP1	*Moniliophthora perniciosa* cerato-platanin	*Moniliophthora perniciosa*	Cacao and tobacco	Elicitor; induced HR (monomer controls activity); total 12 CP genes present in Mp; MpCP2 acts as expansin and facilitates basidiospore germination; MpCP5 blocks NAG6-induced defence response of cacao plants	Zaparoli et al. (2009); Barsottini et al. (2013)
9.	BcSpl1	*Botrytis cinerea* SnodProt like	*Botrytis cinerea;* recombinant protein – *P. pastoris* (rBcSPL1-p)/*E. coli* (rBcSPL1-e)	Tomato, tobacco and *Arabidopsis*	Induces system acquired resistance (SAR); elicitor; HR; required for virulence; rBcSPL1-p induces slightly more defence than rBcSPL1-e	Frias et al. (2011, 2013, 2014); Rathi et al. (2012); Zhang et al. (2017a)
10.	ChEC5	Extracellular	*Colletotrichum higginsianum*	*Arabidopsis*	Putative effector	Kleemann et al. (2012)
11.	CfEPL1	*Colletotrichum falcatum* EPL1	*Colletotrichum falcatum*	Sugarcane and tobacco	Highly abundant PAMP; induces HR and systemic resistance in tobacco and sugarcane, respectively through SA-mediated defence pathway activation; suppresses red rot disease severity	Ashwin et al. (2017a, b, 2020b, 2022)

12.	CgCP1	*Colletotrichum gloeosporioides CP1*	*Colletotrichum gloeosporioides*	Rubber tree and tobacco	Not essential for growth and hyphal development, but involved in conidiation and virulence; induces HR in tobacco and defence activities in rubber tree	Wang et al. (2018)
13.	CtEPL1	*Colletotrichum truncatum EPL1*	*Colletotrichum truncatum*	Tobacco and *Lens culinaris*	Do not induce HR	Bhadauria et al. (2011)
14.	EPL1	Eliciting plant response-like protein	*Trichoderma atroviride*	Tobacco, tomato and maize	Elicitor; not essential for fungal growth; chitin binding; self-assembling property to form highly ordered monolayer at air/water interface	Seidl et al. (2006); Frischmann et al. (2013); Gaderer et al., (2015); Bonazza et al. (2015)
15.	EPL1, Sm1	Eliciting plant response-like protein; small protein	*Trichoderma virens, T. atroviride*	Rice, maize, tomato, cotton and *Arabidopsis*	Elicitor; no toxicity; monomer is the active form that induces defence; dimerisation controls activity; overexpression of EPL1 enhanced disease resistance; Sm1 is not glycosylated but interacts with oligosaccharides	Djonovic et al., (2006, 2007); Vargas et al. (2008); Druzhinina et al. (2011); Nawrocka and Małolepsza (2013); Tai-Long et al. (2014); Crutcher et al. (2015); Salas-Marina et al. (2015); Crutcher and Kenerley (2019)
16.	EPL1, Sm1	Eliciting plant response-like protein; small protein	*Trichoderma harzianum; Trichoderma formosa*	*Phaseolus vulgaris* (common bean), tobacco	More expression in galactose-rich medium; elicitor; essential for mycoparasitism and symbiosis; induces immunity against tomato mosaic virus (ToMV) infection	Freitas et al. (2014); Gomes et al. (2015); Ramada et al. (2016); Cheng et al. (2018)

(continued)

Table 2.1 (continued)

S. No.	CPPs	Protein description	Organism/source	Host/substrate	Biological significance	References
17.	TaEPL1	*Trichoderma asperellum EPL1*	*Trichoderma asperellum* – recombinant protein – *P. pastoris* (rEPL1-p)/*E. coli* (rEPL1-e)	Soybean, *Populus* sp.	rEPL1-p exists in both monomer and dimer; induces chitinase, glucanase, peroxidase and proteinase inhibitors against *Cercosporidium sofinum* infection in soybean; rEPL1-p induces more robust defence than rEPL1-e; triggers both SA- and JA-mediated pathway genes to suppress disease severity against *Alternaria alternata*	Wang et al. (2013); Yu et al. (2018)
18.	EPL1	Eliciting plant response-like protein	*T. harzianum; Trichoderma guizhouense*; recombinant protein – *P. pastoris*	Tomato	Induces tomato defence-related genes while regulating pathogenicity genes of *Botrytis cinerea*; reduces surface hydrophobicity and modifies surfactant properties that facilitate nutritional versatility around the hydrosphere; plays minor role in direct biotic interactions of *Trichoderma* sp. with other pathogens; EPL1 deletion suppresses JA-mediated defence; reduces root colonisation while inducing immunity	Gomes et al. (2017); Gao et al. (2020)
19.	ThCP, TrCP	*Trichoderma harzianum* CP,	*Trichoderma harzianum, Trichoderma reesei*	Apple pomace, coffee silverskin,	Efficient biosurfactant and bioemulsifier; reduces	

		Trichoderma reesei CP		potato peel, bacterial cellulose	surface tension and its pre-treatment increases sugar conversion, minimises process time and increases energy efficiency; CP hinders the activity of bacterial cellulases, thereby affecting their hydrolytic activities, which is in contrast to their enhancing effect on plant cellulose	Pitocchi et al. (2020); Pennacchio et al. (2021); Rovera et al. (2021)
20.	FgCPP1, 2	*Fusarium graminearum CPP1, 2*	Recombinant protein – *P. pastoris*, *Fusarium graminearum*	Soybean, wheat, cellulose substrates	FgCPPs protect fungal cell wall from enzymatic degradation; loosen cellulose substrates and enhance fungal cellulase activity like expansin proteins; Not essential for growth and aggressiveness, but overexpression contributes to virulence	Quarantin et al. (2016, 2019); Eranthodi et al. (2022)
21.	FocCP1	*Fusarium oxysporum f.* sp. *cubense CP1*	*Fusarium oxysporum f.* sp. *cubense*	Tobacco, *Musa acuminata*	Not essential for vegetative growth and conidiation, but essential for penetration and virulence in banana; induces immunity against tomato mosaic virus (ToMV) infection and *Pseudomonas syringae* infection; inhibits the antifungal activity of MaPR1 through direct protein–protein interaction	Li et al. (2019); Liu et al. (2019); Feng et al. (2021)

(continued)

Table 2.1 (continued)

S. No.	CPPs	Protein description	Organism/source	Host/substrate	Biological significance	References
22.	HaCPL2	*Heterobasidion annosum CPL2*	*Heterobasidion annosum* sensu stricto	Tobacco, *Pinus sylvestris*	Defence elicitation and cell death-inducing ability; retarded of apical root growth	Chen et al. (2015)
23.	HiCP	*Heterobasidion irregulare CP*	*Heterobasidion irregulare*	*Pinus* spp., *Juniperus* spp.	HiCPs-not affected by the heterokaryotic stage	Baccelli et al. (2015)
24.	PsCP	*Paradendryphiella salina* CP	*Paradendryphiella salina* (marine fungi)	Seaweed biomass	Hydrophobic protein; self-assembling ability to form amyloid-like aggregates with the properties of conformational plasticity	Landeta-Salgado et al. (2021)
25.	SsSm1, SsCP1	*Sclerotinia sclerotiorum* small protein 1	*Sclerotinia sclerotiorum*	Tobacco, *Arabidopsis*	Induces HR in tobacco; SsSm1 silencing makes the fungus more sensitive to fungicide and reduces salt tolerance; essential for hyphal development, infection cushions, sclerotiorum and virulence; ectopic expression of SsCP1 enhances broad-spectrum resistance against fungal diseases by interacting with PR1 gene of tobacco	Pan et al. (2018); Yang et al. (2018)
26.	VdCP1	*Verticillium dahliae CP1*	*Verticillium dahliae;* recombinant protein – *P. pastoris*	*Gossypium hirsutum*, tobacco	Not essential for growth and sporulation, but knockout affects pathogenicity and becomes more sensitive to chitinase; priming induces	Zhang et al. (2017b)

					ROS, many defence-related genes and confers broad-spectrum resistance against other fungal and bacterial diseases as well in both cotton and tobacco	
27.	DcSnodProt	*Dactylellina cionopaga SnodProt*	*Dactylellina cionopaga* (nematophagous fungi); recombinant protein – *P. pastoris*	Soybean, tomato	SnodProt alters chemotaxis and increases body-bend frequency of *Caenorhabditis elegans*; it may belong to a novel parasitism-related protein of nematophagous fungi	Yu et al. (2012)

2.8 Potential Roles of CPPs During Biocontrol

As a surface-active protein, EPL1 of *Trichoderma* spp. reduces the surface hydrophobicity of the host/environment, which might favour attachment of the fungus and make substrates accessible to enzymes for acquisition of nutrients (Gao et al. 2020). Biophysical studies on EPL1 from *T. atroviride* disclosed the fact that, contrary to hydrophobins, EPL1 increased the polarity of aqueous solutions and exposed surfaces (Frischmann et al. 2013). This property implicates that the phenomena could even enhance the wettability of hyphae, while shielding their hyphae from desiccation by covering their hyphae.

Sm1, a small secreted protein of *Trichoderma virens*, when primed in cotton seedlings induced systemic resistance against the foliar pathogens of *Colletotrichum* sp. (Djonovic et al., 2006). The CPPs facilitate root colonisation of *Trichoderma* spp., which is frequently associated with activation of localised and systemic attack against pathogen invasion. It was further established that the expression of Sm1 is constitutive and ubiquitous, even during interaction with host plant and competing fungal antagonist. Djonovic et al. (2006) reported that the higher expression of the elicitor Sm1 during in planta colonisation indicates its involvement of Sm1 in the plant–fungus interaction.

Global microarray analysis during *Trichoderma*–tomato plant interaction revealed differential expression of transcripts coding for Sm1 protein. Samolski et al. (2009) concluded that Sm1 is one of the key genes involved in early transcriptional response of the interaction and would augment the biocontrol potential of *Trichoderma harzianum*. Homologues of proteins from the CP family, Sm1 and EPL1 from *T. virens* and *T. atroviride*, respectively, were found to trigger plant defence responses and induce systemic resistance via PTI. Ectopic expression of SnodProt of *Dactylellina cionopaga* – a novel parasitism-related CP protein – into the nematophagous fungus was reported to enhance its biocontrol capacity (Yu et al. 2012).

Induction of PR-proteins on soybean leaves primed with EplT4 monomer (CP isolated from *Trichoderma asperellum* T4) resulted in protection against the pathogen *Cercosporidium sofinum* (Wang et al. 2013). Gomes et al. (2015) analysed the expression pattern of Epl1 protein in modulating the mycoparasitism process of *T. harzianum* against phytopathogens. It was reported that absence of Epl1 gene resulted in downregulation of genes related to mycoparasitism against the target fungal pathogen *Sclerotinia sclerotium*. The study concluded that Epl1 plays a critical role in the adaptation of *Trichoderma* during the three-way interaction involving the host plant and the phytopathogenic fungi. Further, Epl1 protein was reported to negatively influence the expression of *Botrytis cinerea* virulence genes during mycoparasitism under in vitro conditions (Gomes et al. 2017). Further, the Epl1-triggered PTI is reported to operate via SA-mediated defence pathway and thereby regulate the priming-based defence response in host tomato plants.

De Lima et al. (2017) reported that the secreted proteins of *T. harzianum* deactivated mycelial growth of *Guignardia citricarpa* causing citrus leaf spot and induced systemic plant disease resistance even in the absence of the pathogen. Among the identified secreted proteins, Epl1 constituted 54% of the total proteins, which is responsible for mycoparasitism and for induction of plant defence. Upon secretion, CPPs induce systemic resistance in maize, for which the Sm1 protein needs to be maintained in an active monomer form (Crutcher and Kenerley 2019). Li et al. (2019) characterised and purified FocCP1 from *Fusarium oxysporum* and analysed the mechanism of triggering PTI against wilt in banana. The PTI thus activated did conferred broad spectrum of resistance accentuated with the accumulation of ROS, callose deposition and necrosis, and expression and accumulation of SA- and JA-pathway-associated genes in tobacco against tobacco mosaic virus (TMV) and *Pseudomonas syringae* infection.

2.9 Biological Targets (Interacting Partners) of CPPs

PRRs localised on cell membranes are either receptor-like kinases (RKs/RLKs) or receptor-like proteins (RLPs) in tandem with RKs that recognise PAMPs and trigger PTI (Zipfel 2014). PTI orchestrates a sequence of events comprising of the activation of mitogen-activated protein kinases (MAPKs), expression of WRKY transcription factors, production of nitric oxide (NO) and reactive oxygen species (ROS), synthesis of pathogenesis-related (PR) proteins, expression of defence-related genes, and activation of signal transduction pathways involving SA, ethylene and JA to restrict the progression of pathogens.

Application of most of the CPPs to plant tissues induced plant defence responses, which implicated them as either MAMPs or PAMPs, that are recognised by the plant via its membrane receptors (Luti et al. 2020a, b). For instance, ectopic expression of SsCP1 showed resistance against many fungal diseases by directly interacting with PR1 protein of tobacco. The direct interaction of SsCP1 with PR1 in the apoplast was confirmed by yeast two-hybrid, co-immunoprecipitation assays and bimolecular florescence complementation studies. The peptide between motif 1 and motif 2 (48–71 aa) of CP1 was found to be essential for SsCP1–AtPR1 interaction (Yang et al. 2018). Similarly, the highly conserved loops β1-β2 and β2-β3 of BcSpl1 were found to be essential motifs for inducing necrosis in plants, which depicts its significance in binding to its interactors (Frias et al. 2014). FocCP1 was reported to physically interact with MaPR1 by direct interaction and inhibit the antifungal activity in plant cell apoplast, attributing to SA-mediated system acquired resistance (SAR) (Feng et al. 2021), which indicated that the SA pathway is essential for CP-associated plant defence (Yang et al. 2018).

CP and Pop1 of *Trichoderma* spp. acted as PAMP, as evidenced by phosphorylation of MAPKs, generation of ROS and NO, and expression of phytoalexin and defence-related genes. Additionally, it also exhibited enhanced expression of PR5 (thaumatin-like proteins) (Lombardi et al. 2013). CP upregulated MAPK4 and WRKY33, and induced camalexin biosynthesis in *Arabidopsis* (Baccelli et al. 2014b). Pop1 dimers activated HR response in *C. populicola* and showed induction of PR1 genes (Martellini et al. 2013). CP from *Ceratocystis fimbriata* was reported to bind to chitin and aid the pathogen to escape from being recognised by plant receptors (Pazzagli et al. 2014). Similarly, VdCP1 was found to protect its cell wall upon chitinase treatment, which in turn prevented chitin oligomers to be recognised by LysM receptors (Zhang et al. 2017b).

Wang et al. (2016) proposed a model of MSP1-induced PTI in rice, hypothesising that MSP1 binds to an unknown membrane receptor in apoplast, inducing kinase and phosphorylation activity of proteins, and the downstream signalling resulted in H_2O_2 burst, expression of pathogenesis-related proteins and cell death. Transcriptome analysis of MSP1-treated rice showed upregulation of 31 RLK genes, predominantly of S-locus glycoprotein like RLKs and leucine-rich repeat (LRR) RLKs (Meng et al. 2018). Since RLKs' recognition marks the activation of MAPK cascades and WRKY transcription factors, the study has also identified upregulation of 14 MAPKKK, 2 MPK genes, 1 MEK and 30 OsWRKY regulating JA- and abscisic acid (ABA)-signalling pathways. Proteomic analysis of MSP1-treated rice leaves revealed induction of RLKs, namely RLCK109, DUF26 and BIP128, and accumulation of four classes of PR-proteins, comprising of OsPR10, PBZ1 and TLP (Meng et al. 2019).

MgSM1 induced defence responses via a coordinated expression of SA-responsive and JA/ET-responsive PRs and PDF1.2 genes, respectively (Yang et al. 2009). Specifically, it induced the expression of OsPR1s and OsPR10 in rice and PR1, PR5 and PDF1.2 genes in *Arabidopsis* (Hong et al. 2017; Yang et al. 2009). It was also reported that Sm1 from *Trichoderma* spp. activated defence responses via JA and green leafy volatile (GLV) signalling pathways in maize (Djonovic et al. 2007; Vargas et al. 2008). BcSpl1 was recognised as PAMP in *Arabidopsis* and it induced HIN1 and HSR203J (HR markers), PR-1 and PR-5 (PR genes) and necrosis via BAK1 protein-mediated signalling cascade, which also implicated that BAK1 in conjunction with LRR-RK family receptor might activate the immune system (Frias et al. 2011, 2013, 2014). Overall, the reports suggested that the probable receptors of CPPs are the proteins from PR family of plants, especially PR1. However, the kind of interaction either direct or indirect is yet to be deciphered categorically.

2.10 Industrial Application Potentials of CPPs as Biosurfactants and Bioemulsifiers

Among the above-discussed physico-chemical properties of CPPs, one of the important and characteristic physical properties of CPPs that has huge potential in industrial applications is their surfactant property and cellulose-loosening ability with strong foaming, formation of abundant microbubbles during pipetting and high propensity of the concentrated protein solution to adhere to pipette tips (Frischmann et al. 2013; Ashwin et al. 2017a, b). For instance, CP from *T. harzianum* was reported to reduce surface tension and stabilise emulsions, demonstrating it to be an efficient biosurfactant and bioemulsifier (Pitocchi et al. 2020). During bioconversion of cellulosic substances for biofuels and during saccharification processes, addition of surfactants is reported to increase the saccharification yields, as surfactants are known to reduce non-specific binding of enzymes on lignins and promote the delignification process (Singhania et al. 2022). In this context, the biosurfactant properties of CPs have huge potential as they promote cellulose-loosening ability and surface-active properties (Pitocchi et al. 2020; Pennacchio et al. 2021). The surfactant properties of CPs were comparable to the commercially available surfactants used for bioconversion processes (Pitocchi et al. 2020).

2.11 Conclusion

Since the CPPs have been identified and characterised in *Ceratocystis* spp. as phytotoxin by Pazzagli et al. (1999), a lot has changed over the years. Many new features with paradoxical functional roles have been either characterised or proposed in many filamentous fungi as elaborated above. All these findings have considerably improved our knowledge on the role of CPPs in biological control and enhanced our understanding on how CPs induce PTI, which altogether would help in developing efficient plant disease management strategies. Further exploration of biological roles of CPPs, especially individual homologues, would lead to meaningful leads in understanding the molecular perspectives of biological control and expedite their biotechnological (industrial) application potential.

Acknowledgements The authors thank Director, ICAR-Sugarcane Breeding Institute, Indian Council of Agricultural Research, India, for providing facilities and continuous encouragement. The work was supported by the grants from DBT, Govt. of India (Sanction no. BT/PR23621/BPA/118/297/2017) and DST-SERB, Govt. of India (Sanction. no. CRG/2022/002835).

Conflict of Interest Statement All the authors declare that they have no conflict of interest.

References

Amaral F, Jácome C, Ângela M, Tillmann A, Ufpel F (2015) Bacterial secretion systems – an overview. Microbiol Spectr 4:1–32. https://doi.org/10.1128/microbiolspec.VMBF-0012-2015. Bacterial

Ashkenazy H, Abadi S, Martz E, Chay O, Mayrose I, Pupko T, Ben-Tal N (2016) ConSurf 2016: an improved methodology to estimate and visualize evolutionary conservation in macromolecules. Nucleic Acids Res 44:W344–W350. https://doi.org/10.1093/NAR/GKW408

Ashwin NMR, Barnabas L, Ramesh Sundar A, Malathi P, Viswanathan R, Masi A, Agrawal GK, Rakwal R (2017a) Comparative secretome analysis of *Colletotrichum falcatum* identifies a cerato-platanin protein (EPL1) as a potential pathogen-associated molecular pattern (PAMP) inducing systemic resistance in sugarcane. J Proteome 169:2–20. https://doi.org/10.1016/j.jprot.2017.05.020

Ashwin NMR, Barnabas L, Ramesh Sundar A, Malathi P, Viswanathan R, Masi A, Agrawal GK, Rakwal R (2017b) Advances in proteomic technologies and their scope of application in understanding plant–pathogen interactions. J Plant Biochem Biotechnol 26:371–386. https://doi.org/10.1007/s13562-017-0402-1

Ashwin NMR, Barnabas L, Ramesh Sundar A, Malathi P, Viswanathan R, Masi A, Agrawal GK, Rakwal R (2018) CfPDIP1, a novel secreted protein of *Colletotrichum falcatum*, elicits defense responses in sugarcane and triggers hypersensitive response in tobacco. Appl Microbiol Biotechnol 102:6001–6021. https://doi.org/10.1007/s00253-018-9009-2

Ashwin NMR, Lakshana KV, Amalamol D, Ramesh Sundar A, Malathi P, Jayakumar V, Viswanathan R (2020a) Tête-à-Tête during plant-pathogen interactions: intricacies involved and beyond. Plant Dis Res 35:89–96. https://doi.org/10.5958/2249-8788.2020.00020.7

Ashwin NMR, Barnabas L, Amalamol D, Lakshana KV, Ramesh Sundar A, Malathi P, Viswanathan R (2020b) Transcriptional reprogramming of major defense - signaling pathways during defense priming and sugarcane - *Colletotrichum falcatum* interaction. Mol Biol Rep 47: 8911–8923. https://doi.org/10.1007/s11033-020-05944-z

Ashwin NMR, Agisha VN, Vinodhini T, Amalamol D, Lakshana KV, Nalayeni K, Barnabas L, Ramesh Sundar A, Malathi P, Viswanathan R (2022) Synergistic integration of sugarcane proteomics with genomics - Proteogenomics to decipher the mechanism of disease resistance in sugarcane. In: Rao GP, Verma KK, Song XP, Rajput VD, Solomon S, Li YR (eds) Agro-industrial Perspectives on Sugarcane Production under Environmental Stress. Springer Nature Singapore, pp 389–401. https://doi.org/10.1007/978-981-19-3955-6_17

Baccelli I (2015) Cerato-platanin family proteins: one function for multiple biological roles? Front Plant Sci 5:769. https://doi.org/10.3389/fpls.2014.00769

Baccelli I, Comparini C, Bettini PP, Martellini F, Ruocco M, Pazzagli L, Bernardi R, Scala A (2012) The expression of the cerato-platanin gene is related to hyphal growth and chlamydospores formation in *Ceratocystis platani*. FEMS Microbiol Lett 327:155–163. https://doi.org/10.1111/j.1574-6968.2011.02475.x

Baccelli I, Luti S, Bernardi R, Scala A, Pazzagli L (2014a) Cerato-platanin shows expansin-like activity on cellulosic materials. Appl Microbiol Biotechnol 98:175–184. https://doi.org/10.1007/s00253-013-4822-0

Baccelli I, Lombardi L, Luti S, Bernardi R, Picciarelli P, Scala A, Pazzagli L (2014b) Cerato-platanin induces resistance in *Arabidopsis* leaves through stomatal perception, overexpression of salicylic acid- and ethylene-signalling genes and camalexin biosynthesis. PLoS One 9: e100959. https://doi.org/10.1371/journal.pone.0100959

Baccelli I, Gonthier P, Bernardi R (2015) Gene expression analyses reveal a relationship between conidiation and cerato-platanin in homokaryotic and heterokaryotic strains of the fungal plant pathogen *Heterobasidion irregulare*. Mycol Prog 14:40. https://doi.org/10.1007/s11557-015-1063-x

Baroni F, Gallo M, Pazzagli L, Luti S, Baccelli I, Spisni A, Pertinhez TA (2021) A mechanistic model may explain the dissimilar biological efficiency of the fungal elicitors cerato-platanin and cerato-populin. Biochem Biophys Acta 1865:129843. https://doi.org/10.1016/j.bbagen.2021.129843

Barsottini MR, De Oliveira JF, Adamoski D, Teixeira PJPL, Do Prado PFV, Tiezzi HO, Sforça ML, Cassago A, Portugal RV, De Oliveira PSL, Ana AC, Dias SMG, Pereira GAG, Ambrosio ALB (2013) Functional diversification of cerato-platanins in *Moniliophthora perniciosa* as seen by differential expression and protein function specialization. Mol Plant-Microbe Interact 26: 1281–1293. https://doi.org/10.1094/MPMI-05-13-0148-R

Bhadauria V, Banniza S, Vandenberg A, Selvaraj G, Wei Y (2011) EST mining identifies proteins putatively secreted by the anthracnose pathogen *Colletotrichum truncatum*. BMC Genomics 12: 327. https://doi.org/10.1186/1471-2164-12-327

Boddi S, Comparini C, Calamassi R, Pazzagli L, Cappugi G, Scala A (2004) Cerato-platanin protein is located in the cell walls of ascospores, conidia and hyphae of *Ceratocystis fimbriata* f. sp. *Platani*. FEMS Microbiol Lett 233:341–346. https://doi.org/10.1016/j.femsle.2004.03.001

Bohm H, Albert I, Fan L, Reinhard A, Nürnberger T (2014) Immune receptor complexes at the plant cell surface. Curr Opin Plant Biol 20:47–54. https://doi.org/10.1016/j.pbi.2014.04.007

Bonazza K, Gaderer R, Neudl S, Przylucka A, Allmaier G, Druzhinina IS, Grothe H, Friedbacher G, Seidl-Seiboth V (2015) The fungal cerato-platanin protein EPL1 forms highly ordered layers at hydrophobic/hydrophilic interfaces. Soft Matter 11:1723–1732. https://doi.org/10.1039/C4SM02389G

Braga RM, Dourado MN, Araújo WL (2016) Microbial interactions: ecology in a molecular perspective. Braz J Microbiol 47:86–98

Buensanteai N, Mukherjee PK, Horwitz BA, Cheng C, Dangott LJ, Kenerley CM (2010) Expression and purification of biologically active *Trichoderma virens* proteinaceous elicitor Sm1 in *Pichia pastoris*. Protein Expr Purif 72:131–138. https://doi.org/10.1016/j.pep.2010.03.006

Carresi L, Pantera B, Zoppi C, Cappugi G, Oliveira AL, Pertinhez TA, Spisni A, Scala A, Pazzagli L (2006) Cerato-platanin, a phytotoxic protein from *Ceratocystis fimbriata*: expression in *Pichia pastoris*, purification and characterization. Protein Expr Purif 49:159–167. https://doi.org/10.1016/j.pep.2006.07.006

Chen H, Kerio S, Asiegbu FO (2013) Distribution and bioinformatic analysis of the cerato-platanin protein family in Dikarya. Mycologia 105:1479–1488. https://doi.org/10.3852/13-115

Chen H, Quintana J, Kovalchuk A, Ubhayasekera W, Asiegbu FO (2015) A cerato-platanin-like protein HaCPL2 from *Heterobasidion annosum sensu stricto* induces cell death in *Nicotiana tabacum* and *Pinus sylvestris*. Fungal Genet Biol 84:41–51. https://doi.org/10.1016/j.fgb.2015.09.007

Cheng C-H, Shen B-N, Shang Q-W, Liu L-Y, Peng K-C, Chen Y-H, Chen F-F, Hu S-F, Wang Y-T, Wang H-C, Wu H-Y, Lo C-T, Lin S-S (2018) Gene-to-gene network analysis of the mediation of plant innate immunity by the eliciting plant response-like 1 (Epl1) elicitor of *Trichoderma formosa*. Mol Plant-Microbe Interact 31:683–691. https://doi.org/10.1094/MPMI-01-18-0002-TA

Comparini C, Carresi L, Pagni E, Sbrana F, Sebastiani F, Luchi N, Santini A, Capretti P, Tiribilli B, Pazzagli L, Cappugi G, Scala A (2009) New proteins orthologous to cerato-platanin in various *Ceratocystis* species and the purification and characterization of cerato-populin from *Ceratocystis populicola*. Appl Microbiol Biotechnol 84:309–322. https://doi.org/10.1007/s00253-009-1998-4

Cosgrove DJ (2017) Microbial expansins. Annu Rev Microbiol 71:479–497. https://doi.org/10.1146/annurev-micro-090816-093315

Crutcher FK, Kenerley CM (2019) Analysis of a putative glycosylation site in the *Trichoderma virens* elicitor SM1 reveals no role in protein dimerization. Biochem Biophys Res Commun 509:817–821. https://doi.org/10.1016/j.bbrc.2019.01.007

Crutcher FK, Moran-Diez ME, Ding S-L, Liu J, Horwitz BA, Mukherjee PK, Kenerley CM (2015) A paralog of the proteinaceous elicitor SM1 is involved in colonization of maize roots by *Trichoderma virens*. Fungal Biol 119:476–486. https://doi.org/10.1016/j.funbio.2015.01.004

de Lima FB, Félix C, Osório N, Alves A, Vitorino R, Domingues P, da Silva Ribeiro RT, Esteves AC (2017) *Trichoderma harzianum* T1A constitutively secretes proteins involved in the biological control of *Guignardia citricarpa*. Biol Control 1(106):99–109

de Oliveira AL, Gallo M, Pazzagli L, Benedetti CE, Cappugi G, Scala A, Pantera B, Spisni A, Pertinhez TA, Cicero DO (2011) The structure of the elicitor Cerato-platanin (CP), the first member of the CP fungal protein family, reveals a double ψβ-barrel fold and carbohydrate binding. J Biol Chem 286:17560–17568. https://doi.org/10.1074/jbc.M111.223644

Djonovic S, Pozo MJ, Dangott LJ, Howell CR, Kenerley CM (2006) Sm1, a proteinaceous elicitor secreted by the biocontrol fungus *Trichoderma virens* induces plant defense responses and systemic resistance. Mol Plant-Microbe Interact 19:838–853. https://doi.org/10.1094/MPMI-19-0838

Djonovic S, Vargas WA, Kolomiets MV, Horndeski M, Wiest A, Kenerley CM (2007) A proteinaceous elicitor Sm1 from the beneficial fungus *Trichoderma virens* is required for induced systemic resistance in maize. Plant Physiol 145:875–889. https://doi.org/10.1104/pp.107.103689

Drew GC, Stevens EJ, King KC (2021) Microbial evolution and transitions along the parasite–mutualist continuum. Nat Rev Microbiol 19:623–638

Druzhinina IS, Seidl-Seiboth V, Herrera-Estrella A, Horwitz BA, Kenerley CM, Monte E, Mukherjee PK, Zeilinger S, Grigoriev IV, Kubicek CP (2011) *Trichoderma*: the genomics of opportunistic success. Nat Rev Microbiol 9:749–759. https://doi.org/10.1038/nrmicro2637

Eranthodi A, González-Peña Fundora D, Montenegro Alonso AP, Bakkeren G, Subramaniam R, Ouellet T, Rampitsch C, Smith M, Goyal RK, Schultz E, Foroud NA (2022) Cerato-platanin protein 1 is not critical for *Fusarium graminearum* growth and aggressiveness, but its overexpression provides an edge to fusarium head blight in wheat. Can J Plant Pathol 44:577. https://doi.org/10.1080/07060661.2022.2044910

Feng Q, Gao X, An B, He C, Wang Q (2021) Two cerato-platanin proteins FocCP1 interact with MaPR1 and contribute to virulence of *Fusarium oxysporum* f. sp. *cubense* to banana. J Plant Interact 16:238–245. https://doi.org/10.1080/17429145.2021.1928305

Fontana F, Santini A, Salvini M, Pazzagli L, Cappugi G, Scala A, Durante M, Bernardi R (2008) Cerato-platanin treated plane leaves restrict *Ceratocystis platani* growth and overexpress defence-related genes. J Plant Pathol 90:295–306

Freitas RS, Steindorff AS, Ramada MHS, de Siqueira SJL, Noronha EF, Ulhoa CJ (2014) Cloning and characterization of a protein elicitor Sm1 gene from *Trichoderma harzianum*. Biotechnol Lett 36:783–788. https://doi.org/10.1007/s10529-013-1410-4

Frias M, González C, Brito N (2011) BcSpl1, a cerato-platanin family protein, contributes to *Botrytis cinerea* virulence and elicits the hypersensitive response in the host. New Phytol 192: 483–495. https://doi.org/10.1111/j.1469-8137.2011.03802.x

Frias M, Brito N, González C (2013) The Botrytis cinerea cerato-platanin BcSpl1 is a potent inducer of systemic acquired resistance (SAR) in tobacco and generates a wave of salicylic acid expanding from the site of application. Mol Plant Pathol 14:191–196. https://doi.org/10.1111/j.1364-3703.2012.00842.x

Frias M, Brito N, González M, González C (2014) The phytotoxic activity of the cerato-platanin BcSpl1 resides in a two-peptide motif on the protein surface. Mol Plant Pathol 15:342–351. https://doi.org/10.1111/mpp.12097

Frischmann A, Neudl S, Gaderer R, Bonazza K, Zach S, Gruber S, Spadiut O, Friedbacher G, Grothe H, Seidl-Seiboth V (2013) Self-assembly at air/water interfaces and carbohydrate binding properties of the small secreted protein EPL1 from the fungus *Trichoderma atroviride*. J Biol Chem 288:4278–4287. https://doi.org/10.1074/jbc.M112.427633

Gaderer R, Bonazza K, Seidl-Seiboth V (2014) Cerato-platanins: a fungal protein family with intriguing properties and application potential. Appl Microbiol Biotechnol 98:4795–4803. https://doi.org/10.1007/s00253-014-5690-y

Gaderer R, Lamdan NL, Frischmann A, Sulyok M, Krska R, Horwitz BA, Seidl-Seiboth V (2015) Sm2, a paralog of the *Trichoderma* cerato-platanin elicitor Sm1, is also highly important for plant protection conferred by the fungal-root interaction of *Trichoderma* with maize. BMC Microbiol 15:1–9. https://doi.org/10.1186/s12866-014-0333-0

Gao R, Ding M, Jiang S, Zhao Z, Chenthamara K, Shen Q, Cai F, Druzhinina IS (2020) The evolutionary and functional paradox of cerato-platanins in fungi. Appl Environ Microbiol 86: e00696–e00620. https://doi.org/10.1128/AEM.00696-20

Gluck-Thaler E, Cerutti A, Perez-Quintero AL, Butchacas J, Roman-Reyna V, Madhavan VN, Shantharaj D, Merfa MV, Pesce C, Jauneau A, Vancheva T (2020) Repeated gain and loss of a single gene modulates the evolution of vascular plant pathogen lifestyles. Sci Adv 6:eabc4516

Gomes EV, Costa MDN, de Paula RG, Ricci de Azevedo R, da Silva FL, Noronha EF, José Ulhoa C, Neves Monteiro V, Elena Cardoza R, Gutiérrez S, Nascimento Silva R (2015) The Cerato-Platanin protein Epl-1 from *Trichoderma harzianum* is involved in mycoparasitism, plant resistance induction and self cell wall protection. Sci Rep 5:17998. https://doi.org/10.1038/srep17998

Gomes EV, Ulhoa CJ, Cardoza RE, Silva RN, Gutiérrez S (2017) Involvement of *Trichoderma harzianum* Epl-1 protein in the regulation of *Botrytis* virulence- and tomato defense-related genes. Front Plant Sci 8:1–11. https://doi.org/10.3389/fpls.2017.00880

Hall N, Keon JPR, Hargreaves JA (1999) A homologue of a gene implicated in the virulence of human fungal diseases is present in a plant fungal pathogen and is expressed during infection. Physiol Mol Plant Pathol 55:69–73. https://doi.org/10.1006/pmpp.1999.0203

Hong Y, Yang Y, Zhang H, Huang L, Li D, Song F (2017) Overexpression of MoSM1, encoding for an immunity-inducing protein from *Magnaporthe oryzae*, in rice confers broad-spectrum resistance against fungal and bacterial diseases. Sci Rep 7:41037. https://doi.org/10.1038/srep41037

Jeong JS, Mitchell TK, Dean R, a. (2007) The *Magnaporthe grisea* snodprot1 homolog, MSP1, is required for virulence. FEMS Microbiol Lett 273:157–165. https://doi.org/10.1111/j.1574-6968.2007.00796.x

Jones JDG, Dangl JL (2006) The plant immune system. Nature 444:323–329. https://doi.org/10.1038/nature05286

Kleemann J, Rincon-Rivera LJ, Takahara H, Neumann U, van Themaat EVL, van der Does HC, Hacquard S, Stüber K, Will I, Schmalenbach W, Schmelzer E, O'Connell RJ, Loren V, van Themaat E, van Themaat EVL, van der Does HC, Hacquard S, Stüber K, Will I, Schmalenbach W, Schmelzer E, O'Connell RJ (2012) Sequential delivery of host-induced virulence effectors by appressoria and intracellular hyphae of the phytopathogen *Colletotrichum higginsianum*. PLoS Pathog 8:e1002643. https://doi.org/10.1371/journal.ppat.1002643

Köhl J, Kolnaar R, Ravensberg WJ (2019) Mode of action of microbial biological control agents against plant diseases: relevance beyond efficacy. Front Plant Sci 845

Kumar A (2022) Microbial biocontrol: sustainable agriculture and Phytopathogen management, vol 1. Springer Nature Chem, Cham, Switzerland, pp 1–369. https://doi.org/10.1007/978-3-030-87512-1

Kumari M, Qureshi KA, Jaremko M, White JF, Singh SK, Sharma VK, Singh KK, Santoyo G, Puopolo G, Kumar A (2022) Deciphering the role of endophytic microbiome in postharvest diseases management of fruits: opportunity areas in commercial up-scale production. Front Plant Sci 13:1026575. https://doi.org/10.3389/fpls.2022.1026575

Landeta-Salgado C, Cicatiello P, Stanzione I, Medina D, Berlanga Mora I, Gomez C, Lienqueo ME (2021) The growth of marine fungi on seaweed polysaccharides produces cerato-platanin and hydrophobin self-assembling proteins. Microbiol Res 251:126835. https://doi.org/10.1016/j.micres.2021.126835

Lapin D, Van den Ackerveken G (2013) Susceptibility to plant disease: more than a failure of host immunity. Trends Plant Sci 18:546–554. https://doi.org/10.1016/j.tplants.2013.05.005

Li S, Dong Y, Li L, Zhang Y, Yang X, Zeng H, Shi M, Pei X, Qiu D, Yuan Q (2019) The novel cerato-platanin-like protein FocCP1 from *Fusarium oxysporum* triggers an immune response in plants. Int J Mol Sci 20:2849. https://doi.org/10.3390/ijms20112849

Linder MB, Szilvay GR, Nakari-Setälä T, Penttilä ME (2005) Hydrophobins: the protein-amphiphiles of filamentous fungi. FEMS Microbiol Rev 29(5):877–896. https://doi.org/10.1016/j.femsre.2005.01.004

Liu S, Wu B, Yang J, Bi F, Dong T, Yang Q, Hu C, Xiang D, Chen H, Huang H, Shao C, Chen Y, Yi G, Li C, Guo X (2019) A cerato-platanin family protein FocCP1 is essential for the penetration and virulence of *Fusarium oxysporum* f. Sp. *cubense* tropical race 4. Int J Mol Sci 20:3785. https://doi.org/10.3390/ijms20153785

Lombardi L, Faoro F, Luti S, Baccelli I, Martellini F, Bernardi R, Picciarelli P, Scala A, Pazzagli L (2013) Differential timing of defense-related responses induced by cerato-platanin and cerato-populin, two non-catalytic fungal elicitors. Physiol Plant 149:408–421. https://doi.org/10.1111/ppl.12041

Lucke M, Correa MG, Levy A (2020) The role of secretion systems, effectors, and secondary metabolites of beneficial rhizobacteria in interactions with plants and microbes. Front Plant Sci 1718:11

Luti S, Martellini F, Bemporad F, Mazzoli L, Paoli P, Pazzagli L (2017) A single amino acid mutation affects elicitor and expansins-like activities of cerato-platanin, a non-catalytic fungal protein. PLoS One 12:e0178337. https://doi.org/10.1371/journal.pone.0178337

Luti S, Bemporad F, Vivoli Vega M, Leri M, Musiani F, Baccelli I, Pazzagli L (2020a) Partitioning the structural features that underlie expansin-like and elicitor activities of cerato-platanin. Int J Biol Macromol 165:2845–2854. https://doi.org/10.1016/j.ijbiomac.2020.10.122

Luti S, Caselli A, Taiti C, Bazihizina N, Gonnelli C, Mancuso S, Pazzagli L (2016) PAMP activity of cerato-platanin during plant interaction: an-omic approach. Int J Mol Sci 17:866. https://doi.org/10.3390/ijms17060866

Luti S, Sella L, Quarantin A, Pazzagli L, Baccelli I (2020b) Twenty years of research on cerato-platanin family proteins: clues, conclusions, and unsolved issues. Fungal Biol Rev 34:13–24. https://doi.org/10.1016/j.fbr.2019.10.001

Lyu D, Msimbira LA, Nazari M, Antar M, Pagé A, Shah A, Monjezi N, Zajonc J, Tanney CA, Backer R, Smith DL (2021) The coevolution of plants and microbes underpins sustainable agriculture. Microorganisms 9:1036

Macho AP, Zipfel C (2014) Plant PRRs and the activation of innate immune signaling. Mol Cell 54: 263–272. https://doi.org/10.1016/j.molcel.2014.03.028

Martellini F, Faoro F, Carresi L, Pantera B, Baccelli I, Maffi D, Tiribilli B, Sbrana F, Luti S, Comparini C, Bernardi R, Cappugi G, Scala A, Pazzagli L (2013) Cerato-populin and Cerato-platanin, two non-catalytic proteins from phytopathogenic fungi, interact with hydrophobic inanimate surfaces and leaves. Mol Biotechnol 55:27–42. https://doi.org/10.1007/s12033-012-9618-4

Mendes R, Garbeva P, Raaijmakers JM (2013) The rhizosphere microbiome: significance of plant beneficial, plant pathogenic, and human pathogenic microorganisms. FEMS Microbiol Rev 37(5):634–663

Meng Q, Gupta R, Kwon SJ, Wang Y, Agrawal GK, Rakwal R, Park S-R, Kim ST (2018) Transcriptomic analysis of *Oryza sativa* leaves reveals key changes in response to *Magnaporthe oryzae* MSP1. Plant Pathol J 34:257–268. https://doi.org/10.5423/PPJ.OA.01.2018.0008

Meng Q, Gupta R, Min CW, Kim J, Kramer K, Wang Y, Park SR, Finkemeier I, Kim ST (2019) A proteomic insight into the MSP1 and flg22 induced signaling in *Oryza sativa* leaves. J Proteome 196:120–130. https://doi.org/10.1016/j.jprot.2018.04.015

Mergaert P (2018) Role of antimicrobial peptides in controlling symbiotic bacterial populations. Nat Prod Rep 35(4):336–356

Miwa H, Okazaki S (2017) How effectors promote beneficial interactions. Curr Opin Plant Biol 38: 148–154. https://doi.org/10.1016/j.pbi.2017.05.011

Moënne-Loccoz Y, Mavingui P, Combes C, Normand P, Steinberg C (2015) Microorganisms and biotic interactions. In: Environmental microbiology: fundamentals and applications. Springer, Dordrecht, pp 395–444

Nawrocka J, Małolepsza U (2013) Diversity in plant systemic resistance induced by *Trichoderma*. Biol Control 67:149–156. https://doi.org/10.1016/j.biocontrol.2013.07.005

Pan Y, Wei J, Yao C, Reng H, Gao Z (2018) SsSm1, a Cerato-platanin family protein, is involved in the hyphal development and pathogenic process of *Sclerotinia sclerotiorum*. Plant Sci 270:37–46. https://doi.org/10.1016/j.plantsci.2018.02.001

Pathak P, Rai VK, Can H, Singh SK, Kumar D, Bhardwaj N, Roychowdhury R, de Azevedo LCB, Kaushalendra VH, Kumar A (2022) Plant-endophyte interaction during biotic stress management. Plan Theory 11(17):2203. https://doi.org/10.3390/plants11172203

Pazzagli L, Cappugi G, Cappugi G, Manao G, Manao G, Camici G, Camici G, Santini A, Santini A, Scala A (1999) Purification, characterization, and amino acid sequence of cerato-platanin, a new phytotoxic protein from *Ceratocystis fimbriata* f. sp. *platani*. Mol Biol 274:24959–24964. https://doi.org/10.1074/jbc.274.35.24959

Pazzagli L, Pantera B, Carresi L, Zoppi C, Pertinhez TA, Spisni A, Tegli S, Scala A, Cappugi G (2006) Cerato-platanin, the first member of a new fungal protein family: cloning, expression, and characterization. Cell Biochem Biophys 44:512–521. https://doi.org/10.1385/CBB:44:3:512

Pazzagli L, Zoppi C, Carresi L, Tiribilli B, Sbrana F, Schiff S, Pertinhez TA, Scala A, Cappugi G (2009) Characterization of ordered aggregates of cerato-platanin and their involvement in fungus-host interactions. Biochim Biophys Acta 1790:1334–1344. https://doi.org/10.1016/j.bbagen.2009.07.014

Pazzagli L, Seidl-Seiboth V, Barsottini M, Vargas WA, Scala A, Mukherjee PK (2014) Cerato-platanins: elicitors and effectors. Plant Sci 228:79–87. https://doi.org/10.1016/j.plantsci.2014.02.009

Pennacchio A, Pitocchi R, Varese GC, Giardina P, Piscitelli A (2021) *Trichoderma harzianum* cerato-platanin enhances hydrolysis of lignocellulosic materials. Microb Biotechnol 14:1699–1706. https://doi.org/10.1111/1751-7915.13836

Pitocchi R, Cicatiello P, Birolo L, Piscitelli A, Bovio E, Cristina Varese G, Giardina P (2020) Cerato-platanins from marine fungi as effective protein biosurfactants and bioemulsifiers. Int J Mol Sci 21:2913. https://doi.org/10.3390/ijms21082913

Quarantin A, Glasenapp A, Schäfer W, Favaron F, Sella L (2016) Involvement of the *Fusarium graminearum* cerato-platanin proteins in fungal growth and plant infection. Plant Physiol Biochem 109:220–229. https://doi.org/10.1016/j.plaphy.2016.10.001

Quarantin A, Castiglioni C, Schäfer W, Favaron F, Sella L (2019) The *Fusarium graminearum* cerato-platanins loosen cellulose substrates enhancing fungal cellulase activity as expansin-like proteins. Plant Physiol Biochem 139:229–238. https://doi.org/10.1016/j.plaphy.2019.03.025

Ramada MHS, Steindorff AS, Bloch C, Ulhoa CJ (2016) Secretome analysis of the mycoparasitic fungus *Trichoderma harzianum* ALL 42 cultivated in different media supplemented with *Fusarium solani* cell wall or glucose. Proteomics 16:477–490. https://doi.org/10.1002/pmic.201400546

Rathi S, McFeeters H, McFeeters RL, Davis MR (2012) Purification and phytotoxic analysis of *Botrytis cinerea* virulence factors: new avenues for crop protection. Agriculture 2:154–164. https://doi.org/10.3390/agriculture2030154

Rovera C, Luti S, Pazzagli L, Heeley EL, Wan C, Romano D, Farris S (2021) Enzymatic hydrolysis of bacterial cellulose in the presence of a non-catalytic cerato-platanin protein. J Appl Polym Sci 139:51886. https://doi.org/10.1002/app.51886

Salas-Marina MA, Isordia-Jasso MI, Islas-Osuna MA, Delgado-Sanchez P, Jimanez-Bremont JF, Rodriguez-Kessler M, Rosales-Saavedra MT, Herrera-Estrella A, Casas-Flores S (2015) The Epl1 and Sm1 proteins from *Trichoderma atroviride* and *Trichoderma virens* differentially modulate systemic disease resistance against different life style pathogens in *Solanum lycopersicum*. Front Plant Sci 6:77. https://doi.org/10.3389/fpls.2015.00077

Samolski I, de Luis A, Vizcaíno JA, Monte E, Suárez MB (2009) Gene expression analysis of the biocontrol fungus *Trichoderma harzianum* in the presence of tomato plants, chitin, or glucose using a high-density oligonucleotide microarray. BMC Microbiol 9(1):1–4

Sbrana F, Bongini L, Cappugi G, Fanelli D, Guarino A, Pazzagli L, Scala A, Vassalli M, Zoppi C, Tiribilli B (2007) Atomic force microscopy images suggest aggregation mechanism in cerato-platanin. Eur Biophys J 36:727–732. https://doi.org/10.1007/s00249-007-0159-x

Seidl V, Marchetti M, Schandl R, Allmaier G, Kubicek CP (2006) Epl1, the major secreted protein of *Hypocrea atroviridis* on glucose, is a member of a strongly conserved protein family comprising plant defense response elicitors. FEBS J 273:4346–4359. https://doi.org/10.1111/j.1742-4658.2006.05435.x

Singh VK, Singh AK, Kumar A (2017) Disease management of tomato through PGPB: current trends and future perspective 3. Biotech 7(4):1–10

Singh M, Srivastava M, Kumar A, Singh AK, Pandey KD (2020) Endophytic bacteria in plant disease management. In Kumar, A., Singh, K.V., (Eds).; Microbial endophytes: prospects for sustainable agriculture Woodhead Publication. Cambridge USA 61–89. https://doi.org/10.1016/B978-0-12-818734-0.00004-8

Singhania RR, Patel AK, Singh A, Haldar D, Soam S, Chen CW, Tsai ML, Dong CD (2022) Consolidated bioprocessing of lignocellulosic biomass: technological advances and challenges. Bioresour Technol 11:127153

Sundar AR, Ashwin NMR, Barnabas L, Nalayeni K, Malathi P, Viswanathan R (2018) A molecular insight into the understanding of major fungal diseases of sugarcane. In: Viswanathan R, Sundar AR, Malathi P, Selvakumar R, Jayakumar V, Nithya K (eds) Research accomplishments in sugarcane pathology at ICAR-SBI. ICAR - Sugarcane Breeding Institute, Coimbatore, India, pp 59–75

Tai-Long Z, Jie-Wei Z, Jie C (2014) SM1 protein produced by *Trichoderma virens* induces defense responses and activates ROS signaling pathway in *Arabidopsis*. J Mycosystem 33:1302–1312. https://doi.org/10.13346/j.mycosystema.140502

Vargas WA, Djonović S, Sukno SA, Kenerley CM (2008) Dimerization controls the activity of fungal elicitors that trigger systemic resistance in plants. J Biol Chem 283:19804–19815. https://doi.org/10.1074/jbc.M802724200

Wang D, Dong X (2011) A highway for war and peace: the secretory pathway in plant–microbe interactions. Mol Plant 4(4):581–587

Wang Y, Wang Y (2018) Trick or treat: microbial pathogens evolved apoplastic effectors modulating plant susceptibility to infection. Mol Plant-Microbe Interact 31:6–12. https://doi.org/10.1094/MPMI-07-17-0177-FI

Wang Y, Song J, Wu Y, Odeph M, Liu Z, Howlett BJ, Wang S, Yang P, Yao L, Zhao L, Yang Q (2013) Eplt4 proteinaceous elicitor produced in *Pichia pastoris* has a protective effect against *Cercosporidium sofinum* infections of soybean leaves. Appl Biochem Biotechnol 169:722–737. https://doi.org/10.1007/s12010-012-0015-z

Wang Y, Wu J, Kim SG, Tsuda K, Gupta R, Park S-Y, Kim ST, Kang KY (2016) *Magnaporthe oryzae*-secreted protein MSP1 induces cell death and elicits defense responses in rice. Mol Plant-Microbe Interact 29:299–312. https://doi.org/10.1094/MPMI-12-15-0266-R

Wang W, An B, Feng L, He C, Luo H (2018) A *Colletotrichum gloeosporioides* cerato-platanin protein, CgCP1, contributes to conidiation and plays roles in the interaction with rubber tree. Can J Microbiol 64:826–834. https://doi.org/10.1139/cjm-2018-0087

Wang Q, Zhong C, Xiao H (2020) Genetic engineering of filamentous fungi for efficient protein expression and secretion. Front Bioeng Biotechnol 8:293

Wilson LM, Idnurm A, Howlett BJ (2002) Characterization of a gene (sp1) encoding a secreted protein from *Leptosphaeria maculans*, the blackleg pathogen of *Brassica napus*. Mol Plant Pathol 3:487–493. https://doi.org/10.1046/j.1364-3703.2002.00144.x

Yang Y, Zhang H, Li G, Li W, Wang X, Song F (2009) Ectopic expression of MgSM1, a Cerato-platanin family protein from *Magnaporthe grisea*, confers broad-spectrum disease resistance in *Arabidopsis*. Plant Biotechnol J 7:763–777. https://doi.org/10.1111/j.1467-7652.2009.00442.x

Yang G, Tang L, Gong Y, Xie J, Fu Y, Jiang D, Li G, Collinge DB, Chen W, Cheng J (2018) A cerato-platanin protein SsCP1 targets plant PR1 and contributes to virulence of *Sclerotinia sclerotiorum*. New Phytol 217:739–755. https://doi.org/10.1111/nph.14842

Yu H, Li L (2014) Phylogeny and molecular dating of the cerato-platanin-encoding genes. Genet Mol Biol 37:423–427

Yu H, Duan J, Wang B, Jiang X (2012) The function of snodprot in the cerato-platanin family from *Dactylellina cionopaga* in nematophagous fungi. Biosci Biotechnol Biochem 76:1835–1842. https://doi.org/10.1271/bbb.120173

Yu W, Mijiti G, Huang Y, Fan H, Wang Y, Liu Z (2018) Functional analysis of eliciting plant response protein Epl1-Tas from *Trichoderma asperellum* ACCC30536. Sci Rep 8:1–13. https://doi.org/10.1038/s41598-018-26328-1

Zaparoli G, Cabrera OG, Medrano FJ, Tiburcio R, Lacerda G, Pereira GG (2009) Identification of a second family of genes in *Moniliophthora perniciosa*, the causal agent of witches' broom disease in cacao, encoding necrosis-inducing proteins similar to cerato-platanins. Mycol Res 113:61–72. https://doi.org/10.1016/j.mycres.2008.08.004

Zhang Y, Liang Y, Qiu D, Yuan J, Yang X (2017a) Comparison of cerato-platanin family protein BcSpl1 produced in *Pichia pastoris* and *Escherichia coli*. Protein Expr Purif 136:20–26. https://doi.org/10.1016/j.pep.2017.06.004

Zhang Y, Gao Y, Liang Y, Dong Y, Yang X, Yuan J, Qiu D (2017b) The *Verticillium dahliae* SnodProt1-like protein VdCP1 contributes to virulence and triggers the plant immune system. Front Plant Sci 8:1880. https://doi.org/10.3389/fpls.2017.01880

Zhang L, Xing J, Lin J (2019) At the intersection of exocytosis and endocytosis in plants. New Phytol 224:1479–1489

Zipfel C (2014) Plant pattern-recognition receptors. Trends Immunol 35:345–351. https://doi.org/10.1016/j.it.2014.05.004

Chapter 3
Molecular Events and Defence Mechanism Against Biotic Stress Induced by Bio-Priming of Beneficial Microbes

Bharani Manoharan, Shobana Narayanasamy, J. Beslin Joshi, Sridharan Jegadeesan, Shanshan Qi, Zhicong Dai, Daolin Du, Senthil Natesan, and Sivakumar Uthandi

Abstract The monotonous and regular use of chemical pesticides or fertilisers not only adversely affects the nutrient and food quality but also affects the health of human beings and surrounding environment. Hence nowadays utilisation of biofertilisers is paramount to support eco-friendly and sustainable agriculture. Modern agriculture should be accompanied by beneficial microorganisms to revamp soil fertility and agricultural productivity. Bio-priming with the efficient microorganisms showed multiple benefit to the plants by modulating plant growth and phytopathogen management. Plant growth-promoting bacteria or fungi (PGPB/PGPF) as priming agents facilitate the acquisition of soil nutrients, and modulate phytohormones. However, it is poorly understood how beneficial microbes (both invading and

Bharani Manoharan and Shobana Narayanasamy contributed equally with all other contributors.

B. Manoharan (✉) · S. Qi · Z. Dai · D. Du
Institute of Environment and Ecology, Academy of Environmental Health and Ecological Security, Jiangsu University, Zhenjiang, P.R. China

School of the Environment and Safety Engineering, Jiangsu University, Zhenjiang, P.R. China
e-mail: ddl@ujs.edu.cn

S. Narayanasamy · S. Uthandi
Biocatalysts Laboratory, Department of Agricultural Microbiology, Tamil Nadu Agricultural University, Coimbatore, India

J. B. Joshi
Deparment of Plant Biotechnology, Centre for Plant Molecular Biology and Biotechnology, Tamil Nadu Agricultural University, Coimbatore, India

S. Jegadeesan
Faculty of Life Sciences, School of Plant Sciences and Food Security, Tel Aviv University, Tel Aviv, Israel

S. Natesan
Deparment of Plant Molecular Biology and Bioinformatics, Centre for Plant Molecular Biology and Biotechnology, Tamil Nadu Agricultural University, Coimbatore, India
e-mail: senthil_natesan@tnau.ac.in

K. K. Bastas et al. (eds.), *Microbial Biocontrol: Molecular Perspective in Plant Disease Management*, Microorganisms for Sustainability 49,
https://doi.org/10.1007/978-981-99-3947-3_3

non-invading) connect successful association with plants by evading host plant immune system. Here, we aimed to discuss the basics of bio-priming agents and describe various molecular events used by beneficial microbes to avoid recognising the immune signalling of plants for their colonisation to 'primed state'. The importance of multi-omics approaches for unravelling bio-priming is also outlined. The chapter may advance our current understanding of bio-priming-induced molecular events and defence mechanisms against biotic stresses and its importance in integrating into modern agriculture.

Keywords Bio-priming · Biotic stress · PGPR · Defence priming · Induced and systemic resistance · Molecular mechanisms · Omics technologies

3.1 Introduction

Increasing world population demands increased food production around the globe. The alarming decrease in arable land due to urbanisation and land pollution has made the task of attaining food security a challenge. In this lieu, the only reliable way to achieve food security is by increasing the crop productivity. Various biotic and abiotic stresses encountered by crop during different growth stages have caused significant decline in yield and productivity (Kumar 2022). Annually, USD$220 billion loss around the globe was recorded due to pest and diseases (Chakraborty and Newton 2011; Kumari et al. 2022). Though using chemicals for stress management is effective, the residual toxicity effect on environment and living being was alarming (Lushchak et al. 2018). With this concern, the public shifted their interest towards biological agents for improving crop health and productivity.

Priming the plants using biological agents to face different stresses, besides improving seed vigour and plant growth, is known as 'bio-priming'. This newly emerging technique sounds like a promising alternative to chemical inputs used in agriculture (Sharma et al. 2018). The physiological process of priming causes the pre-germinative metabolic process to be expedited and increased for quick germination and establishment (Dawood 2018). According to reports, bio-priming can boost crop output by 5–10% and nutrient utilisation effectiveness by 5–25% (European Biostimulants Industry Consortium 2011). It is expected that the global market for bio-stimulants will increase by 12% annually (Calvo et al. 2014).

PGPRs (plant growth-promoting rhizobacteria), which live in plant rhizosphere and promote plant growth and establishment, are a well-known bio-priming agent (Kloepper et al. 1980; Entesari et al. 2013). In addition to promoting growth, the antibiotics produced by PGPR also stifled soil-borne microorganisms and generated systemic resistance (Kavino et al. 2007; Singh et al. 2017; Singh et al. 2020). The plant's resistance to abiotic stress conditions as salt, dehydration and cold was also said to be strengthened by PGPR (Grichko and Glick 2001; Mayak et al. 2004; Alizadeh Forutan et al. 2017; Singh et al. 2018). Additionally, the PGPR preserves and enhances soil structure, which promotes crop development (Jing et al. 2007). Later, numerous symbiotic bacteria and fungi, including *Trichoderma*, *Azospirillum*, *Azotobacter*, *Klebsiella* and *Frankia*, were employed to enhance crop health and yield (Staley and Drahos 1994; Gholami et al. 2008; Savazzini et al. 2009). This method was found to be a cost-effective method for improving the plant productivity and stress resistance overall, in addition to priming the plant against numerous stresses (Rao et al. 2009). The current chapter explores various omics approaches for understanding the bio-priming processes as well as the physiological and molecular events connected to bio-priming against biotic stress.

3.2 Bio-Priming Agents and Their Characteristics

Callan and his co-workers introduced the bio-priming method by mixing the bio-agent with pre-soaked seeds (Callan et al. 1990). The seeds were incubated at 25–35 °C for 48 h under high humidity. By doing so, the bio-agent forms a protective layer on the seed surface (Callan et al. 1990; Lutts et al. 2016), thereby protecting the seed from biotic stresses (Reddy 2013). Bio-priming agents can be of various substances like humic acid, fulvic acid, amino acids, protein hydrolysates, sea weed extracts from biological organisms or microorganisms directly to augment development and growth of plants (du Jardin 2012; Calvo et al. 2014). The best bio-priming agent should have the following characteristics: cheap, durable, compatible, non-toxic and eco-friendly. The different bio-priming agents and their role are discussed below:

3.2.1 PGPR

A single strain of PGPR or a group of PGPRs or plant growth-promoting fungi (PGPF) consortia can be used as bio-priming agents (Yadav et al. 2012; Alizadeh Forutan et al. 2017). Due to their superior genetic diversity and capacity to quickly infiltrate the root zone and occupy the smallest niches, PGPR consortia are chosen over single-strain products (Reddy 2014). As a result, the combination of PGPR strains can withstand a variety of potential plant pathogens under different conditions of plant growth and environment (Palanisamy et al. 2008; Reddy 2014). Additionally, the consumption of fertiliser is reduced by 25% and nitrogen uptake is enhanced when *Bacillus pumilus* T4 and *Bacillus amyloliquefaciens* strain IN937a are combined with arbuscular mycorrhiza fungi *Rhizophagus intraradices* (Adesemoye et al. 2009). According to several studies (Bhaskara Reddy et al. 1999; Mäder et al. 2011; Araújo et al. 2013), local PGPR strains of a host plant had strong physiological and genetic adaptation and co-evolved with other native strains in a typical environment.

3.2.2 PGPR-Based Products

It has been demonstrated in the past that exogenous application of PGPR substances, such as exopolysaccharides, metabolites and phytohormones, encourages the growth of specific beneficial bacteria in the soil (Marks et al. 2013). Wheat plant growth and seed production were stimulated by bio-priming of the bacterial quorum sensing signal *N*-hexanoyl-L-homoserine lactone (C6-HSL) molecule (Moshynets et al. 2019). Depending on the plant species and variety, soil type and environmental factors, PGPR-based products perform differently. The effectiveness and reproducibility of the product are significantly influenced by the commercial formulation type (Calvo et al. 2014).

3.2.3 Elicitors

Elicitor is a signal molecule that plants perceive, and its differentiation as non-self can induce a defensive reaction in them (Vallad and Goodman 2004). Induced resistance is a valuable disease control strategy that offers long-lasting, broad range of pathogens and disease control using its own resistance (Walters et al. 2014). Laminarin and benzo-(1, 2, 3)-thiadiazole-7-carbothioic acid S-methyl ester from algal extract are well-known elicitor products on market (Sobhy et al. 2012). Elicitors from crustacean shells and minerals were also identified (Henry et al. 2012).

It is advantageous to look for new elicitors for usage in agriculture. The three types of elicitors that have been so far identified are damage- or danger-associated molecular patterns (DAMPs) from the plant upon insect attack, microbe-associated molecular patterns (MAMPs), by beneficial/non-pathogenic microorganisms, and pathogen-associated molecular patterns (PAMPs) from potential pathogens (Henry et al. 2012). It has been demonstrated that these warning signs can shield plants from a variety of biotic stresses (Schwessinger and Ronald 2012). Lipopolysaccharides (LPS), bacterial flagellin, siderophores, fungal chitin, ergosterol, oligogalacturonides (OGAs), cyclic lipopeptides (fengycin, surfactin) and volatile organic compounds (VOCs) are among the well-characterised elicitors (Kutschera and Ranf 2019; Seo et al. 2019; Kanchiswamy et al. 2015).

3.2.4 Semiochemicals

Semiochemicals are messenger molecules from living organism that can provoke behavioural and physiological reactions in other individuals. They are divided into two groups: intraspecific (pheromones) and interspecific (allelochemicals) (Vet and Dicke 1992). Semiochemicals are slightly toxic or non-toxic molecules with high specificity and eco-friendliness, and are promising alternative to insecticides

(Rodriguez-Saona et al. 2012). It is based on the push–pull strategy where a repellent or deterrent stimulus will push pests away from plants or some highly attractive stimuli will pull pests towards the trap (Cook et al. 2007). The push–pull strategy was successfully demonstrated against maize and sorghum stem borers in Africa (Khan and Plckett 2001). Methyl salicylate (MeSA) is a volatile compound produced by diverse plant species in response to insect invasions (Pichersky and Gershenzon 2002). Recently, the enticing impact of MeSA on natural foes was demonstrated (Rodriguez-Saona et al. 2011) and is commercially sold as PredaLure® in the USA (AgBio Inc., Westminster, Colorado, USA).

3.2.5 Sea Weed Extracts

Sea weeds are a crucial source of minerals and organic materials. In agriculture, brown, red and green varieties of sea weed are utilised as soil conditioners or bio-stimulants (Norrie and Keathley 2006; Gajc-Wolska et al. 2012; Sharma et al. 2014). Sea weed contains cytokinins and auxins, which function as biostimulants (Hamza and Suggars 2001). Plant development was accelerated by organic substances such as laminarin, fucoidan, alginates and minerals (Battacharyya et al. 2015). Consideration and attention were given to the usage of micro-algae as plant biostimulants (Oancea et al. 2013; Mógor et al. 2018; Chiaiese et al. 2018).

3.3 Biochemical and Physiological Mechanisms of Bio-Priming

Bio-priming enhances the preparatory processes of germination, stress tolerance and nutrient mobilisation (Bradford 1986). The large carbohydrate reserves in bio-primed seeds will support the plant growth under stress conditions when compared to non-primed seeds (Ella et al. 2011). Biochemical changes associated with bio-priming include an increase in hydrolytic enzyme activities, ROS (reactive oxygen species) detoxifying enzyme activities and changes in phenol, flavonoid and plant hormone levels accompanied by differential expression of genes related to enhanced plant growth and stress resistance. Physiological changes associated with bio-priming are improved germination, vigor, root architecture, leaf area, chlorophyll content, lignin content, root and shoot ratio and nutrient utilisation in plants resulting in better crop stand and productivity (O'Toole and Bland 1987; Priya et al. 2016; Rawat et al. 2012; Rubin 2008; Yadav et al. 2012; Namvar and Khandan 2014; Tanwar et al. 2013; Moeinzadeh et al. 2010; Muruli 2013). Understanding the biochemical and physiological changes associated with bio-priming is necessary and innovative research using 'omics' techniques will widen the knowledge of phytostimulation and growth enhancement.

Bio-priming using PGPR will enhance plant growth by the secretion and solubilization of minerals directly in the root zone. For instance, the N-fixing PGPR like *Diazotrophs*, *Rhizobia*, *Azospirillum*, *Azotobacter* and Cyanobacteria by symbiotic or non-symbiotic association converts the atmospheric nitrogen into plant-absorbable form by nitrogenase enzyme coded by *nif* gene (Thakuria et al. 2004; Kim and Rees 1994; Glick et al. 1999; Ahemad and Khan 2012). Similarly, phosphorus-solubilizing microorganism (PSM) released organic ions or phosphatase enzyme to solubilise the insoluble P to soluble form for plant uptake. The well-known P solublizers are *Beijerinckia*, *Azotobacter*, *Bacillus*, *Enterobacter*, *Pseudomonas, Serratia* and *Mycobacterium* (Zaidi et al. 2009; Glick 2012). Few PGPR can efficiently solubilize potassium from mica, illite and orthoclase minerals by organic acid excretion (Sheng and He 2006). In oats and *Arabidopsis*, bio-priming using siderophore-producing PGPR improved the iron uptake (Crowley and Kraemer 2007; Vansuyt et al. 2007; Bhattacharyya and Jha 2012) and suppressed plant disease (Keswani et al. 2014; Jain et al. 2012). Besides the above advantages, the 1-amino cyclopropane-1- carboxylate (ACC) deaminase produced by PGPR under stress condition enhanced the plant growth and development (Saleem et al. 2007; Upadhyay et al. 2012). This has been established in wheat under stress conditions (Shah et al. 2015). In addition, plant growth and root development were also favoured by auxin-producing PGPR (Patten and Glick 1996). Glick (2012) observed cell wall loosening and increased root exudation from the plant primed with indole acetic acid (IAA)-producing PGPR. Plant height and growth were positively regulated by *Azotobacter chroococcum* and *Rhizobium leguminosarum* priming through gibberellic acid (GA_3) and cytokinin production (Noel et al. 1996; Verma et al. 2001). Bio-priming using PGPR increased root to leaves ratio, lignin content in rice (Rêgo et al. 2014) and root and shoot length in soybean (Anitha and Jahagirdar 2015). By comparing bio-primed plants to un-primed plants at various growth stages, it was found that the overall content of amino acids, proteins and phenolic compounds increased (Ahemad and Khan 2012; Aishwath et al. 2012; Dhanya 2014; Warwate et al. 2017; Singh et al. 2003; Sofy et al. 2014). In addition, after bio-priming, the overall amount of soluble sugars also increased. These findings imply that bio-priming activated the sugar and protein molecules that regulate cell homeostasis and function as signal transducers in plant growth and development under both normal and stressful circumstances (Gupta and Kaur 2005; Rolland et al. 2006; Chen 2007). In addition to these, defence-related enzymes such as catalase, superoxide dismutase, peroxidase, ammonia lyases and chitinase, among others, provide resistance against biotic stress in plants.

3.4 Molecular Events of Bio-Priming

Plants face different stresses that adversely affect their productivity. The Plant growth-promoting microorganism (PGPM) helps the plants overcome stressful conditions. For example, the well-known beneficial symbiotic microbes, such as mycorrhizal fungi, and beneficial non-symbiotic or non-invading free-living microbes, such as the diverse genera of PGPR and PGPF, collectively known as PGPM, associate with plants for stress alleviation. These beneficial microbes influence plant growth by releasing phytohormones (phyto-stimulation), increasing the availability of nutrients (biofertilisation) and protecting plants from phytopathogens (biocontrol).

Plant defence genes are primed as a result of inducers or elicitors from beneficial microbes, and is termed induced systemic resistance (ISR) (Pieterse et al. 2014). Upon successful colonization, these beneficial microbes further influence plant growth particularly by reducing abiotic and biotic stress conditions through various defence mechanisms. However, limited studies focused on how these beneficial microbes enter the plants (e.g. intracellular symbioses) or colonise on rhizoplane by evading the recognition of the host plant immune system. Here, we first describe the plant immune system and then explore various molecular events associated with the successful colonisation of beneficial microbes to avoid or evade recognising the host immune system. These include microbe perception or detection, downstream signalling events and the induction or evasion of defence responses in plants by these microbes (Fig. 3.1). These molecular events are not well understood for most beneficial microbes for colonisation with plants (Zipfel and Oldroyd 2017; Choudhary et al. 2016).

Recognition of MAMP signals by plant pattern recognition receptors (PRRs) in the non-host and effectors by nucleotide-binding leucine-rich repeats (NLRs) in the host plants is perceived. Further, various early and multiple signalling events occur after pathogen detection. Specifically, hormonal signals and their cross-talk with other hormones arise depending on the mode of pathogen and insect attack. Finally, upon successful recognition and signalling, various plant defence responses are induced in order to prevent the further growth of pathogens. Pathogen infection in non-primed plants resulted in retarded plant growth and establishment, shown in left (Fig. 3.1). On the right, accommodation of beneficial microbes, i.e. microbe-primed plants, enhances plant defence by evading the host plant immune system. The dotted lines from non-primed plants show the evasion of various molecular events to microbe-primed plants.

3.4.1 The Plant Immune System

Plants have evolved two strategies for detecting microbes (Jones and Dangl 2006; Chisholm et al. 2006). In the first strategy, molecules secreted from the microbes into the extracellular space, i.e. chitin, lipopolysaccharides (LPS) and flagellin, are called pathogen/microbial associated molecular patterns (PAMPs/MAMPs). MAMPs are sensed by plant cell surface pattern recognition receptors (PRRs) and lead to MAMP-triggered immunity (MTI) (Boller and Felix 2009; Schellenberger et al. 2019).

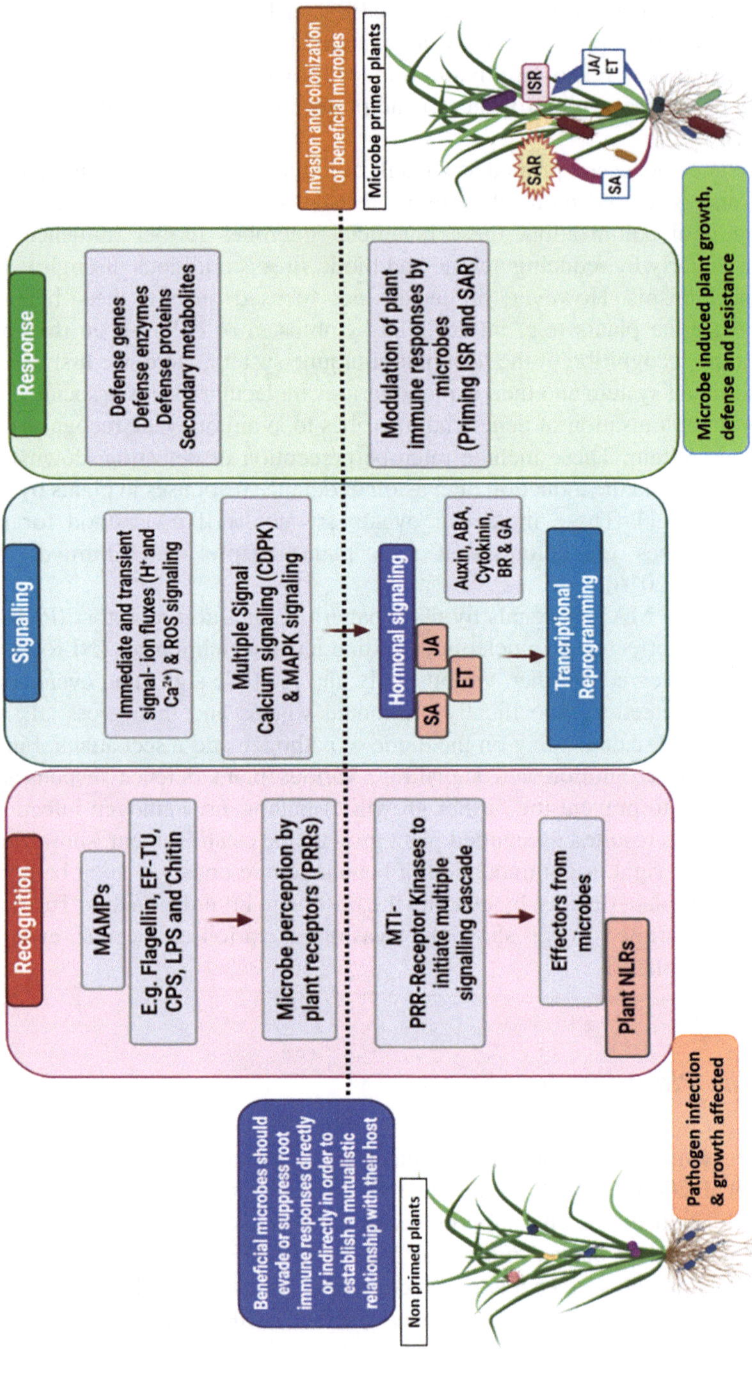

Fig. 3.1 Induction of various molecular events by beneficial microbes in the plants

The second form of perception involves the intracellular receptor-mediated detection of effectors such as microbial virulence or avirulence compounds. To convey effector proteins into the host cell, bacterial pathogens use a type-III secretion system, whereas fungi and oomycetes use haustoria or other intracellular structures. Usually, these intracellular effectors work to lessen pattern triggered immunity (PTI) or microbe triggered immunity (MTI). On the other hand, a few are detected by intracellular nucleotide-binding (NB)-LRR receptors, leading to effector-triggered immunity (ETI). The co-evolutionary dynamics between plants and diseases that result from this type of recognition are described below.

Microbial pathogens of animals and plants have evolved through 'the co-evolutionary arms race', meaning the development of both virulence in microbes and defences in plants (Jones and Dangl 2006; Arnold et al. 2007; Zhou and Chai 2008). The zig-zag theory propounded by Jones and Dangl (2006) explains how plant–pathogens co-evolve with each other. The zig-zag model also explains two components of the plant immune system, i.e. PTI and ETI, but in addition explains ETS (effector-triggered susceptibility) based on effectors that interfere with PTI (stage two) and evolution of effectors and *R* genes by mutation or deletion (stage four) (Jones and Dangl 2006).

Similar signalling pathways and immunological responses are activated by both PTI and ETI; however, ETI activation is typically stronger and lasts longer than PTI (Tsuda and Katagiri 2010). While ETI is frequently active against adapted pathogens known as host resistance, PTI is typically effective against non-adapted infections and is known as non-host resistance. The first barrier that phytopathogens confront is the plant cell wall. Barriers like wax layers or the stiff cell walls of plants must first be breached by pathogens (Zipfel 2008; Spoel and Dong 2012). Through a variety of chemical and physical defence mechanisms, plants defend themselves against pathogens. The synthesis and deposition of callose are the most frequent. A matrix for the deposition of antibacterial substances is provided by the polymer of -l,3-glucan called callose. Papillae are effective barriers that are produced at the sites of attack during the early stages of pathogen invasion. They are callose-containing cell wall appositions (Luna et al. 2011). Additionally, lignin deposition in the cell wall, the creation of reactive oxygen species (ROS) and the manufacture of antimicrobial chemicals are all components of plant immune responses (Tsuda and Katagiri 2010; Gimenez-Ibanez and Rathjen 2010). The signalling mechanisms used by plants to defend themselves include calcium fluxes, transcriptional reprogramming, salicylic acid synthesis and mitogen-activated protein kinase (MAPK) cascades (Sects. 3.4.4 and 3.4.5) (Gimenez-Ibanez and Rathjen 2010; Tsuda and Katagiri 2010). R proteins, which are structurally related and defend the important cellular signalling hubs, constitute the foundation of plant signal-specific immunity. These hubs are disrupted by pathogen effectors, which also activate R proteins to cause programmed cell death (Spoel and Dong 2012).

3.4.2 Molecular Detection/Recognition of Potential Beneficial Partners

The mutual recognition between beneficial microbes and plants occurs in the rhizospheric region. In nature, the plant root exudates recruit potential endophytes and prime them for interaction. The external application of PGPMs directly associates plants above or below ground regions. Although these interactions benefit plants, plants consider any molecules from microbes (pathogens, beneficial organisms and even pests) as 'non-self' or 'foreign agents' as 'warning' or 'alert' to initiate immune response signalling. These molecules from microbes are widespread and conserved in both pathogenic (e.g. PAMPs) and beneficial microorganisms (e.g. MAMPs) described above. Apart from these MAMPs, many metabolites of PGPR were identified as elicitors of plant systemic resistance (Sects. 3.2.2 and 3.2.3) (Pršić and Ongena 2020). Upon a receptor-mediated recognition of these non-self-molecules, plant immune signalling is triggered. The MAMP-induced defence responses are known as MAMP-triggered immunity (MTI) (Boller and Felix 2009; Jones and Dangl 2006) (Fig. 3.1). Signalling events for MTI (mostly in roots) are less explored compared to PTI (which are generally in leaves). MAMPs are highly conserved because MAMPs are essential components of microbial structures that are highly tricky to alter without loss of function. Plant receptors (PRRs) are able to detect these conserved MAMPs from rhizobial symbiosis and root-associated epiphytic PGPR (Zipfel and Oldroyd 2017).

The symbioses with mycorrhiza and rhizobia occur through communication between the rhizosphere microorganisms and host plant root. Both these symbiosis are intracellular associations in the plants. The plant root signals, such as strigolactones (for mycorrhizal fungi) and flavonoids (for rhizobia), are secreted by the plant and in turn mycorrhizal factors (Myc factors) and rhizobia nodulation factors (Nod factors) are produced by the respective symbiotic microbes. The plant recognises these factors to activate common symbiosis signalling (Sym) pathway, then leads to symbiotic associations. In non-symbiotic beneficial microbes, i.e. PGPRs also activate specific signalling components of the Sym pathway, suggesting plant signalling networks triggered by several beneficial microbes are partially converged (Sanchez et al. 2005).

Beneficial microbes must either inhibit MTI (by secreting effectors that interfere with signalling and immune response) or evade MTI (by masking or altering their MAMPs) in order to successfully colonise plants in the first stage. This allows them to establish a mutualistic relationship with their host (Stringlis et al. 2018; Yu et al. 2019). It has been shown that both beneficial and pathogenic microorganisms can get around the MTI process by altering the MAMP or secreting virulence-inducing substances known as effectors. For instance, a MAMP from rhizobia, flg22, is not inducing plant immune responses (not immunogenic) because pathogenic bacteria (flg22, a PAMP) are not conserved; if they were, it would make nitrogen-fixing rhizobial bacteria susceptible to immune detection when they first come into contact with legumes (Lopez-Gomez et al. 2012), indicating that rhizobia have evolved to escape.

3.4.3 Modulation/Regulation of Host Immunity

The successful colonisation of plants and the development of mutualistic plant–microbe interactions depend on PGPM's control of plant immune responses. Hacquard et al. (2017) described that plant employs four distinct barriers to lessen threat of microbes in the endosphere. The first is a rhizosphere-specific ecological barrier that hinders microbial invasion (e.g. PGPM) or phyllosphere (e.g. pathogens). These microbes have highly competitive/cooperative mechanisms against other plant microbiota. The second is the physical barrier that must bypass epidermal cell walls through the secretion of virulence compounds and cell wall degrading enzymes that cause stomata opening (i.e. coronatine). The third barrier (i.e. the first immune layer as described in above paragraph) is sensing of microbe-associated molecular patterns (MAMPs) by pattern recognition receptors (PRRs) of plants, which leads to MAMP-triggered immunity (MTI) therefore restricting entry of microorganisms. The receptors of conserved MAMPs perceive or detect signal from outside the cell. The conserved molecules released from beneficial microbes should avoid recognition by the host plant immune system, which are triggered in roots upon MAMP perception. For example, the successful symbiotic and non-symbiotic beneficial microbes are evolved in different modes to minimise recognition of their MAMPs (Zamioudis and Pieterse 2012). Both rhizobia and arbuscular mycorrhizal fungi secrete effectors that encourage their host plants to colonise (Deakin and Broughton 2009; Kloppholz et al. 2011).

Further, the mechanism to evade MAMPs modification and suppress MTI by effector delivery is described above in Sect. 3.4.2. The final barrier (i.e. the second immune layer) involves nucleotide-binding-site-leucine rich repeats (NBS_LRR) proteins or *R* genes that mainly detect the actions of effectors from either pathogen or beneficial microbes, which leads to ETI (Fig. 3.1). Effectors from beneficial microorganisms were shown to alter plant immune responses in mycorrhizal fungi (Kloppholz et al. 2011; Marchetti et al. 2010), endophytic fungi (Schäfer et al. 2009) and dinitrogen-fixing rhizobia (Berrabah et al. 2015; Berrabah et al. 2014; Domonkos et al. 2013). These microbes are capable of suppressing plant immune responses in order to colonise the host plants using basic mechanism as that of bacterial pathogens. Therefore, it is found that the beneficial microorganisms should evade or suppress various root immune responses. Different evading mechanisms by beneficial microbes were discussed and detailed in several researches (Yu et al. 2019; Zipfel and Oldroyd 2017; Zamioudis and Pieterse 2012). Overcoming all these barriers by a microbe leads to either disease condition (mostly by pathogens in above-ground region) or beneficial association (mostly by PGPM in below-ground region).

3.4.4 Early Signalling Events

The common early signalling events such as ROS and MAPK cascades (which also express transiently) are activated upon recognition of beneficial microbes and rapidly induce the downstream immune signalling and defence genes expression (Fig. 3.1). To colonise successfully with host plants, the beneficial microbes have to eliminate ROS burst and suppress MAPK cascade (Zamioudis and Pieterse 2012; Yu et al. 2019). For instance, NopM (an effector from *Rhizobium* sp. strain NGR234) was previously discovered to prevent flg22 (a MAMP)-induced ROS bursts in *Nicotiana benthamiana* (Xin et al. 2012). Furthermore, in *Nicotiana tabacum* and *Lotus japonicus*, the effector NopL from the *Sinorhizobium* sp. strain NGR234 interfered with host MAPK cascades to inhibit the expression of pathogenesis-related defence proteins (Bartsev et al. 2004). Beneficial bacteria can prevent immune signal transduction and the activation of subsequent immune responses by interfering with host plant MAPK cascades (Fig. 3.1). In a different study, it was discovered that the PGPF *Piriformospora indica* used the jasmonic acid (JA) signalling pathway to suppress both early (ROS generation) and late (mainly salicylic acid (SA) mediated responses) defensive responses in *Arabidopsis* (Millet et al. 2010). JAR1 and MYC2, which are components of the JA signalling pathway, enable this (JIN1).

3.4.5 Hormonal Modulation/Regulation

Among the various defence signalling pathways, hormone signalling networks link signal perception to regulate broad transcriptional reprogramming and induction of defence responses (Ma and Ma 2016). Like pathogen interaction, beneficial interaction with plants leads to manipulation of hormone signalling pathways to suppress immune responses of root and promote beneficial colonisation with plants. In general, SA, JA and ethylene (ET) have consistently been reported to be induced against biotic stresses (Ma and Ma 2016). In contrast, abscisic acid (ABA) plays an important role against abiotic stresses (Dar et al. 2017). In particular, SA signalling triggers resistance against biotrophic and hemi-biotrophic pathogens (e.g. most bacterial pathogens) (Vlot et al. 2009). Both JA and ET are induced against necrotrophic pathogens (e.g. most fungal pathogens) (Glazebrook 2005). JA alone induces defence against herbivores (Ma and Ma 2016). Further, JA and ET pathways are commonly induced for successful association with beneficial microbes in the host. For example, it was found that beneficial microbes induced JA/ET signalling through the secretion of effectors for suppression of host downstream signalling events and defence gene expression (Ma and Ma 2016). A study by Kloppholz et al. (2011) showed that the arbuscular mycorrhizal fungus, *Rhizophagus irregularis*, secretes the effector SP7, which binds directly with the JA/ET inducible-ERF19 transcription factor and inhibits the expression of EFR19-activated defence-related genes in roots of *Medicago truncatula*. It is evident that these effectors foster beneficial colonisation by suppressing immune recognition or signalling.

Additionally, in response to certain pathogens, herbivores and mutualistic interactions, plants can fine-tune to regulate their responses by modifying gene expression via cross-talk (synergistic or antagonistic) across SA, JA and ET pathways (Broekgaarden et al. 2011; Lazebnik et al. 2014; Manoharan et al. 2019; Spoel and Dong 2008). It was established that rhizobia use the Nod-signalling pathway to decrease SA-dependent defensive reactions. Rhizobia are susceptible to defence responses that are controlled by SA, and numerous reports show that SA signalling has a negative impact on the rate and intensity of rhizobial infection and nodulation (Stacey et al. 2006; van Spronsen et al. 2003). In *M. truncatula*, overexpression of *NPR1*, a crucial regulator of SA that is crucial for plant immunity, suppressed root hair deformation in response to *Sinorhizobium meliloti*, as demonstrated by Peleg-Grossman et al. (2009). On the other hand, *NPR1* level reduction brought on by *RNAi* accelerated root hair curling.

3.4.6 Suppression of SA-Responsive Defence Genes

In general, pathogenesis-related (PR) genes are linked to specific signalling pathways, and a variety of plant hormones can cause them to express themselves (Derksen et al. 2013). ISRs are frequently SA-independent, PGPR and PGPF-induced, and do not significantly alter the expression of defence-related genes (Van der Ent et al. 2009). As opposed to this, pathogen-induced systemic acquired resistance (SAR) is heavily dependent on SA signalling and is linked to increased expression of a vast number of PR genes (Durrant and Dong 2004; Vlot et al. 2009). When insects or pathogens invade, PGPR-primed plants show a quicker, frequently JA-dependent defensive response (Pozo et al. 2008; van Hulten et al. 2006). In contrast to the arbuscular mycorrhizal fungi (AMF) *Glomus intraradices*, which is known to colonise tomato plants more commonly, López-Ráez et al. (2010)'s research showed that the SA content and expression of the SA-responsive gene *PR1a* were solely expressed in tomato roots colonized by *Glomus mosseae*.

3.5 Omics: A Holistic Approach to Bio-Priming

In nature, during the interaction of plants with necrotrophic pathogens, beneficial microbes or metabolites from beneficial microbes can render the immune system of plants more reactive, leading to rapid induction of resistance mechanism (Balmer et al. 2015; Pastor et al. 2014; Conrath et al. 2006). This is a distinct physiological state wherein plants are pre-conditioned to mount stronger or more effective defence responses to biotic stresses (Pastor et al. 2014; Balmer et al. 2015; Conrath et al. 2015). Despite the conflict of plant resource allocation to defence vs. development, priming has a tiny fitness cost mostly to ward off possible pathogen invasions (Hilker et al. 2016). Additionally, the primed defence state in a plant can be transmitted epigenetically (Holeski et al. 2012; Jaskiewicz et al. 2011).

Most researches focused only on characterising and defining the physiological and biochemical changes associated with priming processes. Since these techniques are crucial in elucidating the key elements of priming, such as improved physiological stimuli, latent signal transduction enzymes, chromatin remodelling and transcriptional factors (Návarová et al. 2012; Pastor et al. 2013; Conrath et al. 2015; Pastor et al. 2014), there are still discrepancies in our insights on interconnected biological information networks involved in the whole priming event. Notably, the priming elicitors and the patho-system have a significant impact on a particular primed state, making it challenging to establish broad guidelines for understanding how primed plants respond. Several biological processes, such as transcriptional, post-translational, metabolic, physiological and epigenetic reprogramming, are finely regulated by elicitor-induced resistance (Hake and Romeis 2019; Martinez-Medina et al. 2016). The immune systems of plants that have been pre-treated with elicitors can be trans-generationally and temporarily adapted to a 'primed state' (Xu and Charles 2019). When pathogens attack, these 'post-primed' plants permit a greater and quicker defence response than mock plants (de Vega et al. 2018). Thus, omics approaches (interconnected multi-layer comprising genomics, transcriptomics, proteomics and metabolomics) enable the experiential depictions of the physiological responses constituting the 'prime-ome' (Balmer et al. 2015; Hilker et al. 2016; Martinez-Medina et al. 2016; Windram et al. 2014).

3.5.1 Transcriptomic Approaches for Understanding the Primed State in Plants

The transcriptomic approach is widely used to understand and characterise the plant–pathogen interactions by studying the expression of defence genes (Chen et al. 2021; Wang et al. 2022). During the pre-stage primed phase, typical changes in the accumulation of primary metabolites (sugars, derivatives of amino acids and tricarboxylic acid) were investigated and reported (Pastor et al. 2014). However, most of the researches on understanding the changes that occur in primed plants in response to pathogen infection are mainly based on the expression profiling of important plant defence genes associated with early and hormone-specific (SA, JA and ET) signalling (Manoharan et al. 2019; Pieterse et al. 2009). For analysing the target genes in defence primed plants, the transcriptome approach quantitative -reverse transcriptase- polymerase chain reaction (qRT-PCR) is widely used. For instance, *Trichoderma asperellum* primed in *Arabidopsis thaliana* showed defence against *Pseudomonas syringae* by triggering the defence-responsive genes of ET/JA (*eir1*, *eto3*, *lox2*) pathways (Brotman et al. 2012).

Salicylic acid is critical in regulating the plant defence against pathogens at primed state. For example, *Agrobacterium tumefaciens* primed in tobacco plants prior to infection with virulent pathogen *Pseudomonas syringae* (*Pst*) exhibited the primed upregulation of genes for SA and MAPK pathway and subsequent increase in producing ROS and callose deposition (Sheikh et al. 2014). Further, in *T. asperellum* primed *Arabidopsis* infected with *Pst* showed significant increase in SA and Et/JA-responsive genes expression (Brotman et al. 2012). In particular, transcription factors that regulate the expression of glucosinolate pathways and JA-mediated cross-talk and signalling genes were upregulated (Gigolashvili et al. 2007). In *Arabidopsis*, a transcriptome investigation of ISR induced by *Trichoderma hamatum* T382 against *Botrytis cinerea* was carried out in the priming stage resulting in the accumulation of *PR1* gene (Mathys et al. 2012). Petti et al. (2010) evidenced that the *Piriformospora indica* primed barley plants triggered the expression of 22 defence- responsive genes against the *Blumeria graminis* infection. In addition, barley primed with *Pseudomonas fluorescens* MKB158, a non-pathogenic elicitor, exhibited expression of 74 defence genes responsible for protease inhibition, lipid transfer proteins and JA-responsive genes (Petti et al. 2010). Transcriptome analysis on wheat seedlings exposed to H_2O_2 revealed the stimulation of JA/ET genes, which are possibly regulators of resistance to *Blumeria graminis* (Li et al. 2011). Bio-priming of *Trichoderma erinaceum* in tomato challenged with *Fusarium oxysporum* f.sp. *lycopersici* enhanced the upregulation of defence-responsive WRKY transcription factors, and increased the expression genes for ROS enzyme production (Aamir et al. 2019).

3.5.2 Proteomics Approaches in Primed Plants

Identification of the proteins involved in complex biological processes and research into the molecular control mechanisms are both made possible by proteomics (Jia et al. 2020). Proteomics studies have been carried out in recent years to investigate the induced resistance of many elicitors, including H_2O_2, flagellin, SA, chitin and LPS. These studies provided in-depth insights into the complex mechanisms by which elicitors are produced, the components of PTI signalling and the elicitor-induced 'primed state' of plants. MeJA-treated calla lily plants produced more ROS enzymes and secondary metabolites in response to the pathogen *Pectobacterium carotovorum*. Additionally, the expression levels of 215 proteins changed after being infected with the pathogen and primed with MeJA, according to results from a non-targeted label-free mass spectrometry (MS) method used to monitor protein levels. Nucleic acid-binding proteins, ion-binding proteins, cofactor-binding proteins, hydrolases, oxidoreductases and transferases were among the 215 proteins that significantly increased in expression. Additionally, in-plant analysis of oxidoreductases revealed that increased enzymatic activity restricted to areas around the pathogen penetration spots of *Zantedeschia aethiopica* plants against *P. carotovorum* leads to the possibility that proteins are involved in plant defence. Similar to this, the

25 primary metabolism-related proteins were significantly increased in the leaf proteome of potato plants that had been chemically treated with Gamma-aminobutyric acid (GABA), β-aminobutyric acid (BABA), 2,6-dichloroisonicotinic acid and laminarin. The increased defence against *Sclerospora graminicola* in pearl millet seeds primed with *P. fluorescens* and BABA was demonstrated by elucidating the proteins in response to pathogen suppression. In the primed state, primary metabolism, energy and signal transduction are all regulated by 63 differentially expressed proteins. Plants require distinct mechanisms to swiftly modify their metabolism to diverse situations in order to attain a new state of equilibrium in a potentially changing environment. The priming phase, which typically occurs as 'silent' before being challenged to a stress, is characterised by minor changes in primary metabolism (Pastor et al. 2014) that set the plant's defence systems on standby.

It would be ideal to comprehend the circumstances surrounding the beginning, persistence and transmission of the primed state to the progeny. Understanding the underlying metabolic networks that control the adaptive metabolism of priming events in a biosystem that distinguish the primed condition among species is therefore essential for general implications ranging from model plants to industrially significant crop plants.

3.5.3 Metabolomics: The Way Forward to Understand the Drivers in Primed Plants

The numerous layers of induced defence systems that are active at distinct stages of plant–pathogen interaction may be involved in priming. The quantity of descriptive data supplied by metabolic profiling advancements is advancing our understanding of priming processes. These latest efforts have revealed primary metabolic reprogramming and changed secondary metabolite synthesis as the two main mechanisms in priming events (Conrath 2011; Mhlongo et al. 2016; Tenenboim and Brotman 2016). Experimental data suggested that indole metabolites (indole, glucosinolates, IAA, camalexin and indole-3-carboxylic acid) are important in priming events. These investigations of *A. thaliana* primed with bacterial LPS were comparable and untargeted (Finnegan et al. 2016). Additionally, the phenylpropanoid pathway intermediates, tyramine, quinic acid and polyamines or glucose, were altered in tobacco cells exposed to pathogen-derived elicitors such LPS, chitosan and flagellin, which primed the plant's defence against biotrophic stimuli (Mhlongo et al. 2016). In 2018, Kaling et al. investigated the relationship between ectomycorrhizal fungus and poplar roots. A rise in volatiles, chitinases and compounds containing nitrogen was among the defence mechanisms in leaves that *Laccaria bicolor* showed to be systemically regulated. The mycorrhiza-primed condition strengthened the plants' defences against interactions with above-ground insects. Additionally, as demonstrated by Akram et al., tomato plants primed with

Bacillus fortis IAGS162 increased the production of amino acids, sugars, caffeic acid, quinic acid and shikimic acid, which led to plant defence against *Fusarium* wilt (Akram et al. 2016).

An intriguing approach might be to examine the fungal metabolome in response to priming. This should give you a basic understanding of antifungal priming. The question of whether fungi likewise use chemicals to 'prepare' plants for more effective invasion may thus be answered by future omics research on fungal invaders. Although numerous metabolomics studies on plant–fungus interactions have shed light on metabolites associated to defence, our knowledge of metabolites that aid in antifungal priming is still restricted. For instance, applying flavonoid derivatives to *B. cinerea* as a first line of defence against fungus causes early H_2O_2 build-up in *Arabidopsis* (Małolepsza 2005). We looked at the green leaf volatile Z-3-hexenyl acetate as a priming agent against *Fusarium graminearum* in wheat, which is usually associated with defence against herbivores (Ameye et al. 2015). One of the essential metabolites aiding priming against fungal infections is IAA in wheat primed by *P. fluorescens* against *Fusarium culmorum* (Petti et al. 2012). Although the mechanism of action of each of these substances in defence priming is unknown, they share a characteristic.

3.6 Conclusion and Future Prospects

The modern agriculture should integrate the beneficial microbes for alleviating various stresses for eco-friendly and sustainable agriculture. The various bio-priming agents act either directly or indirectly and induce changes in physiological and molecular processes of plants. The molecular changes associated with plants in response to either pathogens or beneficial interactions were investigated. Thus together, plants in association with beneficial microbes against a particular biotic or abiotic stress could lead to elucidate an integrated view of plant immune responses in nature. The recent investigations found the molecular events, such as signal recognition, regulation (common signalling pathways) and defence responses, in plants are mostly similar to both pathogens and beneficial microbes. Thus, the sessile nature of plants in environment is able to distinguish those signals that are beneficial from those that are harmful during various environmental conditions. However, more studies are needed to understand the common and different molecular and signalling pathways associated between the beneficial microbes and pathogens or insect pests with the advancement of 'Omics' tools and in next-generation sequencing approaches for sustainable agricultural production.

References

Aamir M, Kashyap SP, Zehra A, Dubey MK, Singh VK, Ansari WA, Upadhyay RS, Singh S (2019) *Trichoderma erinaceum* bio-priming modulates the WRKYs defense programming in tomato against the *Fusarium oxysporum* f. sp. *lycopersici* (Fol) challenged condition. Front Plant Sci 10. https://doi.org/10.3389/fpls.2019.00911

Adesemoye AO, Torbert HA, Kloepper JW (2009) Plant growth-promoting rhizobacteria allow reduced application rates of chemical fertilizers. Microbial Ecol 58(4):921–929. https://doi.org/10.1007/s00248-009-9531-y

Ahemad M, Khan MS (2012) Evaluation of plant-growth-promoting activities of rhizobacterium *Pseudomonas putida* under herbicide stress. Ann Microbiol 62(4):1531–1540. https://doi.org/10.1007/s13213-011-0407-2

Aishwath OP, Lal G, Kant K, Sharma YK, Ali SF, Naimuddin (2012) Influence of biofertilizers on growth and yield of coriander under typic haplustepts. Int J Seed Spic 2:9–14

Akram W, Anjum T, Ali B (2016) Phenylacetic acid is ISR determinant produced by *Bacillus fortis* IAGS162, which involves extensive re-modulation in metabolomics of tomato to protect against *Fusarium* wilt. Front Plant Sci 7:498. https://doi.org/10.3389/fpls.2016.00498

Alizadeh Forutan M, Pirdashti H, Yaghoubian Y, Babaeizad V (2017) The effect of *Piriformospora indica* seed bio-priming and paclobutrazol foliar spraying on tolerance to chilling stress in green beans (*Phaseolus vulgaris* L.). Environ Stress Crop Sci 10(3):459–474

Ameye M, Audenaert K, De Zutter N, Steppe K, Van Meulebroek L, Vanhaecke L, De Vleesschauwer D, Haesaert G, Smagghe G (2015) Priming of wheat with the green leaf volatile Z-3-hexenyl acetate enhances defense against *Fusarium graminearum* but boosts deoxynivalenol production. Plant Physiol 167(4):1671–1684. https://doi.org/10.1104/pp.15.00107

Anitha MU, Jahagirdar S (2015) Influence of seed priming agents on yield, yield parameters and purple seed stain disease in soybean. J Agril Sci 28(1):20–23

Araújo AES, Baldani VLD, Galisa PS, Pereira JA, Baldani JI (2013) Response of traditional upland rice varieties to inoculation with selected diazotrophic bacteria isolated from rice cropped at the northeast region of Brazil. App Soil Ecol 64:49–55. https://doi.org/10.1016/j.apsoil.2012.10.004

Arnold DL, Jackson RW, Waterfield NR, Mansfield JW (2007) Evolution of microbial virulence: the benefits of stress. Trends Genet 23(6):293–300. https://doi.org/10.1016/j.tig.2007.03.017

Balmer A, Pastor V, Gamir J, Flors V, Mauch-Mani B (2015) The 'prime-ome': towards a holistic approach to priming. Trends Plant Sci 20(7):443–452. https://doi.org/10.1016/j.tplants.2015.04.002

Bartsev AV, Deakin WJ, Boukli NM, McAlvin CB, Stacey G, Malnoë P, Broughton WJ, Staehelin C (2004) NopL, an effector protein of *Rhizobium* sp. NGR234, thwarts activation of plant defense reactions. Plant Physiol 134(2):871–879. https://doi.org/10.1104/pp.103.031740

Battacharyya D, Babgohari MZ, Rathor P, Prithiviraj B (2015) Seaweed extracts as biostimulants in horticulture. Sci Hortic 196:39–48. https://doi.org/10.1016/j.scienta.2015.09.012

Berrabah F, Bourcy M, Eschstruth A, Cayrel A, Guefrachi I, Mergaert P, Wen J, Jean V, Mysore KS, Gourion B, Ratet P (2014) A nonRD receptor-like kinase prevents nodule early senescence and defense-like reactions during symbiosis. New Phytol 203(4):1305–1314. https://doi.org/10.1111/nph.12881

Berrabah F, Ratet P, Gourion B (2015) Multiple steps control immunity during the intracellular accommodation of rhizobia. J Exp Bot 66(7):1977–1985

Bhaskara Reddy MV, Arul J, Angers P, Couture L (1999) Chitosan treatment of wheat seeds induces resistance to *Fusarium graminearum* improves seed quality. J Agric Food Chem 47(3):1208–1216. https://doi.org/10.1021/jf981225k

Bhattacharyya PN, Jha DK (2012) Plant growth-promoting rhizobacteria (PGPR): emergence in agriculture. W J Microbiol Biotechnol 28(4):1327–1350. https://doi.org/10.1007/s11274-011-0979-9

Boller T, Felix G (2009) A renaissance of elicitors: perception of microbe-associated molecular patterns and danger signals by pattern-recognition receptors. Ann Rev Plant Biol 60:379–406. https://doi.org/10.1146/annurev.arplant.57.032905.105346

Bradford K (1986) Manipulation of seed water relations via osmotic priming to improve germination under stress conditions. Hort Sci 21:1105–1112. https://doi.org/10.21273/HORTSCI.21.5.1105

Broekgaarden C, Snoeren TA, Dicke M, Vosman B (2011) Exploiting natural variation to identify insect-resistance genes. Plant Biotechnol J 9(8):819–825. https://doi.org/10.1111/j.1467-7652.2011.00635.x

Brotman Y, Lisec J, Méret M, Chet I, Willmitzer L, Viterbo A (2012) Transcript and metabolite analysis of the *Trichoderma*-induced systemic resistance response to *Pseudomonas syringae* in *Arabidopsis thaliana*. Microbiol (Reading, England) 158(Pt 1):139–146. https://doi.org/10.1099/mic.0.052621-0

Callan NW, Mathre D, Miller JB (1990) Bio-priming seed treatment for biological control of *Pythium ultimum* preemergence damping-off in sh-2 sweet corn. Plant Dis 74:368–372

Calvo P, Nelson L, Kloepper JW (2014) Agricultural uses of plant biostimulants. Plant Soil 383(1):3–41. https://doi.org/10.1007/s11104-014-2131-8

Chakraborty S, Newton AC (2011) Climate change, plant diseases and food security: an overview. Plant Pathol 60:2–14

Chen JG (2007) Sweet sensor, surprising partners. Sci STKE (373):pe7. https://doi.org/10.1126/stke.3732007pe7

Chen XL, Sun MC, Chong SL, Si JP, & , Wu LS (2021) Transcriptomic and metabolomic approaches deepen our knowledge of plant-endophyte interactions. Front Plant Sci 12:700200. doi:https://doi.org/10.3390/ijms20194916

Chiaiese P, Corrado G, Colla G, Kyriacou MC, Rouphael Y (2018) Renewable sources of plant biostimulation: microalgae as a sustainable means to improve crop performance. Front Plant Sci 9

Chisholm ST, Coaker G, Day B, Staskawicz BJ (2006) Host-microbe interactions: shaping the evolution of the plant immune response. Cell 124(4):803–814. https://doi.org/10.1016/j.cell.2006.02.008

Choudhary M, Ghasal PC, Kumar S, Yadav RP, Singh S, Meena VS, Bisht JK (2016) Conservation agriculture and climate change: an overview. In: Bisht JK, Meena VS, Mishra PK, Pattanayak A (eds) Conservation agriculture: an approach to combat climate change in Indian Himalaya. Springer Singapore, Singapore, pp 1–37. https://doi.org/10.1007/978-981-10-2558-7_1

Conrath U (2011) Molecular aspects of defence priming. Trends Plant Sci 16(10):524–531. https://doi.org/10.1016/j.tplants.2011.06.004

Conrath U, Beckers GJ, Flors V, García-Agustín P, Jakab G, Mauch F, Newman MA, Pieterse CM, Poinssot B, Pozo MJ, Pugin A, Schaffrath U, Ton J, Wendehenne D, Zimmerli L, Mauch-Mani B (2006) Priming: getting ready for battle. Mol Plant Microbe Interact 19(10):1062–1071. https://doi.org/10.1094/mpmi-19-1062

Conrath U, Beckers GJ, Langenbach CJ, Jaskiewicz MR (2015) Priming for enhanced defense. Annu Rev Phytopathol 53:97–119. https://doi.org/10.1146/annurev-phyto-080614-120132

Cook SM, Khan ZR, Pickett JA (2007) The use of push-pull strategies in integrated pest management. Annu Rev Entomol 52:375–400. https://doi.org/10.1146/annurev.ento.52.110405.091407

Crowley D, Kraemer S (2007) Function of siderophores in the plant rhizosphere. In: Pinton R, Varanini Z, Nannipierl P (eds) The rhizosphere. CRC Press, pp 173–200. https://doi.org/10.1201/9781420005585.ch7

Dar NA, Amin I, Wani W, Wani SA, Shikari AB, Wani SH, Masoodi KZ (2017) Abscisic acid: a key regulator of abiotic stress tolerance in plants. Plant Gene 11:106–111. https://doi.org/10.1016/j.plgene.2017.07.003

Dawood MG (2018) Stimulating plant tolerance against abiotic stress through seed priming. In: Rakshit A, Singh HB (eds) Advances in seed priming. Springer, Singapore, pp 147–183. https://doi.org/10.1007/978-981-13-0032-5_10

de Vega D, Newton AC, Sadanandom A (2018) Post-translational modifications in priming the plant immune system: ripe for exploitation? FEBS Lett 592(12):1929–1936. https://doi.org/10.1002/1873-3468.13076

Deakin WJ, Broughton WJ (2009) Symbiotic use of pathogenic strategies: rhizobial protein secretion systems. Nat Rev Microbiol 7(4):312–320. https://doi.org/10.1038/nrmicro2091

Derksen H, Rampitsch C, Daayf F (2013) Signaling cross-talk in plant disease resistance. Plant Sci 207:79–87. https://doi.org/10.1016/j.plantsci.2013.03.004

Dhanya BA (2014) Evaluation of microbial seed priming in relation to seed germination, plant growth promotion in *Morinda citrifolia* L. (Noni). University of Agricultural Sciences, Bangalore

Domonkos A, Horvath B, Marsh JF, Halasz G, Ayaydin F, Oldroyd GED, Kalo P (2013) The identification of novel loci required for appropriate nodule development in *Medicago truncatula*. BMC Plant Biol 13(1):157. https://doi.org/10.1186/1471-2229-13-157

du Jardin P (2012) The science of plant biostimulants—A bibliographic analysis. Final Report:27–30

Durrant WE, Dong X (2004) Systemic acquired resistance. Ann Rev Phytopathol 42:185–209. https://doi.org/10.1146/annurev.phyto.42.040803.140421

Ella ES, Dionisio-Sese ML, Ismail AM (2011) Seed pre-treatment in rice reduces damage, enhances carbohydrate mobilization and improves emergence and seedling establishment under flooded conditions. AoB Plants:plr007. https://doi.org/10.1093/aobpla/plr007

Entesari M, Sharifzadeh F, Ahmadzadeh M, Farhangfar M (2013) Seed biopriming with *Trichoderma* species and *Pseudomonas fluorescent* on growth parameters, enzymes activity and nutritional status of soybean. Int J Agron Plant Produc 4(4):610–619

European Biostimulants Industry Consortium (2011) Position on the revision of Reg (EC) 2003/2003

Finnegan T, Steenkamp PA, Piater LA, Dubery IA (2016) The lipopolysaccharide-induced metabolome signature in *Arabidopsis thaliana* reveals dynamic reprogramming of phytoalexin and phytoanticipin pathways. PLoS One 11(9):e0163572. https://doi.org/10.1371/journal.pone.0163572

Gajc-Wolska J, Spiżewski T, Grabowska A (2012) The effect of seaweed extracts on the yield and quality parameters of broccoli (*Brassica oleracea* var. *cymosa* L.) in open field production. Acta Hortic 1009:83–89. https://doi.org/10.17660/ActaHortic.2013.1009.9

Gholami A, Shahsavani S, Nezarat S (2008) The effect of plant growth promoting rhizobacteria (PGPR) on germination, seedling growth and yield of maize. World Acad Sci Eng Technol 49

Gigolashvili T, Berger B, Mock HP, Müller C, Weisshaar B, Flügge UI (2007) The transcription factor *HIG1/MYB51* regulates indolic glucosinolate biosynthesis in *Arabidopsis thaliana*. Plant J 50(5):886–901. https://doi.org/10.1111/j.1365-313X.2007.03099.x

Gimenez-Ibanez S, Rathjen JP (2010) The case for the defense: plants versus *Pseudomonas syringae*. Microbes Infec 12(6):428–437. https://doi.org/10.1016/j.micinf.2010.03.002

Glazebrook J (2005) Contrasting mechanisms of defense against biotrophic and necrotrophic pathogens. Annu Rev Phytopathol 43:205–227. https://doi.org/10.1146/annurev.phyto.43.040204.135923

Glick BR (2012) Plant growth-promoting bacteria: mechanisms and applications. Scientifica 2012:963401. https://doi.org/10.6064/2012/963401

Glick BR, Patten CL, Holguin G, Penrose DM (1999) Biochemical and genetic mechanisms used by plant growth promoting bacteria. Published by Imperial College Press and Distributed by World Scientific Publishing Co. https://doi.org/10.1142/p130

Grichko VP, Glick BR (2001) Amelioration of flooding stress by ACC deaminase-containing plant growth-promoting bacteria. Plant Physiol Biochem 39(1):11–17. https://doi.org/10.1016/S0981-9428(00)01212-2

Gupta AK, Kaur N (2005) Sugar signalling and gene expression in relation to carbohydrate metabolism under abiotic stresses in plants. J Biosci 30(5):761–776. https://doi.org/10.1007/bf02703574

Hacquard S, Spaepen S, Garrido-Oter R, Schulze-Lefert P (2017) Interplay between innate immunity and the plant microbiota. Annu Rev Phytopathol 55:565–589. https://doi.org/10.1146/annurev-phyto-080516-035623

Hake K, Romeis T (2019) Protein kinase-mediated signalling in priming: immune signal initiation, propagation, and establishment of long-term pathogen resistance in plants. Plant Cell Environ 42(3):904–917. https://doi.org/10.1111/pce.13429

Hamza B, Suggars A (2001) Biostimulants: myths and realities. TurfGrass Trends 8:6–10

Henry G, Thonart P, Ongena M (2012) PAMPs, MAMPs, DAMPs and others: an update on the diversity of plant immunity elicitors. BASE 16(2):257–268

Hilker M, Schwachtje J, Baier M, Balazadeh S, Bäurle I, Geiselhardt S, Hincha DK, Kunze R, Mueller-Roeber B, Rillig MC, Rolff J, Romeis T, Schmülling T, Steppuhn A, van Dongen J, Whitcomb SJ, Wurst S, Zuther E, Kopka J (2016) Priming and memory of stress responses in organisms lacking a nervous system. Biol Rev Camb Philos Soc 91(4):1118–1133. https://doi.org/10.1111/brv.12215

Holeski LM, Jander G, Agrawal AA (2012) Transgenerational defense induction and epigenetic inheritance in plants. Trends Ecol Evol 27(11):618–626. https://doi.org/10.1016/j.tree.2012.07.011

Jain A, Singh S, Kumar Sarma B, Bahadur Singh H (2012) Microbial consortium-mediated reprogramming of defence network in pea to enhance tolerance against *Sclerotinia sclerotiorum*. J Appl Microbiol 112(3):537–550. https://doi.org/10.1111/j.1365-2672.2011.05220.x

Jaskiewicz M, Conrath U, Peterhänsel C (2011) Chromatin modification acts as a memory for systemic acquired resistance in the plant stress response. EMBO Rep 12(1):50–55. https://doi.org/10.1038/embor.2010.186

Jia X, Qin H, Bose SK, Liu T, He J, Xie S, Ye M, Yin H (2020) Proteomics analysis reveals the defense priming effect of chitosan oligosaccharides in *Arabidopsis*-Pst DC3000 interaction. Plant Physiol Biochem 149:301–312. https://doi.org/10.1016/j.plaphy.2020.01.037

Jing YD, He ZL, Yang XE (2007) Role of soil rhizobacteria in phytoremediation of heavy metal contaminated soils. J Zhejiang Univ Sci B 8(3):192–207. https://doi.org/10.1631/jzus.2007.B0192

Jones JDG, Dangl JL (2006) The plant immune system. Nature 444(7117):323–329. https://doi.org/10.1038/nature05286

Kaling M, Schmidt A, Moritz F, Rosenkranz M, Witting M, Kasper K, Janz D, Schmitt-Kopplin P, Schnitzler J-P, Polle A (2018) Mycorrhiza-triggered transcriptomic and metabolomic networks impinge on herbivore fitness. Plant Physiol 176(4):2639–2656. https://doi.org/10.1104/pp.17.01810

Kanchiswamy CN, Malnoy M, Maffei ME (2015) Chemical diversity of microbial volatiles and their potential for plant growth and productivity. Front Plant Sci 6

Kavino M, Harish S, Kumar N, Saravanakumar D, Damodaran T, Soorianathasundaram K, Samiyappan R (2007) Rhizosphere and endophytic bacteria for induction of systemic resistance of banana plantlets against bunchy top virus. Soil Biol Biochem 39(5):1087–1098. https://doi.org/10.1016/j.soilbio.2006.11.020

Keswani C, Mishra S, Sarma BK, Singh SP, Singh HB (2014) Unraveling the efficient applications of secondary metabolites of various *Trichoderma* spp. Appl Microbiol Biotechnol 98(2):533–544. https://doi.org/10.1007/s00253-013-5344-5

Khan Z, Plckett JA (2001) Habitat management strategies for the control of cereal stemborers and striga in maize in Kenya. Insect Sci Appl 21:375–380

Kim J, Rees DC (1994) Nitrogenase and biological nitrogen fixation. Biochemistry 33(2):389–397. https://doi.org/10.1021/bi00168a001

Kloepper JW, Leong J, Teintze M, Schroth MN (1980) Enhanced plant growth by siderophores produced by plant growth-promoting rhizobacteria. Nature 286(5776):885–886

Kloppholz S, Kuhn H, Requena N (2011) A secreted fungal effector of *Glomus intraradices* promotes symbiotic biotrophy. Curr Biol 21(14):1204–1209. https://doi.org/10.1016/j.cub.2011.06.044

Kumar A (2022) Microbial biocontrol: sustainable agriculture and phytopathogen management, vol 1. Springer Nature Chem, Cham, Switzerland, pp 1–369. https://doi.org/10.1007/978-3-030-87512-1

Kumari M, Qureshi KA, Jaremko M, White JF, Singh SK, Sharma VK, Singh KK, Santoyo G, Puopolo G, Kumar A (2022) Deciphering the role of endophytic microbiome in postharvest diseases management of fruits: opportunity areas in commercial up-scale production. Front Plant Sci 13:1026575. https://doi.org/10.3389/fpls.2022.1026575

Kutschera A, Ranf S (2019) The multifaceted functions of lipopolysaccharide in plant-bacteria interactions. Biochimie 159:93–98. https://doi.org/10.1016/j.biochi.2018.07.028

Lazebnik J, Frago E, Dicke M, van Loon JJ (2014) Phytohormone mediation of interactions between herbivores and plant pathogens. J Chem Ecol 40(7):730–741. https://doi.org/10.1007/s10886-014-0480-7

Li A, Zhang R, Pan L, Tang L, Zhao G, Zhu M, Chu J, Sun X, Wei B, Zhang X, Jia J, Mao L (2011) Transcriptome analysis of H_2O_2-treated wheat seedlings reveals a H_2O_2-responsive fatty acid desaturase gene participating in powdery mildew resistance. PLoS One 6(12):e28810. https://doi.org/10.1371/journal.pone.0028810

Lopez-Gomez M, Sandal N, Stougaard J, Boller T (2012) Interplay of flg22-induced defence responses and nodulation in *Lotus japonicus*. J Exp Bot 63(1):393–401. https://doi.org/10.1093/jxb/err291

López-Ráez JA, Verhage A, Fernández I, García JM, Azcón-Aguilar C, Flors V, Pozo MJ (2010) Hormonal and transcriptional profiles highlight common and differential host responses to arbuscular mycorrhizal fungi and the regulation of the oxylipin pathway. J Exp Bot 61(10):2589–2601. https://doi.org/10.1093/jxb/erq089

Luna E, Pastor V, Robert J, Flors V, Mauch-Mani B, Ton J (2011) Callose deposition: a multifaceted plant defense response. Mol Plant Microbe Interact 24(2):183–193. https://doi.org/10.1094/mpmi-07-10-0149

Lushchak VI, Matviishyn TM, Husak VV, Storey JM, Storey KB (2018) Pesticide toxicity: a mechanistic approach. EXCLI J 17:1101–1136. https://doi.org/10.17179/excli2018-1710

Lutts S, Benincasa P, Wojtyla L, Kubala S, Pace R, Lechowska K, Quinet M, Garnczarska M (2016) Seed priming: new comprehensive approaches for an old empirical technique. New challenges in seed biology-basic and translational research driving seed technology, pp 1–46

Ma KW, Ma W (2016) Phytohormone pathways as targets of pathogens to facilitate infection. Plant Mol Biol 91(6):713–725. https://doi.org/10.1007/s11103-016-0452-0

Mäder P, Kaiser F, Adholeya A, Singh R, Uppal HS, Sharma AK, Srivastava R, Sahai V, Aragno M, Wiemken A, Johri BN, Fried PM (2011) Inoculation of root microorganisms for sustainable wheat–rice and wheat–black gram rotations in India. Soil Biol Biochem 43(3):609–619. https://doi.org/10.1016/j.soilbio.2010.11.031

Małolepsza U (2005) Spatial and temporal variation of reactive oxygen species and antioxidant enzymes in o-hydroxyethylorutin-treated tomato leaves inoculated with Botrytis cinerea. Plant Pathol 54(3):317–324. https://doi.org/10.1111/j.1365-3059.2005.01167.x

Manoharan B, Qi SS, Dhandapani V (2019) Gene expression profiling reveals enhanced defense responses in an invasive weed compared to its native congener during pathogenesis. Int J Mol Sci 20(19). https://doi.org/10.3390/ijms20194916

Marchetti M, Capela D, Glew M, Cruveiller S, Chane-Woon-Ming B, Gris C, Timmers T, Poinsot V, Gilbert LB, Heeb P, Médigue C, Batut J, Masson-Boivin C (2010) Experimental evolution of a plant pathogen into a legume symbiont. PLoS Biol 8(1):e1000280. https://doi.org/10.1371/journal.pbio.1000280

Marks BB, Megías M, Nogueira MA, Hungria M (2013) Biotechnological potential of rhizobial metabolites to enhance the performance of *Bradyrhizobium* spp. and *Azospirillum brasilense* inoculants with soybean and maize. AMB Express 3:21. https://doi.org/10.1186/2191-0855-3-21

Martinez-Medina A, Flors V, Heil M, Mauch-Mani B, Pieterse CMJ, Pozo MJ, Ton J, van Dam NM, Conrath U (2016) Recognizing plant defense priming. Trends Plant Sci 21(10):818–822. https://doi.org/10.1016/j.tplants.2016.07.009

Mathys J, De Cremer K, Timmermans P, Van Kerckhove S, Lievens B, Vanhaecke M, Cammue BP, De Coninck B (2012) Genome-wide characterization of ISR induced in *Arabidopsis thaliana* by *Trichoderma hamatum* T382 against *Botrytis cinerea* infection. Front Plant Sci 3: 108. https://doi.org/10.3389/fpls.2012.00108

Mayak S, Tirosh T, Glick BR (2004) Plant growth-promoting bacteria that confer resistance to water stress in tomatoes and peppers. Plant Sci 166(2):525–530. https://doi.org/10.1016/j.plantsci.2003.10.025

Mhlongo MI, Steenkamp PA, Piater LA, Madala NE, Dubery IA (2016) Profiling of altered metabolomic states in *Nicotiana tabacum* cells induced by priming agents. Front Plant Sci 7

Millet YA, Danna CH, Clay NK, Songnuan W, Simon MD, Werck-Reichhart D, Ausubel FM (2010) Innate immune responses activated in arabidopsis roots by microbe-associated molecular patterns. Plant Cell 22(3):973–990. https://doi.org/10.1105/tpc.109.069658

Moeinzadeh A, Sharifzadeh F, Ahmadzadeh M, Heidari Tajabadi F (2010) Biopriming of sunflower (*Helianthus annuus* L.) seed with *Pseudomonas fluorescens* for improvement of seed invigoration and seedling growth. Aust J Crop Sci 4

Mógor ÁF, Ördög V, Lima GPP, Molnár Z, Mógor G (2018) Biostimulant properties of cyanobacterial hydrolysate related to polyamines. J Appl Phycol 30(1):453–460. https://doi.org/10.1007/s10811-017-1242-z

Moshynets OV, Babenko LM, Rogalsky SP, Iungin OS, Foster J, Kosakivska IV, Potters G, Spiers AJ (2019) Priming winter wheat seeds with the bacterial quorum sensing signal N-hexanoyl-L-homoserine lactone (C_6-HSL) shows potential to improve plant growth and seed yield. PLoS One 14(2):e0209460. https://doi.org/10.1371/journal.pone.0209460

Muruli CN (2013) Studies on seed priming and primed seed longevity in onion (*Allium cepa* L.) and capsicum (*Capsicum annuum* L.). University of Agricultural Sciences, Bangalore

Namvar A, Khandan T (2014) Biopriming and mineral fertilizers effects on agronomical performance of rapeseed (*Brassica napus* L.). Ekologija 60. https://doi.org/10.6001/ekologija.v60i3.2981

Návarová H, Bernsdorff F, Döring AC, Zeier J (2012) Pipecolic acid, an endogenous mediator of defense amplification and priming, is a critical regulator of inducible plant immunity. Plant Cell 24(12):5123–5141. https://doi.org/10.1105/tpc.112.103564

Noel TC, Sheng C, Yost CK, Pharis RP, Hynes MF (1996) *Rhizobium leguminosarum* as a plant growth-promoting rhizobacterium: direct growth promotion of canola and lettuce. Can J Microbiol 42(3):279–283. https://doi.org/10.1139/m96-040%M8868235

Norrie J, Keathley JP (2006) Benefits of *Ascophyllum nodosum* marine-plant extract applications to â'thompson seedlessâ´ grape production. International Society for Horticultural Science (ISHS), Leuven, Belgium, pp 243–248. https://doi.org/10.17660/ActaHortic.2006.727.27

Oancea F, Velea S, Fătu V, Mincea C, Ilie L (2013) Micro-algae based plant biostimulant and its effect on water stressed tomato plants. Rom J Plant Prot 6:104–117

O'Toole JC, Bland WL (1987) Genotypic variation in crop plant root systems. In: Brady NC (ed) Advances in agronomy, vol 41. Academic Press, pp 91–145. https://doi.org/10.1016/S0065-2113(08)60803-2

Palanisamy G, Theerthagiri A, Prakasam V, Manoharan B, Ragupathi N, Ramasamy S (2008) Plant growth promoting rhizobacterial (PGPR) bioconsortia mediates induction of defense-related proteins against infection of root rot pathogen in mulberry plants. J Plant Interac 3:233–244

Pastor V, Luna E, Mauch-Mani B, Ton J, Flors V (2013) Primed plants do not forget. Environ Exp Bot 94:46–56. https://doi.org/10.1016/j.envexpbot.2012.02.013

Pastor V, Balmer A, Gamir J, Flors V, Mauch-Mani B (2014) Preparing to fight back: generation and storage of priming compounds. Front Plant Sci 5:295. https://doi.org/10.3389/fpls.2014.00295

Patten CL, Glick BR (1996) Bacterial biosynthesis of indole-3-acetic acid. Can J Microbiol 42(3): 207–220. https://doi.org/10.1139/m96-032

Peleg-Grossman S, Golani Y, Kaye Y, Melamed-Book N, Levine A (2009) NPR1 protein regulates pathogenic and symbiotic interactions between *Rhizobium* and legumes and non-legumes. PLoS One 4(12):e8399. https://doi.org/10.1371/journal.pone.0008399

Petti C, Khan M, Doohan F (2010) Lipid transfer proteins and protease inhibitors as key factors in the priming of barley responses to fusarium head blight disease by a biocontrol strain of *Pseudomonas fluorescens*. Funct Integ Genomics 10(4):619–627. https://doi.org/10.1007/s10142-010-0177-0

Petti C, Reiber K, Ali SS, Berney M, Doohan FM (2012) Auxin as a player in the biocontrol of *fusarium* head blight disease of barley and its potential as a disease control agent. BMC Plant Biol 12(1):224. https://doi.org/10.1186/1471-2229-12-224

Pichersky E, Gershenzon J (2002) The formation and function of plant volatiles: perfumes for pollinator attraction and defense. Cur Opin Plant Biol 5(3):237–243. https://doi.org/10.1016/s1369-5266(02)00251-0

Pieterse CMJ, Leon-Reyes A, Van der Ent S, Van Wees SCM (2009) Networking by small-molecule hormones in plant immunity. Nat Chem Biol 5(5):308–316. https://doi.org/10.1038/nchembio.164

Pieterse CM, Zamioudis C, Berendsen RL, Weller DM, Van Wees SC, Bakker PA (2014) Induced systemic resistance by beneficial microbes. Annu Rev Phytopathol 52:347–375. https://doi.org/10.1146/annurev-phyto-082712-102340

Pozo MJ, Van Der Ent S, Van Loon LC, Pieterse CMJ (2008) Transcription factor *MYC2* is involved in priming for enhanced defense during rhizobacteria-induced systemic resistance in *Arabidopsis thaliana*. New Phytol 180(2):511–523. https://doi.org/10.1111/j.1469-8137.2008.02578.x

Priya P, Patil V, Kumar BA (2016) Characterization of grain sorghum [*Sorghum bicolor* (L.) *Moench*] for root traits associated with drought tolerance. Res Environ Life Sci 9(2):163–165

Pršić J, Ongena M (2020) Elicitors of plant immunity triggered by beneficial bacteria. Front Plant Sci 11:594530. https://doi.org/10.3389/fpls.2020.594530

Rao MS, Ramachandran N, Sowmya DS (2009) Biological control of nematode induced disease complex in certain vegetable crops. Paper presented at the international conference on horticulture. Bangalore.

Rawat L, Singh Y, Shukla N, Kumar J (2012) Seed biopriming with salinity tolerant isolates of trichoderma harzmnumalleviates salt stress in rice: growth, physiological and biochemical characteristics. J Plant Pathol 94:353–365

Reddy PP (2013) Bio-priming of seeds. In: Reddy PP (ed) Recent advances in crop protection. Springer, New Delhi, pp 83–90. https://doi.org/10.1007/978-81-322-0723-8_6

Reddy P (2014) Plant growth promoting rhizobacteria for horticultural crop protection, 1st edn. Springer, New Delhi

Rêgo MCF, Ilkiu-Borges F, de Filippi MCC, Gonçalves LA, de Silva GB (2014) Morphoanatomical and biochemical changes in the roots of rice plants induced by plant growth-promoting microorganisms. J Bot: 818797. doi:https://doi.org/10.1155/2014/818797

Rodriguez-Saona C, Kaplan I, Braasch J, Chinnasamy D, Williams L (2011) Field responses of predaceous arthropods to methyl salicylate: a meta-analysis and case study in cranberries. Biol Control 59(2):294–303. https://doi.org/10.1016/j.biocontrol.2011.06.017

Rodriguez-Saona C, Blaauw B, Isaacs R (2012) Manipulation of natural enemies in agroecosystems: habitat and semiochemicals for sustainable insect pest control. In: Soloneski S, and Larramendy ML (eds) Integrated pest management and pest control—Current and future tactics. Intechpen, pp: 89–126. doi:https://doi.org/10.5772/30375

Rolland F, Baena-Gonzalez E, Sheen J (2006) Sugar sensing and signaling in plants: conserved and novel mechanisms. Ann Rev Plant Biol 57:675–709. https://doi.org/10.1146/annurev.arplant.57.032905.105441

Rubin EM (2008) Genomics of cellulosic biofuels. Nature 454(7206):841–845. https://doi.org/10.1038/nature07190

Saleem M, Arshad M, Hussain S, Bhatti AS (2007) Perspective of plant growth promoting rhizobacteria (PGPR) containing ACC deaminase in stress agriculture. J Indus Microbiol Biotechnol 34(10):635–648. https://doi.org/10.1007/s10295-007-0240-6

Sanchez L, Sp W, Arnould C, Bernard AR, Gianinazzi S, Gianinazzi-Pearson V (2005) *Pseudomonas fluorescens* and *Glomus mosseae* trigger *DMI3*-Dependent Activation of genes related to a signal transduction pathway in roots of *Medicago truncatula*. Plant Physiol 139(2): 1065–1077. https://doi.org/10.1104/pp.105.067603

Savazzini F, Longa CMO, Pertot I (2009) Impact of the biocontrol agent *Trichoderma atroviride* SC1 on soil microbial communities of a vineyard in northern Italy. Soil Biol Biochem 41(7): 1457–1465. https://doi.org/10.1016/j.soilbio.2009.03.027

Schäfer P, Pfiffi S, Voll LM, Zajic D, Chandler PM, Waller F, Scholz U, Pons-Kühnemann J, Sonnewald S, Sonnewald U, Kogel KH (2009) Manipulation of plant innate immunity and gibberellin as factor of compatibility in the mutualistic association of barley roots with *Piriformospora indica*. Plant J 59(3):461–474. https://doi.org/10.1111/j.1365-313X.2009.03887.x

Schellenberger R, Touchard M, Clément C, Baillieul F, Cordelier S, Crouzet J, Dorey S (2019) Apoplastic invasion patterns triggering plant immunity: plasma membrane sensing at the frontline. Mol Plant Pathol 20(11):1602–1616. https://doi.org/10.1111/mpp.12857

Schwessinger B, Ronald PC (2012) Plant innate immunity: perception of conserved microbial signatures. Ann Rev Plant Biol 63:451–482. https://doi.org/10.1146/annurev-arplant-042811-105518

Seo JS, Diloknawarit P, Park BS (2019) *ELF18*-induced long noncoding RNA 1 evicts fibrillarin from mediator subunit to enhance *Pathogenesis-related gene 1* (*PR1*) expression. New Phytol 221(4):2067–2079. https://doi.org/10.1111/nph.15530

Shah Z-u-h, Ansari TS, Shah AN, Jamro GM, Rajpar I (2015) Biopriming of wheat seeds with rhizobacteria containing acc-deaminase and phosphate solubilizing activities increases wheat growth and yield under phosphorus deficiency. PJAAEVS 31:24–32

Sharma HSS, Fleming C, Selby C, Rao JR, Martin T (2014) Plant biostimulants: a review on the processing of macroalgae and use of extracts for crop management to reduce abiotic and biotic stresses. J Appl Phycol 26(1):465–490. https://doi.org/10.1007/s10811-013-0101-9

Sharma P, Bhatt A, Jyoti B (2018) Effect of seed bio-priming with microbial inoculants on plant growth, yield and yield contributing characters in soybean [*Glycine max* (L.) Merril]. Int J Econ Plants 5:053–058. https://doi.org/10.23910/IJEP/2018.5.2.0214

Sheikh AH, Raghuram B, Eschen-Lippold L, Scheel D, Lee J, Sinha AK (2014) Agroinfiltration by cytokinin-producing *Agrobacterium* sp. strain GV3101 primes defense responses in *Nicotiana tabacum*. Mol Plant Microbe Inter 27(11):1175–1185. https://doi.org/10.1094/mpmi-04-14-0114-r

Sheng XF, He LY (2006) Solubilization of potassium-bearing minerals by a wild-type strain of *Bacillus edaphicus* and its mutants and increased potassium uptake by wheat. Can J Microbiol 52(1):66–72. https://doi.org/10.1139/w05-117

Singh UP, Sarma BK, Singh DP (2003) Effect of plant growth-promoting rhizobacteria and culture filtrate of *Sclerotium rolfsii* on phenolic and salicylic acid contents in chickpea (*Cicer arietinum*). Cur Microbiol 46(2):131–140. https://doi.org/10.1007/s00284-002-3834-2

Singh VK, Singh AK, Kumar A (2017) Disease management of tomato through PGPB: current trends and future perspective. 3 Biotech 7(4):1–10

Singh VK, Singh AK, Singh PP, Kumar A (2018) Interaction of plant growth promoting bacteria with tomato under abiotic stress: a review. Agric Ecosyst Environ 15(267):129–140

Singh M, Srivastava M, Kumar A, Singh AK, Pandey KD (2020) Endophytic bacteria in plant disease management. In: Kumar A, Singh KV (eds) Microbial endophytes: prospects for sustainable agriculture. Woodhead Publication, Cambridge USA, pp 61–89. https://doi.org/10.1016/B978-0-12-818734-0.00004-8

Sobhy I, Erb M, Sarhan A, El-Hussieni M, Mandour N, Turlings T (2012) Less is more: treatment with BTH and laminarin reduces herbivore-induced volatile emissions in maize but increases parasitoid attraction. J Chem Ecol 38:348–360. https://doi.org/10.1007/s10886-012-0098-6

Sofy A, Salah Attia M, Sharaf A, El Dougdoug K (2014) Potential impacts of seed bacterization or Salix extract in faba bean for enhancing protection against bean yellow mosaic disease. Nat Sci 12:67–82

Spoel SH, Dong X (2008) Making sense of hormone crosstalk during plant immune responses. Cell Host Microbe 3(6):348–351. https://doi.org/10.1016/j.chom.2008.05.009

Spoel SH, Dong X (2012) How do plants achieve immunity? Defence without specialized immune cells. Nat Rev Immunol 12(2):89–100. https://doi.org/10.1038/nri3141

Stacey G, Libault M, Brechenmacher L, Wan J, May GD (2006) Genetics and functional genomics of legume nodulation. Curr Opin Plant Biol 9(2):110–121. https://doi.org/10.1016/j.pbi.2006.01.005

Staley TE, Drahos DJ (1994) Marking soil bacteria with lacZY. In: Methods of Soil analysis. SSSA Book Series, pp 689–706. https://doi.org/10.2136/sssabookser5.2.c33

Stringlis IA, Yu K, Feussner K, de Jonge R, Van Bentum S, Van Verk MC, Berendsen RL, Bakker PAHM, Feussner I, Pieterse CMJ (2018) MYB72-dependent coumarin exudation shapes root microbiome assembly to promote plant health. PNAS 115(22):E5213–E5222. https://doi.org/10.1073/pnas.1722335115

Tanwar A, Aggarwal A, Kaushish S, Chauhan S (2013) Interactive effect of AM fungi with *Trichoderma viride* and *Pseudomonas fluorescens* on growth and yield of broccoli. Plant Protec Sci 49(3):137–145

Tenenboim H, Brotman Y (2016) Omic relief for the biotically stressed: metabolomics of plant biotic interactions. Trends Plant Sci 21(9):781–791. https://doi.org/10.1016/j.tplants.2016.04.009

Thakuria D, Talukdar NCT, Goswami C, Hazarika S, Boro R, Khan M (2004) Characterization and screening of bacteria from rhizosphere of rice grown in acidic soils of Assam. Curr Sci 86:978–985

Tsuda K, Katagiri F (2010) Comparing signaling mechanisms engaged in pattern-triggered and effector-triggered immunity. Cur Opin Plant Biol 13(4):459–465. https://doi.org/10.1016/j.pbi.2010.04.006

Upadhyay SK, Singh JS, Saxena AK, Singh DP (2012) Impact of PGPR inoculation on growth and antioxidant status of wheat under saline conditions. Plant Biol 14(4):605–611

Vallad GE, Goodman RM (2004) Systemic acquired resistance and induced systemic resistance in conventional agriculture. Crop Sci 44(6):1920–1934. https://doi.org/10.2135/cropsci2004.1920

Van der Ent S, Van Wees SCM, Pieterse CMJ (2009) Jasmonate signaling in plant interactions with resistance-inducing beneficial microbes. Phytochemistry 70(13):1581–1588. https://doi.org/10.1016/j.phytochem.2009.06.009

van Hulten M, Pelser M, van Loon LC, Pieterse CM, Ton J (2006) Costs and benefits of priming for defense in *Arabidopsis*. PNAS (USA) 103(14):5602–5607. https://doi.org/10.1073/pnas.0510213103

van Spronsen PC, Tak T, Rood AM, van Brussel AA, Kijne JW, Boot KJ (2003) Salicylic acid inhibits indeterminate-type nodulation but not determinate-type nodulation. Mol Plant Microbe Interac 16(1):83–91. https://doi.org/10.1094/mpmi.2003.16.1.83

Vansuyt G, Robin A, Briat JF, Curie C, Lemanceau P (2007) Iron acquisition from Fe-pyoverdine by *Arabidopsis thaliana*. Mol Plant Microbe Interac 20(4):441–447. https://doi.org/10.1094/mpmi-20-4-0441

Verma ADP, Kukreja KU, Pathak DV, Suneja S, Narula N (2001) *In vitro* production of plant growth regulators (PGRs) by *Azotobacter chroococcum*. Ind J Microbiol 41:305–307

Vet LEM, Dicke M (1992) Ecology of infochemical use by natural enemies in a tritrophic context. Annu Rev Entomol 37(1):141–172. https://doi.org/10.1146/annurev.en.37.010192.001041

Vlot AC, Dempsey DA, Klessig DF (2009) Salicylic acid, a multifaceted hormone to combat disease. Annu Rev Phytopathol 47:177–206. https://doi.org/10.1146/annurev.phyto.050908.135202

Walters DR, Newton AC, Lyon GD (2014) Induced resistance for plant defense: a sustainable approach to crop protection. John Wiley & Sons

Wang Y, Guo H, Wu X, Wang J, Li H, Zhang R (2022) Transcriptomic and physiological responses of contrasting maize genotypes to drought stress. Front Plant Sci 13

Warwate S, Kandoliya U, Bhadja N, Golakiya BA (2017) The effect of plant growth promoting rhizobacteria (PGPR) on biochemical parameters of coriander (*Coriandrum sativum* L.) seedling. Int J Curr Microbiol App Sci 6:1935–1944. https://doi.org/10.20546/ijcmas.2017.603.220

Windram O, Penfold CA, Denby KJ (2014) Network modeling to understand plant immunity. Annu Rev Phytopathol 52:93–111. https://doi.org/10.1146/annurev-phyto-102313-050103

Xin D-W, Liao S, Xie Z-P, Hann DR, Steinle L, Boller T, Staehelin C (2012) Functional analysis of *NopM*, a novel E3 ubiquitin ligase (NEL) domain effector of *Rhizobium* sp. Strain NGR234. PLoS Pathogens 8(5):e1002707. https://doi.org/10.1371/journal.ppat.1002707

Xu Y, Charles MT (2019) Ultraviolet-C priming of strawberry leaves against subsequent *Mycosphaerella fragariae* infection involves the action of reactive oxygen species, plant hormones, and terpenes. Plant Cell Environ 42(3):815–831. https://doi.org/10.1111/pce.13491

Yadav S, Dave A, Sarkar A, Singh H, Sarma B (2012) Co-inoculated biopriming with *Trichoderma*, *Pseudomonas* and *Rhizobium* improves crop growth in *Cicer arietinum* and *Phaseolus vulgaris*. Int J Agric Environ Biotechnol 6

Yu K, Pieterse CMJ, Bakker PAHM, Berendsen RL (2019) Beneficial microbes going underground of root immunity. Plant Cell Environ 42(10):2860–2870. https://doi.org/10.1111/pce.13632

Zaidi A, Khan MS, Ahemad M, Oves M (2009) Plant growth promotion by phosphate solubilizing bacteria. Acta Microbiol Immunol Hung 56(3):263–284. https://doi.org/10.1556/AMicr.56.2009.3.6

Zamioudis C, Pieterse CM (2012) Modulation of host immunity by beneficial microbes. Mol Plant Microbe Interac 25(2):139–150. https://doi.org/10.1094/mpmi-06-11-0179

Zhou JM, Chai J (2008) Plant pathogenic bacterial type III effectors subdue host responses. Curr Opin Microbiol 11(2):179–185. https://doi.org/10.1016/j.mib.2008.02.004

Zipfel C (2008) Pattern-recognition receptors in plant innate immunity. Curr Opin Immunol 20(1):10–16. https://doi.org/10.1016/j.coi.2007.11.003

Zipfel C, Oldroyd GED (2017) Plant signalling in symbiosis and immunity. Nature 543(7645):328–336. https://doi.org/10.1038/nature22009

Chapter 4
Plant Associated Endophytes as Potential Agents for the Protection of Crops from Phytopathogens

S. Harish, V. Sendhilvel, L. Rajendran, S. Parthasarathy, and T. Raguchander

Abstract Sustainable plant disease management can be achieved by using beneficial microorganisms isolated from the plant epiphytic or endophytic region. Endophytic microorganisms reside inside the host plant without causing any damage and facilitate plant growth and development in all possible ways. In nature, the resistance and susceptibility of the plants depend on phylloplane or endophytic microbiome, which deter the pathogens. The antimicrobial and growth-promoting substances secreted by the endophytes play a promising role in combating pathogenic microbes in the plants. The interaction of the pathogen with the endophytes kindles defense-related genes/proteins to restrain the pathogen. In this context, the commercialization of endophytic biocontrol agents through suitable formulation technology is essential. Thus, understanding plant, endophyte, and phytopathogen interactions will open up a new avenue for exploring the disease-resistant mechanism in plants.

Keywords Biological control · Endophytes · Growth promotion · Pathogens · Plant diseases

4.1 Introduction

Plant-associated endophytes to increase productivity are a highly demanded catalyst in agriculture. The global population may increase by over 9 billion by 2050, necessitating a doubling of the food supply by 60% (Ganeshan et al. 2021; Kumar 2022). The intensive farming system and green revolution provided distinctive grain

S. Harish (✉) · V. Sendhilvel · L. Rajendran · T. Raguchander
Department of Plant Pathology, Tamil Nadu Agricultural University, Coimbatore, India
e-mail: harish.s@tnau.ac.in; sendhilvel@tnau.ac.in; rajendran.l@tnau.ac.in

S. Parthasarathy
Department of Plant Pathology, Amrita School of Agricultural Sciences, Amrita Vishwa Vidyapeetham, Coimbatore, India
e-mail: s_parthasarathy@cb.amrita.edu

K. K. Bastas et al. (eds.), *Microbial Biocontrol: Molecular Perspective in Plant Disease Management*, Microorganisms for Sustainability 49,
https://doi.org/10.1007/978-981-99-3947-3_4

and vegetable production achievements, resulting in new pests and diseases outbreak (Kumari et al. 2022). Plant diseases account for a sizable portion of the losses in crop production, resulting in economic loss due to lower yields and the failure of variety, increased mitigation costs, and adverse effects on human health (Ristaino et al. 2021). Global losses caused by diseases are estimated to reach 12% of the potential agricultural production. The challenge of boosting crop yield while maintaining a sustainable phytosanitary state has been met using resistant crop varieties and chemical-based management strategies. However, the chemical management of diseases has disrupted the delicate biological balance of the soil, resulting in groundwater contamination and residue persistence in agricultural and horticultural goods.

Furthermore, pathogens are continually becoming resistant to pre-existing defense in the host and repeated fungicides. Besides lowering crop productivity, fungal pathogens frequently decrease the quality of crops by producing human-harming toxins. Instead, our attention has been on utilizing advanced chemistry and molecular biology to substitute less harmful chemicals or harmless biologically based products for conventional pesticides (Singh et al. 2017). The resilience of agriculture depends mainly on maintaining the ecological balance to sustain the farming system. To keep sustainability, eco-friendly approaches have to be explored to overcome threats caused by the pests and diseases that create an economic impact on farmers.

In such circumstances, plant protection research should look beyond simply substituting more ecologically and biologically based treatments for hazardous compounds. This includes the utilization of antagonistic microbes, which are not only non-toxic, beneficial, economical and inhibit the plant pathogens but also enhance the plant growth (Nega 2014; Harish et al. 2019; Singh et al. 2020). Biological control of plant diseases is focused extensively on using endophytic fungi and bacteria as biocontrol agents. Thus far, the diversity of microbial endophytic communities has been understudied.

Endophytes are microorganisms that reside within the higher plants without producing any diseases. The complex community of microorganisms, composed mostly of bacterial and fungal taxa, dwells ubiquitously throughout the plant system, neither causing harm nor reaping benefits beyond its occupancy (Kobayashi and Palumbo 2000; Kumar et al. 2016; Singh et al. 2022). A distinct endophyte species may exhibit marked differences in behavior based on the host and external environment, including physicochemical soil properties and microorganisms associated with the diverse cropping system. These endophytes possess a foreseeable part in diverse nutrient acquisition and transfer mechanisms, atmospheric nitrogen metabolism, siderophore production, solubilization of phosphorous, induction of phytohormone synthesis, and biocontrol mechanisms (Fig. 4.1).

Endophytic microorganisms have enormous biological control and biotechnological significance, and their properties are used in many fields. They are especially effective in sustainable agriculture, where they are used to stimulate plant growth, eradicate biological pests, and increase host resistance to stress conditions (Ikram et al. 2018; Li et al. 2018). Numerous studies show that endophytes reduce insect pests and pathogenic fungal attacks on the plant species (Landum et al. 2016; Jaber

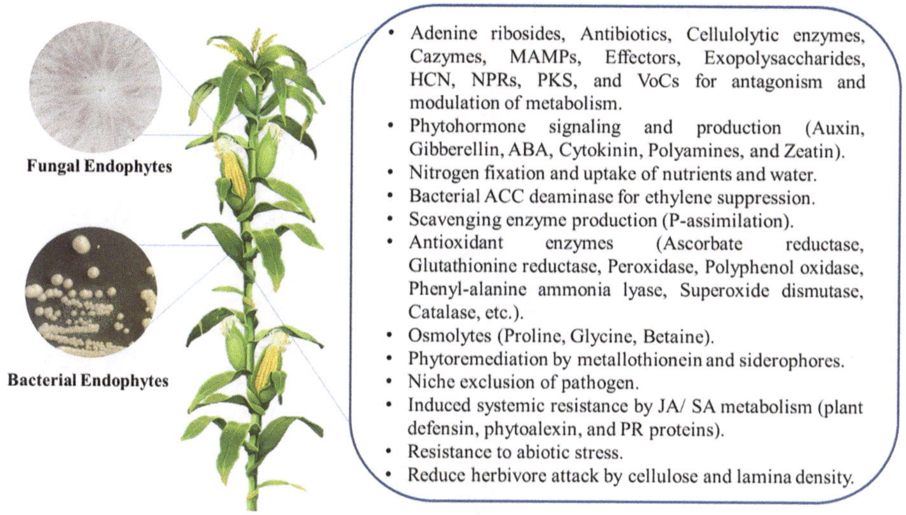

Fig. 4.1 Beneficial effects of endophytes in plant system

and Ownley 2018; Kumari et al. 2022). In addition, the interaction between endophytes and plants results in the creation of a variety of bioactive chemicals that are important for industry in the regulation of phytopathogens and other biological processes (Ikeda et al. 2019).

The agricultural production, research, and innovation sectors are experiencing tremendous strides in isolating, identifying, and categorizing potentially beneficial endophytic microbes for plants. Furthermore, the growth of these biopesticide and biofertilizer industries must overcome real-world obstacles such as well-organized fermentation, scale-up production, storage life, and formulations for large numbers of defined beneficial microbes used to produce succeeding bio-products (Nadarajah and Abdul Rahman 2021). The end goal is to build a well-structured and organized database so that new information can be used to make precision farming technologies that use beneficial or desirable microbial endophytes.

4.2 Diversity of Endophytes

Endophytes have garnered considerable study during the last few years, yet many concerns remain unresolved about their biodiversity patterns at local to global scales. Endophytes and their plant hosts have co-evolved when plants originally colonized the earth, following the wide assortment of distinct symbiotic interactions that we observe nowadays. Endophytes are extremely complex and abundant in species diversity; still a minor subsection of known endophytes has been defined (Suman et al. 2016; Kumar et al. 2020). Within, there are dynamic, complex functional and behavioral groups of endophytes that can be formed, like endopaths (pathogens),

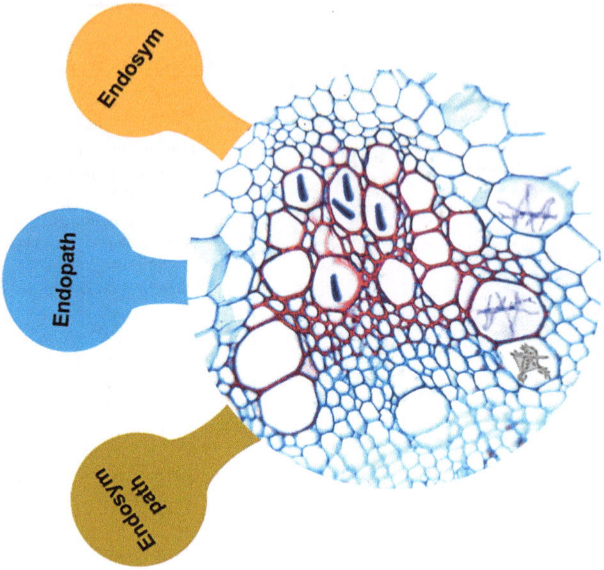

Fig. 4.2 Functional and behavioral groups of endophytes

endosyms (endosymbionts), and endosympaths (endosymbionts are intermittents of mutualists and parasites) (Fig. 4.2). Based on various factors and environments, these groups can change their roles and habitats, such as mutualism, commensalism, and parasitism (Murphy and Hodkinson 2018). Based on phylogeny, these endophytes are grouped as bacteria, fungi, and protists and their respective species. They are found residing in the inter- and intracellular regions of host tissues, with which they are connected and transferred in horizontal or vertical mode. Primarily in tropical and temperate habitats, endophyte diversity is reasonably high. Almost all terrestrial plants are connected with complex and diverse endophytic organisms, and it has been expected that there are over one million species of endophytic fungi on Earth (Deshmukh et al. 2015). These endophytes are differentiated according to their mode of entry into plants. They are classed as "obligate," "facultative," or "passive" endophytes. Murphy et al. (2019) explored a study in which they found a lot of different types of mycorrhizal symbiont groups in a landrace of barley. Krishnaraj and Pasha (2017) characterized all of this microbiota in a particular niche, both functionally and structurally, through metagenomic investigations. Fungal endophytes have been identified in at most one host among 30% of the embryophyte family, but only 10% of endophytic bacteria have now been investigated. The growth behavior of the host determines the variation in endophyte communities in above-ground tissues (Harrison and Griffin 2020). There are two types of endophytes depending on the organism. It has been found that fungal endophytes primarily play a crucial role in the stress resistance of plants and are highly suited to environmental conditions. They have been collected from plant species from many climatic areas, including the Arctic tundra, arid deserts, farmlands, pastures, mangroves, and temperate and tropical forests. Based on their taxonomy, fungal endophytes are classified as clavicipitaceous associated with grasses or non-clavicipitaceous angiosperm, conifers, ferns, and non-vascular plants based on their phylogenetic relations, habitat, and host range. Clavicipitaceous endophytes include fungal species such as *Atkinsonella*, *Balansia*, *Balansiopsis*, *Claviceps*, *Epichloe*, and *Neotyphodium*, mostly dependent on their host over their life period as commensals. Non-clavicipitaceous endophytes such as *Acremonium*, *Aspergillus*, *Chaetomium*, *Cladosporium*, *Colletotrichum*, *Curvularia*, *Gaeumannomyces*, *Gliocladium*, *Guignardia*, *Diaporthe*, *Fusarium*, *Nigrospora*, *Penicillium*, *Pestalotiopsis*, *Phomopsis*, *Piriformospora*, *Stemphylium*, *Trichoderma*, and *Xylaria* are common on terrestrial plants and do not require any hosts to fulfill their life cycles (Modi et al. 2020; Kumar and Dara 2021). Also, many important fungi play different roles in the environment, such as human pathogens from plants and soil (*Coccidioides posadasii*). *Chaetomium globosum* is well known for its antagonistic, fungal endophytic, saprophytic, and infectious properties (Arunkumar et al. 2021). However, bacterial endophytes are more prevalent in plants, which are conquered by the α, β, and γ-Proteobacteria classes, besides Actinobacteria and Firmicutes phyla, also being routinely detected as endophytic microbes. Further bacterial classes, such as Acidobacteria, Bacteroidetes, Planctomycetes, and Verrucomicrobia, are found as endophytes less frequently. *Acinetobacter*, *Achromobacter*, *Ascocoryne*, *Azoarcus*, *Azotobacter*, *Azospirillum*, *Bacillus*, *Brevibacterium*, *Burkholderia*, *Enterobacter*, *Gluconacetobacter*, *Herbaspirillum*, *Klebsiella*, *Microbacterium*, *Micrococcus*, *Pantoea*, *Paenibacillus*, *Phialocephala*, *Pseudomonas*, *Rahnella*, *Rhizobium*, *Rhodotorula*, *Serratia*, *Stenotrophomonas*, and *Streptomyces* are the genera of bacterial endophytes that are most commonly observed (Da et al. 2012; Walitang et al. 2017). These species, classified as bacterial endophytes, are also frequent rhizosphere dwellers. As a result, it is postulated that the endophyte microbiome is a subgroup of rhizosphere-dwelling bacteria.

4.3 Mutualisms of Endophytes and Host Plant

The relationship between plants and endophytes is mutualistic, in which both organisms benefit. In this case, endophytic microbes help plants acquire more nutrients, promote growth, gain more biomass, and be more resistant to disease through the metabolites they render. In turn, the plants support the microbes' evolution and survival. The microbial community and their associated plants engage in complex and diverse interactions in all ecosystems, such as holobionts or metaorganisms (Nadarajah and Abdul Rahman 2021; Pathak et al. 2022). While most endophytic microbes function commensally with their plant hosts, multiple microflorae form mutualistic relationships, ranging from commensalism to dormant or moderate hostility.

The host plants give the endophytes food and a place to live in exchange for the endophytes' ability to promote plant development and fend against pests and illnesses. Under the right conditions, such as environmental factors, plant health, soil nutrient availability, pathogens, and abiotic stresses, plants try to recruit the best microbial community possible because the appearance and health of plants depend on the associated microbiome (Kumar et al. 2021).

These aspects are not only influenced by the plant microbiome but also impact the soil and surrounding microbiome and, more precisely, the plant tissues associated with endophytes. Bacterial and fungal endophytes use the plant tissues as an exclusive protecting life niche that provides a stable and safe environment unaffected by the inconstant environmental conditions that impact rhizosphere and epiphyte microbes. In addition, most endophytic microorganisms have biphasic development, which means they live within a plant and in soil conditions. Endophytes are considered a subtype of rhizospheric microbes, recognized for entering the endorhiza, generally identified as plant growth-promoting (PGP) microbes (Afzal et al. 2019). For instance, rhizobia-legume relationship evolved 60 million years ago, one of the most-known endophytic interactions. The bacterial endosymbiont regulates the host nitrogen intake. The dispersion of endophytes between the soil rhizosphere across the seeds is determined by the nutritional source accessible within the tissue to enable its growth and development. These endophytic microorganisms can live inside the plant host, which includes above- and below-ground plant tissues as well as seeds, and they can have a beneficial effect on plant development (Nadarajah and Abdul Rahman 2021). However, their proliferation may be controlled by the natural immune system in the associated plants. Within a single crop, niche selection can result endophyte populations, as numerous researchers have found that endophyte abundances differ according to tissue type (Harrison and Griffin 2020).

Consequently, endophyte diversity was more minor in stems and leaves than in the phyllosphere communities of natural ecosystems. Conversely, for bacterial diversity, the opposite tendency was observed. There are discernible taxonomic differences between endophytes and phyllosphere populations. In particular, a study on tomato roots, stems, leaves, fruits, and seeds revealed a dominant *Acinetobacter* community in stems and leaves. Bacterial endophytes were enriched in several parts of the fruits, with *Acinetobacter* found in the placenta, *Enterobacter* intense in the seeds and pericarp, and *Weissella* extended in the jelly (Dong et al. 2019).

Similarly, a significant group of fungal endophytes is the dark septate endophytes, which penetrate root systems and create distinctive intracellular structures such as melanized dark mycelium and microsclerotia, referred to as the mutualism-parasitism-continuum paradigm. The conception of the "plant microbiome" has drastically transformed the ecology landscape. Thus, the symbiosis between plants and their endophytes must be investigated to establish the factors leading to co-habitation and quantify the interaction's potential benefits. Rodriguez et al. (2009) reported that a phylloplane is mostly colonized by fungal endophytes exhibiting restricted growth. One or two species dominate, while the remaining taxa have low colonization frequencies as satellite species (Vaz et al. 2018; Suryanarayanan et al. 2018). The nature of interactions between co-occurring bacterial and fungal endophytes within leaves remains unknown. In order to sustain the presence of endophytes within plant tissues, there is a phenomenon of antagonism observed among the different endophytic organisms (Vijaya et al. 2020). Schulz et al. (2015) found a balancing antagonism between the numerous endophytes in plant tissue to preserve the endophyte community.

4.4 Endophyte Biochemical Substances and Growth Promotion

Endophytes stimulate plant growth by stimulating or secreting plant hormones and enzymes, enhancing nutrient uptake through two-way nutrient diffusion, and enhancing plant health by resisting pathogens and pests. Numerous endophytes produce and activate the synthesis of growth-promoting hormones, such as auxins, cytokinins, ethylene, and gibberellic acid. Wang et al. (2013) described the capacity of endophytic bacteria to stimulate peanut development by producing indole-acetic acid (IAA) and siderophore. Tariq et al. (2014) tested bacterial endophytes in pea, and they revealed that five bacterial isolates produce indole-acetic acid (IAA) ranging from 0.86 to 16.16 $\mu g\ mL^{-1}$and phosphate solubilization capacity, which varied from 5.57 to 11.73 $\mu g\ mL^{-1}$. In addition, when the PGPB endophyte synthesizes and secretes IAA, it is subsequently absorbed by plant tissue and combined with plant-synthesized IAA, promoting plant development or activating 1-aminocyclopropyl-1-carboxylate synthase (ACC) transcription. Ultimately, this enhances ethylene production. Elevated ethylene response restricts IAA signaling pathways, hence limiting IAA-catalyzed plant development. Adding an endophyte expressing the enzymatic ACC deaminase, however, reduces the quantity of ethylene released by the plant, hence lessening the inhibitory effects described (del Carmen Orozco-Mosqueda et al. 2020).

Under various stress conditions, the metabolic activity of ACC deaminase produces ketobutyrate and ammonia, which by lowering ACC levels, reduces excessive elevations in the formation of ethylene (Glick 2014). Accordingly, it is presumed that the ACC deaminase enzymatic role is to prevent an increase in ethylene levels that would restrict the growth of roots (Glick et al. 1999), and the role of ACC deaminase has been observed in many plants to improve plant growth and health (del Carmen Orozco-Mosqueda et al. 2020). Yan et al. (2018) searched for the endophytic bacteria that are found in tea that produce siderophore, IAA, and phosphate solubilization, among other phytostimulation activities. The findings showed that eight *Herbaspirillum* isolates produced IAA, siderophore formation, and P-solubilization, three phyto-stimulation activities. *Bradyrhizobium* isolates produced both siderophore and IAA, whereas 10 *Herbaspirillum* and three *Methylobacterium* isolates only produced IAA. The maximum siderophore production, P-solubilization, and IAA production rates were seen in *C. roseus* (Karthikeyan et al. 2012).

A high level of IAA-like molecule synthesis and siderophore production by the bacterial endophytes isolated from tomato plants was examined by Rashid et al. (2012). Three strains belong to *Microbacterium* spp. showed high IAA production and low siderophore production, whereas all the *Pseudomonas* spp. displayed high siderophore production and medium IAA activity, these strains are among the most potent biological control agents for enhancing the host plant resistance to both biotic and abiotic pressures. Sessitsch et al. (2004) reported the production of IAA and iron scavenging siderophore production ability of endophytes in potatoes. Quantitative analysis revealed that the high-level IAA-producing strains in peanut (*Bacillus megaterium*, *Pantoea agglomerans*, *Enterobacter asburiae*, and *Rhizobium* spp.) were from the harvested maturity stage. Bacterial endophytes in date palm having the capability to solubilize the phosphorous from insoluble phosphate and higher affinity iron-chelating siderophore production have been reported by Yaish et al. (2015). Bacteria endophytes have the ability to stimulate the plant growth metabolism through IAA production, P-solubilization, and siderophore biosynthesis was reported by Jasim et al. (2013) in *Piper nigrum*. Among all the endophytes tested, the isolates of *Bacillus* sp., *Klebsiella pneumoniae*, and *Enterobacter* sp. could utilize the phosphate in the phosphate solubilization test. Herrera et al. (2016) examined that the endophytes isolated from the wheat belong to an Enterobacteriaceae of *Pantoea* genus, which showed active IAA production and phosphate solubilization siderophore production. Walitang et al. (2017) assessed the growth-promoting activity due to rice endophytes and described that 73% of isolates could solubilize phosphate, and 65% have siderophore production. Among them, *Flavobacterium* spp. has been considered an IAA producer and phosphate solubilizer. Two strains isolated from *Phyllanthus amarus*, which belongs to *Bacillus* spp. PVMX4 (1472.7 ng mL^{-1}) and *Acinetobacter* spp. ACMS25 (1294.3 ng mL^{-1}) showed the highest IAA production and siderophore production (4.6 g mL^{-1}) (Joe et al. (2016). Passari et al. (2016) estimated the endophytes obtained from *Clerodendrum colebrookianum*, and the results showed that it could solubilize the phosphate that ranged from 1.25 to 10.5 mm. Archana et al. (2020) reported that EGN 4 (*Acinetobacter* spp.) has IAA and phosphate solubilization activity in peanuts. Similarly, Ijazet al. (2020) identified *Acinetobacter* sp. with an ability of IAA production and phosphate solubilization in groundnut. It helps improve nutrient availability by enhancing nutrient absorption and afford a better environment, which improves plant growth development and higher yields in groundnut.

Hassan (2017) reported the highest concentration of IAA from fungal and bacterial endophytes registered, suggesting its role in stimulating root development and promoting plant growth of *Teucrium polium*. The sunflower growth characteristics were promoted in fungal endophyte-treated plants against *Sclerotium rolfsii* (Waqas et al. 2015). Suleman et al. (2018) described that *Enterobacter* spp. MS32 and *Pseudomonas* spp. MS16 inoculated in the root tissues of the wheat seedlings augmented the surface area of root (5.4–7.5 cm^2) and root tips (26.5–58 cm^2), respectively, compared to control as observed using rhizoscanner. Son et al. (2012) also assessed the architecture of the *Oenothera odorata*, (Evening Primrose) root using rhizo scanner (SNAPScan 1236s, AGFA, Belgium). Archana et al. (2020)

reported that groundnut plants treated with seed endophyte EGN 1 (*Pseudomonas* spp.) isolated from peanut showed increased elongation of root, including the lateral roots, increased root length (3370 mm), forks (425 root tips), and surface area (501 forks).

4.5 Endophytes as a Bio-Shield Against Plant Pathogens

Lacking endophytes, a plant would be vulnerable to environmental stress and forfeit its capacity to overcome pathogens. The endophytes (fungi or bacteria) that colonize embryos and endosperm are gaining importance. These endophytes facilitate plant development by hindering phytopathogens and improving the structure of the soil, and help in bioremediation by sequestering dangerous metals from soil. Endophytes exert diverse biocontrol potentials operating simultaneously in the hosts, directly by suppressing pathogens or indirectly by strengthening the local or systemic plant defense systems (Fig. 4.3). The production of antimicrobial chemicals primarily accomplishes the direct suppression of diseases (Brader et al. 2017). Harish et al. (2009) showed the production of pathogenesis-related proteins and defense-related

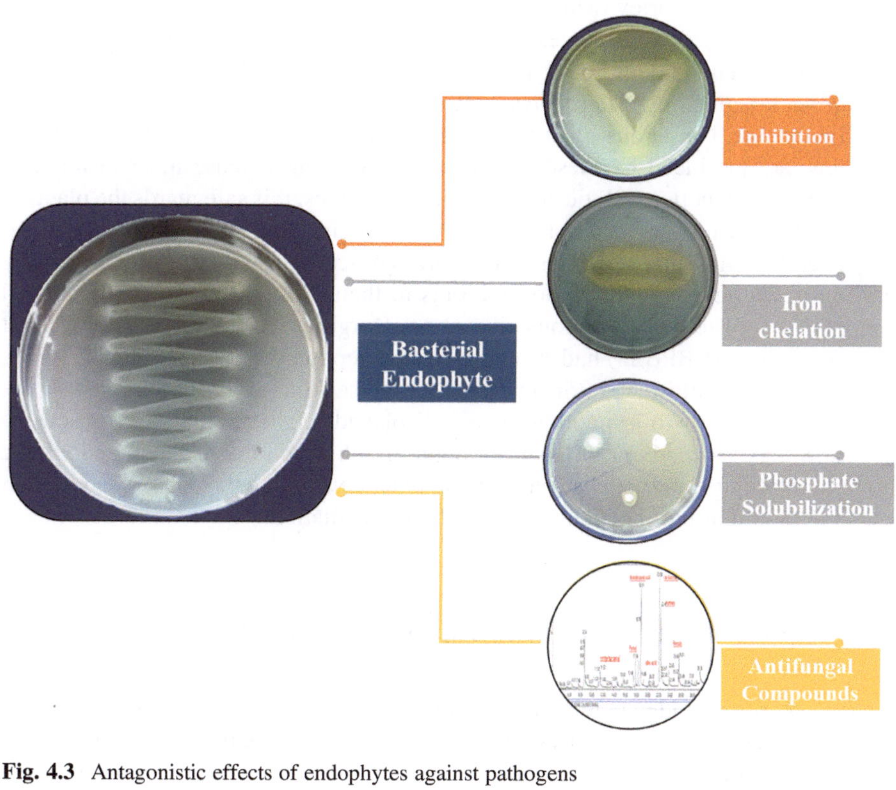

Fig. 4.3 Antagonistic effects of endophytes against pathogens

enzymes in banana plantations inoculated with endophytic *P. fluorescens* and exposed with black aphids carried *Banana bunchy top virus* (BBTV). Another natural method is quorum quenching signals by degrading pathogen autoinducer signals (Kusari et al. 2014). Endophytic bacteria employ indirect biocontrol approaches, such as systemic resistance development in host plants to inhibit wider phytopathogens (Harish et al. 2009; Seethapathy et al. 2016).

Endophytic fungi either colonize target infections directly or synthesizing plant cell wall degrading enzymes (CWDEs) that allow them to drive holes into pathogenic fungi and collect resources for growth. This process highlights the agricultural potential of endophytes in various field applications. Endophytes are transmitted vertically through the seed and can even be vital for the microbial population's evolution in seedlings (Zvinavashe et al. 2021). Additionally, Frank et al. (2017) demonstrated that the vertical transfer of endophytes reduces the capacity of microbial pathogenic strength to support plant growth and seed development. According to Cao et al. (2008), three endophytic fungi that produce chitinases and 1,3-glucanases, such as *Choiromyces aboriginum*, *Cylindrocarpon* sp., and *Stachybotrys elegans*, are antagonistic to *R. solani*, a fungus that infects pine trees. Through the creation of papillae, local cell death, and increased resistance stimulation of defense gene expression, in vitro investigations showed that the *Serendipita indica*, a root endophyte associated in the interior of a wide range of hosts, prevented powdery mildew in barley (Molitor et al. 2011).

In a review of seed endophytes, Truyens et al. (2015) showed that the majority of the bacterial endophytes discovered in seeds belonged to genera, namely *Acinetobacter*, *Bacillus*, *Micrococcus*, *Paenibacillus*, *Pantoea*, *Pseudomonas*, and *Staphylococcus*. Numerous plant hormones, including auxins, cytokinins, and gibberellins, are produced by these bacteria. By triggering defense mechanisms and boosting the production of lytic enzymes and antibiotics, this safeguards the plant. In addition to playing a crucial part in plant growth and defense, these bacterial and fungal endophytes that are transmitted through seeds benefit the host plants by passing down advantageous endosymbionts to their offspring. Out of 80 bacterial endophytes isolated from various rice parts. Durgadevi et al. (2021) found that *Bacillus subtilis* (EBPBS4) had 13 antimicrobial peptide genes, including bacilysin, ericin, fengycin, iturin A, iturin D, iturin C, mersacidin, mycosubtilin, subtilosin, surfactin, and growth-promoting genes displayed highest antagonistic activity against *Rhizoctonia solani.* It has been regularly observed that isolated endophytic bacteria introduced into plants can promote plant development. This has led to their use in agricultural bio-inoculants and/or phytoremediation.

4.6 Tripartite Interaction of Endophytes–Host–Pathogens

A major barrier to understanding plant–endophyte–pathogen interactions is a lack of understanding of how host physiology and ecological variation affect observable traits such as differences in microbial mechanisms and physiology in vitro or planta circumstances, and how laboratory-derived traits can be used to assess functional outcomes in the environment. For example, how endophyte variation in micro-climatic tolerance affects their interactions with pathogenic organisms in the host is unknown. However, the complexity of endophyte–pathogen–host connections remains unclear. Recent research on foliar fungal endophyte–host interactions revealed that environmental influences plant development directly but have a smaller effect on plant–endophyte interactions (da Silva Santos et al. 2022). In another study, genome sequencing followed by confocal laser scanning microscopy (CLSM) analyses was employed to discern all the possible mechanisms underlying the interactions between a beneficial mycorrhiza-like fungus (*Serendipita indica*) and a variety of endophytic bacteria from the genera *Methylobacterium*, *Tardiphaga*, *Rhodanobacter*, and *Trinickia*, as well as their implications on tomato development and biocontrol of root diseases (del Barrio-Duque et al. 2020). It is essential to look at various environmental factors to understand the three-way interactions between host plants, diseases, and the microbes that live inside them (Whitaker and Bakker 2019).

In light of growing evidence that the plant–endophytic interaction is beneficial, promoting host plant growth while also assisting with specific abiotic or biotic stress alleviation, understanding the mechanisms is the need of the hour. To identify the molecular pathways driving plant–endophyte–pathogen interaction, omics tools, including comparative genomics, high-throughput genomics, transcriptomics, microarray, next-generation sequencing (NGS), proteomics, rooteomics, and metabolomics, play an important role (Gouda et al. 2016). The currently available high-throughput nucleotide sequencing platforms, for example Illumina (e.g., HiSeq and MiSeq), Ion Torrent/Ion Proton, and Roche 454 GS FLX+, have been extensively used in investigations of interactions among plants and their associated beneficial and harmful microbes.

The most often used methods for determining the beneficial effects of plant endophytes depend on laboratory assays. If functional activities are context-dependent, laboratory analyses might not accurately reflect functional results in varying environmental situations unless multi-omics techniques are applied. Combining omics methods, including as genomics, transcriptomics, and proteomics, is indeed better to a single strategy for elucidating the genetic process underpinning host–microbiome interactions (del Carmen Orozco-Mosqueda et al. 2020). Additional reliable approaches for determining the capabilities of microbial endophytes in the plants include meta-genomic, meta-transcriptomic, and meta-proteomic analyses, which will also disclose the influence of the environment and alterations in plant genome functions. These evaluations can also be useful in elucidating the response of plants to beneficial or harmful microbial activity. For example, the transcriptomic investigation may provide insight into plant genes that may be exploited by pathogenic microbes in disease development, such as the downregulation of functional and resistance-related genes consistent with the development of specific disease symptoms or the gene expressions, which are involved in other physiological functions.

Although Brader et al. (2014) evaluated the environmental and epigenetic traits that separate pathogenic from non-pathogenic endophytes and emphasized the importance of regulatory/secretory proteins T3SSs and T6SSs, which could also activate the immune responses in plants, secretion systems also played a role in the downregulation of such plant defense mechanisms. According to the results, many of these genes, including T3SSs, may actually be lacking or absent in endophytic bacteria as opposed to their presence in the genomes of plant–pathogens, which is consistent with earlier findings. An investigation of the dual interactions between the pathogen, endophytes, and host plant is used in another study to optimize the selection and application of endophytes. Using in vitro plant–fungus co-cultures of European ash and manna ash trees as well as a fungus–fungus co-culture methodology in three biological control frameworks, the biocontrol potential of two ash endophytes, *Minimidochium* sp. and *Thielavia basicola*, against *Hymenoscyphus fraxineus* was examined (Nawrot-Chorabik et al. 2021).

The capacity of pathogenic strains to disrupt immunological signaling and apoptosis is most likely because they have developed a range of effectors that are unique to their host. When compared to pathogens, endophytes have a lower number of effector genes, which allows them to be distinguished from one another. The application of an endophytic strain of F. oxysporum the application of Fo47 on tomato plants has been found to reduce wilt disease by inhibiting the growth of the pathogenic strain F. oxysporum f. sp. lycopersici. However, the proteomic investigation shows that the intial infestation of the pathogen in sieve elements has minimal impact on the composition of the xylem sap (De Lamo and Takken 2020). They reported a tripartite interface, wherein the synthesis of glucanases, and PR proteins, NP24 (a PR-5), was altered (Constantin et al. 2020). This tripartite interaction demonstrates how endophytes might aid tomato in developing resistance to a virulent pathogen. Durgadevi et al. (2021) used 2D-PAGE-MALDI-TOF-MS to examine the defense protein expression during the tripartite interaction of endophytic *Bacillus subtilis* in rice against *R. solani*. The investigation revealed 15 defense relative proteins and putative NBS-LRR type disease resistance protein (RGA1), which are found to be differentially regulated during the antagonistic interaction in rice. The detection of a spectrum of metabolites produced during the triadic interaction, such as resistance molecules, activated following endophyte colonization to reduce pathogen growth, has also aided in deciphering endophyte–pathogen interactions in the host environment.

4.7 Application of Endophytes in Disease Management

Both fungal and bacterial endophytes are un-questionably essential parts of the world biosphere. Endophytes can influence plant characteristics by reducing vulnerability to disease, increasing resistance to abiotic stresses, altering phytochemical patterns, and modulating the development of functional characteristics in plants. A recent study has demonstrated that the various effects of endophytes may influence ecosystem functions. Endophytes were identified as possible antagonists of a variety of plant diseases. Endophyte's ability to colonize interior host tissues for growth promotion and disease suppression has made them a critical component of crop improvement. Endophytic microbes have been widely advocated as a biological control that could be employed in place of pesticides because they typically occupy the very same ecosystems as fungal and bacterial phytopathogens. Endophytic microorganisms are protected from environmental influences and variations that could endanger their lives and diminish their efficacy. Endophytes are excellent biological control agents due to their ability to have a lasting effect on plants. Two ways have been reported to help manage plant pathogens: 1. They are preventing the pathogen from getting inside the plant and competition over the ecological niche, and 2. Producing metabolites against harmful to the disease-causing agents or boosting the host defenses. Endophytic bacteria introduced in the same environment and niche where the pathogen is colonized would be more effective at controlling the disease. The closely related endophytic genus, such as *Agrobacterium radiobacter*, *P. fluorescens*, and *Enterobacter* sp., will be more effective in managing *Erwinia amylovora* and *A. tumefaciens*, respectively.

Antagonism of crop diseases by endophytes is arbitrated by a range of diverse processes, including direct competition, mycoparasitism, antibiosis, the synthesis of hydrogen cyanide, siderophores, and cellulolytic enzymes, and induction of systemic resistance in its associated host plant. Vijaya et al. (2020) found that bacterial endophytes have a significant role in triggering host defense against soil-borne diseases and agricultural environment productivity overall. Sraphet and Javadi (2022) stated that tissue culture potato plants were protected by endophytic bacteria against *E. carotovora* sub sp. *atroseptica* causing black leg disease. Thangavelu et al. (2003) detected that among 40 strains of endophytic bacteria isolated from banana, the strain EB 22 demonstrated the maximum suppression of *F. oxysporum* f. sp. *cubense* producing banana vascular wilt. They found a combinatorial application of bacterial strains EP 22 and Pf1 was effective in managing the *F. oxysporum* f. sp. *cubense* in soil than individual strains. Rajendran and Samiyappan (2008) observed that two *Bacillus* spp. isolates from cotton, EPCO 102 and EPCO 16, demonstrated the best inhibitory activity against *R. solani* in vitro.

The endophytic bacteria reported to control fusarium wilt diseases on cotton (Parris et al. 2022), on pea (Gupta et al. 2022), on potato (Montes-Osuna et al. 2022), *Sclerotium rolfsii* rot on sugar beet (Farhaoui et al. 2022), and *Helminthosporium oryzae* brown spot on rice (Putri 2022). Endophytic bacteria found in potato tissue were antagonistic against common scab of potatoes (Shuang et al. 2022), potato soft rot and *Fusarium* wilt of tomato (Choi and Ahsan 2022). The mechanism of antibiosis has been shown in the endophytic bacteria isolated from potato against diverse fusarium species viz., *Fusarium avenaceum*, *F. oxysporum*, *F. sambuctnum*, *R. solani*, and from maize against *F. moniliforme* (Slama et al. 2019). Some endophytic bacteria effectively controlled the bacterial pathogen *P. savastanoi* pv. *savastanoi* in olive trees (Filiz Doksöz and Bozkurt 2022). *Bacillus* species were found to colonize the plants endophytically and effectively against vascular pathogens (Rajendran and Samiyappan 2008). The antagonistic efficiency of various endophytes in combating plant pathogens is listed in Table 4.1.

Table 4.1 Biocontrol activity of endophytes against fungal pathogens

Host	Pathogen	Endophytes	References
Ashwagandha or winter cherry	*F. Oxysporum* var. *ciceri* and *Rhizoctonia solani*	*Pseudomonas stutzeri*	Kumar et al. (2022)
Tomato	*Fusarium oxysporum*, and *Alternaria solani*	*Xylaria feejeensis*	Brooks et al. (2022)
Ash trees	*Hymenoscyphus fraxineus*	*Thielavia basicola*	Nawrot-Chorabik et al. (2021)
Rice	*Rhizoctonia solani*	*Bacillus subtilis*	Durgadevi et al. (2021)
Gerbera	*F. Oxysporum* f. sp. *gerberae*	*Bacillus subtilis*	Ramyabharathi et al. (2020)
Tomato	*F. Oxysporum* f. sp. *lycopersici*	*Fusarium oxysporum*	de Lamo and Takken (2020)
Cucumber	*F. Oxysporum* f. sp. *cucumerinum*	*Eupenicillium javanicum, Guignardia mangiferae, Hypocrea* sp., *Lasiodiplodia theobromae, Neurospora* sp., and *Penicillium* sp.	Abro et al. (2019)
Groundnut	*Verticillium dahliae*	*Phomopsis liquidambri*	Xie et al. (2017)
Cotton	*Rhizoctonia solani*	*Bacillus subtilis, Pseudomonas aeruginosa* and *Stenotrophomonas maltophila*	Selim et al. (2017)
Mango	*Lasiodiplodia theobromae*	*Bacillus subtilis*	Seethapathy et al. (2016)
Chestnut	*Fusarium solani*	*Bacillus cereus*	Cheng et al. (2015)
Banana	*F. Oxysporum* f. sp. *cubense*	*Bacillus subtilis*	Thangavelu and Gopi (2015)
Sunflower	*Sclerotium rolfsii*	*Aspergillus terreus* and *Penicillium citrinum*	Waqas et al. (2015)
Rice	*Rhizoctonia solani*	*Bacillus subtilis* var. *amyloliquefaciens*	Nagendran et al. (2014)
Tomato	*F. Oxysporum* f. sp. *lycopersici*	*Bacillus subtilis*	Ramyabharathi and Raguchander (2014)
Chilli	*Fusarium solani*	*Bacillus subtilis*	Sundaramoorthy et al. (2012)
Hot pepper	*Phytophthora capsici*	*Trichoderma* sp.	Bae et al. (2011)
Barley	*Blumeria graminis* f.sp. *hordei*	*Serendipita indica*	Molitor et al. (2011)
Tomato	*Verticillium dahliae*	*Piriformospora indica*	Fakhro et al. (2010)
Grapevine	*Botrytis cinerea*	*Pseudomonas* sp.	Fakhro et al. (2010)

(continued)

Table 4.1 (continued)

Host	Pathogen	Endophytes	References
Pine	*Rhizoctonia solani*	*Choiromyces aboriginum*	Cao et al. (2008)
Cotton	*Rhizoctonia solani*	*Bacillus* spp.	Rajendran and Samiyappan (2008)
Banana	*Banana bunchy top virus*	*P. fluorescens*	Harish et al. (2008)
Groundnut	*Sclerotinia sclerotiorum*	*Pseudomonas aeruginosa*	Gupta et al. (2006)
Banana	*F. Oxysporum* f. sp. *cubense*	*Pseudomonas fluorescens*	Thangavelu et al. (2003)
Grapevine	*Botrytis cinerea*	*Pseudomonas* sp.	Barka et al. (2002)
Groundnut	*Macrophomina phaseolina*	*Pseudomonas fluorescens*	Gupta et al. (2002)

4.8 Commercial Formulation of Endophytes in Disease Management

The increasing importance and demand for plant endophytes have necessitated the development of formulations and commercialization of beneficial endophytes for agricultural use (White et al. 2019). This necessitates the availability of an eco-friendly alternative technique; hence, numerous combinations of these valuable endophytes are being intensively researched. Formulation plays a crucial part in determining the success or failure of a commercial product derived from an experimentally effective microbial endophyte throughout the biopesticide sector. The novel endophyte should be included in the formulation in the required concentration as an active ingredient, along with appropriate careers for developing a commercial formulation for plant application. Commercial formulations should comply with applicable local and national farm legislation and farmers' expectations for consistently favorable results, reasonable rates, and processability (Bashan et al. 2014). Scientifically, successful endophyte formulations rely on the crop, environmental conditions, and the target pathogen.

Recent registration and licensing processes for endophyte-based formulations for use in agricultural production as "bio-fertilizers" or "bio-inoculants" or "bio-pesticides" are significantly less onerous; as a result, several endophyte-based products or consortial formulations are commercially available to farmers globally (Vurukonda et al. 2018). Endophyte formulations for commercial use can be developed and commercialized in four different forms: aqueous suspension, granules, slurry, and talc or powder (Bashan et al. 2014). Numerous experiments on various types of carrier materials have been undertaken. Its shelf life and operating requirements are evaluated to develop formulations of microbial inoculants that are suitable for use (Sallam et al. 2013). With advancements, microbial formulations face a unique challenge confronted by microorganisms: The formulation's shelf life is deteriorated during long-term storage in the ranging temperature of 5–30 °C, which is the optimal temperature used in the warehouses to store their bio-products.

Appropriate formulations of endophytic microbes must be chosen to maximize biocontrol ability against targeted pathogens. Powder formulations are made available to attain greater viability and effective disease inhibition. Recent literature has revealed the effectiveness of endophytic formulations against various diseases (De Silva et al. 2019). Further, it was reported that the application of talc-based *B. subtilis* bioformulations (Cotton endophyte EPCO16 and Coconut endophyte EPC5 strains) was effective in reducing *Fusarium* wilt in chili pepper (Sundaramoorthy et al. 2012), tomato (Ramyabharathi and Raguchander 2014), banana (Thangavelu and Gopi 2015), and gerbera (Ramyabharathi et al. 2020).

Soil drenching of talc formulations of endophytic strains such as *Bacillus subtilis* - EPCO16, EPCO102 (Rajendran and Samiyappan 2008), and fungal strains *Cladorrhinum foecundissimum* - S8, A32 (Gasoni and Stegman De Gurfinkel 2009), as well as a combinations of endophytic bacteria, *Bacillus subtilis*, *Pseudomonas aeruginosa*, and *Stenotrophomonas maltophila*, was reported to be most potent against *R. solani* in cotton (Selim et al. 2017). To standardize the efficacy of endophytic formulations and delivery systems against the sheath blight pathogen *R. solani*, a *Bacillus subtilis* var. *amyloliquefaciens* active ingredient based corn starch-formulation has been developed and applied to rice (Nagendran et al. 2014). The wettable powder formulations of freeze-dried cells of *Bacillus cereus* strain CE3 had an extended shelf life and lasting efficiency against *Endothia parasitica. Fusarium solani* in chestnut and other fruit rots. *B. cereus* wettable powder formulations with diatomite (28.9%) as a carrier, sulfonate (6%) as disperser, alkyl naphthalene sulfonate (6%) as a wetting agent, sodium lignin (4%), d ipotassium phosphate (1%) as a stabilizer, β-cyclodextrin (0.1%) as an UV protectant would be more effective and widely suitable for field applications (Cheng et al. 2015). The endophyte *Piriformospora indica* was developed as a talc formulation using different carrier ingredients: bio-boost (an organic supplement), clay, sawdust, and talc powder, and Tripathi et al. (2015) revealed that a talc-based formulation provided the best shelf life in terms of microbial composition, storage conditions, growth performance, and biological properties. Similarly, in talc-based formulations, the formulated endophyte *Bacillus* sp. CaB5 was effective in growth promotion activities of cowpea and bhendi when the formulation was designed with CaB5 cells (9×10^8 cfu mL^{-1}) with sterile talc (1 kg), calcium carbonate powder (15 g), and carboxymethyl cellulose powder (10 g). Basheer et al. (2019) used a standard plate count to determine its viability, and microbial count for 45 days.

Remarkably, most of the endophyte formulations developed for the management of plant diseases are bacterial or fungal in origin, owing to their ability to colonize plant systems, stability or persistence in career, introduced environment, co-existence with the native microbiome, toxicity, and collateral reactions in plants, and adaptation to adverse environmental conditions. Additionally, they adapt to micro-environmental conditions such as pH, ions, minerals, metabolite activity, and the presence of high osmotic pressure in tissues. Once the beneficial effects of

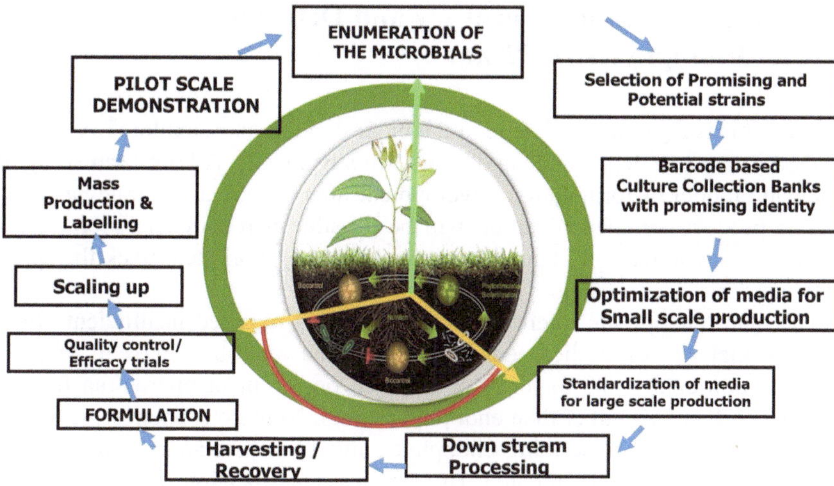

Fig. 4.4 A generic fermentation scale-up plan

identified endophytic microbes on agronomic improvements are established, attention switches to processing streamlining and wide-scale manufacturing (Fig. 4.4). The most important steps in developing commercial endophyte products are the selection of the right endophyte, compatibility, field testing, mass production, carrier optimization, grounding, mixing, packing, post-production field testing, and maintaining appropriate population, and strict quality control. These industries are becoming increasingly sensitive to the use of superior strains as microbial inoculants and biopesticides in agricultural production. Though, scaling up fermentation and discovering low-cost measures remain the most difficult hurdles in biotechnology innovation. The scaling-up plant endophyte production technique is still largely unexplored (Ganeshan et al. 2021). The endophyte-based bio-products commercially available were mostly based on the bacteria; specifically, Avogreen™, a *B. subtilis*-based product, was used in South Africa to control avocado *Cercospora* spot. In Spain, a *Pantoea agglomerans* CPA-2™ was effective against pome and citrus diseases. Additionally, Amylo-X™ is a product derived from *B. amyloliquefaciens* used in Italy to control bacterial and fungal diseases of vegetables. Multi-endophytic strain mixes, also known as microbial consortiums, tend to promote plant development even more effectively than single strains. The commercial product Micosat F® (CCS Aosta srl, Aosta, Italy), developed in Italy, already contains *Trichoderma harzianum* TH01, *Pichia pastoris* PP59, and *Bacillus* sp. BV84. Many endophytic, Gram-positive bacteria of the genus *Bacillus* produce long-lived endospores in nature, and it is suggested that these endospores target infectious oomycetes and fungi in plant spheres, soil, and tissues.

4.9 Post-Application Recovery and Detection of Endophytes *In Planta*

Because this is a paradigm change in the biopesticide sector, exploratory research is necessary to prove dependable and long-term advantages. A long-term study of the kinematic model of formulation development after intervention strategies and barriers to product commercialization will be possible if researchers keep an eye on introduced endophytes in plants, figure out how they work, and track the spread of inoculants.

The plant endophytic microbial community has received insufficient attention, with the majority of studies concentrating on the microbial makeup of the crop sphere. Additionally, the majority of research on the plant microbiota focuses on bacterial endophytes rather than endophytic fungi. Numerous phenotypic and genotypic approaches for describing endophytic microbes associated with plants have been developed and implemented. Biochemical classification, fatty acid profiling, and multilocus electrophoresis are used for phenotypic characterization, though the phenotypic characteristics of microbial populations in the plants are less variable. Thus, beneficial endophytes' biological, biochemical, physiological, immunological, and genomic characteristics capable of suppressing crop diseases are determined to provide reliable and precise identification. Molecular techniques include DNA sequence-based typing and characterization, both widely useful for bacterial species identification. Correlations between microorganisms' genetic types and their antagonistic capacity will aid in the deployment of molecular characterization tools for the quick documentation of actual microorganisms and tracking of commercially significant endophytic microflora.

The lack of a rapid detection process in assessing the post-application endophytic microbial populations in plant tissues has to be focused more on advancing research into applications of endophytes in crops. With this concern, we discussed the recovery of endophytes after colonization and established effects and the genome mining and tagging of introduced endophytes that spread in different parts of plants. This will help us better understand whether these introduced endophytes interact with plants and the ecosystem that occur over time, space, and function.

In recent decades, research focusing on the distribution and behavior of endophyte colonization, population dynamics, and antagonistic potential was carried out by recovering the inoculated strains from the targeted plant tissues. Thus, culture-independent approaches were critical since they disclosed greater species diversity and efficacy of isolation methods. However, culture-dependent methods will remain the only approach to obtaining strains for future research. Most studies indicate that endophytic bacteria colonize plant roots more frequently than plant stems. Endophytes associated with introduced plants were isolated frequently using a customized dilution-to-extinction process. Endophytes are most frequently detected and counted through isolation from ground host plant tissue. Isolation and characterization of isolates are required for these culture-dependent methods. Endophytes can be isolated using any media having the necessary carbon and nitrogen, together with minimal mineral supplementation and the needed pH. To optimize endophyte isolation efficiency, Prior et al. (2014) tested seven distinct synthetic mediums consisting of a wide range of leaf extracts.

Additionally, they created a method for increasing the extraction efficiency of species diversity through enzymatic digestion of plant tissue. They found that additional leaf extracts had a negligible effect on yeast and malt extract media. Still, isolations involving enzymatic digestion of leaf tissue combined with a 1:10 concentration of nutrients indicated significantly more efficacy and species diversity. This is evident in the substantial number of individual endophytes obtained when studying the variety of endophytes using culture-independent techniques.

Around 99% of microorganisms in the environment are unculturable; as a result, substantially more work is necessary to make them culturable and ultimately detectable. Likewise, numerous culture-independent assays using fluorescence in-situ hybridization (FISH) cum CLSM or fluorescence microscope analyses have recently enabled the quantification and localization of microbial cells in the plant tissues (Thomas and Franco 2021). These techniques detect variations in the overall endophytic community and its associated functional changes in the host tissues (Gouda et al. 2016). The spontaneous mutation approach will be largely adopted for in situ detection of introduced microbial agents in the plants. Spontaneous chloramphenicol mutants and BOX-PCR are fingerprinting monitored endophytic colonization of antagonistic endophytic *Pseudomonas* species in the introduced host plant vascular system.

In practice, numerous methods are utilized to detect endophytes in planta: microscopy; isolation; biochemical; physiological; histopathological; radio-immunolabeling; and omics-based approaches (Raja et al. 2016). Polymerase chain reaction (PCR) fingerprinting and fatty acid methyl ester (FAME) have proven reliable methods for detecting inoculated bacterial endophytes. According to transmission electron microscopy (TEM) studies, *Bacillus subtilis* colonized the intercellular gaps of plant cortical tissues, with those next to the vascular elements of seedlings. Gold labeling was used to detect endophytes, and the bacteria were identified in the fibrillar substance that filled the intercellular gaps (Wulff et al. 2003). Denaturing gradient gel electrophoresis (DGGE), coupled with 16S rRNA gene sequence analysis or terminal restriction fragment length polymorphism (T-RFLP) analysis, enables the early identification of introduced or resident microbial communities in the plant tissues (Sessitsch et al. 2004). Additionally, laser capture microdissection pressure catapulting is a valuable tool. It enables the specific procurement of target tissue under microscopical picturing and the fast single-step acquisition of inhabitants from a complex, heterogeneous population of a host tissue (Balestrini and Bonfante 2008).

It is feasible to detect the archetype composition and population of endophytes in plant tissues using spectroscopic techniques such as fluorescence microscopy, flow cytometry, and spectrofluorometry (Chow and Ting 2019). Several strategies of this method are as follows: the interaction of wheat germ agglutinin Alexa Fluor with fungal chitin enables the detection of respective fungal mycelium through the formation of pale green color. Subsequently, using confocal laser scanning microscopy (CLSM), micrographs of cellular-level endophytic colonization were acquired after 4 weeks (Hussin et al. 2017). The CLSM procedure enables access to pictorial data about cellular elements and macromolecules localization. It also facilitates the capture of 3-D images and access to evidence about the relationship within the arrangements and utilities of endophytes. These findings show that the technique has the potential to detect the progression of endophyte colonization, which may help us understand this post-application biological process, mobility, and existence. The gradual development of research over the last few decades has demonstrated that post-application restoration of endophytes from inoculated plants and *in-planta* detection using multi-omic tools, including metagenomics, metabolomics (microfluidics-coupled mass spectrometry devices), metatranscriptomics, metaproteomics, and SIPomics, our understanding of endophyte colonization and function has exploded (Gouda et al. 2016). The following approach is the stable isotope probing (SIP) technique, enabling the functional activity of introduced microbiota to be connected with their structure. A stable isotope atom from a substratum is detected in the endophyte cell constituents that offer insight into the molecular, including DNA, RNA, proteins, enzymes, lipids, and ion exchange. Through their novel integration, SIP and metagenomics have also aided in a greater consideration of active microbial populations' functional distribution and abundance. Nucleic acid can be isolated from tissues and targeted genes amplified using an appropriate polymerase chain reaction. The nucleic acid-based approach was used in conjunction with a strong surface disinfection process designed for endophytes. The characterization of gene sequences such as 16S rDNA, GyrA, and GyrB for endophytic bacteria and 18s rRNA, Actin, and Tubulin gene sequences for fungi has been widely used in taxonomy and identification at the genus or species level. The 16S-ARDRA is based on the straightforward amplified restriction analysis of 16S rDNA technique (ARDRA). The differentiation capability of ARDRA depends on the restriction enzymes utilized. Some PCR profiling techniques, like ERIC, REP, and BOX-PCR, amplify DNA sequences found in prokaryotic genomes.

Following this, the host tissue was treated with propidium monoazide to eliminate surface DNA contaminants from the subsequent reaction. A unique primer set constructed from the region of interest can be employed for a nested PCR/RT-qPCR procedure to detect short regions. Numerous fingerprinting techniques, including DGGE, T-RFLP, and capillary electrophoresis, can be employed to detect amplified genes. Some researchers are already working to elucidate the effects on plant expression of genes after contact with a certain endophyte. During the investigation, endophytic *Paraburkholderia phytofirmans* PsJN inoculation led to DNA cytosine methylation alterations. With, 30 plant proteins changed their methylation status after interacting with the endophytic bacteria (Da et al. 2012). Conversely, computational biology methods for analyzing inoculated microbiome data with the plant carposphere, endosphere, phyllosphere, rhizosphere, and spermosphere for the period are a rapidly growing tool called "microbiomics" (Berg et al. 2020).

4.10 Conclusion

Endophytic bacteria have tremendous agricultural potential, suggesting a possible alternative to the agrochemicals currently in use. Understanding the specific molecular underpinnings of its interaction with plants, particularly with regard to advantageous plant traits, would be helpful in determining the cellular foundation of that relationship. The creation of biopesticides from these endophyte cultures may benefit from these characteristics. Understanding plant endophytic interactions and their effects on disease and growth may therefore have a big impact on sustainable farming practices.

Acknowledgments The authors are grateful to the DST-SERB, New Delhi, India, for funding the endophytes research (Grant No. SR/FT/LS-105/2010).

References

Abro MA, Sun X, Li X, Jatoi GH, Guo LD (2019) Biocontrol potential of fungal endophytes against fusarium oxysporumf. Sp. cucumerinum causing wilt in cucumber. Plant Pathol J 35:598–608

Afzal I, Shinwari ZK, Sikandar S, Shahzad S (2019) Plant beneficial endophytic bacteria: mechanisms, diversity, host range and genetic determinants. Microbiol Res 221:36–49

Archana T, Rajendran L, Manoranjitham SK, Krishnan VS, Paramasivan M, Karthikeyan G (2020) Culture-dependent analysis of seed bacterial endophyte, pseudomonas spp. EGN 1 against the stem rot disease (Sclerotium rolfsii Sacc.) in groundnut. Egypt J Biol Pest Cont 30(1):1–13

Arunkumar K, Parthasarathy S, Harish S, Rajendran L, Raguchander T (2021) Evaluating an isotonic aqueous formulation of Chaetomium globosum Kunze for the management of potato black scurf disease caused by Rhizoctonia solani Kuhn in India. J Plant Pathol 12:1–2

Bae H, Roberts DP, Lim HS, Strem MD, Park SC, Ryu CM et al (2011) Endophytic Trichoderma isolates from tropical environments delay disease onset and induce resistance against Phytophthora capsici in hot pepper using multiple mechanisms. Mol Plant Microbe Interact 24:336–351. https://doi.org/10.1094/MPMI-09-10-0221

Balestrini R, Bonfante P (2008) Laser microdissection (LM): applications to plant materials. Plant Biosyst 142(2):331–336

Barka EA, Gognies S, Nowak J, Audran JC, Belarbi A (2002) Inhibitory effect of endophyte bacteria on Botrytis cinerea and its influence to promote the grapevine growth. Biol Control 24: 135–142. https://doi.org/10.1016/s1049-9644(02)00034-8

Bashan Y, de Bashan LE, Prabhu SR, Hernandez JP (2014) Advances in plant growth-promoting bacterial inoculant technology: formulations and practical perspectives (1998-2013). Plant Soil 378:1–33. https://doi.org/10.1007/s11104-013-1956-x

Basheer J, Ravi A, Mathew J, Krishnankutty RE (2019) Assessment of plant-probiotic performance of novel endophytic bacillus sp. in talc-based formulation. Prob Antimicro Prot 11:256–263. https://doi.org/10.1007/s12602-018-9386-y

Berg G, Rybakova D, Fischer D, Cernava T, Vergès M-CC, Charles T, Chen X, Cocolin L, Eversole K, Corral GH, Kazou M, Kinkel L, Lange L, Lima N, Loy A, Macklin JA, Maguin E, Mauchline T, McClure R, Mitter B, Ryan M, Sarand I, Smidt H, Schelkle B, Roume H, Kiran GS, Selvin J, de Souza RSC, van Overbeek L, Singh BK, Wagner M, Walsh A, Sessitsch A, Schloter M (2020) Microbiome definition re-visited: old concepts and new challenges. Microbiome 8:1–22. https://doi.org/10.1186/s40168-020-00905-x

Brader G, Compant S, Vescio K, Mitter B, Trognitz F, Ma LJ, Sessitsch A (2017) Ecology and genomic insights into plant-pathogenic and plant-nonpathogenic endophytes. Annu Rev Phytopathol 55:61–83

Brader G, Compant S, Mitter B, Trognitz F, Sessitsch A (2014) Metabolic potential of endophytic bacteria. Curr Opin Biotechnol 27:30–37

Brooks S, Klomchit A, Chimthai S, Jaidee W, Bastian AC (2022) Xylaria feejeensis, SRNE2BP a fungal endophyte with biocontrol properties to control early blight and fusarium wilt disease in tomato and plant growth promotion activity. Curr Microbiol 79(4):1–15

Cao R, Liu X, Gao K, Mendgen K, Kang Z (2008) Antagonism of Choiromyces aboriginum Mü1w1c6, an endophytic fungus isolated from common reed against phytopathogenic fungi and its potential of biocontrol. Zhiwu-baohu-xuebao: jikan= Acta phytophylacica Sinica 35(2): 107–112

Cheng H, Li L, Hua J, Yuan H, Cheng S (2015) A preliminary preparation of endophytic bacteria CE3 wettable powder for biological control of postharvest diseases. Not Bot Horti Agrobot Cluj Napoca 43:159–164. https://doi.org/10.15835/nbha4319699

Choi HW, Ahsan SM (2022) Biocontrol activity of aspergillus terreus ANU-301 against two distinct plant diseases, tomato fusarium wilt and potato soft rot. Plant Pathol J 38(1):33

Chow YY, Ting ASY (2019) Influence of fungal infection on plant tissues: FTIR detects compositional changes to plant cell walls. Fungal Ecol 37:38–47. https://doi.org/10.1016/j.funeco.2018.10.004

Constantin ME, Vlieger BV, Takken FL, Rep M (2020) Diminished pathogen and enhanced endophyte colonization upon coinoculation of endophytic and pathogenic fusarium strains. Microorganisms 8(4):544

da Silva Santos SD, da Silva AA, Polonio JC, Polli AD, Orlandelli RC, dos Santos Oliveira JA, Brandão Filho JU, Azevedo JL, Pamphile JA (2022) Influence of plant growth-promoting endophytes Colletotrichum siamense and Diaporthe masirevici on tomato plants (Lycopersicon esculentum mill.). Mycology 13:257–270

Da K, Nowak J, Flinn B (2012) Potato cytosine methylation and gene expression changes induced by a beneficial bacterial endophyte, Burkholderia phytofirmans strain PsJN. Plant Physiol Biochem 50:24–34. https://doi.org/10.1016/j.plaphy.2011.09.013

De Lamo FJ, Takken FL (2020) Biocontrol by fusarium oxysporum using endophyte-mediated resistance. Front Plant Sci 11:37

De Silva NI, Brooks S, Lumyong S, Hyde KD (2019) Use of endophytes as biocontrol agents. Fungal Biol Rev 33:133–148. https://doi.org/10.1016/j.fbr.2018.10.001

del Carmen Orozco-Mosqueda M, Glick BR, Santoyo G (2020) ACC deaminase in plant growth-promoting bacteria (PGPB): an efficient mechanism to counter salt stress in crops. Microbiol Res 235:126439

Deshmukh SK, Verekar SA, Bhave SV (2015) Endophytic fungi: a reservoir of antibacterials. Front Microbiol 5:715

Dong C-J, Wang L-L, Li Q, Shang Q-M (2019) Bacterial communities in the rhizosphere, phyllosphere and endosphere of tomato plants. PLoS One 14(11):e0223847

Durgadevi D, Harish S, Manikandan R, Prabhukarthikeyan SR, Alice D, Raguchander T (2021) Proteomic profiling of defense/resistant genes induced during the tripartite interaction of *Oryza sativa*, *Rhizoctonia solani* AG1-1A, and *Bacillus subtilis* against rice sheath blight. Physiol Mol Plant Pathol 115:101669

Fakhro A, Andrade-Linares DR, von Bargen S, Bandte M, Büttner C, Grosch R, Schwarz D, Franken P (2010) Impact of Piriformospora indica on tomato growth and on interaction with fungal and viral pathogens. Mycorrhiza 20:191–200. https://doi.org/10.1007/s00572-009-0279-5

Farhaoui A, Adadi A, Tahiri A, El Alami N, Khayi S, Mentag R, Ezrari S, Nabil R, Mokrini F, Belabess Z, Lahlali R (2022) Biocontrol potential of plant growth-promoting rhizobacteria (PGPR) against Sclerotiorum rolfsii diseases on sugar beet (Beta vulgaris L.). Physiol Mol Plant Pathol 101829:101829

Filiz Doksöz S, Bozkurt IA (2022) Biological control of pseudomonas savastanoi pv. Savastanoi causing the olive knot disease with epiphytic and endophytic bacteria. J Plant Pathol 104(1): 65–78

Frank AC, Saldierna Guzmán JP, Shay JE (2017) Transmission of bacterial endophytes. Microorganisms 5(4):1–21

Ganeshan S, Kim SH, Vujanovic V (2021) Scaling-up production of plant endophytes in bioreactors: concepts, challenges and perspectives. Biores Bioprocess 8(1):1–16. https://doi.org/10.1186/s40643-021-00417-y

Gasoni L, Stegman De Gurfinkel B (2009) Biocontrol of Rhizoctonia solani by the endophytic fungus Cladorrhinum foecundissimum in cotton plants. Australas Plant Pathol 38:389–391. https://doi.org/10.1071/AP09013

Glick BR (2014) Bacteria with ACC deaminase can promote plant growth and help to feed the world. Microbiol Res 169(1):30–39

Gouda S, Das G, Sen SK, Shin HS, Patra JK (2016) Endophytes: A treasure house of bioactive compounds of medicinal importance. Front Microbiol 7:1–8. https://doi.org/10.3389/fmicb.2016.01538

Gupta C, Dubey R, Maheshwari D (2002) Plant growth enhancement and suppression of Macrophomina phaseolina causing charcoal rot of peanut by fluorescent pseudomonas. Biol Fertil Soils 35(6):399–405

Gupta CP, Kumar B, Dubey RC, Maheshwari DK (2006) Chitinase-mediated destructive antagonistic potential of Pseudomonas aeruginosa GRC1 against Sclerotinia sclerotiorum causing stem rot of peanut. BioControl 51(6):821–835

Gupta S, Pandey S, Sharma S (2022) Decoding the plant growth promotion and antagonistic potential of bacterial endophytes from Ocimum sanctum Linn. Against root rot pathogen fusarium oxysporum in Pisum sativum. Front Plant Sci 13:813686

Harish S, Kavino M, Kumar N, Saravanakumar D, Soorianathasundaram K, Samiyappan R (2008) Biohardening with plant growth promoting rhizosphere and endophytic bacteria induces systemic resistance against banana bunchy top virus. Appl Soil Ecol 39(2):187–120

Harish S, Kavino M, Kumar N, Samiyappan R (2009) Differential expression of pathogenesis-related proteins and defense enzymes in Banana: interaction between endophytic bacteria, banana bunchy top virus and Pentalonia nigronervosa. Biocontrol Sci Tech 19(8):843–857

Harish S, Parthasarathy S, Durgadevi D, Anandhi K, Raguchander T (2019) Plant growth-promoting rhizobacteria: harnessing its potential for sustainable plant disease management. In: Plant growth promoting Rhizobacteria for agricultural sustainability. Springer, Singapore, pp 151–187

Harrison JG, Griffin EA (2020) The diversity and distribution of endophytes across biomes, plant phylogeny and host tissues: how far have we come and where do we go from here? Environ Microbiol 22(6):2107–2123

Hassan SE (2017) Plant growth-promoting activities for bacterial and fungal endophytes isolated from medicinal plant of Teucrium polium L. J Adv Res 8(6):687–695. https://doi.org/10.1016/j.jare.2017.09.001

Herrera SD, Grossi C, Zawoznik M, Groppa MD (2016) Wheat seeds harbor bacterial endophytes with potential as plant growth promoters and biocontrol agents of fusarium graminearum. Microbiol Res 186:37–43

Hussin S, Khalifa W, Geissler N, Koyro HW (2017) Influence of the root endophyte Piriformospora indica on the plant water relations, gas exchange and growth of Chenopodium quinoa at limited water availability. J Agron Crop Sci 203:373–384. https://doi.org/10.1111/jac.12199

Ijaz M, Perveen S, Nawaz A, Ul-Allah S, Sattar A, Sher A, Ahmad S, Nawaz F, Rasheed I (2020) Eco-friendly alternatives to synthetic fertilizers for maximizing peanut (Arachis hypogea L.) production under arid regions in Punjab, Pakistan. J Plant Nutr 43(5):762–772

Ikeda K, Park P, Nakayashiki H (2019) Cell biology in phytopathogenic fungi during host infection: commonalities and differences. J Gen Plant Pathol 85(3):163–173

Ikram M, Ali N, Jan G, Jan FG, Rahman IU, Iqbal A, Hamayun M (2018) IAA producing fungal endophyte Penicillium roqueforti Thom., enhances stress tolerance and nutrients uptake in wheat plants grown on heavy metal contaminated soils. PLoS One 13(11):e0208150

Jaber LR, Ownley BH (2018) Can we use entomopathogenic fungi as endophytes for dual biological control of insect pests and plant pathogens? Biol Control 116:36–45

Jasim B, John Jimtha C, Jyothis M, Radhakrishnan EK (2013) Plant growth promoting potential of endophytic bacteria isolated from Piper nigrum. Plant Growth Regul 71(1):1–11

Joe MM, Devaraj S, Benson A, Sa T (2016) Isolation of phosphate solubilizing endophytic bacteria from Phyllanthus amarus Schum & Thonn: evaluation of plant growth promotion and antioxidant activity under salt stress. J Appl Res Med Aromat Plants 3(2):71–77

Karthikeyan B, Joe MM, Islam MR, Sa T (2012) ACC deaminase containing diazotrophic endophytic bacteria ameliorate salt stress in Catharanthus roseus through reduced ethylene levels and induction of antioxidative defense systems. Symbiosis 56(2):77–86

Kobayashi DY, Palumbo JD (2000) Bacterial endophytes and their effects on plants and uses in agriculture. In: Bacon CW, White JF (eds) Microbial endophytes. Dekker, New York, pp 199–236

Krishnaraj PU, Pasha MA (2017) Metagenome of rhizosphere and endophytic ecosystem. In: Understanding host-microbiome interactions-an omics approach. Springer, Singapore, pp 125–156

Kumar A (2022) Microbial biocontrol: sustainable agriculture and Phytopathogen management, vol 1. Springer Nature Chem, Cham, Switzerland, pp 1–369. https://doi.org/10.1007/978-3-030-87512-1

Kumar KK, Dara SK (2021) Fungal and bacterial endophytes as microbial control agents for plant-parasitic nematodes. Int J Environ Res Public Health 18(8):4269

Kumar A, Singh R, Yadav A, Giri DD, Singh PK, Pandey KD (2016) Isolation and characterization of bacterial endophytes of Curcuma longa L. 3 Biotech 6(1):1–8

Kumar A, Droby S, Singh VK, Singh SK, White JF (2020) Entry, colonization, and distribution of endophytic microorganisms in plants. In: Kumar A, Radhakrishnan EK (eds) Microbial endophytes: functional biology and applications. Woodhead Publishing, Cambridge USA, pp 1–33. https://doi.org/10.1016/B978-0-12-819654-0.00001-6

Kumar A, Zhimo Y, Biasi A, Salim S, Feygenberg O, Wisniewski M, Droby S (2021) Endophytic microbiome in the Carposphere and its importance in fruit physiology and pathology. In: Spadaro D, Droby S, Gullino ML (eds) Postharvest pathology. Plant pathology in the 21st century, vol 11. Springer, Cham, pp 73–88. https://doi.org/10.1007/978-3-030-56530-5_5

Kumar A, Kumar SJ, Chintagunta AD, Agarwal DK, Pal G, Singh AN, Simal-Gandara J (2022) Biocontrol potential of *Pseudomonas stutzeri* endophyte from *Withania somnifera* (Ashwagandha) seed extract against pathogenic *Fusarium oxysporum* and *Rhizoctonia solani*. Arch Phytopathol Plant Protect 55(1):1–18

Kumari M, Qureshi KA, Jaremko M, White JF, Singh SK, Sharma VK, Singh KK, Santoyo G, Puopolo G, Kumar A (2022) Deciphering the role of endophytic microbiome in postharvest diseases management of fruits: opportunity areas in commercial up-scale production. Front Plant Sci 13:1026575. https://doi.org/10.3389/fpls.2022.1026575

Landum MC, Félix MR, Alho J, Garcia R, Cabrita MJ, Rei F, Varanda CMR (2016) Antagonistic activity of fungi of Olea europaea L. against Colletotrichum acutatum. Microbiol Res 183:100–108

Li H, Guan Y, Dong Y, Zhao L, Rong S, Chen W, Lv M, Xu H, Gao X, Chen R, Li L, Xu Z (2018) Isolation and evaluation of endophytic bacillus tequilensis GYLH001 with potential application for biological control of Magnaporthe oryzae. PLoS One 13(10):e0203505

Modi A, Kanani P, Kumar A (2020) Fungal endophytes-induced gene expression studies in biotic and abiotic stress management. In: Microbial endophytes. Woodhead Publishing, pp 251–271

Molitor A, Zajic D, Voll LM, Pons-Kühnemann J, Samans B, Kogel KH, Waller F (2011) Barley leaf transcriptome and metabolite analysis reveals new aspects of compatibility and

Piriformospora indica–mediated systemic induced resistance to powdery mildew. Mol Plant-Microbe Interact 24(12):1427–1439
Montes-Osuna N, Cernava T, Gómez-Lama Cabanás C, Berg G, Mercado-Blanco J (2022) Identification of volatile organic compounds emitted by two beneficial endophytic pseudomonas strains from olive roots. Plan Theory 11(3):318
Murphy BR, Hodkinson TR (2018) Endophyte ecology, diversity and utilisation. Plant Ecol Div 11(5–6):551–554
Murphy BR, Doohan FM, Hodkinson TR (2019) Prospecting crop wild relatives for beneficial endophytes. In: Endophytes for a growing world, pp. 390–410
Nadarajah K, Abdul Rahman NS (2021) Plant–Microbe Interaction: aboveground to belowground, from the Good to the Bad. Int J Mol Sci 22(19):10388
Nagendran K, Karthikeyan G, Mohammed Faisal P, Kalaiselvi P, Raveendran M, Prabakar K, Raguchander T (2014) Exploiting endophytic bacteria for the management of sheath blight disease in rice. Biol Agric Hortic 30:8–23. https://doi.org/10.1080/01448765.2013.841099
Nawrot-Chorabik K, Marcol-Rumak N, Latowski D (2021) Investigation of the biocontrol potential of two ash endophytes against Hymenoscyphus fraxineus using invitro plant–fungus dual cultures. Forests 12(12):1750
Nega A (2014) Review on concepts in biological control of plant pathogens. J Biol Agric Healthcare 4(27):33–54
Parris SM, Jeffers SN, Olvey JM, Olvey JM, Adelberg JW, Wen L, Udall JA, Coleman JJ, Jones DC, Saski CA (2022) An in vitro co-culture system for rapid differential response to Fusarium oxysporum f. sp. vasinfectum race 4 in three cotton cultivars. Plant Disease 106(3):990–995
Passari AK, Mishra VK, Leo VV, Gupta VK, Singh BP (2016) Phytohormone production endowed with antagonistic potential and plant growth promoting abilities of culturable endophytic bacteria isolated from Clerodendrum colebrookianum Walp. Microbiol Res 193:57–73
Pathak P, Rai VK, Can H, Singh SK, Kumar D, Bhardwaj N, Roychowdhury R, de Azevedo LCB, Kaushalendra, Verma H, Kumar A (2022) Plant-endophyte interaction during biotic stress management. Plan Theory 11(17):2203. https://doi.org/10.3390/plants11172203
Prior R, Görges K, Yurkov A, Begerow D (2014) New isolation method for endophytes based on enzyme digestion. Mycol Prog 13(3):849–856
Putri (2022) Endophytic fungi as potential agents of Helminthosporium sp. in Rice Plant (Oryza sativa L.). J Trop Plant Prot 2(2):50
Raja S, Subhashini P, Thangaradjou T (2016) Differential methods of localisation of fungal endophytes in the seagrasses. Mycology 7:112–123. https://doi.org/10.1080/21501203.2016.1218966
Rajendran L, Samiyappan R (2008) Endophytic bacillus species confer increased resistance in cotton against damping off disease caused by Rhizoctonia solani. Plant Pathol J 7:1–12
Ramyabharathi SA, Raguchander T (2014) Mode of action of bacillus subtilis EPCO16 against tomato fusarium wilt. Biochem Cell Arch 14:47–50
Ramyabharathi S, Meena KS, Rajendran L, Raguchander T, Jonathan EI (2020) Potential of a rhizobacterium Bacillus subtilis (Bbv 57) on Fusarium oxysporum f. sp. gerberae and Meloidogyne incognita infecting gerbera grown in protected cultivation. Eur J Plant Pathol 158:615–632. https://doi.org/10.1007/s10658-020-02087-6
Rashid S, Charles TC, Glick BR (2012) Isolation and characterization of new plant growth-promoting bacterial endophytes. Appl Soil Ecol 61:217–224
Ristaino JB, Anderson PK, Bebber DP, Brauman KA, Cunniffe NJ, Fedoroff NV, Finegold C, Garrett KA, Gilligan CA, Jones CM, Martin MD (2021) The persistent threat of emerging plant disease pandemics to global food security. Proc Natl Acad Sci 118(23):e2022239118
Rodriguez RJ, White JF Jr, Arnold AE, Redman RS (2009) Fungal endophytes: diversity and functional roles. New Phytol 182:314–330. https://doi.org/10.1111/j.1469-8137.2009.02773.x
Sallam NA, Riad SN, Mohamed MS, El-Eslam AS (2013) Formulations of bacillus spp. and Pseudomonas fluorescens for biocontrol of cantaloupe root rot caused by fusarium solani. J Plant Prot Res 53:295–300. https://doi.org/10.2478/jppr-2013-0044

Schulz B, Haas S, Junker C, Andrée N, Schobert M (2015) Fungal endophytes are involved in multiple balanced antagonisms. Curr Sci 109:39–45

Seethapathy P, Gurudevan T, Subramanian KS, Kuppusamy P (2016) Bacterial antagonists and hexanal-induced systemic resistance of mango fruits against Lasiodiplodia theobromae causing stem-end rot. J Plant Interact 11(1):158–166

Selim HMM, Gomaa NM, Essa AMM (2017) Application of endophytic bacteria for the biocontrol of Rhizoctonia solani (Cantharellales: ceratobasidiaceae) damping-off disease in cotton seedlings. Biocontrol Sci Tech 27:81–95. https://doi.org/10.1080/09583157.2016.1258452

Sessitsch A, Reiter B, Berg G (2004) Endophytic bacterial communities of field-grown potato plants and their plant-growth-promoting and antagonistic abilities. Can J Microbiol 50(4): 239–249

Shuang M, Wang Y, Teng W, Jin G (2022) Isolation and identification of an endophytic bacteria bacillus sp. K-9 exhibiting biocontrol activity against potato common scab. Res Square. https://doi.org/10.21203/rs.3.rs-1221321/v1

Singh VK, Singh AK, Kumar A (2017) Disease management of tomato through PGPB: current trends and future perspective. 3 Biotech 7(4):1–10

Singh M, Srivastava M, Kumar A, Singh AK, Pandey KD (2020) Endophytic bacteria in plant disease management. In: Kumar A, Singh KV (eds) Microbial endophytes: prospects for sustainable agriculture. Woodhead Publication, Cambridge USA, pp 61–89. https://doi.org/10.1016/B978-0-12-818734-0.00004-8

Singh R, Pandey KD, Singh M, Singh SK, Hashem A, Al-Arjani AB, Abd-Allah EF, Singh PK, Kumar A (2022) Isolation and characterization of endophytes bacterial strains of Momordica charantia L. and their possible approach in stress management. Microorganisms 10(2):290

Slama HB, Cherif-Silini H, Chenari Bouket A, Qader M, Silini A, Yahiaoui B, Alenezi FN, Luptakova L, Triki MA, Vallat A, Oszako T, Rateb ME, Belbahri L (2019) Screening for fusarium antagonistic bacteria from contrasting niches designated the endophyte bacillus halotolerans as plant Warden against fusarium. Front Microbiol 9:3236. https://doi.org/10.3389/fmicb.2018.03236

Son KH, Kim DY, Koo N, Kim KR, Kim JG, Owens G (2012) Detoxification through phytochelatin synthesis in Oenothera odorata exposed to cd solutions. Environ Exp Bot 75:9–15

Sraphet S, Javadi B (2022) Unraveling techniques for plant microbiome structure analysis. Diversity 14(3):206

Suleman M, Yasmin S, Rasul M, Yahya M, Atta BM, Mirza MS (2018) Phosphate solubilizing bacteria with glucose dehydrogenase gene for phosphorus uptake and beneficial effects on wheat. PLoS One 13(9):1–28

Suman A, Yadav AN, Verma P (2016) Endophytic microbes in crops: diversity and beneficial impact for sustainable agriculture. In: Microbial inoculants in sustainable agricultural productivity. Springer, New Delhi, pp 117–143

Sundaramoorthy S, Raguchander T, Ragupathi N, Samiyappan R (2012) Combinatorial effect of endophytic and plant growth promoting rhizobacteria against wilt disease of Capsicum annum L. caused by fusarium solani. Biol Control 60:59–67. https://doi.org/10.1016/j.biocontrol.2011.10.002

Suryanarayanan TS, Devarajan PT, Girivasan KP, Govindarajulu MB, Kumaresan V, Murali TS et al (2018) The host range of multi-host endophytic fungi. Curr Sci 115:1963–1969. https://doi.org/10.18520/cs/v115/i10/1963-1969

Tariq M, Hameed S, Yasmeen T, Zahid M, Zafar M (2014) Molecular characterization and identification of plant growth promoting endophytic bacteria isolated from the root nodules of pea (Pisum sativum L.). World J Microbiol Biotechnol 30(2):719–725

Thangavelu R, Gopi M (2015) Combined application of native Trichoderma isolates possessing multiple functions for the control of fusarium wilt disease in banana cv. grand naine. Biocontrol Sci Technol 25:1147–1164. https://doi.org/10.1080/09583157.2015.1036727

Thangavelu R, Palaniswami A, Doraiswamy S, Velazhahan R (2003) The effect of Pseudomonas fluorescens and Fusarium oxysporum f. sp. cubense on induction of defense enzymes and phenolics in banana. Biol Plant 46(1):107–112

Thomas P, Franco CMM (2021) Intracellular bacteria in plants: elucidation of abundant and diverse cytoplasmic bacteria in healthy plant cells using in vitro cell and callus cultures. Microorganisms 9:1–27. https://doi.org/10.3390/microorganisms9020269

Tripathi S, Das A, Chandra A, Varma A (2015) Development of carrier-based formulation of root endophyte Piriformospora indica and its evaluation on Phaseolus vulgaris L. World J Microbiol Biotechnol 31:337–344. https://doi.org/10.1007/s11274-014-1785-y

Truyens S, Weyens N, Cuypers A, Vangronsveld J (2015) Bacterial seed endophytes: genera, vertical transmission and interaction with plants. Environ Microbiol Rep 7(1):40–50

Vaz ABM, Fonseca PLC, Badotti F, Skaltsas D, Tomé LMR, Silva AC et al (2018) A multiscale study of fungal endophyte communities of the foliar endosphere of native rubber trees in eastern Amazon. Sci Rep 8:16151. https://doi.org/10.1038/s41598-018-34619-w

Vijaya CR, Suryanarayanan TS, Nataraja KN, Prasad SR, Oelmüller R, Shaanker RU (2020) Fungal endophyte-mediated crop improvement: the way ahead. Front Plant Sci 11:561007. https://doi.org/10.3389/fpls.2020.561007

Vurukonda SSKP, Giovanardi D, Stefani E (2018) Plant growth promoting and biocontrol activity of streptomyces spp. as endophytes. Int J Mol Sci 19. https://doi.org/10.3390/ijms19040952

Walitang DI, Kim K, Madhaiyan M, Kim YK, Kang Y, Sa T (2017) Characterizing endophytic competence and plant growth promotion of bacterial endophytes inhabiting the seed endosphere of rice. BMC Microbiol 17(1):1–13

Wang S, Wang W, Jin Z, Du B, Ding Y, Ni T, Jiao F (2013) Screening and diversity of plant growth promoting endophytic bacteria from peanut. Afr J Microbiol Res 7(10):875–884

Waqas M, Khan AL, Hamayun M, Shahzad R, Kang SM, Kim JG, Lee IJ (2015) Endophytic fungi promote plant growth and mitigate the adverse effects of stem rot: an example of Penicillium citrinum and aspergillus terreus. J Plant Interact 10(1):280–287

Whitaker BK, Bakker MG (2019) Bacterial endophyte antagonism toward a fungal pathogen in vitro does not predict protection in live plant tissue. FEMS Microbiol Ecol 95:fiy237

White JF, Kingsley KL, Zhang Q, Verma R, Obi N, Dvinskikh S, Elmore MT, Verma SK, Gond SK, Kowalski KP (2019) Review: endophytic microbes and their potential applications in crop management. Pest Manag Sci 75:2558–2565. https://doi.org/10.1002/ps.5527

Wulff EG, Van Vuurde JW, Hockenhull J (2003) The ability of the biological control agent Bacillus subtilis, strain BB, to colonise vegetable brassicas endophytically following seed inoculation. Plant Soil 255(2):463–474

Xie XG, Fu WQ, Zhang FM, Shi XM, Zeng YT, Li H et al (2017) The endophytic fungus Phomopsis liquidambari increases nodulation and N2 fixation in Arachis hypogaea by enhancing hydrogen peroxide and nitric oxide signalling. Microb Ecol 74:427–440. https://doi.org/10.1007/s00248-017-0944-8

Yaish MW, Antony I, Glick BR (2015) Isolation and characterization of endophytic plant growth-promoting bacteria from date palm tree (Phoenix dactylifera L.) and their potential role in salinity tolerance. Antonie Van Leeuwenhoek 107(6):1519–1532

Yan X, Wang Z, Mei Y, Wang L, Wang X, Xu Q, Peng S, Zhou Y, Wei C (2018) Isolation, diversity, and growth-promoting activities of endophytic bacteria from tea cultivars of Zijuan and Yunkang-10. Front Microbiol 9:1–11

Zvinavashe AT, Mardad I, Mhada M, Kouisni L, Marelli B (2021) Engineering the plant microenvironment to facilitate plant-growth-promoting microbe association. J Agric Food Chem 69(45): 13270–13285

Thangavelu R, Palaniswami A, Doraiswamy S, Velazhahan R (2003) The effect of Pseudomonas fluorescens and Fusarium oxysporum f. sp. cubense on induction of defense enzymes and phenolics in banana. Biol Plant 46(1):107–112

Thomas P, Sekhar AC [illegible]

[illegible]

Chapter 5
Plant Nematode Management Using Beneficial Endophytic Microbes

A. Ramalakshmi, M. Mythili, and U. Sivakumar

Abstract Plant parasitic nematodes are group of plant disease causing organisms which accounts for more than 40% of the yield loss of major crops globally. In this chapter, biological control of plant parasitic nematodes by endophytic microorganisms is discussed in detail. Endophytic microorganisms like nematodes live inside the plant parts and help the plants in nutrient update and growth promotion. Endophytes and nematodes enter the plants through same path, and application of plant endophytes through seed, plants, and others would reduce the colonization of plant parasitic nematodes through various mechanisms. In this chapter, bacterial, fungal endophytes, and their mechanisms of nematode control are discussed in detail. AM fungi and root fungus have symbiotic relationship with almost all angiosperms can contol nematode infection. Trichoderma also possesses the anti-nematicidal compounds to control nematodes. In this chapter, the mechanism of action of fungal endophytes is also discussed in detail.

Keywords Plant parasitic nematodes · Bacterial endophytes · Fungal endophytes · Mechanism of action · Biological control

5.1 Introduction

Endophytic microorganisms invade plant cells and form symbiotic relationships with the host plants. Endophytes invade plants through seeds, seedlings, or vegetative material and are present in the rhizosphere and phylloplane (Pathak et al. 2022; Kumar et al. 2020). These microorganisms prevent plant-parasitic nematode multiplication by using trapping structures or releasing toxic substances (Kumari et al. 2022). Bacteria, fungi, archaea, and protists are examples of endophytes; out of which fungi and bacteria are the most dominant and well-studied group. As these

A. Ramalakshmi (✉) · M. Mythili · U. Sivakumar
Department of Agricultural Microbiology, Tamil Nadu Agricultural University, Coimbatore, Tamil Nadu, India
e-mail: ramalakshmi.a@tnau.ac.in; usiva@tnau.ac.in

K. K. Bastas et al. (eds.), *Microbial Biocontrol: Molecular Perspective in Plant Disease Management*, Microorganisms for Sustainability 49,
https://doi.org/10.1007/978-981-99-3947-3_5

fungi and bacteria appear to be efficient in suppressing nematodes, there is growing interest in using these endophytic microorganisms as bionematicides, particularly for sedentary parasitic nematodes because both the endophytes and nematodes inhabit the same part of the plant, it is easy for endophytes to suppress nematodes tissues, and the former has a better opportunity of suppressing the latter (Sikora et al. 2007). Endophytic microorganisms produce secondary metabolites with the pesticidal ability, and act as great choices for bionematicides (Yadav 2019; Singh et al. 2017; Singh et al. 2020). Bionematicides from bacteria include *Bacillus* spp., *Pseudomonas fluorescens*, and *P. penetrans*, as well as the fungi include *Trichoderma* spp., *Purpureocillium lilacinum*, Pochonia *chlamydosporia*, *Verticillium* spp., and *Arthrobotrys oligospora* were used in a variety of crops namely tomato-lettuce, tomato-potato, carrot, wheat, sugarcane, and cucumber (Mazzuchelli et al. 2020).

Nematodes are generally minute soil-borne pathogens that attack different plant parts (roots, seeds, leaves, stems, and flowers), though most species feed on roots. These nematodes use stylet to penetrate plant cells for feeding. This stylet is connected to pharyngeal glands that secrete effector molecules to facilitate penetration, internal migration, and parasitism. The nematodes mainly affect the rootsthrough formation of sites of feed like syncytia, single giant cells, and coenocytes that produce a protective feeding environment (Palomares-Rius et al. 2017). The nematode feeding causes root deformation, plant stunting, leaf yellowing, and yield loss. To date, a total of 4100 plant parasitic nematode species were identified, of which a few genera are regarded as important pathogens, while other species are confined to a narrow group of plants. Both vital pathogens and narrow spectrum speciescombinely would cause significant influence on major crops. An estimate suggests that plant nematodes can cause yield loss of 12.3% globally which equates to ($157 billion; [Singh et al. 2015]), the economic loss due to nematode damage is more than half of what is caused by the damage due to exotic insects ($70 billion; [Bradshaw et al. 2016]). Siddique and Grundler (2018) suggested that the total degree of global damage caused by nematodes is likely underreported because farmers are unaware of nematode presence due to non-specific symptoms in plants, making it difficult to trace crop losses caused by nematodes. Further losses could occur due to poor quality of food and visual deformities connected with disease symptoms (Palomares-Rius et al. 2017). Plant-parasitic nematodes are classified as ectoparasitic or endoparasitic, depending on their feeding habits. Two major crop damaging nematodes are root-knot nematode (*Meloidogyne* sp.) and cyst nematodes (*Heterodera* and *Globodera* sp.).

A variety of nematode-control measures have been developed, including the use of synthetic nematicides. Since these synthetic nematicides are dangerous to the health of humans and a pollutant to the environment, there is a demand to adopt an alternate method. Biocontrol agents are the most viable method for managing parasitic nematodes. The word "biocontrol" refers to the application of live microorganisms to reduce the numbers and effect of a concerned pest. The biocontrol agents can interact directly or indirectly with pathogens (Poinar and Jansson 2018). Bacteria, fungi, viruses, protists, and other invertebrates are some of the organisms known to function as biological control agents against various species of plant-parasitic nematodes.

5.2 Endophytic Bacteria

Based on the occurrence of endophytes, bacteria are the second most ones inhabitating intracellular and vascular plants (Zinniel et al. 2002). Most of antagonistic pathogens belong to Gram positive and Gram negative bacteria (Kobayashi and Palumbo 2000) have the ability to control plant parasitic nematode. *Pseudomonas fluorescens*, *Burkholderia cepacian*, and *Agrobacterium radiobacter* are the examples of Gram-negative endophytes, and *Bacillus* spp. are the examples of Gram-positive endophytes. (Yadav et al. 2017). The endophytic bacteria which are efficient in controlling plant parasitic nematodes in different crops are highlighted in Table 5.1.

5.2.1 Bacillus Thuringiensis

Endophytic *Bacillus* such as *Bacillus thuringiensis*, *B. amyloliquifaciens*, and *B. substilis* are found to provide different benefits to the plants, including plant growth promotion and pest control (insects, nematodes, and microorganisms), without doing any visible harm to the plant. *Bacillus thuringiensis* is the most commonly used Bacillus sp. in biological pest control, accounting for about 53% of the biopesticide market (Polanczyk et al. 2017).

During sporulation, *Bacillus thuringiensis* produces a crystalline protein called the parasporal body. This is encoded by a group of genes called *cry* genes. Parasporal body located in sporangium onside the endospore is found to be a protoxin, which in an alkaline condition of insect gut will be converted into toxin form. These crystal proteins are highly specific. Not all organism possess the same receptors imparting high species specificity for the cry proteins. Among the 70 different *cry* genes groups, *cry5*, *cry 6*, *cry12*, *cry13*, *cry14*, *cry15*, *cry21*, and *cry55* are found to exhibit nematicidal properties (Bravo et al. 2013). Other nematicidal compounds such as thuringiensin, chitinase, and metalloproteinase are also produced by Bt (Jouzani et al. 2017). The various mechanisms of *Bacillus thuringiensis* against plant-parasitic nematodes are described in Table 5.2.

These compounds are found to act individually or in synergistically with each other for controlling nematodes. For the enhanced nematicidal activity of *cry5* and *cry6*, the presence of collagenase protein called metalloprotein ColB is indispensable (Peng et al. 2016).

Table 5.1 Endophytic bacteria against plant-parasitic nematodes (PPN)

Crop	Nematode species	Endophytic bacteria	Effect of bacterial endophytes on PPN	References
Rice	*M. graminicola*	*Bacillus megaterium*	Reduced nematode invasion and gall formation	Padgham and Sikora (2007)
Tomato	*Meloidogyne incognita*	*Pseudomonas putida, P. fluorescens, bacillus megaterium*	Reduced nematode infestation and number of galls	Munif et al. (2000)
Bhendi	*M. incognita*	*Pseudomonas* sp., *Bacillus* sp., *Methylobacterium* sp	Reduced the egg mass, adult female nematode population, and lowered root gall index	Vetrivelkalai et al. (2010)
Banana	*M. incognita, R. similis, Helicotylenchusmulticinctus*	*Bacillus subtilis*	Reduced nematode population	Jonathan and Umamaheswari (2006)
Potato	*Globoderarostochiensis*	*P. fluorescens, P. putida, P. aurantiacea*	Reduction in nematode multiplication rate	Trifonova et al. (2014)
Black pepper	*R. similis, Meloidogyne*sp.	*B. Megaterium*	Suppress egg hatching and reduces nematode population	Tran et al. (2019)
Coffee	*Meloidogyne* sp., *Xiphinema* sp., *Rotylenchulus* sp.	*Bacillus* sp., *Enterobacter* sp. and *Streptomyces* sp.	Inhibit egg hatching and increase mortality rate	Hoang et al. (2020)
Pine trees	*Bursaphelenchusxylophilus*	*Bacillus* sp.	Inhibit growth of nematode	Ponpandian et al. (2019)

Table 5.2 Mode of action of *Bacillus thuringiensis* against nematodes

Compound	Mode of action	References
Crystal protein	Destroys the intestinal lining following the germination of the spore	Salehi Jouzani et al. (2017)
Other Bt compounds		
• Metalloproteinase	• Leads to the digestion of intestinal tissue	Shiomi et al. (2010)
• Chitinase	• It is a cell wall degrading enzyme that helps in the digestion of body wall and intestinal tissue of nematode	Martínez-Zavala et al. (2020)
• Exotoxin (thuringiensin)	• It acts as an analogue to nucleotide and RNA polymerase inhibitors, thereby inhibiting the replication and transcriptional process	Salehi Jouzani et al. (2017)

5.2.2 Genus Pseudomonas

Species of the genus *Pseudomonas* are present in a broad range of terrestrial and aquatic ecosystems, including soil, plants, and water. They are physiologically multifunctional. *Pseudomonas* species produce a variety of secondary metabolites that are effective against various plant diseases(Troxler et al. 2012). *P. putida* inhibits the growth of *M. incognita* by inducing systemic acquired resistance and promoting the defense enzyme (peroxidase, polyphenol oxidase, and phenylalanine ammonia-lyase) activities (Zhai et al. 2018). The production of cell wall lytic enzymes can reduce the mobility of nematodes in soil. The volatile organic compounds produced by *P. putida* actively act against various nematode species. *P. fluorescens* increased the activity of PPO, PAL, PO, and phenol by inducing ISR and thus preventing the nematode attack. *P. fluorescens* induces the activity of chitinase and peroxidase in plants, which are essential for ISR (Thiyagarajan and Kuppusamy 2014).

5.3 Fungal Endophytes

One of the most popular, well-established, and well-studied group of microorganism is fungal endophytic organism that offers resistance to biotic stress in plants (Yan et al. 2019). Clavicipitaceous and non-clavicipitaceous are two groups of fungal endophytes classified based on the taxonomy, ecology, evolutionary relations, and range of hosts (Santangelo et al. 2015). The major host plants to complete life cycle of clavicipitaceous endophytes are grasses. Horizontal and vertical distribution of endophytes in the above ground plant parts happen by multiplication of these endophytes in intercellular gaps.. They include *Atkinsonella*, *Balansia*, *Neotyphodium*, *Echinodothis*, *Balansiopsis*, *Myriogenospora*, *Parepichloe*, and *Epichloe* (De Silva et al. 2016). Terrestrial plants serve as the host for non-clavicipitaceous endophytes, but theses endophytes do not depend on host for

Table 5.3 Endophytic fungi with plant parasitic nematode in different plant species

Crop	Nematode species	Endophytic fungi	Effect of fungal endophytes on PPN	References
Tomato	*Meloidogyne incognita*	*Fusariumoxysporum*	Reduced nematode invasion	Bogner et al. (2016)
Cucumber	*M. incognita*	*Chaetomium, Paecilomyces, Phyllosticta*	Reduction in number of galls by 25 to 58%	Yan et al. (2011)
Banana	*Radopholussimilis*	*F. Oxysporum*	Reduced the number of J2	Van Dessel et al. (2011)
Banana	*Pratylenchusgoodeyi*	*F. Oxysporum*	Increased mortality and paralysis of nematodes	Waweru et al. (2013)
Melon	*M. incognita*	*F. Oxysporum*	Reduced penetration of J2 in roots	Menjivar et al. (2011)
Rice	*M. graminicola*	*Fusarium*sp.	Reducethe formation of gall and increase root weight	Le et al. (2009)
Rice	*M. graminicola*	*Fusariummoniliforme*	Reduced penetration of J2 into roots	Le et al. (2016)
Cotton	*M. incognita*	*Chaetomiumglobosum*	Reduced nematode infection and female production	Zhou et al. (2016)

its completion of life cycle. These include *Colletotrichum*, *Fusarium*, *Phomopsis*, and *Xylaria* (Jayawardena et al. 2016). Common enophytic fungi *Fusarium oxysporum* generally isolated from various plants serves as a competitor of insects, plant pathogens, and parasitic nematodes. Several researchers have revealed that *F. oxysporum* is antagonistic to *Meloidogyne incognita*, *M. graminicola*, *Pratylenchusgoodeyi*, *Helicotylenchusmulticinctus*, and *R. similis* in tomato, melons, and banana (Bogner et al. 2016). Interaction of different endophytic fungi with plant parasitic nematode (PPN) in different plant species is presentedin Table 5.3.

5.3.1 AM Fungi

AM fungi are obligate symbionts, forming symbiotic relationships with 80% of the vascular plants. The major benefits of AMF interaction on plants are increased nutrient uptake, improved soil quality, and structure, enhanced plant tolerance toward different biotic and abiotic stresses with improved plant growth promotion and development of disease (Meir et al. 2010). AM fungi have been extensively studied as biocontrol agents against nematodes in different crops, including maize (Alvarado-Herrejón et al. 2019), switchgrass, and miscanthus (Emery et al. 2017). The major functions of Rhizosphere are increasing water and nutrient uptake, modifying root morphology by enhancing root growth or altering rhizosphere interactions (Wani et al. 2017). Nematode control by AM fungi has not been described. Activation of JA-mediated systemic acquired resistance by AM fungi is known as mycorrhizal-induced resistance (MIR; Pozo and Azcón-Aguilar 2007) which happens due to the continuous detection of AM fungi inside colonized cells. In mycorrhizal plants, increased accumulation of various defense components such as ROS, phytoalexins, phenolics, suberin, lignin, Pathogen Response proteins, and sulfur amino acids has been observed (Hill et al. 2018). Marro et al. (2018) confirmed that the application of AM fungal species (*Rhizophagus intraradices* and *Funneliformis mosseae*) reduced the penetration of root knot nematode *Nacobbus aberrans* in tomato root. Similar to plant pathogens, mycorrhizal fungi-derived plant defense mechanisms act initially, afterward AM fungi modify plant response to ensure successful colonization. In this way, salicylic acid and Jasmonic acid pathyways is activated and boosted or enhanced by AM fungal colonization in response to the attack by plant pathogens (Hohmann and Messmer 2017).

5.3.2 Genus Trichoderma

The *Trichoderma* is filamentous fungi that grow in different environments and act as natural root symbionts. Several modes of action of *Trichoderma* in nematode management are competition with nematodes, mycoparasitism, antibiosis, plant growth stimulation, increased plant tolerance to abiotic stressors, and activation of pathogen defenses. The colonization of *Trichoderma* species in the plant root blocked the functioning of nematodes at various phases of parasitism, such as reproduction, galling, and invasion (Martínez-Medina et al. 2017). *Trichoderma harzianum* had a major effect on the infection induced by the nematode *Meloidogyne javanica* by influencing their establishment, reproduction, and development in the tomato plant. They also noticed a considerable drop in the hatched egg numbers. Both the eggs and larvae of nematodes are affected by the genus *Trichoderma*. Nematode egg shell components and hatching capability are reduced by enhanced secretion of cellulose enzyme by the fungi. The exudates produced by the *Trichoderma* species directly impacted *M. incognita* by increasing the mortality rate of J2 and decreasing the egg hatching (Khan et al. 2018).

Trichoderma can cause resistance in most different plant species, resulting in plant metabolomic, transcriptomic, and proteomic changes (Mukherjee et al. 2012). Stimulating systemic defense enhances the plant's immune response, providing a quicker response following priming against nematode attack and therefore minimizing the risk of disease transmission (Mendoza-Mendoza et al. 2018). This ISR is managed by JA/ET signaling. Leonetti et al. (2014) noticed that during primary stage of *M. javanica* infection, SA signaling was downregulated, whereas the JA/ET-mediated response is induced in *Trichoderma*-treated tomato roots, indicating that the presence of fungus stimulates the ISR in the plants. *Trichoderma*-induced defenses occur in two phases: 1. *T. harzianum* promotes salicylic acid-mediated defense reactions preserve the roots from parasitic nematode infection during the first phase. 2. In the next phase, *Trichoderma* enhances the activation of jasmonic acid-mediated defenses when nematodes affect the JA-mediated defenses in the plant roots. Consequently, nematode *M. incognita*-mediated defense *was* suppressed effectively, resulting in a reduction in nematode reproduction and development. Once parasitism is developed, the activation of salycylcic acid-mediated defense is promoted by fungus, most likely by detection of eggs, this can help improve defense against invasion of juvenile i (Martínez-Medina et al. 2017).

5.4 Mode of Action of Endophytes

Endophytic microbes help to enhance plant health by suppressing the growth and multiplication of parasitic nematodes at different stages through various processes. Directly or indirectly the control strategies are divided (Fig. 5.1).

5.4.1 Direct Mechanism

Endophytic organisms can directly attack, kill, or resist parasitic nematodes when they discover the host, compete for space, release toxic compounds and some

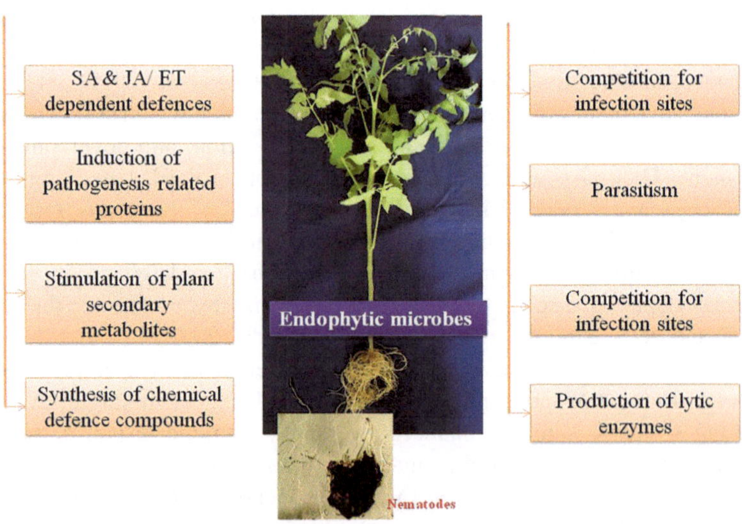

Fig. 5.1 Mechanisms of endophytic microbes against plant parasitic nematodes

chemicals against nematodes. Metabolites like alkaloids, flavonoids, terpenoids, peptides, phenols, quinones, steroids, polyketones, and cellulolytic enzymes, hemicellulases, and other hydrolyses are produced by endophytes (Fadiji and Babalola 2020). These molecules prevent the growth of nematodes by antibiosis. By colonizing the plant tissues, endophytes restrict the nematode via niche competition (Schouten 2016).

5.4.2 Indirect Mechanism

Upregulation of genes related to secretion of Volatile organic compounds by endophytes, phytoalexins, phytohormones, and PR proteins activates jasmonic acid, salicylic acid, and ET pathway, which give protection to plants from stresses. A few of these defense mechanisms can antagonize nematodes, while the others, like phytohormones, stimulate the growth of plant and compensate for the damage caused by nematodes. Application of *Rhizobium etli* and *Bacillus sphaericus* significantly reduced the *M. incognita* in tomato by Induced Systemic Resistance (Schäfer 2007).

5.5 Conclusion

Biocontrol approaches for parasitic nematode management are feasible alternatives to toxic chemical nematicides. Endophytic microorganisms are becoming more widely recognized for their role in protecting plants from insects, nematodes, pathogenic bacteria, and fungi. They also alter the physiological composition of their hosts, making them more tolerant to different pest diseases and climatic changes. Screening of endophytes having nematicidal properties and mass multiplication and commercialization of efficient endophytes are promising field for the long-term management of nematodes. It generates induced systemic resistance against various stressors and antagonizes different kinds of pests. Hence, the endophytes can be an effective tool for enhancing overall crop yield and health and minimizing crop care costs.

References

Alvarado-Herrejón M, Larsen J, Gavito ME et al (2019) Relation between arbuscular mycorrhizal fungi, root-lesion nematodes and soil characteristics in maize agroecosystems. Appl Soil Ecol 135:1–8

Bogner CW, Kariuki GM, Elashry A et al (2016) Fungal root endophytes of tomato from Kenya and their nematode biocontrol potential. Mycol Prog 15:30

Bradshaw CJA, Leroy B, Bellard C et al (2016) Massive yet grossly underestimated global costs of invasive insects. Nat Commun 7:1–8

Bravo A, Gómez I, Porta H et al (2013) Evolution of *Bacillus thuringiensis* cry toxins insecticidal activity. Microb Biotechnol 6:17–26

De Silva NI, Lumyong S, Hyde KD et al (2016) Mycosphere essays 9: defining biotrophs and hemibiotrophs. Mycosphere 7:545–559

Emery SM, Reid ML, Bell-Dereske L et al (2017) Soil mycorrhizal and nematode diversity vary in response to bioenergy crop identity and fertilization. GCB Bioenergy 9:1644–1656

Fadiji AE, Babalola OO (2020) Elucidating mechanisms of endophytes used in plant protection and other bioactivities with multifunctional prospects. Front Bioeng Biotechnol 8:467

Hill EM, Robinson LA, Abdul-Sada A et al (2018) Arbuscular mycorrhizal fungi and plant chemical defence: effects of colonisation on aboveground and belowground metabolomes. J Chem Ecol 44:198–208

Hoang H, Tran LH, Nguyen TH et al (2020) Occurrence of endophytic bacteria in Vietnamese Robusta coffee roots and their effects on plant-parasitic nematodes. Symbiosis 80:75–84

Hohmann P, Messmer MM (2017) Breeding for mycorrhizal symbiosis: focus on disease resistance. Euphytica 213:113

Jayawardena RS, Hyde KD, Damm U et al (2016) Notes on currently accepted species of *Colletotrichum*. Mycosphere 7:1192–1260

Jonathan EI, Umamaheswari R (2006) Biomanagement of nematodes infesting banana by bacterial endophytes (*Bacillus subtilis*). Indian J Nematol 36:213–216

Jouzani GS, Valijanian E, Sharafi R (2017) *Bacillus thuringiensis*: a successful insecticide with new environmental features and tidings. Appl Microbiol Biotechnol 101:2691–2711

Khan MR, Ahmad I, Ahamad F (2018) Effect of pure culture and culture filtrates of Trichoderma species on root-knot nematode, *Meloidogyne incognita* infesting tomato. Indian Phytopathol 71: 265–274

Kobayashi DY, Palumbo JD (2000) Bacterial endophytes and their effects on plants and uses in agriculture. In: Microbial endophytes. CRC Press, pp 213–250

Kumar A, Droby S, Singh VK, Singh SK, White JF (2020) Entry, colonization, and distribution of endophytic microorganisms in plants. In: Kumar A, Radhakrishnan EK (eds) Microbial endophytes: functional biology and applications. Woodhead Publishing, Cambridge USA, pp 1–33. https://doi.org/10.1016/B978-0-12-819654-0.00001-6

Kumari M, Qureshi KA, Jaremko M, White JF, Singh SK, Sharma VK, Singh KK, Santoyo G, Puopolo G, Kumar A (2022) Deciphering the role of endophytic microbiome in postharvest diseases management of fruits: opportunity areas in commercial up-scale production. Front Plant Sci 13:1026575. https://doi.org/10.3389/fpls.2022.1026575

Le HTT, Padgham JL, Sikora RA (2009) Biological control of the rice root-knot nematode *Meloidogyne graminicola* on rice, using endophytic and rhizosphere fungi. Int J Pest Manag 55:31–36

Le HTT, Padgham JL, Hagemann MH et al (2016) Developmental and behavioural effects of the endophytic *Fusarium moniliforme* Fe14 towards *Meloidogyne graminicola* in rice. Ann Appl Biol 169:134–143

Leonetti P, Costanza A, Zonno M et al (2014) How fungi interact with nematode to activate the plant defence response to tomato plants. Commun Agric Appl Biol Sci 79:357–362

Marro N, Caccia M, Doucet ME et al (2018) Mycorrhizas reduce tomato root penetration by false root-knot nematode Nacobbus aberrans. Appl Soil Ecol 124:262–265

Martínez-Medina A, Fernandez I, Lok GB et al (2017) Shifting from priming of salicylic acid-to jasmonic acid-regulated defences by *Trichoderma* protects tomato against the root knot nematode *Meloidogyne incognita*. New Phytol 213:1363–1377

Martínez-Zavala SA, Barboza-Pérez UE, Hernández-Guzmán G et al (2020) Chitinases of *Bacillus thuringiensis*: phylogeny, modular structure, and applied potentials. Front Microbiol 10:3032–3032

Mazzuchelli RDCL, Mazzuchelli EHL, De Araujo FF (2020) Efficiency of Bacillus subtilis for root-knot and lesion nematodes management in sugarcane. Biol Control 143:104185
Meir D, Pivonia S, Levita R et al (2010) Application of mycorrhizae to ornamental horticultural crops: lisianthus (*Eustoma gradiflorum*) as a test case. Spanish J Agric Res 8:5–10
Mendoza-Mendoza A, Zaid R, Lawry R et al (2018) Molecular dialogues between *Trichoderma* and roots: role of the fungal secretome. Fungal Biol Rev 32:62–85
Menjivar RD, Hagemann MH, Kranz J et al (2011) Biological control of *Meloidogyne incognita* on cucurbitaceous crops by the non-pathogenic endophytic fungus *Fusarium oxysporum* strain 162. Int J Pest Manag 57:249–253
Mukherjee M, Mukherjee PK, Horwitz BA et al (2012) *Trichoderma*–plant–pathogen interactions: advances in genetics of biological control. Indian J Microbiol 52:522–529
Munif A, Hallmann J, Sikora RA (2000) Evaluation of the biocontrol activity of endophytic bacteria from tomato against Meloidogyne incognita. Mededelingen-Faculteit Landbouwkundige en Toegepaste Biologische Wetenschappen, Universiteit Gent 65:471–480
Padgham JL, Sikora RA (2007) Biological control potential and modes of action of *Bacillus megaterium* against *Meloidogyne graminicola* on rice. Crop Prot 26:971–977
Palomares-Rius JE, Escobar C, Cabrera J et al (2017) Anatomical alterations in plant tissues induced by plant-parasitic nematodes. Front Plant Sci 8:1987
Pathak P, Rai VK, Can H, Singh SK, Kumar D, Bhardwaj N, Roychowdhury R, de Azevedo LCB, Kaushalendra VH, Kumar A (2022) Plant-endophyte interaction during biotic stress management. Plan Theory 11(17):2203. https://doi.org/10.3390/plants11172203
Peng D, Lin J, Huang Q et al (2016) A novel metalloproteinase virulence factor is involved in *Bacillus thuringiensis* pathogenesis in nematodes and insects. Environ Microbiol 18:846–862
Poinar GO, Jansson HB (2018) Diseases of nematodes. CRC Press, Boca Raton
Polanczyk RA, Van Frankenhuyzen K, Pauli G (2017) The american *Bacillus thuringiensis* based biopesticides market. In: *Bacillus thuringiensis* and *Lysinibacillus sphaericus*. Springer, pp 173–184
Ponpandian LN, Rim SO, Shanmugam G et al (2019) Phylogenetic characterization of bacterial endophytes from four Pinus species and their nematicidal activity against the pine wood nematode. Sci Rep 9:1–11
Pozo MJ, Azcón-Aguilar C (2007) Unraveling mycorrhiza-induced resistance. Curr Opin Plant Biol 10:393–398
Salehi Jouzani G, Valijanian E, Sharafi R (2017) *Bacillus thuringiensis*: a successful insecticide with new environmental features and tidings. Appl Microbiol Biotechnol 101:2691–2711
Santangelo JS, Turley NE, Johnson MTJ (2015) Fungal endophytes of *Festuca rubra* increase in frequency following long-term exclusion of rabbits. Botany 93:233–241
Schäfer K (2007) Dissecting rhizobacteria-induced systemic resistance in tomato against *Meloidogyne incognita*: the first step using molecular tools. Aust Plant Pathol 36:124–134
Schouten A (2016) Mechanisms involved in nematode control by endophytic fungi. Annu Rev Phytopathol 54:121–142
Shiomi T, Lemaître V, D'armiento J et al (2010) Matrix metalloproteinases, a disintegrin and metalloproteinases, and a disintegrin and metalloproteinases with thrombospondin motifs in non-neoplastic diseases. Pathol Int 60:477–496
Siddique S, Grundler FMW (2018) Parasitic nematodes manipulate plant development to establish feeding sites. Curr Opin Microbiol 46:102–108
Sikora RA, Schäfer K, Dababat AA (2007) Modes of action associated with microbially induced in planta suppression of plant-parasitic nematodes. Australas Plant Pathol 36:124–134
Singh S, Singh B, Singh AP (2015) Nematodes: a threat to sustainability of agriculture. Procedia Environ Sci 29:215–216
Singh VK, Singh AK, Kumar A (2017) Disease management of tomato through PGPB: current trends and future perspective. 3 Biotech 7(4):1–10
Singh M, Srivastava M, Kumar A, Singh AK, Pandey KD (2020) Endophytic bacteria in plant disease management. In: Kumar A, Singh KV (eds) Microbial endophytes: prospects for

sustainable agriculture. Woodhead Publication, Cambridge USA, pp 61–89. https://doi.org/10.1016/B978-0-12-818734-0.00004-8

Thiyagarajan SS, Kuppusamy H (2014) Biological control of root-knot nematodes in chillies through *Pseudomonas fluorescens*'s antagonistic mechanism. J Plant Sci 2:152–158

Tran TPH, Wang S-L, Nguyen VB et al (2019) Study of novel endophytic bacteria for biocontrol of black pepper root-knot nematodes in the central highlands of Vietnam. Agronomy 9:714

Trifonova Z, Tsvetkov I, Bogatzevska N et al (2014) Efficiency of *Pseudomonas* spp. for biocontrol of the potato cyst nematode *Globodera rostochiensis* (Woll.). Bulgarian J Agric Sci 20:666–669

Troxler J, Svercel M, Natsch A et al (2012) Persistence of a biocontrol pseudomonas inoculant as high populations of culturable and non-culturable cells in 200-cm-deep soil profiles. Soil Biol Biochem 44:122–129

Van Dessel P, Coyne D, Dubois T et al (2011) In vitro nematicidal effect of endophytic *Fusarium oxysporum* against *Radopholus similis, Pratylenchus goodeyi* and *Helicotylenchus multicinctus*. Nematropica 41:154–160

Vetrivelkalai P, Sivakumar M, Jonathan EI (2010) Biocontrol potential of endophytic bacteria on Meloidogyne incognita and its effect on plant growth in bhendi. J Biopest 3:452

Wani KA, Manzoor J, Shuab R et al (2017) Arbuscular mycorrhizal fungi as biocontrol agents for parasitic nematodes in plants. In: Mycorrhiza-nutrient uptake, biocontrol, ecorestoration. Springer, pp 195–210

Waweru BW, Losenge T, Kahangi EM et al (2013) Potential biological control of lesion nematodes on banana using Kenyan strains of endophytic *Fusarium oxysporum*. Nematology 15:101–107

Yadav AN (2019) Endophytic fungi for plant growth promotion and adaptation under abiotic stress conditions. Acta Sci Agric 3:91–93

Yadav AN, Verma P, Kour D et al (2017) Plant microbiomes and its beneficial multifunctional plant growth promoting attributes. Int J Environ Sci Nat Resour 3:1–8

Yan X-N, Sikora RA, Zheng J-W (2011) Potential use of cucumber (*Cucumis sativus* L.) endophytic fungi as seed treatment agents against root-knot nematode Meloidogyne incognita. J Zhejiang Univ Sci B 12:219–225

Yan L, Zhu J, Zhao X et al (2019) Beneficial effects of endophytic fungi colonization on plants. Appl Microbiol Biotechnol 103:3327–3340

Zhai Y, Shao Z, Cai M et al (2018) Multiple modes of nematode control by volatiles of *Pseudomonas putida* 1A00316 from Antarctic soil against Meloidogyne incognita. Front Microbiol 9:253

Zhou W, Starr JL, Krumm JL et al (2016) The fungal endophyte *Chaetomium globosum* negatively affects both above-and belowground herbivores in cotton. FEMS Microbiol Ecol 92:fiw158

Zinniel DK, Lambrecht P, Harris NB et al (2002) Isolation and characterization of endophytic colonizing bacteria from agronomic crops and prairie plants. Appl Environ Microbiol 68:2198–2208

Chapter 6
Impact of Persistence and Movement of Gliotoxin Produced by *Trichoderma virens* in Agricultural Soil and Crop Plants

R. Oviya, G. Sobanbabu, S. T. Mehetre, R. Kannan, M. Theradimani, and V. Ramamoorthy

Abstract Gliotoxin is a short non-ribosomal peptide synthesized by Q strains of *Trichoderma virens*, *Aspergillus fumigatus* and certain *Aspergillus* and *Penicillium* species. Isolating gliotoxin-producing *T. virens* is tedious. Hence, a selective medium is used for isolating gliotoxin-producing *T. virens*. Any basal medium with suitable carbon and nitrogen sources or complex medium such as potato dextrose agar amended with gliotoxin serves as a selective medium that specifically allows the growth of gliotoxin-producing *T. virens*. Production of gliotoxin is positively correlated with disease suppression and mycelial growth reduction of various soil-borne pathogens and diseases caused by them. Thus, gliotoxin produced by *T. virens* is considered an antibiotic/biopesticide and a beneficial factor for the suppression of phytopathogenic organisms. Gliotoxin-producing *T. virens* strain (GL 20) was the first biocontrol agent which was commercially formulated and marketed as SoilGard (formerly GlioGard) for the management of various soil-borne diseases. Gliotoxin shows a phytotoxic effect when applied to the seed or aerial parts

R. Oviya · G. Sobanbabu
Department of Plant Pathology, Agricultural College and Research Institute, Tamil Nadu Agricultural University, Madurai, Tamil Nadu, India

S. T. Mehetre
Nuclear Agriculture and Biotechnology Division, Bhabha Atomic Research Centre, Mumbai, India
e-mail: smehetre@barc.gov.in

R. Kannan
Department of Plant Pathology, Tamil Nadu Agricultural University, Coimbatore, Tamil Nadu, India

M. Theradimani
Agricultural College and Research Institute, Tamil Nadu Agricultural University, Killikulam, Tamil Nadu, India

V. Ramamoorthy (✉)
Department of Plant Pathology, Agricultural College and Research Institute, Tamil Nadu Agricultural University, Eachangkottai, Thanjavur, Tamil Nadu, India

K. K. Bastas et al. (eds.), *Microbial Biocontrol: Molecular Perspective in Plant Disease Management*, Microorganisms for Sustainability 49,
https://doi.org/10.1007/978-981-99-3947-3_6

of the plants at higher concentrations. However, there is no phytotoxic effect when it is applied to plants at the inhibitory concentration required for the inhibition of fungal and bacterial pathogens. Thus, gliotoxin-producing *T. virens* has potential application for the management of soil-borne pathogens.

Keywords *Trichoderma virens* · Gliotoxin · Phytopathogen management · Biocontrol

6.1 Introduction

Trichoderma sp. is an effective mycoparasite that produces a variety of secondary metabolites, including epipolythiodioxopiperazines (ETPs), peptaibols, polyketides, pyrones, terpenes, and siderophores (Leylaie and Zafari 2018). Secondary metabolites are not essential for the survival of the producing organisms, but they are important for human society. Antimicrobial secondary metabolites are generally low-molecular weight compounds, and they are secreted to the external environment. Antimicrobial secondary metabolites of *Trichoderma* spp. are divided into volatile and non-volatile antibiotics based on their volatility (Dennis and Webster 1971a, b).

Only *T. virens* produces gliotoxin, whereas other species of *Trichoderma* do not produce gliotoxin. Certain strains of *T. virens* produce gliotoxin and they are called Q-type, whereas other strains of *T. virens* that do not produce gliotoxin are called P-type. Q-type strain of *T. virens* produces an enormous amount of gliotoxin compared to other secondary metabolites produced by it. Gliotoxin is a disulfide thiodioxopiperazine (ETP) type antibiotic containing a disulfide bridge (Dolan et al. 2015). It is considered as second antibiotic next to penicillin (Weindling 1934). It exhibited antifungal and antibacterial activity against various phytopathogenic fungi and bacteria, respectively (Brian and Hemming 1945). From the 1980s onwards, several research works on gliotoxin production and its importance in the biological control of phytopathogens have been described. The highest gliotoxin-producing strain of *T. virens* (GL 20) is the first commercial biocontrol agent that is marketed as SoilGard in the USA for the management of various soil-borne pathogens. In addition to gliotoxin production, it also showed mycoparasitic activity. Unfortunately, the production of gliotoxin by more than 90% of environmental isolates and clinical isolates of *A. fumigatus* has been reported. Human severe aspergillosis and the amount of gliotoxin produced by *A. fumigatus* are strongly associated. Thus, gliotoxin is regarded as a virulence factor and a mycotoxin in animal system (Sugui et al. 2017). Gliotoxin, which is produced by *T. virens* in the agricultural ecosystem, is recognized as a biopesticide and antibiotic that helps crops grow. Whereas in the animal system, gliotoxin produced by *A. fumigatus* is recognized as a mycotoxin and a lethal factor in causing aspergillosis (Scharf et al. 2016). Thus, it is essential to address and reaffirm the role of gliotoxin in the agricultural ecosystem and its biosafety. This review paper describes various fungi producing gliotoxin, the efficacy of gliotoxin on the suppression of various phytopathogens, various edaphic factors affecting the stability of gliotoxin in agricultural eco-system and its phytotoxicity effects in crop plants.

6.2 Gliotoxin-Producing Fungi

Fungi produce several secondary metabolites. Gliotoxin is one of them which shows antimicrobial activity. However, only a few fungal species produce gliotoxin. Weindling (1934) first described gliotoxin production in *Trichoderma virens*. However, only a limited strain of *Trichoderma* produces gliotoxin. Based on the ability to produce gliotoxin, various strains of *T. virens* are grouped into Q-type stains (gliotoxin producers) and P-type strains (gliotoxin non-producers) (Howell 1999; Park et al. 1992). Comparative genomic analysis of several gliotoxin-non-producing *Trichoderma* spp. and gliotoxin-producing *T. virens* revealed that GliH, a key protein, is required for gliotoxin synthesis. Thus, GliH can be used as a genetic marker for screening the gliotoxin-producing *T. virens* strains (Bulgari et al. 2020). In addition to Q strains of *T. virens*, gliotoxin is produced by *Eurotium chevalieri*, *Aspergillus fumigatus*, *Neosartorya pseudofischeri* and some *Acremonium* and *Penicillium* species (Scharf et al. 2016).

6.3 Isolation of Gliotoxin-Producing *T. virens*

Trichoderma is one of the most common fungal genera, isolated from various sources such as soil and plant organic matter (Kumar et al. 2012). *Trichoderma*-selective agar medium (TSM) has been developed for the isolation of *Trichoderma* spp. It contains chloramphenicol and Rose-Bengal as bacterial inhibitors and sucrose at low concentrations for faster growth and rapid sporulation (Elad and Chet 1983; Elad et al. 1981). The main disadvantage of TSM is that it allows the growth of green-colored *Aspergillus* and *Penicillium* which appear phenotypically similar to *Trichoderma.* In such conditions, the colonies need to be verified by microscope which is a cumbersome process. In addition, other fungi such as *Mucor* and *Rhizopus* can also grow on the TSM medium.

Since gliotoxin is particularly produced by *T. virens*, the use of TSM medium will support the growth of other *Trichoderma* spp. Therefore, specific growth medium or the basal medium or common medium, for example, PDA, amended with gliotoxin at 20 mg L^{-1}, has been used for the isolation of gliotoxin-producing *T. virens* strain specifically the Q -type strains. These media amended with gliotoxin are referred as selective media because they allow the growth of gliotoxin-producing organisms. Potato dextrose agar medium amended with gliotoxin can also be used for the selective isolation of gliotoxin-producing *T. virens*. The selective media allow the growth of particular microorganisms and inhibit others. Such selective medium was highly useful for the selective isolation of gliotoxin-producing *T. virens* because the presence of gliotoxin allows the growth of the gliotoxin-producing *T. virens* but inhibit the growth of other fungi (Park et al. 1992; Howell 1999). Gliotoxin-producing fungi have a self-immune system against the external gliotoxin present in the medium.

6.4 Identification and Differentiating *T. virens* from Other *Trichoderma* spp.

Actively growing *Trichoderma* isolates are characterized by a definite coconut smell because of the production of volatile metabolites (Bisby 1939). Basically, in biological studies, an organism is preliminarily identified by observing its morphology under a microscope. Similarly, screening and identification of gliotoxin-producing *Trichoderma* spp. are done by observing the conidiophore morphology of the culture under the light microscope. Gliotoxin-producing *T. virens* has typical *Gliocladium* type conidiophore and phialides. Conidiophores of *T. virens* are unbranched and formed singly. On the conidiophores, the phialides are grouped at an acute angle in a whorl and clustered together at the end of the conidiophore. The conidia are borne in a ball. Since several conidia are formed from the clustered phialides, they are all glued together. Because of *Gliocladium*-like conidiophores, the *T. virens* was earlier named as *Gliocladium virens*. Whereas the conidiophore of other *Trichoderma* spp. is well-branched into primary, secondary and tertiary conidiophores. The phialides (usually three phialides) are produced at a point in the tertiary conidiophore that arises from the secondary conidiophore that branches from the primary conidiophore. Phialides are positioned at a right angle to each other and to the tertiary conidiophore. A few conidia are formed on the phialides, and conidia are freely arranged. In appearance, the entire conidiophore system appears as a triangular form from the base to the end of the conidiophore arrangement (Webster and Lomas 1964).

Phenotypically, *T. virens* can be differentiated from other *Trichoderma* spp. based on the conidiophore morphology (conidiophore phenotype of *T. virens* looks like *Gliocladium)*. However, for species-level confirmation, certain molecular methods such as analyzing the internal transcribed spacer (ITS) region of the fungal rDNA are commonly used (Bruns et al. 1991; Hibbett 1992). Modern taxonomy of living organisms has been classified based on the rDNA sequences that encode ribosomal genes (Woese et al. 1990). Chaverri et al. (2001) analyzed the ITS1, ITS2 and 5.8 s regions of rDNA of *Hypocrea virens* which was found identical to *T. virens* isolates. Thus, *T. virens* can be confirmed by analyzing the ITS1 and ITS2 regions of the rDNA.

6.5 Production and Stability of Gliotoxin in Bioformulation

The time of production of any antimicrobial compounds by the biocontrol agents plays an important role in the suppression of phytopathogens. The earlier the antibiotic production by the *Trichoderma* spp., the better would be the disease management. Generally, the secondary metabolites are produced at and after the stationary phase. Conversely, the production of gliotoxin in *T. virens* starts at the early stage of the logarithmic phase and attains maximum accumulation at the stationary phase.

Gliotoxin is generally produced in 2 days after inoculation in the culture medium and its production and accumulation reach the maximum level in 6–8 days of incubation and thereafter it is degraded in the culture medium (Jayalakshmi et al. 2021). In another study, it has been described that gliotoxin synthesis started on the second day after inoculation and continued to accumulate up to the tenth day of incubation in the liquid culture medium (liquid formulation). Thereafter, its accumulation declined. In yet another study, it was shown that gliotoxin biosynthesis started 32 h after the inoculation of *T. virens* in the culture medium and attained the maximum level at 38 h. Thereafter its accumulation levels declined and completely stopped at 48 hours of incubation (Wilhite and Straney 1996). Thus, all these studies depicted that gliotoxin synthesis in *T. virens* occurs at the early growth phase. The rate of gliotoxin production was slower when *T. virens* was grown in C:N ratio of 80: 1 compared to that grown in other C:N ratios such as 18:1, 31:1 and 42:1 (Park et al. 1991). Generally, gliotoxin is synthesized in the initial growth period in a culture medium by *T. virens*. Later, *T. virens* synthesizes an extracellular enzyme called S-methyl transferase which converts the gliotoxin into bis-thiomethyl gliotoxin through methylation of disulfide bridge (Scharf et al. 2016; Sugui et al. 2017). Gliotoxin is a toxic compound and shows antimicrobial activity, whereas the modified bis-thiomethyl gliotoxin is non-toxic and does not show antimicrobial activity (Lumsden et al. 1992a, b; Domingo et al. 2012; Scharf et al. 2016).

Gliotoxin production by *T. virens* varies according to the growth conditions in the culture medium such as culture media type (culture media) with appropriate nutrients, temperature, pH conditions and the number of days it is cultured. Weindling medium, used for culturing the *T. virens*, contains dextrose 25.0 g, ammonium tartrate 2.0 g, potassium dihydrogen phosphate 2.0 g, magnesium sulfate 1–0 g, ferrous sulfate 0.01 g and water 1 L and pH adjusted to 3.5 with phosphoric acid before autoclaving (Weindling 1934). It is used for the production and analysis of gliotoxin for the first time. Brian and Hemming (1945) tested various media such as Weindling medium, Czapek–Dox medium, corn steep medium and Raulin-Thom medium for the production of gliotoxin and found that Weindling and Raulin-Thom media supported the higher level of gliotoxin production compared to other media tested. The reason for the maximum production of gliotoxin in the Weindling and Raulin-Thom media could be the presence of an ammoniacal form of nitrogen which would be superior to the nitrate form of nitrogen that was present in the Czapek–Dox medium and corn steep medium. Among the various simple and complex sugars tested for gliotoxin production, galactose supported the maximum level of gliotoxin production. Among the various culture media such as molasses-yeast medium, PDA and Weindling medium tested, molasses-yeast medium supported the maximum production of gliotoxin (Jayalakshmi et al. 2021). Carbon and nitrogen sources in the growth medium influence the gliotoxin production. Glucose, galactose and starch, as a carbon source, favor the higher gliotoxin production. Phenylalanine and ammoniacal form of nitrogen favor the gliotoxin production. A defined medium amended with glucose and phenylalanine as carbon and nitrogen sources, respectively, resulted in the highest production of gliotoxin (Park et al. 1991). In another study, it was shown that galactose and starch supported the maximum level of gliotoxin production (Brian and Hemming 1945).

Production, accumulation and stability of gliotoxin are influenced mainly by the pH and temperature of the culture medium. Gliotoxin is degraded under alkaline conditions and high temperature in a liquid culture medium and oppositely, it is stable in acidic conditions and low temperature. *T. virens* produces gliotoxin on the second day after inoculation. When the adsorbing agent such as animal charcoal is added together with *T. virens*, the levels of gliotoxin in the culture medium are very low indicating gliotoxin is adsorbed onto the charcoal (Weindling 1934). In another study, it was shown that gliotoxin was stable at acidic pH and quickly degraded in the neutral and basic pH in the culture medium (Brian and Hemming 1945). The optimum growth temperature for gliotoxin production is 25–30 °C (Wilhite and Straney 1996). In another study, gliotoxin production was found to be stable at 30 °C up to 10 days of incubation (Oviya 2019; Premalatha 2020).

6.6 Effect of Gliotoxin on Suppression of Phytopathogens

For the first time, Weindling (1934) noticed that gliotoxin was toxic to *R. solani*. It showed moderately toxic to various bacteria such as *Bacillus subtilis*, *B. lactis aerogenes*, *E. coli*, *Micrococcus lysodeikticus*, *Salmonella typhi* and *Staphylococcus aureus*. It is very highly toxic to *Colletotrichum lini*, *Fusarium caeruleum*, *Botrytis alli*, *Trichothecium roseum* and *Macrosporium sarcinaeforme* and moderately inhibitory against *Aspergillus niger* and *Penicillium digitatum*. However, gliotoxin-producing *T. virens* is resistant to gliotoxin. Generally, the concentration of gliotoxin required to inhibit spore germination is lower than that required to inhibit the mycelial growth (Brian and Hemming 1945).

Wheat seed dressing with talc-based gliotoxin formulation brought about significant disease reduction of covered smut of barley, leaf spot of oats, and bunt of wheat without phytotoxicity (Brian and Hemming 1945). Many gliotoxin-producing *T. virens* did not show a suppressive effect against zinnia damping-off. However, gliotoxin-producing *T. virens* isolate G 20 (previously designated as GL 21) showed a suppressive effect up to three subsequent plantings (replanting). For effective control of damping-off of zinnia in *R. solani* infected soil-less mix, prior application of *T. virens* at least 3 days before planting of zinnia seeds is effective in controlling the damping-off disease. Because *T. virens* needs a certain time period to contact and myco-parasitize *R. solani*, it should be applied before planting for effective control of zinnia damping-off in the *R. solani* infected soil-less medium. This experiment indicates that prior application and incubation of *T. virens* is needed in planting medium (pot mixture for raising crop) already infested with *R. solani* because antagonists need to grow, multiply, contact, parasitize and kill the resting bodies of *R. solani* (Lumsden and Locke 1989).

For the control of soil-borne fungal diseases of crop plants, *T. virens* G20 (Q type strain) is commercially sold as a biopesticide. In culture media, cropping soil and soilless cultivating medium, it produces massive amounts of gliotoxin, and the prevention of disease has a positive correlation with gliotoxin production. Mutant strains lacking gliotoxin production were created using *T. virens* stain G20 by chemical mutagenesis to demonstrate whether gliotoxin production was responsible for the suppression of soil-borne fungal diseases of crop plants. The isogenic mutant lacking gliotoxin production was ineffective in inhibiting the growth of *P. ultimum* mycelia under in vitro conditions and also had decreased efficacy in suppressing zinnia damping-off (Wilhite et al. 1994). Although it grows more quickly than the wild-type strain of *T. virens*, the mutant is useless as a mycoparasite. In comparison to the wild-type strain, biocontrol experiments in soil revealed that the mutant was unable to defend cotton seedlings against infection caused by *P. ultimum*, and it also showed reduced ability to attack (mycoparasitize) the sclerotia of *S. sclerotiorum*. This study concluded that synthesis of gliotoxin by *T. virens* is indirectly related to its mycoparasitism and biocontrol abilities (Vargas et al. 2014). However, we hypothesize that gliotoxin immediately weakens and halts the pathogen's growth resulting in effective mycoparasitic effect by *T. virens* on the pathogen's propagules such as mycelia and sclerotia.

T. virens is synergistic with metalaxyl in the suppression of the damping-off disease. Metalaxyl does not show fungicidal effect on *T. virens*. A combination of *T. virens* with a reduced dosage of metalaxyl (6.25% of the recommended dose) was more effective and synergistic in the management of cotton pythium damping-off than when they were used alone. *T. virens* is compatible with metalaxyl, PCB and carboxin. It is also noticed that seed treatment with *T. virens* is superior to soil application. A long storage period of the formulation containing *T. virens* is not effective because of late germination and loss of gliotoxin and gliovirin (Howell 1991). Cotton seed treated with *T. virens* together with metalaxyl brought about suppression of disease incidence than untreated controls (Howell et al. 1997).

The sporangial germination and growth of *P. ultimum*, the mycelial growth of *R. solani*, and the sclerotial germination and growth of *Sclerotium rolfsii* were all severely suppressed by the gliotoxin produced by *T. virens* strain G20. Gliovirin was very effective in inhibiting *P. ultimum* and *S. rolfsii* but not *R. solani*. Gliovirin, according to Howell and Stipanovic (1983), prevented the growth of *P. ultimum* but not *R. solani*. Gliovirin is inhibitory only on oomycetes fungi. Viridin had a substantial inhibitory effect against *R. solani* and *S. rolfsii*, whereas it showed weak inhibitory effect against *P. ultimum* (Lumsden et al. 1992a, b). This study demonstrated that, in contrast to the other antibiotics, gliotoxin has broad-spectrum antifungal activity. Other experiments have demonstrated the antifungal and antibacterial activity of gliotoxin against *P. ultimum* and *Bacillus cereus*, respectively (Park et al. 1991; Vargas et al. 2014). The antibacterial effects of gliotoxin and gliovirin against *Colletotrichum gloeosporioides* and *R. solani* were investigated by Jang et al. (2001). The amount of gliotoxin produced by *T. virens* is positively correlated with the suppression of damping-off incidence (Wilhite and Straney 1996; Vinale et al. 2014).

6.7 Factors Affecting the Stability of Gliotoxin in Agricultural Soil-Ecosystem and Irrigation Water

The study on various factors influencing the production and stability of gliotoxin in the culture medium and liquid bioformulation has been well described. However, there is no detailed study on the fate of gliotoxin especially its production by *T. virens* in soil and plant systems, its movement, stability and degradation in soil and plant ecosystems. The edaphic conditions such as soil pH, temperature, light and moisture content could be the major factors that determine the stability of gliotoxin in the soil ecosystem. Recently, we analyzed various edaphic factors on gliotoxin stability in the soil ecosystem and noticed that several edaphic conditions influence the stability of gliotoxin (Jayalakshmi et al. 2021). Thus, a detailed review on the influence of pH, soil moisture and native microflora on the stability/dissipation of gliotoxin was given in this section.

Soil incorporation with wheat straw and inoculation with a spore suspension of gliotoxin-producing *T. virens* resulted in its growth and colonization on wheat straw, and production of gliotoxin was recorded at 18 μg of gliotoxin per gram of straw. The soil immediately surrounding these straws contained one microgram of gliotoxin per gram of soil. The colonization by *T. virens* and production of gliotoxin are more in soil with autoclaved straw when compared to that in soil with unautoclaved straw. However, in soil with neutral pH incorporated with wheat straw and inoculated with the same *T. virens*, the presence of gliotoxin is very low. This study indicated that autoclaving of straw leads to a reduction in pH and favors more gliotoxin accumulation. Increasing the pH of the acidic soil leads to decreased accumulation of gliotoxin and vice versa. Among the various supplements tested, all supplements increased the yield of gliotoxin, but ammonium sulfate was the best supplement and potassium nitrate was the least effective in gliotoxin production (Wright 1956b).

In another study, the same author (Wright 1956a) reported that mustard seed treated with *T. virens* produced gliotoxin on mustard seed coats starting from the fifth day but not before the fourth day after sowing. Similarly, it started to produce gliotoxin (0.5 μg of GT/seed coat) on the pea seed coat from the third day after sowing and accumulation of gliotoxin was the maximum on the sixth day (2–4 μg of gliotoxin/seed coat) after germination and thereafter its accumulation declined gradually to low levels (0.5 μg of gliotoxin/seed coat) on the 12th day after sowing. The author concluded that the reason for the declined amount of gliotoxin during a longer period of incubation could be due to reduction in nutrients/food available on the seed coat for the growth and survival of *T. virens* leading to the arrest of its growth and multiplication. Secondly, it could be due to the enzymatic conversion of gliotoxin to non-toxic bis-thiomethyl gliotoxin by *T. virens* itself and thirdly, it could be due to the degradation of gliotoxin by soil microflora (Wright 1956b).

An earlier study found that *T. virens* produced gliotoxin in the culture media during initial logarithmic growth stage and gliotoxin was enzymatically transformed to bis-thiomethyl gliotoxin in the stationary growth phase. In our lab, we recently discovered that *T. virens* produced gliotoxin at the early logarithmic stage of growth, and it was transformed into bis-thiomethyl gliotoxin both intracellularly in the mycelium and extracellularly in the culture media. Our research demonstrated that alkaline pH, wet soil conditions, and unsterile soil promoted gliotoxin degradation. Gliotoxin dissipation was greater in non-sterile soil than that was in sterile soil; it was greater in wet soil than in dry soil and it was greater in alkaline soil than acidic soil. Alkaline soil pH increased gliotoxin dissipation in wet soil, but not in dry soil, under both sterile and unsterile soil. However, under alkaline pH conditions, high soil moisture significantly accelerated the dissipation of gliotoxin in unsterile soil compared to that in sterile soil. Thus, three factors namely high soil moisture (over 90% water holding capacity), the soil microbial community and soil alkaline pH (pH above 7) decreased the stability of gliotoxin, and these three factors show additive effect in degradation of gliotoxin (Jayalakshmi et al. 2021).

T. virens Q strain (GL 20) produced gliotoxin in peat–moss vermiculites (PV) soil-less medium, composted mineral soil, clay soil and sandy soil. Gliotoxin was not detectable in the peat–moss vermiculite with charcoal. This could be due to the adsorption of gliotoxin onto the charcoal. Gliotoxin accumulation was most abundant at the place of *T. virens* application in the soil-less potting medium. However, gliotoxin was detected 4–5 cm away from the inoculation point of *T. virens* or the area/place where the advancing mycelia growth of *T. virens* was associated with the soil. *T. virens* produced gliotoxin in potting medium on 1–2 days after inoculation. The maximum amount of gliotoxin was produced and accumulated between 1–4 days and then diminished to low levels between 8–18 days. This study showed that the trend in gliotoxin production in the potting medium/planting medium by *T. virens* is almost similar to that produced in the culture medium. In other words, gliotoxin is produced by the *T. virens* at the early growth stage and later it could be converted into bis-thiomethyl gliotoxin or degraded by soil microflora as described by Jayalakshmi et al. (2021). Among the various growth temperature conditions from 15 to 30 °C tested, *T. virens* produced GT at low levels at 15 °C and increased when the temperature is raised and the maximum level is observed at 30 °C (Lumsden et al. 1992b).

Similar to gliotoxin, the stability of gliovirin was gradually declined in unsterile soil while that remained higher in sterile soil. This study shows that soil microflora is responsible for inactivation/degradation of gliovirin. Gliovirin is sensitive to degradation by soil microflora but is neither absorbed nor rendered inactive by the soil. In damp, wet and unsterile soil, its persistence is less. Therefore, gliovirin-producing *T. virens* has limited use as a direct seed treatment to prevent *P. ultimum* causing damping-off in moist nursery beds. It would be beneficial if it is formulated as a slow-release formulation to reduce deterioration of gliovirin (Howell et al. 1993; Howell and Stipanovic 1983).

6.8 Edaphic Conditions on Antimicrobial Activity of Soil-Gliotoxin

It has been well-described that *T. virens* produces gliotoxin at the early growth stage. Gliotoxin production is observed immediately at 18 hrs after sowing of seeds that are coated with *T. virens* (Wright 1956a, b; Lumsden et al. 1992a, b). Early gliotoxin production by *T. virens* is essential for the immediate and quick suppression of germination of the pathogen's propagules, growth and differentiation of the phyto-pathogenic fungi. Since its production starts at 18 h after inoculation and continues to accumulate for a period of 5–8 days, we assume that pathogenic propagules can be effectively suppressed or killed based on the amount of gliotoxin production in the rhizosphere soil. However, production and accumulation of gliotoxin decline in the late stationary growth phase due to the enzymatic degradation. Thus, gliotoxin-producing *T. virens* can be formulated in the slow-releasing form to maintain an active growing phase for several days in the soil ecosystem or suitable mutated *T. virens* stains, having a mutation on S-methyl transferase gene that is responsible for conversion of gliotoxin to bis-thiomethyl gliotoxin, can be developed so that gliotoxin would be stable for a longer period and its conversion into bis-thiomethyl gliotoxin enzymatically can be avoided.

Since gliotoxin stability is mainly influenced by soil moisture in various types of soil (Jayalakshmi et al. 2021), gliotoxin and gliotoxin-producing *T. virens* could be effective when it is applied in moderately wet and dry soils and good for garden land and dry land crop and not suitable for wetland conditions. In addition to soil moisture, its stability is also influenced by soil pH. Thus, it would be highly effective when it is used in acidic soil with moderate moisture levels. Since high soil moisture and alkaline pH have an additive effect on the degradation of soil gliotoxin produced by *T. virens*, it would be better to apply the *T. virens* in garden land soil with acidic pH conditions.

6.9 Phytotoxic Effect of Gliotoxin

Any biomolecule that is used as a biopesticide should not cause phytotoxic effects such as inhibition of germination, growth retardation, yellowing and scorching of leaves or any aerial plant parts. For the development of gliotoxin or gliotoxin-producing *T. virens* as a biopesticide for agricultural purposes, it has to be tested for its phytotoxicity and plant growth-promoting activity on crop plants. Gliotoxin did not show any phytotoxic effect when seeds were treated with gliotoxin (Brian and Hemming 1945). Gliotoxin-producing strains have not shown any harm to plants under field conditions (Howell 2006). After thorough toxicological testing, it was determined that SoilGard, which contains a Q type strain of *T. virens*, is safe under field conditions (Lumsden and Walter 2003).

But some investigations found that gliotoxin has a phytotoxic effect on the development of plant roots and aerial parts (Wright 1951; Furuta et al. 1984; Haraguchi et al. 1992; Haraguchi et al. 1996; Haraguchi et al. 1997). Gliotoxin moderately affects the germination of wheat, clover and mustard seed and root growth (Wright 1951). Haraguchi et al. (1996) reported that pea seeds treated with gliotoxin showed a reduction in root and shoot length. From our lab, it has been noticed that gliotoxin at the inhibitory concentration of 100 ppm (inhibitory to most of the fungi) did not show any phytotoxic effect in black gram and gingelly. However, it showed a phytotoxic effect when it is applied at the higher concentration (Oviya 2019; Premalatha 2020). From these various studies, it has been noticed that gliotoxin has been found to be non-phytotoxic when it is applied at lower/inhibitory concentration, and it could be phytotoxic when it is applied at higher concentrations in crop plants.

6.10 Conclusions and Future Perspectives

Since gliotoxin shows broad-spectrum antimicrobial activity against numerous phytopathogenic fungi and bacteria, it is possible to use either gliotoxin or gliotoxin-producing *T. virens* as a potential bio-pesticide for the control of crop diseases. In addition to gliotoxin production, *T. virens* also shows a mycoparasitic effect parasitizing fungal propagules such as sclerotia and mycelia. Interestingly, gliotoxin is produced at the early logarithmic growth stage of *T. virens*. Thus, early production of gliotoxin immediately arrests and weakens the germinating propagules of the pathogen and also damages the germinated propagules. Further, the weakened and damaged pathogenic propagules could be effectively myco-parasitized. If the concentration of gliotoxin is high, it directly kills the pathogenic propagules. Importantly, the stability of gliotoxin is mainly influenced by the pH conditions of the culture medium, especially it is stable at acidic pH conditions. Thus, care should be taken to prepare buffered acidic culture medium for the growth and multiplication of *T. virens* so that gliotoxin could be stable for a longer period. Gliotoxin is converted into non-toxic bis-thiomethyl gliotoxin enzymatically by S-methyl transferases in *T. virens* itself. Thus, future study should be directed at developing *T. virens* having mutation or deletion of gene encoding methyl transferase. Thus, gliotoxin gets accumulated in the bio-formulation which can be directly applied for plant disease management. In the agricultural soil eco-system, its stability is mainly influenced by soil pH, soil moisture and microbiome. It is degraded by high soil pH and high soil moisture especially in saturated water conditions. Thus, care should be taken to use the gliotoxin and gliotoxin-producing *T. virens* in garden land and dry land soil having acidic pH for effective management of plant diseases. Because nitrogenous fertilizers are nitrogen source for growth of *T. virens* and plants, the soil pH would be changed at the rhizosphere regions upon taking up the ammoniacal or nitrate form of nitrogen by plants and *T. virens*. The pH change would influence the stability of gliotoxin that in turn affect the suppression of the pathogens in the rhizosphere. Thus, further studies should also be conducted on the analysis of acidifying nitrogenous fertilizers (ammonium-based) and non-acidifying nitrogenous fertilizers (nitrate-based) on the growth, multiplication and gliotoxin production by *T. virens* in soil.

References

Bisby GR (1939) *Trichoderma viride* Pers. ex fries, and notes on *Hypocrea*. Trans Brit Mycol Soc 23(2):149–168. https://doi.org/10.1016/S0007-1536(39)80020-1

Brian PW, Hemming HG (1945) Gliotoxin, a fungistatic metabolic product of *Trichoderma viride*. Ann Appl Biol 32(3):214–220. https://doi.org/10.1111/j.1744-7348.1945.tb06238.x

Bruns TD, White TJ, Taylor JW (1991) Fungal molecular systematics. Annu Rev Ecol Syst 22(1): 525–564. https://www.jstor.org/stable/i310052

Bulgari D, Fiorini L, Gianoncelli A, Bertuzzi M, Gobbi E (2020) Enlightening gliotoxin biological system in agriculturally relevant *Trichoderma* spp. Front Microbiol 11:200. https://doi.org/10.3389/fmicb.2020.00200

Chaverri P, Samuels GJ, Stewart EL (2001) *Hypocrea virens* sp. nov., the teleomorph of *Trichoderma virens*. Mycologia 94:1113–1124. https://doi.org/10.1080/00275514.2001.12063245

Dennis C, Webster J (1971a) Antagonistic properties of species-groups of *Trichoderma*: I. production of non-volatile antibiotics. Trans Brit Mycol Soc 57(1):25–39. https://doi.org/10.1016/S0007-1536(71)80077-3

Dennis C, Webster J (1971b) Antagonistic properties of species-groups of *Trichoderma*: II. Production of volatile antibiotics. Trans Brit Mycol Soc 57(1):41–48. https://doi.org/10.1016/S0007-1536(71)80078-5

Dolan SK, O'Keeffe G, Jones GW, Doyle S (2015) Resistance is not futile: gliotoxin biosynthesis, functionality and utility. Trends Microbiol 23(7):419–428. https://doi.org/10.1016/j.tim.2015.02.005

Domingo MP, Colmenarejo C, Martínez-Lostao L, Müllbacher A, Jarne C, Revillo MJ, Delgado P, Roc L, Meis JF, Rezusta A, Pardo J, Gálvez EM (2012) Bis (methyl) gliotoxin proves to be a more stable and reliable marker for invasive aspergillosis than gliotoxin and suitable for use in diagnosis. Diag Microbiol Infect Dis 73(1):57–64. https://doi.org/10.1016/j.diagmicrobio.2012.01.012

Elad Y, Chet I (1983) Improved selective media for isolation of *Trichoderma* spp. or *Fusarium* spp. Phytoparasitica 11(1):55. https://doi.org/10.1007/BF02980712

Elad Y, Chet I, Hetii Y (1981) Biological control of *Rhizoctonia solani* in strawberry fields by *Trichoderma harzianum*. Plant Soil 60:245–254. https://doi.org/10.1007/BF02374109

Furuta T, Kigane T, Suganoya Y (1984) Isolation of gliotoxin as lettuce seedling growth inhibit substance from the culture broth of *Aspergillus fumigatus* Fres. Bull Fac Agric 24:16–25

Haraguchi H, Hamatani Y, Shibata K, Hashimoto K (1992) An inhibitor of acetolactate synthase from a microbe. Biosci Biotechnol Biochem 56(12):2085–2086. https://doi.org/10.1271/bbb.56.2085

Haraguchi H, Hamatani Y, Hamada M, Fujii-Tachino A (1996) Effect of gliotoxin on growth and branched-chain amino acid biosynthesis in plants. Phytochemistry 42(3):645–648. https://doi.org/10.1016/0031-9422(95)00982-5

Haraguchi H, Yamano K, Kusunoki N, Fukuda A (1997) Effect of gliotoxin and related compounds on acetolactate synthase. J Agric Food Chem 45(7):2784–2787. https://doi.org/10.1021/jf960842a

Hibbett DS (1992) Ribosomal RNA and fungal systematics. Trans Mycol Soc Jpn 33:533–556

Howell CR (1991) Biological control of Pythium damping-off of cotton with seed-coating preparations of *Gliocladium virens*. Phytopathology 81:738–741. https://doi.org/10.1094/Phyto-81-738

Howell CR (1999) Selective isolation from soil and separation in vitro of P and Q strains of *Trichoderma virens* with differential media. Mycologia 91(6):930–934. https://doi.org/10.2307/3761624

Howell CR (2006) Understanding the mechanisms employed by *Trichoderma virens* to effect biological control of cotton diseases. Phytopathology 96(2):178–180. https://doi.org/10.1094/PHYTO-96-0178

Howell CR, Stipanovic RD (1983) Gliovirin, a new antibiotic from *Gliocladium virens*, and its role in the biological control of *Pythium ultimum*. Can J Microbiol 29:321–324. https://doi.org/10.1139/m83-053

Howell CR, Stipanovic RD, Lumsden RD (1993) Antibiotic production by strains of *Gliocladium virens* and its relation to the biocontrol of cotton seedling diseases. Biocontrol Sci Tech 3(4):435–441. https://doi.org/10.1080/09583159309355298

Howell CR, DeVay JE, Garber RH, Batson WE (1997) Field control of cotton seedling diseases with *Trichoderma virens* in combination with fungicide seed treatments. J Cotton Sci 1:15–20

Jang KS, Kim HM, Chung BK (2001) Purification and antifungal activities of an antibiotic produced by *Gliocladium virens* G1 against plant pathogens. Plant Pathol J 17(1):52–56

Jayalakshmi R, Oviya R, Premalatha K, Mehetre ST, Paramasivam M, Kannan R, Theradimani M, Pallavi MS, Mukherjee PK, Ramamoorthy V (2021) Production, stability and degradation of *Trichoderma* gliotoxin in growth medium, irrigation water and agricultural soil. Sci Rep 11:16536. https://doi.org/10.1038/s41598-021-95907-6

Kumar K, Amaresan N, Bhagat S, Madhuri K, Srivastava RC (2012) Isolation and characterization of *Trichoderma* spp. for antagonistic activity against root rot and foliar pathogens. Indian J Microbiol 52(2):137–144. https://doi.org/10.1007/s12088-011-0205-3

Leylaie S, Zafari D (2018) Antiproliferative and antimicrobial activities of secondary metabolites and phylogenetic study of endophytic *Trichoderma* species from Vinca plants. Front Microbiol 9:1484. https://doi.org/10.3389/fmicb.2018.01484

Lumsden RD, Locke JC (1989) Biological control of damping-off caused by *Pythium ultimum* and *Rhizoctonia solani* with *Gliocladium virens* in soilless mix. Phytopathology 79:361–366. https://doi.org/10.1094/Phyto-79-361

Lumsden RD, Walter JF (2003) *Gliocladium virens*: risk assessment. Biol Control Benefits Risks 4:263

Lumsden RD, Locke JC, Adkins ST, Walter JF, Ridout CJ (1992a) Isolation and localization of the antibiotics gliotoxin produced by *Gliocladium virens* from alginate prill in soil and soilless media. Phytopathology 82:230–235. https://doi.org/10.1094/Phyto-82-230

Lumsden RD, Ridout CJ, Vendemia ME, Harrison DJ, Waters RM, Walter JF (1992b) Characterization of major secondary metabolites produced in soilless mix by a formulated strain of the biocontrol fungus *Gliocladium virens*. Can J Microbiol 38(12):1274–1280. https://doi.org/10.1139/m92-210

Oviya R (2019) Effects of gliotoxin producing and non-producing *Trichoderma* spp. for the management of dry root rot of black gram. M. Sc Thesis, Tamil Nadu Agricultural University, Coimbatore
Park YH, Stack JP, Kenerley CM (1991) Production of gliotoxin by *Gliocladium virens* as a function of source and concentration of carbon and nitrogen. Mycol Res 95(10):1242–1248. https://doi.org/10.1016/S0953-7562(09)80018-X
Park YH, Stack JP, Kenerley CM (1992) Selective isolation and enumeration of *Gliocladium virens* and *G. roseum* from soil. Plant Dis 76:230–235. https://doi.org/10.1094/PD-76-0230
Premalatha K (2020) Studies on gliotoxin producing *Trichoderma* spp. on the management of dry root rot of sesame. M. Sc Thesis, Tamil Nadu Agricultural University, Coimbatore
Scharf DH, Brakhage AA, Mukherjee PK (2016) Gliotoxin–bane or boon? Environ Microbiol 18(4):1096–1109. https://doi.org/10.1111/1462-2920.13080
Sugui JA, Rose SR, Nardone G, Swamydas M, Lee CCR, Kwon-Chung KJ, Lionakis MS (2017) Host immune status-specific production of gliotoxin and bis-methyl-gliotoxin during invasive aspergillosis in mice. Sci Rep 7(1):1–13. https://doi.org/10.1038/s41598-017-10888-9
Vargas WA, Mukherjee PK, Laughlin D, Wiest A, Moran-Diez ME, Kenerley CM (2014) Role of gliotoxin in the symbiotic and pathogenic interactions of *Trichoderma virens*. Microbiology (Reading) 160(10):2319–2330. https://doi.org/10.1099/mic.0.079210-0
Vinale F, Sivasithamparam K, Ghisalberti EL, Woo SL, Nigro M, Marra R, Lombardi N, Pascale A, Ruocco M, Lanzuise S, Manganiello G, Lorito M (2014) *Trichoderma* secondary metabolites active on plants and fungal pathogens. Open Mycol J 8(1):127–139. https://doi.org/10.2174/1874437001408010127
Webster J, Lomas N (1964) Does *Trichoderma viride* produce gliotoxin and viridin? Trans Brit Mycol Soc 47(4):535–540. https://doi.org/10.1016/S0007-1536(64)80031-0
Weindling R (1934) Studies on a lethal principle effective in the parasitic action of *Trichoderma lignorum* on *Rhizoctonia solani* and other soil fungi. Phytopathology 24:1153–1179
Wilhite SE, Straney DC (1996) Timing of gliotoxin biosynthesis in the fungal biological control agent *Gliocladium virens* (*Trichoderma virens*). Appl Microbiol Biotechnol 45(4):513–518. https://doi.org/10.1007/BF00578464
Wilhite SE, Lumsden RD, Straney DC (1994) Mutational analysis of gliotoxin production by the biocontrol fungus *Gliocladium virens* in relation to suppression of Pythium damping-off. Phytopathology 84:816–821. https://doi.org/10.1094/Phyto-84-816
Woese CR, Kandler O, Wheelis ML (1990) Towards a natural system of organisms: proposal for the domains archaea, bacteria, and Eucarya. PNAS 87(12):4576–4579. https://doi.org/10.1073/pnas.87.12.4576
Wright JM (1951) Phytotoxic effects of some antibiotics. Ann Bot 15(4):493–499. https://doi.org/10.1093/OXFORDJOURNALS.AOB.A083294
Wright JM (1956a) Biological control of a soil-borne *Pythium* infection by seed inoculation. Plant Soil 8(2):132–140. https://www.jstor.org/stable/42931704
Wright JM (1956b) The production of antibiotics in soil III. Production of gliotoxin in wheat straw buried in soil. Ann App Biol 44(3):461–466. https://doi.org/10.1111/j.1744-7348.1956.tb02140.x

Chapter 7
Biological Control of Water Hyacinth (*Eichhornia crassipes*(C.Mart) Solms. Using Fungal Pathogens as Mycoherbicides: An Overview

Kannan Rengasamy, Pavithra Raj, Nivetha Andichamy, Ramamoorthy Vellaisamy, Sabarinathan Kuttalingam Gopalasubramanian, and Uma Sankareswari Rengasamy

Abstract Water hyacinth is one of the common aquatic weeds infesting most of the water bodies all over the world. It reproduces mainly through stolons irrespective of seeds. A single mother plant produces four daughter plants which can reproduce within 2 weeks and grows up into mat of 2 m thickness. It has a rapid rate of reproduction and can invade a variety of freshwater environments, which has a negative impact on the environment. An acre of medium-sized water hyacinth plants can yield over 45 million seeds. It results in a drop in oxygen levels, which lowers the quality of the water, obstructs sunlight, serves as a breeding ground for mosquitoes that spread diseases like malaria, filariasis and encephalitis that affect humans, as well as having a negative impact on aquatic life and the fisheries, irrigation, navigation, water supply and overall ecology of the infected water bodies.

Water hyacinths can be controlled using a variety of techniques, such as mechanical or physical, chemical and biological control. The best alternative to physical, mechanical and chemical techniques of controlling aquatic weeds without harming the aquatic ecology is biological control. Plant pathogenic fungi are the most promising biocontrol agent for the management of weeds. This chapter gives brief reports on ecology, habitat, problems caused by water hyacinth, pathogens,

K. Rengasamy (✉)
Department of Plant Pathology, Tamil Nadu Agricultural University, Coimbatore, India
e-mail: kannan.r@tnau.ac.in

P. Raj · N. Andichamy · S. K. Gopalasubramanian
Agricultural College and Research Institute, Killikulam, Tuticurin, India

R. Vellaisamy · U. S. Rengasamy
Agricultural College and Research Institute, Madurai, India
e-mail: ramamoorthy.v@tnau.ac.in

K. K. Bastas et al. (eds.), *Microbial Biocontrol: Molecular Perspective in Plant Disease Management*, Microorganisms for Sustainability 49,
https://doi.org/10.1007/978-981-99-3947-3_7

symptomatology, host range of pathogens, mass production and formulation of the potential fungal biocontrol agent.

Keywords Biocontrol · Water hyacinth · *Eichhornia crassipes* · Plant pathogens

7.1 Introduction

Eichhornia crassipes (C. Mart) Solms, also known as water hyacinth, is a poisonous aquatic weed that is found all over the world. It is one of the top 10 worst weeds that have invaded the entire world, and one of the 100 most troublesome invasive weed species, according to the International Union for Conservation of Nature (IUCN) (Patel 2012). It is a monocotyledonous, perennial, free-floating, aquatic herb that belongs to the family *Pontederiaceae*. The average height of the plant is 40 cm but also it can attain a height of about 1 m. Roots are fibrous and elongated. Leaves are broad, oblong with narrow ends, glossy green in colour and bulbous petioles. The spiked inflorescence contains 6–10 lily-like flowers which are 4–7 cm in diameter and pale violet or blue colour. Stems and leaves consist of air-filled tissues that give the ability to float on the water (buoyancy).

The water hyacinth came originally from South America, but it was introduced as a horticultural plant to Egypt between 1879 and 1892. The plant was distributed over the world as an attractive and botanical garden plant at the end of the nineteenth century. Its lovely flowers and distinctive leaf form led to its introduction as an ornamental plant in Bengal regions of India in 1896 (Fig. 7.1).

It was considered as a terror of Bengal because the plant colonizes the water surface, blocks light and decreases the level of oxygen which leads to death of the fishes and other aquatic organisms. Now it occupies 0.2 million hectares of water bodies including many rivers, lakes, ponds and canals throughout the country. It has been found infested in the Veeranam Lake and its distributaries in Tamil Nadu, Sundarbans mangrove forest of Bangladesh, wetlands of Kaziranga National Park and Deepor Beel Lake of Assam (Gnanavel and Kathiresan 2007; Biswas et al. 2007; Patel 2012). Gopal and Sharma (1981) stated biological control as 'regulating the

Fig. 7.1 Global spread of water hyacinth. (https://www.cabi.org/isc/datasheet/20544#toDistributionMaps)

Fig. 7.2 *Eichhornia crassipes* floating on water

excessive growth of an organism by employing its natural enemies'. By 'actions aimed at lowering the population of an aquatic weed to acceptable levels by use of a living organism or virus', Pieterse (1990) defined biological management of aquatic weeds. The biological control of weeds by exploiting their natural enemies, such as insects and microbes, has gained more attention in recent years. The organism was safer, more environmentally friendly, more effective and more targeted in their action while acting as a biocontrol agent.

7.2 Ecology and Habitat

Water hyacinth is a plant that can grow in both alkaline and acidic waters, according to Gopal (1987), although neutral water bodies are where the plant grows the fastest. Penetration of sunlight into the water surface was reduced by a dense mat that affects the photosynthesis of phytoplankton and submerged plants (Howard and Harley 1997). Water hyacinth can be located in different habitats like shallow ponds, lakes, rivers and canals all over the world (Gopal 1987). It grows well and rapidly spread in freshwater, slow moving water (Fig. 7.2). It also grows in toxic water and can withstand minimal nutrient supply, adverse pH and temperature conditions. But it does not grow in the water system when average salinity goes beyond 15% of seawater (Jones 2009).

7.3 Problems Caused by Water Hyacinth

Pysek and Richardson (2010) stated that biological alien invasions constitute a major destruction of biodiversity all over the world. Water hyacinth poses a threat to aquatic biodiversity and results in competition among all other nearby species and also affects microbes, growth of phytoplankton and fishes (Villamagna and Murphy 2010; Patel 2012; Gichuki et al. 2012). Transfer of oxygen from the air into the water was prevented by the profusely grown and dense floating water hyacinth mats and also affects the oxygen production (Villamagna and Murphy 2010). The dead and decayed biomass affects the water quality and amount of potable water, increases the cost of treatment for drinking water and decreases dissolved oxygen levels which results in the release of phosphorus triggers eutrophication. (Bicudo et al. 2007; Mironga et al. 2011; Ndimele et al. 2011; Patel 2012). Large dense floating mats of water hyacinth provide the breeding ground for many harmful human pests like anopheles mosquitoes which cause malaria, cholera and mansonoides mosquitoes which act as a vector for human lymphatic filariasis causing nematode Brugia (Chandra et al. 2006; Minakawa et al. 2008; Varshney and Sushilkumar 2008). Heavy infestation of water hyacinth results in increased levels of crocodile attacks and provides shelter for poisonous snakes and reptiles (Ndimele et al. 2011; Patel 2012). Due to its rapid rate of reproduction, water hyacinth grows densely and forming mats that Conceals the water bodies. The dense mats can often create problems to socio-economic activities like ship and boat navigation, hydropower generation, fisheries, water flow, irrigation channels and decrease the amount of water for agriculture and access to recreation (Fig. 7.3). In Brahmaputra River of India, the weed affects navigation, blocks irrigation channels and interferes water flow to the crop fields (Ndimele et al. 2011; Patel 2012).

7.4 Pathogens

Water hyacinth is infected by various fungal pathogens causing various levels of damages as described in Table 7.1. Pathogen that causes highest damage could be used as a potential bioherbicide.

7.5 Symptomatology

As described earlier, various fungal pathogens infect the water hyacinth and nature of symptoms varies as presented in Table 7.2.

7.6 Pathogenicity

Elwakil et al. (1990) sprayed wounded and unwounded water hyacinth plants with sterile distilled water before application of mycelial and conidial suspension of fungal pathogens for proving pathogenicity. Praveena and Naseema (2004) proved pathogenicity of fungal isolates by pin prick method. Fungal discs of seven-day old

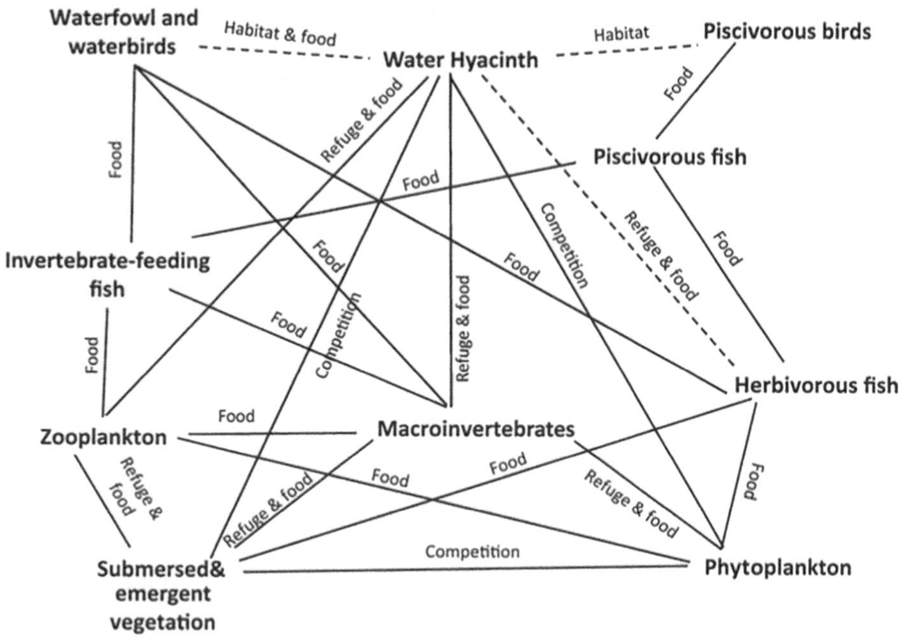

Fig. 7.3 Socio-economic impacts of water hyacinth (*Eichhornia crassipes*) (Villamagna and Murphy 2010)

culture were placed on pin-pricked area along with small wet cotton wool to cover the discs and protected with polyethene bags to maintain relative humidity. Ray and Hill (2012) proved pathogenicity of isolated fungal pathogens from water hyacinth by two steps, namely Petri plate bioassay and whole plant bioassay. For first bioassay, the water hyacinth leaf fragments, placed in Petri plates with moist filter paper, were inoculated with 2 mm fungal disc of actively growing culture and incubated at 26 ± 2 °C for 5–7 days. In the whole plant bioassay, healthy water hyacinth plants were sprayed with 50 mL of spore suspension collected from 21 day old culture along with 0.05 mL of Tween-20. Dagno et al. (2012) sprayed 20 mL culture suspension of fungal pathogens at a concentration of 2×10^6 spores/mL with 5% Tween-20 to the detached leaves placed in Petri plates. They also sprayed 20 mL spore suspension (5×10^6 spores/mL) with 5% Tween 20 on healthy water hyacinth plants in plastic pots. Singh et al. (2016) sprayed culture suspension made from a 10-day-old fungal culture that contained 1106 spores per millilitre with 0.05% Tween-20. They discovered that the isolate #19BJSS caused 100% plant death during the whole plant bioassay and caused 98.8% damage to the leaves under the in-vitro detached leaf assay. Yirefu et al. (2017) used a hand sprayer to inoculate 20 mL of mycelium and spore suspension at a concentration of 1106 propagule/mL with Tween-80 on the healthy plants in order to evaluate 25 fungal pathogens that were isolated from water hyacinth. The development of symptoms was then monitored, while the inoculated plants were covered in polythene bags.

Table 7.1 List of fungal pathogens associated with water hyacinth

S. No.	Pathogens	Country	References
1.	*Mycoleptodiscus terrestis*	USA	Charudattan and Conway (1976)
2.	*Cercospora rodmanii*	Florida	Conway (1976)
3.	*Helminthosporium*, *Myrothecium* and *Chaetomiella* species	Malaysia	Caunter (1984)
4.	*Rhizoctonia solani*	India	Srivastava and Verma (1987)
5.	*Cercospora rodmanii*	India	Aneja et al. (1988)
6.	*Fusarium chlamydosporum*, *Epicoccum nigram* and *Phoma sorghina*	India	
7.	*Alternaria alternata*	Egypt	Elwakil et al. (1990)
8.	*Cercospora piaropi*	South Africa	Morris (1990)
9.	*Fusarium equiseti*, *F. pallidoroseum* and *Colletotrichum gloeosporioides*	India	Santhi (1994)
10.	*Alternaria eichhorniae* and *Myrothecium roridum*	Malaysia	Caunter et al. (1996)
11.	*Acremonium zonatum*, *Alternaria* sp., *Cercospora piaropi*, *Fusarium* sp., and *Verticillium* species	Mexico	Jimenez and Charudattan (1998)
12.	*Cercospora piaropi*, *C. rodmanii*, *Acremonium zonatum* and *A. eichhorniae*	South Africa	Alana-den-Breeyen (2001)
13.	*Alternaria alternata*, *Drechslera hawaiiensis* and *Ulocladium atrum*	Egypt	El-Morsy (2004)
14.	*Myrothecium advena* and *Fusarium pallidoroseum*	India	Praveena and Naseema (2004)
15.	*Colletotrichum* and *Alternaria* species	China	Ding et al. (2008)
16.	*Alternaria jacinthicola*	West Africa	Dagno (2011)
17.	*Acremonium zonatum*, *Alternaria eichhorniae*, *Bipolaris hawaiiensis*, *Fusarium*, *Myrothecium roridum* and *Ulocladium* species	South Africa	Ray and Hill (2012)
18.	*Gibberella sacchari*, *Cadophora malorum* and *Alternaria* species	West Africa	Dagno et al. (2012)
19.	*Alternaria geophila*, *Ascochyta chartarum*, *Fusarium chlamydosporium*, *F. equiseti* and *Pythium ultimum*	East Africa	Tegene et al. (2012)
20.	*Alternaria japonica*	India	Dutta et al. (2015)
21.	*Diplodia mutila*	India	Singh et al. (2016)
22.	*Myrothecium roridum*	Thailand	Piyaboon et al. (2016)

Table 7.2 Symptoms produced by various fungal pathogens isolated from water hyacinth

Sl. No.	Pathogen	Symptoms	References
1.	*Alternaria eichhorniae*	Leaf spot and severe leaf blight	Nag Raj and Ponnappa (1970)
2.	*Myrothecium roridum*	Leaf spot	Ponnappa (1970)
3.	*Cercospora rodmanii*	Leaf spot	Freeman and Charudattan (1974)
4.	*Curvularia lunata*	Leaf spot	Chandra (1974)
5.	*Fusarium solani*	Reddish brown leaf spot	Jamil et al. (1983)
6.	*Rhizoctonia solani*	Foliar blight	Srivastava and Verma (1987)
7.	*Alternaria alteranta*	Leaf spot	Aneja and Singh (1989)
8.	*Fusarium chlamydosporum, Epicoccum nigram* and *Phoma sorghina*	Leaf spot	Aneja and Srinivas (1990)
9.	*Fusarium equiseti, F. pallidoroseum and Colletotrichum gloeosporioides*	Leaf spot	Santhi (1994)
10.	*Drechslera hawaiiensis* and *Ulocladium atrum*	Leaf blight	El-Morsy (2004)
11.	*Alternaria jacinthiciola*	Leaf blight	Dagno (2011)
12.	*Acremonium zonatum* and *Bipolaris hawaiiensis*	Zonate leaf spot and leaf spot	Ray and Hill (2012)
13.	*Gibberella sacchari* and *Cadophora malorum*	Leaf blight	Dagno et al. (2012)
14.	*Neofisicoccum parvum*	Leaf spot	Yirefu et al. (2017)
15.	*Diplodia mutila*	Leaf spot	Singh et al. (2016)

7.7 Morphological and Molecular Characterization of Virulent Fungal Pathogens

Dagno (2011) obtained fungal isolate Mlb684 from diseased water hyacinth plants that caused leaf blight and identified by both morphological and molecular characterization. Based on the characterization, the isolate Mlb684 belonged to the genus Alternaria and named as *A. jacinthicola*; it was different from known species of *Alternaria*. Singh et al. (2016) isolated 30 fungal pathogens and obtained an isolate #19BJSS that caused severe damage on water hyacinth plants. Based on both morphological and molecular analysis, the isolate belonged to a family Botryosphaeriaceae and found to be a novel species *Diplodia mutila*, a new aquatic weed pathogen. Yirefu et al. (2017) isolated different fungal pathogens from diseased water hyacinth plants and they were identified based on morphological and molecular characteristics. In identification of fungal pathogens, morphological examination plays a major role which includes description of mycelium, size, colour, fruiting structures and spore characters. They isolated 25 fungal pathogens and identified as following nine genera, namely *Alternaria*, *Aspergillus*, *Botryosphaeria*, *Curvularia*, *Fusarium*, *Mucor*, *Neofusicoccum*, *Penicillium* and *Phoma* based on both morphological and molecular characteristics.

7.8 Effect of Virulent Fungal Pathogens on Non-target Crops (Host Range)

S. No.	Pathogen	Hosts	References
1.	*Alternaria eichhornia*	*Monochoria vaginalis*	Nag Raj and Ponnappa (1970)
2.	*Mycoleptodiscus terrestris*	Legumes	Charudattan and Conway (1976)
3.	*F. sporotrichoides*	*Myriophyllum spicatum*	Andrews and Hecht (1981)
4.	*Fusarium lateritium*	Velvet leaf and prickly sida	Boyette and Walker (1985)
5.	*F. oxysporum Senna,*	Sickle pod and hemp sesbania	Boyette et al. (1993a)
6.	*F. equiseti*, *F. pallidoroseum* and *F. solani*	*Monochoria vaginalis*	Santhi (1994)
7.	*Fusarium equiseti* and *F. pallidoroseum*	*Amaranthus viridis*, *Commelina benghalensis*, *C. joobi*, *Pennisetum purpureum* and *Monochoria vaginalis*	Naseema and Balakrishnan (1999)
8.	*Cercospora piaropi* and *Acremonium zonatum*	Water lettuce	Jimenez and Lopez (2001)
9.	*Cadophora malorum isolate Mln715 and Alternaria* sp. *isolate Mlb684*	Aquatic fern	Dagno et al. (2012)
10.	*Alternaria japonica*	Goosefoot	Dutta et al. (2015))
11.	*Myrothecium roridum*	Water lettuce and duckweed	Piyaboon et al. (2016)
12.	*Alternaria alternata*, *A. tenuissima*, *Alternaria* sp.	Water lettuce	Yirefu et al. (2017)
	Fusarium equiseti	Water lettuce, noug, cabbage, mustard, haricot bean, soybean, pea, onion, sesame, tomato and pepper	
	F. oxysporum	Water lettuce, cabbage, papyrus, chickpea, faba bean, pea, lentil, fenugreek, sesame and pepper	
	Neofusicoccum parvum	Water lettuce, tomato and roscoae	

7.9 Effect of Virulent Fungal Pathogens on Water Quality

Yirefu et al. (2017) investigated the relationship between the presence of fungal species and environmental factors such as wave action, air temperature, season, altitude, rainfall, nitrogen, phosphorous, pH and electrical conductivity. They discovered that environmental and water quality factors had an impact on the prevalence of fungal pathogens and the degree of disease severity.

7.10 Mass Production of Fungal Pathogens

7.10.1 Whole Grain Media

Several fungal bioinoculants are mass multiplied using cereal grains as substrates. Sahayaraj and Namasivayam (2008) made the utilization of various agricultural and by-products for the mass culture of entomopathogenic fungi. Moreover, wheat grains were found suitable for the growth and spore production of *Beauveria bassiana,* whereas sorghum grains were found suitable for *Paecilomyces fumosoroseus* and *Verticillium lecanii.*

7.10.2 Solid Substrate Media

Crop residues and agricultural wastes products containing high amount of cellulose and low levels of lignin are suitable for mass multiplication of several ascomycetous fungi. Mathai et al. (1988) attempted the utilization of locally available and cheap substrates such as rice bran, wheat bran, paddy straw, tapioca bits, vegetable waste and their different combinations. Among them, wheat bran and rice bran with tapioca bits were suitable for the growth and sporulation of *F. pallidoroseum*. Santhi (1994) concluded that wheat and rice bran were good carrier materials for the mass production and storage of *Fusarium* spp. and *Colletotrichum gloeosporioides*. Ciotola et al. (1995) reported that sorghum straw was suitable for the growth and sporulation of *F. oxysporum* isolate M12-4A and it suppressed the weed *Striga hermonthica* by inhibiting germination and attachment of weed to sorghum roots. Faizal and Mathai (1996) reported that wheat and rice bran were suitable substrates for the growth and sporulation of *F. pallidoroseum* and spores were harvested within 8–10 days of incubation. Guinea grass straw was reported as a suitable substrate for the growth and maximum spore production of *F. pallidoroseum* and *F. equiseti* within 20 and 40 days of incubation respectively (Susha and Naseema 2002).

7.10.3 Liquid Substrate Media

In addition to solid substrates, bacterial and fungal bio agents are cultured and multiplied in liquid medium. Ibrahim and Low (1993) reported that coconut water was found to be a suitable liquid substrate for the growth and spore production of *B. bassiana* and *P. fumosoroseus*. *Manisegarame* and *Letchoumanne* (1996) attempted the utilization of different liquid media for the mass production of *F. pallidoroseum,* namely coconut water, parboiled rice water, parboiled rice gruel, raw rice water and raw rice gruel. They reported that coconut water yielded the maximum spore production than other substrates. Rejirani (2001) observed maximum growth and spore production of *F. pallidoroseum* cultured in coconut water when compared to boiled rice water and raw rice water.

7.11 Formulation of Fungal Pathogens

For field level application, fungal bio-agents need to be formulated in appropriate carrier for their easy application. Boyette and Walker (1985) developed a granular formulation of *F. lateritium* which consisted of mycelium and spores mixed with sodium alginate and kaolin clay. ABG-5003 was a wettable powder formulation containing mycelium and spores of *Cercospora rodmanii* developed by Abbott laboratories against water hyacinth (Freeman and Charudattan 1984). Collego was wettable powder formulation of *Colletotrichum gloeosporioides* f. sp. *aeschynomene* developed against northern joint vetch in rice and soybean fields (Bowers 1986). Faizal (1992) prepared wettable powder formulation of *F. pallidoroseum* with diatomaceous earth containing 3.5×10^6 spores/mL that was effective against aphids. An invert emulsion formulation of *Colletotrichum truncatum* was developed by Boyette et al. (1993b) that was effective against hemp sesbania. Shabana et al. (1997) studied the bioherbicidal efficacy of different alginate formulations of *Alternaria eichhornia* isolate Ae5 on water hyacinth, namely mycelium alone, mycelium with culture filtrate and culture filtrate only with and without a hydrophilic humectant (Evergreen 500). Among these, alginate formulations with humectant were effective in disease production than the other formulations.

7.12 Assessment of Shelf Life

The carrier materials used in the formulation should support the shelf life of the bio-agents that need to be assessed periodically. Faizal (1992) observed that water, talc and diatomaceous earth formulations of *F. pallidoroseum* retained the viability for 4 days of storage and after that there was a decrease in their virulence. A wettable powder formulation of *F. pallidoroseum* with diatomaceous earth retained viable for 10 months under refrigeration (Sunitha 1997). Booth et al. (2000) observed the viability of dry mycelial formulation of *Metarhizium* retained for more than 1 year when stored at 4 °C.

7.13 Concluding Remarks

Water hyacinth is used for paper manufacturing, fibreboard production, yarn and rope making, basket making, charcoal briquetting, biogas production, bioremediation, animal fodder, fish feed, fertilizers, industrial uses and household articles. Different countries like Bangladesh. Papua New Guinea and Sri Lanka are striving hard to use water hyacinth as a substrate for paper making. Despite its benefits, it has significant negative social, economic and environmental effects. A variety of techniques, including mechanical or physical, chemical, and biological control, can be used to manage water hyacinth. The two most expensive, short-term and damaging to aquatic life and unintended crops of the three methods are mechanical and chemical control. Therefore, using natural enemies to control water hyacinth is an efficient and affordable method of biological control. Fungal pathogens are found to be more suitable biocontrol agents for the management of water hyacinth as they are easy to isolate, maintain, identify, grow and also for mass production. They have less harmful effects compared to chemicals, non-toxic to humans and animals. A wettable powder formulation of *F. pallidoroseum* with diatomaceous earth retained viable for 10 months under refrigeration (Sunitha 1997). They are environmentally and economically feasible, safer and non-pathogenic effects to non-target crops. They can be formulated as a commercial mycoherbicide and also used in integrated pest management (IPM) approaches.

References

Alana-den-Breeyen (2001) The international Mycoherbicide programme for water hyacinth in Africa. ARC Plant Protection Research Institute, South Africa. Plant Protection News Bulletin No 59, pp. 8–11

Andrews JH, Hecht EP (1981) Evidence for pathogenicity of fusarium sporotrichioides to Eurasian water milfoil, *Myriophyllum spicatum*. Can J Bot 59(6):1069–1077

Aneja KR, Singh K, Bhan R (1988) A new leaf spot disease of water hyacinth. Indian Phytopathol 41(1):160

Aneja KR, Singh K (1989) *Alternaria alternata* (Fr.) Keissler, a pathogen of water hyacinth with biocontrol potential. Int J Pest Manag 35(4):354–356

Aneja KR, Srinivas B (1990) Three new pathogenic fungi of water hyacinth from India. Trop Pest Manag 36(1):76p

Bicudo DDC, Fonseca BM, Bini LM, Crossetti LO, Bicudo CEDM, Jesus TA (2007) Undesirable side-effects of water hyacinth control in a shallow tropical reservoir. Freshw Biol 52(6): 1120–1133

Biswas SR, Choudhury JK, Nishat A, Rahman MM (2007) Do invasive plants threaten the Sundarbans mangrove forest of Bangladesh? For Ecol Manag 245(3):1–9

Booth SR, Tanigoshi L, Dewes I (2000) Potential of a dried mycelium formulation of an indigenous strain of *Metarhizium anisopliae* against subterranean pests of cranberry. Biocontrol Sci Tech 10(5):659–668

Bowers RC (1986) Commercialization of Collego™–an Industrialist's view. Weed Sci 34(1):24–25

Boyette CD, Abbas HK, Connick WJ Jr (1993a) Evaluation of *Fusarium oxysporum* as a potential bioherbicide for sickle pod (*Cassia obtusifolia*), coffee senna (*C. occidentalis*) and hemp sesbania (*Sesbania exaltata*). Weed Sci 41:678–681

Boyette CD, Quimby PC Jr, Bryson CT, Egley GH, Fulgham FE (1993b) Biological control of hemp sesbania (*Sesbania exaltata*) under field conditions with *Colletotrichum truncatum* formulated in an invert emulsion. Weed Sci 41:497–500

Boyette CD, Walker HL (1985) Factors influencing biocontrol of velvetleaf (*Abutilon theophrasti*) and prickly sida (*Sida spinosa*) with *Fusarium lateritium*. Weed Sci 32:209–211

Caunter IG (1984) Potential for biological control of water hyacinth with indigenous fungal pathogens. In: Proceedings of international conference on plant protection in tropics, School of Biological Sciences, University of Sians, Malaysia, 16–19 June, pp. 489–492

Caunter IG Lee KC and Moran VC (1996) Initiating the use of fungi for biocontrol of weeds in Malaysia. In: Proceedings of ninth international symposium on biological control of weeds. University of Capetown, Republic of South Africa, 19–26 January, pp. 249–252

Chandra S (1974) Some new leaf spot disease from Allahabad, India. Nova Hedwigia Beihefte 47: 335–341

Chandra G, Ghosh A, Biswas D, Chatterjee SN (2006) Host plant preference of Mansonia mosquitoes. J Aquat Plant Manag 44:142–144

Charudattan R, Conway KE (1976) *Mycoleptodiscus terrestris* leaf spot on water hyacinth. Plant Dis Rep 60(1):77–80

Ciotola M, Watson AK, Hallett SG (1995) Discovery of an isolate of *Fusarium oxysporum* with potential to control *Striga hermonthica* in Africa. Weed Res 35(4):303–309

Conway KE (1976) *Cercospora rodmanii*, a new pathogen of water hyacinth with biological control potential. Can J Bot 54(10):1079–1083

Dagno K (2011) *Alternaria jacinthicola*, a new fungal species causing blight leaf disease on water hyacinth [*Eichhornia crassipes* (Martius) Solms-Laubach]. J Yeast Fungal Res 2(7):99–105

Dagno K, Lahlali R, Diourté M, Jijakli H (2012) Fungi occurring on water hyacinth (*Eichhornia crassipes* [Martius] Solms-Laubach) in Niger River in Mali and their evaluation as mycoherbicides. J Aquat Plant Manag 50:25–32

Ding Y, Zhao N, Chu J (2008) Nine pathogenic fungi of water hyacinth isolated in China. J Shanghai Jiaotong Univ (Sci) 13(5):617–622

Dutta W, Ray D, Ray P (2015) Molecular characterization and host range studies of indigenous fungus as prospective mycoherbicidal agent of water hyacinth. Indian J Weed Sci 47(1):59–65

El-Morsy EM (2004) Evaluation of microfungi for the biological control of water hyacinth in Egypt. Fungal Divers 16(1):35–51

Elwakil MA, Sadik EA, Fayzalla EA, Shabana YM (1990) Biological control of water hyacinth with fungal plant pathogens in Egypt. In: Proceedings of the eighth international symposium on biological control of weeds. Canterbury, New Zealand, 2–7 February, 497p

Faizal MH (1992) Studies on the entomogenous fungus associated with cowpea aphid. M.Sc. (Agriculture) Thesis, Department of Agricultural Entomology, Kerala Agriculture University, Thrissur

Faizal MH, Mathai S (1996) Evaluation of different substrates for mass production of *Fusarium pallidoroseum* an entomogenous fungus. J Entomol 58(2):99–102

Freeman TE, Charudattan R (1974) Occurrence of *Cercospora piaropi* on water hyacinth in Florida. Plant Dis Rep 58:277–278

Freeman TE, Charudattan R (1984) *Cercospora rodmanii* Conway, a biocontrol agent for water hyacinth. Florida Agriculture Experiment Station Technical Bulletin 18:842

Gichuki J, Omondi R, Boera P, Okorut T, Matano AS, Jembe T, Ofulla A (2012) Water hyacinth *Eichhornia crassipes* (Mart.) Solms-Laubach dynamics and succession in the Nyanza gulf of Lake Victoria (East Africa): implications for water quality and biodiversity conservation. Sci World J 2012:1–10

Gnanavel I, Kathiresan RM (2007) Impact of integrated biological control of water hyacinth (*Eichhonnia crassipes* (Mart.) Solms) on water quality and fish mortality. Res J Agric Biol Sci 3(1):21–23

Gopal B (1987) Water hyacinth. Elseveir Science Publishers, Amsterdam, Netherland, p 471p

Gopal B, Sharma KP (1981) Water-hyacinth (*Eichornia crassipes*): the most troublesome weed of the world. Hindasia Publishers, New Delhi, India, p 128p

Howard GW, Harley KLS (1997) How do floating aquatic weeds affect wetland conservation and development? How can these effects be minimised? Wetl Ecol Manag 5(3):215–225

Ibrahim YB, Low W (1993) Potential of mass-production and field efficacy of isolates of the entomopathogenic fungi *Beauveria bassiana* and *Paecilomyces fumosoroseus* against *Plutella xylostella*. Int J Pest Manag 39(3):288–292

Jamil K, Narasaiah J, Thyagarajan G (1983) Studies on the evaluation of naturally occurring fungal pathogens of water hyacinth for biological control of the weed. Indian J Bot 6(2):185–189

Jimenez MM, Charudattan R (1998) Survey and evaluation of Mexican native fungi for potential biocontrol of water hyacinth. J Aquat Plant Manag 36:145–148

Jimenez MM, Lopez EG (2001) Host range of *Cercospora piaropi* and *Acremonium zonatum*, potential fungal biocontrol agents for water hyacinth in Mexico. Phytoparasitica 29(2):175–177

Jones RW (2009) The impact on biodiversity, and integrated control, of water hyacinth, *Eichhornia crassipes* (Martius) Solms-Laubach (Pontederiaceae) on the Lake Nsezi-Nseleni River System. M.Sc. Thesis. Department of Zoology and Entomology, Rhodes University, South Africa, 115p

Manisegarame S, Letchoumanane S (1996) *Fusarium pallidoroseum* a pathogen on the rice leaf roller *Cnaphalocrocis medinalis*. Indian J Entomol 58:364–368

Mathai S, Geetha D, Mohandas N (1988) Use of different locally available and cheaper substrates for the mass multiplication of *Fusarium pallidoroseum* (Cooke) Sacc., an effective fungal pathogen against pea aphid, *Aphis craccivora* Koch. In: New trends in biotechnology, pp. 249–252

Minakawa N, Sonye G, Dida GO, Futami K, Kaneko S (2008) Recent reduction in the water level of Lake Victoria has created more habitats for *Anopheles funestus*. Malar J 7(1):1–6

Mironga JM, Mathooko JM, Onywere SM (2011) The effect of water hyacinth (*Eichhornia crassipes*) infestation on phytoplankton productivity in Lake Naivasha and the status of control. J Environ Sci Eng 5(10):1252–1261

Morris MJ (1990) *Cercospora piaropi* recorded on the aquatic weed, *Eichhornia crassipes*. South Africa Phytophylactica 22(2):255–256

Nag Raj TR, Ponnappa KM (1970) Blight of water hyacinth caused by *Alternaria eichhorniae* sp. Trans Br Mycol Soc 55(1):123–130

Naseema, A, Balakrishnan S (1999) Bioherbicidal potential of fungal pathogens of water hyacinth. In: Proceedings on alien weeds in moist tropical zones; banes and benefits, Kerala Forest Research Institute, Peechi, India, 2–4 November, pp. 115–121

Ndimele PE, Kumolu-Johnson CA, Anetekhai MA (2011) The invasive aquatic macrophyte, water hyacinth {*Eichhornia crassipes* (Mart.) Solm-Laubach: Pontedericeae}: problems and prospects. Res J Environ Sci 5(6):509

Patel S (2012) Threats, management and envisaged utilizations of aquatic weed *Eichhornia crassipes:* an overview. Rev Environ Sci Biotechnol 11(3):249–259

Pieterse AH (1990) Introduction of aquatic weeds. Oxford University Press, North California, pp 88–110

Piyaboon O, Pawongrat R, Unartngam J, Chinawong S, Unartngam A (2016) Pathogenicity, host range and activities of a secondary metabolite and enzyme from *Myrothecium roridum* on water hyacinth from Thailand. Weed Biol Manag 16(3):132–144

Ponnappa KM (1970) On the pathogenicity of *Myrothecium roridum-Eichhornia crassipes* isolate. Hyacinth Control J 8(1):18–20

Praveena R, Naseema A (2004) Fungi occurring on water hyacinth (*Eichhornia crassipes* [Mart.] Solms) in Kerala. J Trop Agric 42(1):21–23
Pysek P, Richardson DM (2010) Invasive species, environmental change and management and health. Annu Rev Environ Resour 35:25–55
Ray P, Hill MP (2012) Fungi associated with *Eichhornia crassipes* in South Africa and their pathogenicity under controlled conditions. Afr J Aquat Sci 37(3):323–331
Rejirani OP (2001) Production and evaluation of the fungus *Fusarium pallidoroseum* (Cooke) Sacco as a biopesticide against pea aphid Aphis craccivora Koch. Ph.D., Thesis, Kerala Agricultural University, Thrissur
Sahayaraj K, Namasivayam SKR (2008) Mass production of entomopathogenic fungi using agricultural products and by products. Afr J Biotechnol 7(12):1907–1910
Santhi KS (1994) Screening of fungal pathogens for biocontrol of water hyacinth. M.Sc. (Agriculture) Thesis, Department of Plant Pathology, Kerala Agricultural University, Thrissur
Shabana YM, Baka ZAM, Abdel-Fattah GM (1997) *Alternaria eichhorniae*, a biological control agent for water hyacinth: mycoherbicidal formulation and physiological and ultrastructural host responses. Eur J Plant Pathol 103(2):99–111
Singh B, Saxena S, Meshram V, Kumar M (2016) Diplodia mutila as a new pathogen on water hyacinth *Eichhornia crassipes* in Harike wetland. J Biopest 9(2):180–188
Srivastava LS, Verma RN (1987) New records of *Rhizoctonia* blight diseases of coffee and water hyacinth from India. Bangladesh J Bot 16(2):214–215
Sunitha VS (1997) *Fusarium pallidoroseum* (Cooke) Sacc as a biocontrol agent for the pea aphid Aphis craccivora Koch. M.Sc. (Agriculture) Thesis, Department of Agricultural Entomology, Kerala Agricultural University, Thrissur
Susha TS, Naseema A (2002) Mass production of *Fusarium* species, potent biocontrol agents of water hyacinth. J Mycol Plant Pathol 32(1):94–95
Tegene S, Hussein T, Tessema T, Yirefu F, Carmen B, Gossmann M (2012) Exploration of fungal pathogens associated with water hyacinth (*Eichhornia crassipes* (Mart.) Solms Laubach) in Ethiopia. Afr J Agric Res 7(1):11–18
Varshney GJ, Sushilkumar JS (2008) Current status of aquatic weeds and their management in India. In: Proceedings of Taal 2007: the 12th world lake conference, pp. 1039–1045
Villamagna AM, Murphy BR (2010) Ecological and socio-economic impacts of invasive water hyacinth (*Eichhornia crassipes*): a review. Freshw Biol 55(2):282–298
Yirefu F, Struik PC, Lantinga EA, Taye T (2017) Occurrence and diversity of fungal pathogens associated with water hyacinth and their potential as biocontrol agents in the Rift Valley of Ethiopia. Int J Pest Manag 63(4):355–363

Chapter 8
Plant Growth-Promoting Microorganisms as Phytoprotectants and Suitable Nano Delivery Systems

Haripriya Shanmugam, Shobana Narayanasamy, and Sivakumar Uthandi

Abstract Plant diseases are controlled with microbial biological control agents (MBCAs) in crop plants via different mechanisms. These interactions are regulated through a cascade of complex metabolic activities that frequently combine several mechanisms of action. Conventional BCAs formulations have bottlenecks like shorter shelf life, poor stability, decreased microbial load, and high dosage required for covering per unit area. Compared to chemical fertilizers and conventional biocontrol agents, nanoformulation based BCAs provide excellent ecofriendly approaches to plant disease management. The design and delivery of nanoformulation based BCAs is a valuable step towards augmenting disease resistance in host plants. This chapter highlights the role of different BCAs in managing plant diseases, mechanisms of action against plant pathogens, different nanoformulation techniques suitable for its delivery, and possible release mechanisms. It also addresses the application of nanoformulated BCAs, their current status, drawbacks, and potential approaches to combat biotic stresses.

Keywords Biocontrol agents (BCAs) · Mode of action · Nanoformulations · Plant protectants · Phytopathogens

H. Shanmugam (✉)
Centre for Agricultural Nanotechnology, Tamil Nadu Agricultural University, Coimbatore, India
e-mail: haripriya.s@tnau.ac.in

S. Narayanasamy
Biocatalysts Laboratory, Department of Agricultural Microbiology, Tamil Nadu Agricultural University, Coimbatore, India

S. Uthandi
Department of Agricultural Microbiology, Tamil Nadu Agricultural University, Coimbatore, India

K. K. Bastas et al. (eds.), *Microbial Biocontrol: Molecular Perspective in Plant Disease Management*, Microorganisms for Sustainability 49,
https://doi.org/10.1007/978-981-99-3947-3_8

8.1 Introduction

The current scenario of changing climatic conditions and wastage of agricultural crops due to plant diseases or pathogen attack during the developmental stages and postharvest storage conditions is a severe challenge in achieving food security due to rising global populations. Approximately 40% of fresh produces or crops are thought to be lost each year due to pest or pathogen attacks. Most farming population still rely on chemical pesticides to control/manage plant disease or pathogen attacks. However, use of synthetic chemicals has several negative impacts on human health and the environment, which is a global concern. Therefore, the latest strategy is focussed towards developing environmentally sound approach to prevent plant diseases using biocontrol agents by employing naturally occurring antagonistic microbes, or biocontrol agents (BCAs).

In certain instances, pathogen invasion and infection are viable alternatives (Adivitiya and Khasa 2017; Cook 1993). An antagonistic microorganism plays a significant part in pathogen control through a wide range of mechanisms, such as synthesis of inhibitory chemicals from proliferation of other microorganisms. Rather than adopting a single antagonistic strain, application of multiple microorganisms in a single unit as a consortium results in widespread rhizosphere colonization and desirable defense response in various environments (Bashan 1998; Duffy 1995). Aside from plant pathogens, biological control agents (BCAs) have shown to be effective in suppressing insect pests and nematodes (Keswani et al. 2014). A list of biocontrol agents and their mode of action on plants for effective management of plant disease in crops are furnished in Table 8.1. However, nowadays, BCAs-based nanobioformulations have started gaining significance as plant protection agents in managing biotic stresses for sustainable agriculture. Nanoformulated BCAs have the ability to serves as "nutrient booster" and "immuno-modulator," allowing prolonged and targeted delivery of BCAs to the plants throughout its crop development cycle that helps in subduing invading plant pathogens (Sood et al. 2020). Hence, this chapter gives an overview and mechanism of BCAs as well as nanoformulations suitable for delivering BCAs to the host plants, aiding in synergistic management of plant pathogens.

8.2 Mechanisms of Action by Microbial BCAs in Suppressing Phytopathogens

A disease triangle, which consists of the pathogen, host, and environment, interacts to cause plant diseases. A BCA combines with these factors to prevent disease occurrence (Junaid et al. 2013). Effective management of plant diseases requires an understanding of the BCA's mode of action in suppressing plant diseases. The direct and indirect methods of action of BCAs can be separated into two main types, which are shown in Fig. 8.1. Having a thorough understanding of these systems and

Table 8.1 Microbial biocontrol agents and their mode of action against phytopathogens

Biocontrol agents	Mode of action	Crop	Pathogen	References
Bacteria as BCA				
Azospirillum brasilence	Siderophore production	Strawberry	Anthracanose/*Colletotrichum acutatum*	Tortora et al. (2011)
Azotobacter chroococcum	Production of antimicrobial metabolites	Rice, cotton, and tomato	*Rhizoctonia solani*	Chauhan et al. (2012)
Bacillus subtilis BY2	–	Rape seeds	*Sclerotinia sceloritium*	Hu et al. (2019)
Burkholderia cepacia	Hemolytic enzymes	Tomato	Damping off/*R. solani*	Szczech and Shoda (2004)
Pseudomonas chlororaphis	Inhibition of ascospores germination	Canola	*S. sclerotium*	Savchuk and Dilantha Fernando (2004)
Paenibacillus polymyxa BRF1	Induced systemic resistance	Soybean	Root rot/*Phialophora gregata*	Zhou et al. (2008)
Pantoea agglomerans	–	Wheat	Rot/*Rhizoctonia solani*	Barnett et al. (2006)
P. fluorescens 2–79 and 30–84	Antibiotics Phenazine	Wheat	Take all/*Gaeumannomyces graminis* var. *tritici*	Mavrodi et al. (2012)
B. subtilis QST713	Iturin A	Cucumber	Damping off/*Botrytis cinerea* and *R. solani*	Paulitz and Bélanger (2001)
Pseudomonas and *Burkholderia*	Suppression of sporangium formation	Cucumber	*Phytophthora capsici*	Khatun et al. (2018)
Pseudomonas fluorescens	Production of mVOC	Apple	Mucor rot/*Mucor piriformis*	Wallace et al. (2018)
Pseudomonas putida BP25	Induced systemic resistance by triggering PR genes	Rice	Blast/*Magnaporthe oryzae*	Ashajyothi et al. (2020)
Rhizobium japonicum	–	Soybean	Root rot/*Fusarium solani*; *Macrophomina phaseolina*	Al-Ani et al. (2012)
Bacillus thuringiensis		*Brassica campestris* L	Sclerotiniose/*Sclerotinia sclerotiorum*	Wang et al. (2020)

(continued)

Table 8.1 (continued)

Biocontrol agents	Mode of action	Crop	Pathogen	References
	Induced systemic resistance by triggering SA/JA/ET signalling			
Brevibacillus brevis	Inhibition of conidial germination and germ-tube extension	Tomato	*Fusarium oxysporum* f.sp. *lycopersici*	Chandel et al. (2010)
Bacillus subtilis GB03 and *Bacillus amyloliquefaciens* IN937a	Production of VOC	Arabidopsis	*Erwinia carotovora* subsp. *carotovora*	(Ryu et al. (2003), Ryu et al. (2004)
P. fluorescens GRP3	Siderophore production	Rice	*Rhizoctonia* sp.	Pathak et al. (2004)
B. pumilus SE34	Induced systemic resistance (SA signalling)	*Arabidopsis* Tobacco	*P. syringae* pv. *Maculicola* *Peronospora tabacina*	Ryu et al. (2003), Zhang et al. (2002)
Ochrobactrum pseudintermedium CB361-80	Induced systemic resistance	*Cucumber*	*Angular leaf spot*	Akbaba and Ozaktan (2018)
P. putida WCS 358	Production of lipopolysaccharides (LPS) and Siderophore	Beans	*Pseudomonas syringae*	Meziane et al. (2005)
Bacillus amyloliquefaciens FZB42	Antibiotics Bacillomycin, fengycin	Rhizosphere	Wilt/*Fusarium oxysporum*	Koumoutsi et al. (2004)
Fungal biocontrol agents				
Paecilomyces lilacinus	Mycoparasitism and competition	Tomato	Root-knot disease/*Meloidogyne javanica*	Akhtar and Azam (2012)
Trichoderema hamatum	Parasitism	Cabbage	*S. sclerotiorum apothecia*	Jones et al. (2014)
Trichoderma asperellum	Triggers systemic resistance	Onion	*Sclerotium cepivorum*	Rivera-Méndez et al. (2020)
Trichoderma virens	Parasitism and competition	Okra	Root-knot disease/*Meloidogyne incognita*	Mukhtar (2018)
Trichoderma harzianum	Antifungal metabolites production	Rice	Brown spot/*Biopolaris oryzae*	Khalili et al. (2012)

Trichoderma harzianum	Competition for nutrient and space	Grape vine	*Rhizoctonia solani*, *Fusarium* sp., *Sclerotinia sclerotiorum*, *Sclerotium rolfsii*	Ram et al. (2018)
Ampelomyces quisqualis, *Acrodontium crateriforme*	Antibiosis, Mycoparasitism and induced resistance	Fruit, vegetable, and ornamental crops	Powdery mildew fungi (*Erysiphe* sp., *Uncinula* sp.)	Siozios et al. (2015)
Trichoderma viridi	Mycoparasitism, competition and antibiosis		*F. solani*	Zaghloul et al. (2007)
Microdochium dimerum	Induced systemic resistance	Tomato	Gray mold *Botrytis cinerea*	Bardin et al. (2008)
Fusarium solani CEF559		Cotton	*Verticillium dahliae*	Wei and Zhang (2019)
Anthracocystis floculossa PIP1, Penicillium olsonii ML37 Sarocladium strictum C113L, and *A. floculossa F63P*	Antibiosis, Mycoparasitism	Wheat	*Fusarium* head blight (FHB)	Rojas et al. (2020)
Laetisaria arvalis *T. harzianum* and *Trichoderma viride*	Antibiosis, production of hydrolytic enzymes	Tomato, chickpea, soybean, Redgram and tomato	*Pythium ultimum*	Konappa et al. (2022)
T. koningii	Mycoparasitism, competition and antibiosis	Peas	Damping-off/*P. aphanidermatum*	Nelson et al. (1988)
T. harzianum	Antibiosis and parasitism	Groundnut	Stem rot/*Sclerotium rolfsii*	Karthikeyan et al. (2006)
T. longibrachiatum EF5	Competition antimicrobial metabolites, mVOC and hyperparasitism	Rice	*Sclerotium rolfsii* and *Macrophomina phaseolina*	Palanisamy et al. (2020)
Yeasts as biocontrol agents				
Rhodotorula rubra	Competition for niche and nutrients	Tomato	*Botrytis cinerea*	Dal Bello et al. (2008)
Aureobasidium pullulans	Siderophore and toxin production	Pear	*Penicillium expansum*	Wang et al. (2009), Robiglio et al. (2011)
		Apple		Mari et al. (2012)

(continued)

Table 8.1 (continued)

Biocontrol agents	Mode of action	Crop	Pathogen	References
			Botrytis cinerea, *Colletotrichum acutatum* and *Penicillium expansum*	
Candida oleophila	Induction of systemic resistance	Banana	*Colletotrichum musae*, *Fusarium moniliforme* and *Cephalosporium* sp.	Lassois et al. (2008)
Metschnikowia fructicola	Induction of systemic resistance	Grapefruit	*Penicillium digitatum*	Hershkovitz et al. (2013)
Candida (Pichia) guilliermondii	Induction of, secretion of enzymes, volatiles and toxin production	Chili	*Colletotrichum capsici*	Chanchaichaovivat et al. (2007)
		Tomato	*Rhizopus nigricans* *Rhizopus stolonifer*	Zhao et al. (2008), Zambrano et al. (2014)
Cryptococcus laurentii	Hydrolytic enzymes and toxin production	Strawberry	*Botrytis cinerea*	Wei et al. (2014)
Saccharomycopsis schoenii	Secretion of enzymes, volatile production	Citrus	*C. acutatum* Citrus black spot/*Guignardia citricarpa*	Lopes et al. (2015) Fialho et al. (2010)

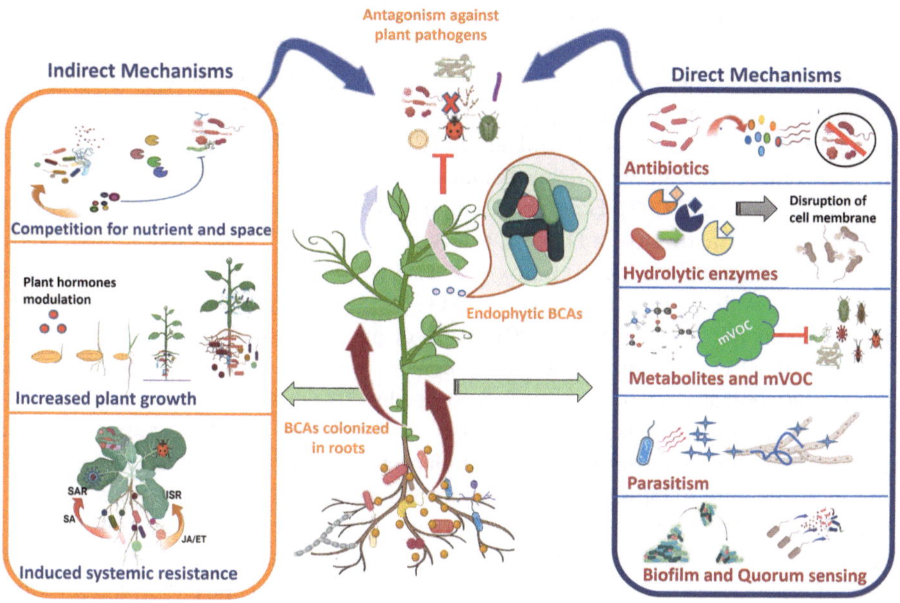

Fig. 8.1 Mechanism of action of biocontrol agents (BCAs) against plant pathogens

the surrounding environment can help BCAs to be an effective solution. This can be accomplished by improving the environment, wherein BCAs can operate by formulating microbial strains with appropriate delivery system as plant disease management tactics.

8.3 Direct Mechanisms

In direct mechanism, the BCAs involved were in physical contact with the pathogens. Different mechanism such as hyperparasitism, mycoparasitism, antibiosis, and enzyme production have been reported to be exhibited by BCAs during plant pathogen management were discussed in detail in the following sections.

8.4 Antibiosis

Microbes can directly impede the growth of other species with their secretory products like antibiotics, which are a low molecular weight compounds that can be effective even at the lower concentration. The capacity of BCAs to generate antibiotics is a crucial trait (Keswani et al. 2017; Maksimov et al. 2011). However, the function of antibiotics differ based on their biochemical make-up and nature, since they might operate as metabolic blockers or interfere with the pathways that lead to translational protein synthesis (Babbal and Khasa 2017; Keswani et al. 2014).

The use of antibiotics as biocontrol agents through parasitism or competition is essential for control and management of plant diseases(Raguchander et al. 2011). Antinematicidal, antibacterial, and antifungal activity of a BCA is demonstrated through synthesis of general or particular metabolites, as well as antibiosis activity (De La Fuente et al. 2002). Haas and Défago (2005) claim that majority of the PGPR BCAs, such as Bacillomycin D and Iturin A, are produced by *Bacillus subtilis*, while *Trichoderma virens* produces gliotoxin, *Pseudomonas fluorescence* produces phenazine, and 2,4-diactylphloroglucinol (DAPG) and *Agrobacterium radiobacter* produce agrocin 84 (Moyne et al. 2001; Ram et al. 2018; Wilhite et al. 2001). The capacity of BCA to control plant pathogens is further improved by the fact that several biocontrol strains have proven to produce a variety of antibiotics that can block one or more diseases. The *Bacillus cereus* synthesizes both antibiotics like Zwittermycin and Kanosamine (Pal and Gardener 2006). This bacterium is important in controlling damping-off in alfalfa brought on by *Phytophthora medicaginis*. According to Glandorf et al. (2001), the genetically altered *Pseudomonas putida* strain generates phenazine and DAPG improved plant disease-inhibitory properties. One of the effective method used by PGPR to avoid occurrence of soil-borne diseases in a number of crops is through use of microbial antibiotics (Handelsman and Stabb 1996). By preventing generation of zoospores, antibiotics help pathogenic *Pythium* species rupture their cell membrane (de Souza et al. 2003). It is well known that *Pseudomonas fluorescens* strains produce the antibiotics, phenazine-1-carboxylic acid to treat the take-all disease in wheat (Weller 2007).

In addition to *Pseudomonas*, *Bacillus* can also act as a potent BCA for controlling soil borne phyto-pathogens in several agricultural crops like beans, rice, maize, wheat, groundnut, chili, tomato, and citrus (Ashwini and Srividya 2014). Certain rhizobacteria secretes or exerts hydrogen cyanide (HCN), which assists the bacteria to establish a beneficial environment for plant development by killing the harmful or non-beneficial organisms (Tian et al. 2007). *P. fluorescens* combined with neem cake decreased the nematode growth by hindering the worm's development (Rizvi et al. 2012). Interestingly, BCA producing antibiotics should have its own defense mechanism against the produced antibiotics to prevent self-suicide, while delivering its toxin (Slininger et al. 2003). When a population of microorganisms in an environment is repeatedly exposed to specific dose(s) of antibiotics, and those organisms with inherent built-in natural resistance can thrive by developing antibiotic resistance. In contrast, those microbes that lack antibiotic resistant happen to perish (Fernando et al. 2005).

8.5 Mycoparasitism or Hyperparasitism

Mycoparasitism, having direct physical interaction with the host mycelium, is the most prevalent type of antagonistic relationship (Pal and Gardener 2006). In Mycoparasitism, fungal mycelium of biocontrol agent rapidly spreads over the target pathogen, followed by substantial coiling and production of hydrolytic enzymes, resulting in disintegration of cell wall membrane in target pathogen. Mycoparasitism is a four-step process. Recognizing the phytopathogenic fungus is the initial stage, followed by chemotropic development of hostile fungal mycelium. The third and fourth phases involve direct attachment and disintegration of phytopathogenic fungus cell wall, preceded by invasion of the host fungal cell. *Trichoderma* sp. employs mycoparsitism as one of the main strategies for eradicating phytopathogenic fungus (Sharma 1996). Altomare et al. (1999) reported that *Trichoderma harzianum* has a strong mycoparasitic activity against the plant pathogen, *Rhizoctonia solani.* Facultative parasites, hypovirus, predators, and obligatory bacterial pathogens are the four main categories of hyperparasites. *Cryphonectria parasitica* fungus is the root cause of chestnut blight infected by hypovirus, a hypoparasite (Milgroom and Cortesi 2004). However, several mycoparasites can attack a single fungus pathogen. For example, the fungus *Acremonium alternatum*, *Ampelomyces quisqualis*, and *Gliocladium virens* can parasitize powdery mildew pathogen in crops like cucurbits, vineyards, tomato, beans, peas, and okra (Kiss 2003). Mycoparasitism of rust fungus, such as *Puccinia* and *Uromyces*, by *Sphaerellopsis filum*, is another example in this line of work (Gordon and Pfender 2012).

8.6 Microbial Enzymes

Microbial enzymes aid in survival of microorganisms in a specific niche, apart from acting as a biocatalyst in crucial metabolic events. Microbes in the rhizosphere are well recognized for their capacity to stimulate plant development and manage phytopathogens. However, it has been well studied that various microbial biocontrol agents released several enzymes including lytic enzymes, which play an essential role in phytopathogen management. In response to phytopathogenic invasion, rhizosphere microorganisms secrete chitinases, cellulases, proteases, and glucanases. Crop plant survival and establishment are indirectly supported by these biocontrol enzymes, which employ several mechanisms for eradication of phytopathogens. The ability of such enzymes to suppress phytopathogens makes them an excellent choice as biocontrol agents.

Chitinase enzyme produced by *Serratia marcescens* combats *Sclerotium rolfsii* that causes southern blight (Ordentlich et al. 1988). The enzyme, β-1,3-glucanase, is necessary for *Lysobacter enzymogenes* C3 to function as a BCA. These enzymes force the phytopathogen's complex cell wall components to degrade in order to get nourishment and ultimately inhibit the activity of the pathogens. Several enzymes produced by *Paenibacillus* species, including chitinase, protease, cellulase, and amylase, are crucial in suppression of pathogenic fungi (Mishra et al. 2020). *Streptomyces vinaceusdrappus* S5 MW2 strain showed antifungal efficacy against *Rhizoctonia solani*(Yandigeri et al. 2015). *Xanthomonas campestris*, a fungus that causes black-rot disease has been destroyed by the extracellular chitinase derived from *P. aeruginosa* (Mishra and Arora 2012). It has been reported that *Pseudomonas* strains isolated from greengram and chickpea rhizospheres produce hydrolytic chitinases and cellulases that act in opposition to *R. solani* and *P. aphanidermatum.*

Likewise, the enzyme cellulases synthesized by *Trichoderma* spp. showed effective antagonistic behavior against the pathogen, *clerotium rolfsii* and *Fusarium cicero* of crop plants (Anand and Reddy 2009). Overexpression of the gene for 1,4- endoglucanase produced by *T. longibrachiatum* is essential to control/management of *P. ultimatum* in cucumbers (Migheli et al. 1998). In a study, protease-mediated activity was demonstrated by an antagonistic strain of *Bacillus amyloliquefaciens* against several pathogen including *Macrophomina phaseolina*, *F. oxysporum*, and *Fusarium semitectum* (Majumdar 2017). Similarly, Essghaier et al. (2009) reported salt-tolerant protease produced by *B. pumilus* M3–16 strain that served as BCA against the phytopathogenic fungus, *B. cinerea.* However, mass multiplication of these enzymes for industrial purposes might also be beneficial for producing high-quality biocontrol products against phytopathogens.

8.7 Microbial Volatile Organic Compounds (mVOCs) as BCA

Volatile organic compounds are low molecular weight, small diffusible compounds produced by different microorganisms as a result of their metabolic process. Microbial volatile organic compounds (mVOCs) have been implicated in having an extensive network of interactions between inter- and intra-kingdom communications. Such interactions play a variety of ecological roles, from providing beneficial cooperation (such as mutualism, symbiosis, and host resistance induction) to hostile relationships, when one of the interacting species engages in microbicidal activity (Kanchiswamy et al. 2015; Maffei et al. 2011)

Recently, various authors are reporting about the VOCs and their potential application as plant protection agents. For example, the VOCs released by *T. viride* significantly enhances biomass production and chlorophyll content of *Arabidopsis thaliana* seedlings as well as by providing protection from fungal pathogen attacks (Kottb et al. 2015; Hung et al. 2013). Similarly, the volatiles compounds, 2,3-butanediol and acetoin produced by *Bacillus amyloliquefaciens* and *B. subtilis respectively* elicited systemic resistance responses against the pathogen, *Erwinia carotovora*, thereby improved the plant growth (Kloepper et al. 2004; Ryu et al. 2003).

8.8 Indirect Mechanism

8.8.1 Induction of Systemic Resistance

Suppression of plant diseases through stimulation of plant defense systems using microbial antagonistics is one of the most common method (Silva et al. 2004). However, the activated resistance may be systemic or local, depending on the origin, nature, and strength of the stimulus. Salicylic acid and *Non-expressor of Pathogenesis-Related Gene 1*, or *NPR1*, are the principal initiators of systemic acquired resistance (SAR). A long-lasting defense against encroaching pathogen infections is provided by SAR, which is brought on by the local infection and persists systemically throughout the plant tissues. Another resistance mechanism controlled by ethylene and jasmonic acid is induced systemic Resistance (ISR). The most studied rhizobacteria that induces ISR are *Pseudomonas* and *Bacillus* spp. (Kloepper et al. 2004; Van Wees et al. 2008). Regardless of the signaling route implicated in this process, De Vleesschauwer and Höfte (2009) coined the name ISR to describe induced systemic resistance fostered by non-pathogenic rhizobacteria or plant-growth promoting rhizobacteria (PGPR). Additionally, salicylic acid-dependent induced resistance brought on by a localized infection is referred to as SAR. Plants are known to develop resistance to a variety of metabolites. For instance, enzyme-active proteins like xylanases, cellulases, and endochitinase can stimulate production of defense proteins in plants. The hydrophobin-like protein SM1 produced by *T. virens* can trigger production of phytoalexins, which serve as a weapon against pathogenic attacks (Mukherjee et al. 2013). Another class of proteins that can make plants resistant is the avirulence (Avr) gene products. They endow plants with unique R (resistance) genes that activate the hypersensitive response and other defense-related responses for a particular Avr gene. A few examples of microbial substances that can cause plants to mount a defense response include gram-negative bacteria produced lipopolysaccharide (LPS) and flagellin, cold-shock proteins of several bacteria, chitin and ergosterol from fungi, as well as transglutaminase and alpha-glucanase from oomycetes (De Vleesschauwer and Höfte 2009; Van Loon and Bakker 2007). As biocontrol agents colonize the roots of plants, resistance mechanisms are imparted. Deposition of callose, cell wall thickening through lignification, production of lytic enzymes such as chitinases and glucanases, synthesis of pathogenesis-related (PR) proteins and peroxidases, as well as production of low molecular weight antimicrobial substances like phytoalexins are typical results of inducing the SAR defense response (Pieterse et al. 2014). According to Sharma et al. (2019), salt-tolerant rhizobacteria called *Klebsiella* sp. produced systemic resistance in peanuts against a variety of soil-borne fungal diseases. According to Verhagen et al. (2010), *P. fluorescens* CHA0 and *P. aeruginosa* 7NSK2 caused grapevines to develop a resistance to *Botrytis cinerea*. By inducing the expression of SA and ET-responsive genes, *Bacillus velezensis* established systemic resistance to crown gall disease (Chen et al. 2020). Thus, the non-pathogenic microorganisms-induced systemic resistance is also an efficient way of approach for controlling numerous phytopathogens under field conditions.

8.9 Competition

On soil and plant surfaces, microorganisms have nutrient-restricted habitats, which cause fierce competition among resident microbial communities for limited resources and habitat. Competition is regarded as an indirect interaction among the disease-causing agents and ecological exclusion that drives the pathogens out of the environment (Lorito et al. 1998). Pathogens and beneficial bacteria must compete for nutritional resources in order to lower the frequency and severity of disease incidence. In contrast to pathogens that directly germinate and infect on plant surfaces through infection pegs and appressoria, *Fusarium* and *Pythium* species spread by mycelial contact are more vulnerable to competition (Prajapati et al. 2020).

In general, plant-associated microorganisms are assumed to defend the plants by rapidly colonizing, thereby depleting limitedly available substrates, leaving none for pathogens to thrive on. For instance, efficient nutrient catabolism in the spermosphere by *Enterobacter cloacae* has been reported to suppress *Pythium ultimum* (Kageyama and Nelson 2003; van Dijk and Nelson 2000). Simultaneously, these microorganisms produce metabolites and bioactive compounds that can inhibit pathogens. Micronutrients like iron and manganese are scarce, so BCA exhibit special transportation mechanism through formation of siderophores to solubilize and chelate these micronutrients (Kloepper et al. 2004). Kloepper et al. (2004) first shown the significance of siderophore formation as a biological regulatory mechanism in *Erwinia carotovora* through different *Pseudomonas fluorescence* strains that promote plant development. The enhanced capacity of commensal bacteria to absorb iron is anticipated to play a critical role in efficient removal of phytopathogens from potential infection sites, owing to its ability to effectively colonize plant roots.

8.10 Conventional vs Nano Formulated BCA

Plant protection synthetic chemicals have been widely used to control pathogen incidence in plants to increase agricultural crop production. However, during the past decade, experts have been more concerned about the negative impacts of excessive use of such chemicals due to reported environmental toxicity and long-term residual toxicity in the soil as well as in the produces. This fueled the need to find non-toxic, environmentally friendly alternatives to meet the intended aim of increasing agricultural production, while avoiding the associated detrimental effects in plant disease protection. To ensure safety of agriculture produces, biocontrol agents are being replaced with synthetic chemicals (Dhir 2017). Biocontrol agents are mainly composed of live or latent cells of antagonistic organisms formulated in granular carrier materials (talc, lignite, vermiculite, night soil, FYM, etc.) or liquid (emulsion, suspensions) and lyophilized formulations, so as to deliver the BCAs to crop plants for suppressing pathogenic infection and enhance plant growth (Itelima et al. 2018). However, the mentioned conventional formulations have some key limitations, including a shorter storage life, poor on-field stability, poor efficacy under changeable environmental conditions (temperature, radiation, pH sensitive), inability to be used for more extended periods of time, a decrease in microbial load, susceptibility to desiccation, and, most importantly, a high dose was required for broad area coverage (Mishra et al. 2017). BCA-based nanoformulations could circumvent all of these limitations and can be used to increase crop productivity (El-Ghamry et al. 2018). Owing to the unique high surface area-to-volume ratio, higher adsorption capacity, and regulated release kinetics, nanoformulations could be efficient in delivering BCAs for plant disease suppression and plant growth promotion. Advantages of BCA-based nanoformulation over conventional formulations are depicted in Fig. 8.2. Nano-based delivery systems for delivery of biocontrol agents mostly range in the size of 100–500 nm with polymer-based nanomaterials (Shanmugam 2022).

Nanoformulation of BCA benefits farmers in an extensive way, as it is a green and renewable approach that can enrich the soil's beneficial microbial population and increase soil fertility by enhancing the host plant disease resistance. Additionally, it extends the shelf life of BCAs, lessens microbial load loss (CFU), and facilitates dispensing by providing a precise delivery system with sustained release at the intended location. Gatahi et al. (2015) looked into the effectiveness of nanoformulated BCA against the tomato plant wilt disease caused by *Ralstonia solanacearum.* Gouda et al. (2018) have noted the protective function of nanoformulation-based BCAs such as *Pseudomonas fluorescens*, *Bacillus subtilis*, *Paenibacilluselgii*, and *Pseudomonas putida* against a variety of harmful diseases in the rhizosphere of leguminous plants. Using *Trichoderma* sp. and *Pseudomonas* sp. in nanoclay-coat formulation, fungal-nematode disease in crops was successfully suppressed (Mukhopadhyay and De 2014).

8.11 BCA-Based Nanoformulation for Crop Plants

Polysaccharides, lipids and protein-based polymers of natural, and synthetic sources can be used as nanocarriers to develop nanoformulations to deliver BCAs. Chitosan, zeolite, and synthetic polymers are commonly used in developing nanoformulations that aid in sustained delivery of target BCA to plants. Various nanoformulation techniques like nanoemulsion nanospheres, and nanogels, may be suitable for delivering BCAs. All of these techniques pave way for effective and eco-friendly delivery of BCAs to plants in order to combat epidemic plant pathogens. The nanotechniques suitable for delivery of BCAs for managing plant disease conditions are discussed.

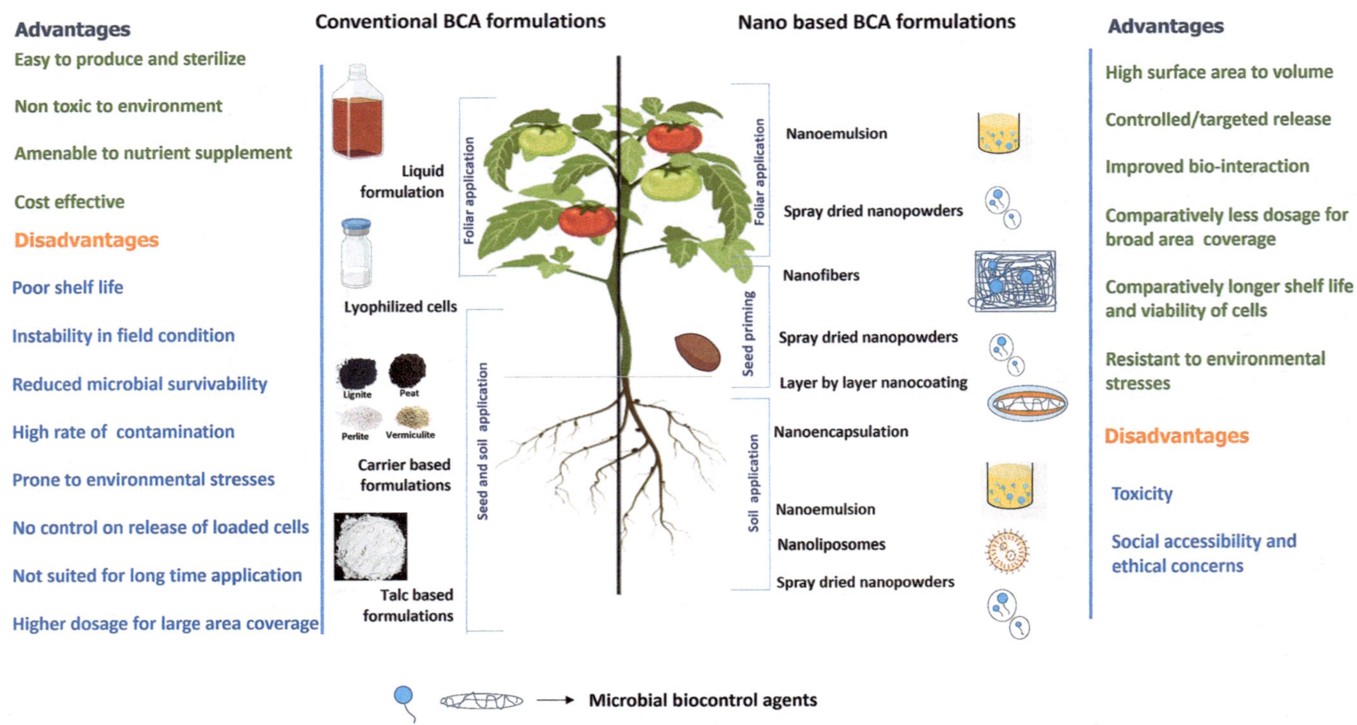

Fig. 8.2 Conventional formulation of BCAs over nano-based formulations suitable for managing plant diseases

8.12 Nanoemulsion

Nanoemulsions are liquid-in-liquid dispersions that are highly stable and have droplet sizes ranged from 10 to 100 nm. The reduced size confers advantageous characteristics such as formulation stability, optical transparency with varied rheological properties for ease in field application. Numerous industries, including biomedicine, food, cosmetics, and biopharmaceuticals, use nanoemulsion technique for product formulation. The system serves as a model for comprehending tiny spheres. To get stable nanoemulsions, high pressure homogenization, ultrasonication, phase inversion temperature, and emulsion inversion point are used in addition to recently reported method like bubble bursting method (Gupta et al. 2016; Shanmugam 2022). The main components of a nanoemulsion are oil, water, and an emulsifier.

An emulsifier lowers the surface energy per unit area between the water and oil phases of an emulsion, which is crucial for formation of smaller droplets. Due to steric hindrance and repellent electrostatic interactions, emulsifiers aid in stabilizing the formed nanoemulsions (Mason et al. 2006). Although proteins and lipids have also shown to be influential in developing nanoemulsions, surfactants are still frequently used as emulsifiers. There are two distinct ways to make nanoemulsions: high-energy and low-energy methods. High-energy techniques that produce microscopic droplets with high-pressure homogenization (HPH) and ultrasonication (Fryd and Mason 2012) consume more energy ($\sim 10^8$–10^{10} W kg^{-1}). However, low-energy techniques take advantage of the certain system to form tiny droplets without using much energy ($\sim 10^3$ W kg^{-1}). Phase inversion temperature (PIT) and emulsion inversion point (EIP) are low energy approaches for developing nanoemulsions (Forgiarini et al. 2000, 2001). Lately, new methods for developing nanoemulsions, such as evaporative ripening and bubble bursting at the oil/water interface, have also been reported. A recent formulation of silica nanoparticles encapsulated *Metarhizium brunneum* controls *Spodoptera littoralis* infection in *Ricinus communis* (Yaakov et al. 2018).

8.13 Nanoencapsulation

Encapsulation is the process of enclosing an active ingredient in a liquid or solid phase inside a matrix, typically with a polymer. The coated ingredient (BCAs or antibiotics) can be preserved within an enclosed wall, increasing its stability in the applied environment. Encapsulation can improve the stability of organisms by regulating their release at the target site (Nedovic et al. 2011; Zuidam and Shimoni 2010). This technique has a broad range of applications in commercial industries like cosmetics, biomedical applications, medicines, food, and agriculture (Aphibanthammakit and Kasemwong 2021). Nanoencapsulation produces particles of 100 nm size, and the encapsulated particles can be categorized into nanocapsules

(<0.2 μm), microcapsules (0.2–5000 μm), and macrocapsules (>5000 μm). One of the key advantages of using nanoencapsulation technique is the homogeneity, which leads to improved encapsulation efficiency with desirable physio-chemical properties (Jafari 2017). Saberi-Riseh and Pour (2019) reported that nanoencapsulated *B. subtilis* Vru1 in sodium alginate (NaAlg), starch and bentonite showed higher inhibition of *R. solani* in beans (*Phaseolusvulgaris*). Similarly, nanoencapsulation of *B. velezensis* in sodium alginate-gelatin microcapsules augmented with SiO_2 nanoparticles and carbon nanotubes showed enhanced bioactivity against *Pistachio gummosis* with a cell viability of 10^7 CFU mL^{-1} maintained even after a year of storage (Moradi Pour et al. 2022). Krell et al. (2018)encapsulated *Metarhizium brunneum* BIPESCO5 mycelium in calcium alginate/starch beads to prevent fungus drying, thereby enabling its growth on various types of soils, and encouraging endophytism in tomato plants. Improved seed germination, early plant growth, as biocontrol agent, and biosensor fabrication for diagnostics are few important uses of carbon nanotubes (CNTs) in agriculture (Khodakovskaya et al. 2012; Zaytseva and Neumann 2016). Alginate-gelatin and nanoparticle (MWNCT, SiO_2) system has been synthesized for encapsulation of antagonistic bacteria such as *Pesudomonas fluorescence* VUPF5 and *Bacillus subtilis* VRU1 that significantly enhanced plant growth and resistance against phytopathogens (Pour et al. 2019).

8.14 Nanocoats

Most BCAs are living entities and must be handled by employing non-destructive formulation techniques to retain their capacity to multiply. In this perspective, researchers suggest nanoformulation techniques with surface modification might be beneficial in rendering healthy living cells by improving their nutritional pathways, protecting them from predators, and allowing them to combat both biotic and abiotic stresses (Fakhrullin and Lvov 2012). Self-assembly of proteins, or polyelectrolytes, enables layer-by-layer (LbL) encapsulation of BCAs. The LbL encapsulation technique uses electrostatically bonded polycations and polyanions that are sequentially deposited, first for forming planar films, then for encapsulating colloids, including microbial cells (Donath et al. 2002). Generally, in a swollen condition, the thickness of a bilayer comprising of a polycation/polyanion is about 4–5 nm size, while the shell thicknesses are 30–100 nm (Bédard et al. 2010). The capacity to form nanocapsules of any composition, comprises of nanosized layers of proteins, polymers, and nanoparticles in a preset order, which controls the capsule characteristics like pH and temperature sensitivity, structural stability, and permeability. Thus, the intrinsic properties of such surface functionalized BCAs can be improved upon or altered. Franz et al. (2010) showed that *A. vinosum* encapsulated in a variety of polyelectrolyte layers developed a physical barrier between the cells and the crop environment using electrostatic LbL nano self-assembly. By adhering to the surfaces of bacteria, *Pseudomonas stutzeri* NRCB010 nanocoated with modified N-hydroxysuccinimide (NHS)-poly (−glutamic acid) and calcium ions form ionic and covalent bonds with them. *P. stutzeri* 's survival rate was increased by 2.38 times, therefore plant growth was also improved.

8.15 Spray-Dried Nanopowder

To form solid dry powder formulation from a liquid feed, spray drying is a quick, continuous, reliable, economical, and adaptable process. Generally speaking, it is an efficient one-step procedure to turn a variety of liquid formulations (such as aqueous and organic solutions, emulsions, and suspensions) into dry powders with a particle size range of 1–5 μm. Using a nano spray dryer that uses a piezoelectric component in the spray head to lower the droplet size of the feed liquids, it is possible to nanoencapsulate temperature tolerant microbes. In order to improve the formulation and delivery of biocontrol agents comprising of solid colloidal particles in the submicron size range, nano spray drying technology has evolved due to the quick development in nanoencapsulation techniques. Nano spray dryer makes it easier to enclose BCA in polymeric wall materials, thereby imparting environmental protection (against oxidation, light, and temperature), stability, handling, and storage (Arpagaus 2018). The major setback of this approach is that it employs high temperatures, which decreases the viability and durability of bacterial cells(Guerin et al. 2017)and is not highly suitable for small-scale applications (Jantzen et al. 2013). Encapsulation of *S. fulvissimus* Uts22 in chitosan and gellan microcapsules synthesized using spray drying technique showed an increase in bacterial survivability and enhanced plant growth by controlling *G. graminis* in wheat (Saberi Riseh and Moradipour 2021).

8.16 Nanofibers

Nanofibers are linear polymeric threads with excellent porosity, large surface area to volume ratio, higher loading efficiency, and controllable topology, with chemical and thermal endurance. Among the several techniques used for synthesizing nanofibers, electrospinning is the most preferred method and relies on an electric driving force, allowing continuous synthesis of nanofibers. Additionally, it enables production of nanofibers from a wide range of organic, inorganic, and hybrid polymers with various physical, chemical, and mechanical characteristics (Leena et al. 2021). Nanoencapsulating heat-sensitive microbial cells and enzymes with suitable viscoelastic polymers for efficient delivery are made possible by the electrospinning method by customizing peripheral working conditions. It is now being exploited for seed treatment with microbial agents to increase seedling establishment and suppress various disease causing soil-borne pathogenic microorganisms (Castañeda et al. 2014; Farias et al. 2019; Krishnamoorthy et al. 2016). Encapsulation of two PGPR strains, *Pantoea agglomerans* and *Burkholderia caribensis* in polyvinyl alcohol-based electrospun nanofibers used as seed priming in soyabean seeds, showed successful colonization in both seeds and plants. Further, reduction in cell viability in nanofiber-immobilized PGPR was minimized up to 30 days of storage (De Gregorio et al. 2017). Campaña and Arias (2020) reported that seed priming of *Phaseolus vulgaris* with polyethylene oxidepor (PEO)-based nanofiber as a delivery system for mycorrhizal fungi showed improved colonization and lowered the inoculation dose compared to conventional formulations. Thus, nanofibers might be a promising alternative for seed priming of BCAs for ease in field application and targeted delivery of BCAs to suppress plant pathogens.

8.17 Nanoliposomes

Liposomes are vesicles with aqueous core consisting of polar lipid bilayers, primarily phospholipids. To form a liposomal structure, scientists depend on the self-assembly mechanism of amphiphilic lipids in aqueous solutions. A reservoir for encapsulation of hydrophilic substances is produced in the inner aqueous phase by concentrated stacking of amphipathic lipids (Briuglia et al. 2015). Natural and reasonably priced ingredients, such as crude lecithin, can form nanoliposomes. They can encapsulate and release both hydrophilic and hydrophobic metabolites formed or synthesized by BCAs separately or simultaneously (Assis et al. 2014; Mosquera et al. 2014). Nanoliposomes have a mean diameter that varies from 50 to 150 nm and are thermodynamically unstable. For best outcomes, it is also important to take into account the nature of the enclosed materials, ionic strength, composition, pH, temperature, oxygen, and exposure to light while forming nanoliposomes (Liu et al. 2015). Atienza et al. (2021) reported the formation of a nano-biofungicide by encapsulating crude extracts from plant growth-promoting bacteria (PGPB) into nanoliposomes that showed inhibitory action against *Fusarium oxysporum* f.sp. *cubense*.

8.18 Application methods of Nanoformulated BCAs

Nanoformulation techniques provide a promising advancement in active delivery of BCAs to the host plant for successful colonization and management of invading plant pathogens (Vinothini et al. 2020). The majority of nanoformulation systems suitable for plant systems include nanoliposomes, nanofibers, nanoemulsion, nanoencapsulation, multilayered nanocoatings, and freeze as well as spray-dried nanopowders (Shanmugam 2022). Primarily, three modes of application of microbial formulations reported for delivery of BCAs to the host plants, includes spray or foliar application (Duhan et al. 2017), seed priming (De Gregorio et al. 2017), and soil application (Hafeez et al. 2015). The crop characteristics, cultivation method, cultivable environment, and characteristics of the BCAs determine the desirable delivery method. Colonization of delivered BCAs under nutrient-deprived conditions is required to improve the plants' resistance to pathogen attack. BCAs restrict the amount of available nutrients at the wound site, making it difficult for pathogens to germinate, develop, and invade their hosts (Li et al. 2013; Zhao et al. 2011). Generally, freeze and spray-dried nanopowders, as well as core-shell nanoparticles-based BCA formulations can be used as soil applications. Pour et al. (2022) reported that encapsulation of *Bacillus velezensis* in sodium alginate–gelatin-based nanoformulations as soil application synergistically controls *Phytophthora drechsleri* in Pistachio.

For seed priming and coating, electrospun nanofiber, multilayered nanostructures, and nanofilms encapsulated with BCAs may provide a better environment for maintaining viable cells for longer period. For example, seeds primed with *P. agglomerans* ISIB55 immobilized nanofibers exhibited a significant increase in plant growth and improved survivability of bacterial cells on seeds for longer storage (De Gregorio et al. 2017). Campaña and Arias (2020) studied the effect of polyethylene oxide (PEO) nanofibers as delivery system for AMF (Arbuscular Mycorrhizal Fungi) application through seed coating in common bean (*Phaseolus vulgaris*). In addition, BCAs-loaded nanoemulsions, spray-dried, and lyophilized nanopowder can also be used as foliar application (Shanmugam et al. 2017; Shanmugam et al. 2022) to control plant pathogens.

8.19 Biosafety of Nanofomulated BCAs

Any novel crop protection methods need to be assessed for potential risks and ethical considerations before being released for commercial application in the field condition. A range of products have been produced due to advances in nanotechnology. In plant disease management, improper usage of plant protection agents-based products can have detrimental effects on the environment and humans as well. Analysis and characterization of environmental persistence, exposure, ecotoxicity, and absorption by biota due to nanomaterials in aquatic and terrestrial environments should be understood for specific formulation types with recommended doses, frequency of application, and microbial load (Deshpande 2019). The proposed changes to test research methodology, and suggestions might help to build regulatory framework for nanoformulated BCAs. Regulatory authorities in the United States (Environmental Protection Agency (EPA)) and Europe (EC 1107/2009) have established rules and regulations for nano-based products for application with environmental safety (Shanmugam 2022; Shanmugam et al. 2017; Shanmugam et al. 2022). The physio-chemical and biological features of risk assessment standards for nano-based products (Singh et al. 2018) are being framed to suit the specific country. To employ BCA-encapsulated nano-vehicles effectively in the near future, its release kinetics and mechanism of absorption must be understood at the molecular level. Globally, researchers are making great strides in commercializing nanoformulated BCAs for controlling pathogens and insects, but little is known about how

BCA-loaded nanoformulations interact with the host plant and invading pathogens with regard to colonization, distribution, antagonism, and metabolism.

8.20 Conclusion and Future Perspectives

Biocontrol agents (BCAs) are getting prominence as a better approach in managing plant diseases. They have several mechanisms of action and target hyperparasitism, which cause synthesis of lytic enzymes that break down plant pathogens' cell membranes. Plants and biocontrol agents engage in mutualistic and commensal interactions due to their contact. BCAs-loaded nanoformulations are found beneficial in combating the target pathogens due to their minuscule size, high surface area, enhanced efficacy, dispersibility, adaptability, and the capability to transform global agricultural system in terms of food safety and security. However, nanobioformulations of BCAs remain less explored. Hence, systematic studies need to be conducted to unravel the interaction of nano delivery systems suitable for host-pathogen interaction and predict their ultimate fate in the agricultural ecosystem. To manage the delivery and release of BCAs at the targeted site, the mutual interaction of nanocarriers and BCAs must be understood. Further, developed nanoformulated BCAs must be technologically advanced and cost-effective for the farmers or end users. Therefore, to ensure their effective use for the benefit of agricultural sustainability, numerous aspects of nano-BCAs, including their current state, prospects, limitations and regulatory network, must be understood before product commercialization for sustainable agricultural production.

References

Adivitiya B, Khasa Y (2017) Microbes as biocontrol agents. In: Probiotics and plant health, pp 507–552. https://doi.org/10.1007/978-981-10-3473-2_24

Akbaba M, Ozaktan H (2018) Biocontrol of angular leaf spot disease and colonization of cucumber (*Cucumis sativus* L.) by endophytic bacteria. Egypt J Biol Pest Control 28(1):14. https://doi.org/10.1186/s41938-017-0020-1

Akhtar MS, Azam T (2012) Effect of *Paecilomyces lilacinus* on the growth of tomato, population of root-knot nematode (*Meloidogyne incognita*) and their histological interactions. J Biotechnol Biomat 2:332. https://doi.org/10.4172/2155-952X.S1.020

Al-Ani RA, Adhab MA, Mahdi MH, Abood HM (2012) *Rhizobium japonicum* as a biocontrol agent of soybean root rot disease caused by *fusarium solani* and *Macrophomina phaseolina*. Plant Protect Sci 48(4):149–155

Altomare C, Norvell WA, Bjorkman T, Harman GE (1999) Solubilization of phosphates and micronutrients by the plant-growth-promoting and biocontrol fungus *Trichoderma harzianum* rifai 1295-22. Appl Environ Microbiol 65(7):2926–2933. https://doi.org/10.1128/aem.65.7.2926-2933.1999

Anand S, Reddy J (2009) Biocontrol potential of *Trichoderma* Sp. against plant pathogen. Int J Agric Sci 1:30. https://doi.org/10.9735/0975-3710.1.2.30-39

Aphibanthammakit C, Kasemwong K (2021) Nanoencapsulation in agricultural applications. In: Lau WJ, Faungnawakij K, Piyachomkwan K, Ruktanonchai UR (eds). Elsevier, Handbook of nanotechnology applications, pp 359–382. https://doi.org/10.1016/B978-0-12-821506-7.00015-6

Arpagaus C (2018) A short review on nano spray drying of pharmaceuticals. J Nanomed Nanosci Res: 1–5

Ashajyothi M, Kumar A, Sheoran N, Ganesan P, Gogoi R, Subbaiyan GK, Bhattacharya R (2020) Black pepper (*Piper nigrum* L.) associated endophytic *Pseudomonas putida* BP25 alters root phenotype and induces defense in rice (*Oryza sativa* L.) against blast disease incited by *Magnaporthe oryzae*. Biol Control 143:104181. https://doi.org/10.1016/j.biocontrol.2019.104181

Ashwini N, Srividya S (2014) Potentiality of *Bacillus subtilis* as biocontrol agent for management of anthracnose disease of chilli caused by *Colletotrichum gloeosporioides* OGC1. 3 Biotech 4(2):127–136. https://doi.org/10.1007/s13205-013-0134-4

Assis LM, Machado AR, Motta AS, Costa JAV, Soares LAS (2014) Development and characterization of nanovesicles containing phenolic compounds of microalgae *Spirulina* strain LEB-18 and Chlorella pyrenoidosa. Adv Mat Phys Chem 4:6–12

Atienza MTJA, Magpantay MDA, Santos KLT, Mora NB, Balaraman RP, Gemeinhardt ME, Dela Cueva FM, Paterno ES, Fernando LM, Kohli P (2021) Encapsulation of plant growth-promoting bacterial crude extract in nanoliposome and its antifungal property against *Fusarium oxysporum*. ACS Agric Sci Tec 1(6):691–701. https://doi.org/10.1021/acsagscitech.1c00188

Babbal A, Khasa YP (2017) Microbes as biocontrol agents. In: Kumar V, Kumar M, Sharma S, Prasad R (eds) Probiotics and plant health. Springer, Singapore, pp 507–552. https://doi.org/10.1007/978-981-10-3473-2_24

Bardin M, Fargues J, Nicot P (2008) Compatibility between biopesticides used to control grey mould, powdery mildew and whitefly on tomato. Biol Control 46:476–483. https://doi.org/10.1016/j.biocontrol.2008.05.012

Barnett SJ, Roget DK, Ryder MH (2006) Suppression of *Rhizoctonia solani* AG-8 induced disease on wheat by the interaction between *Pantoea*, *Exiguobacterium*, and *Microbacteria*. Soil Res 44(4):331–342. https://doi.org/10.1071/SR05113

Bashan Y (1998) Inoculants of plant growth-promoting bacteria for use in agriculture. Biotechnol Adv 16(4):729–770. https://doi.org/10.1016/S0734-9750(98)00003-2

Bédard MF, De Geest BG, Skirtach AG, Möhwald H, Sukhorukov GB (2010) Polymeric microcapsules with light responsive properties for encapsulation and release. Adv Colloid Interf Sci 158(1–2):2–14. https://doi.org/10.1016/j.cis.2009.07.007

Briuglia M-L, Rotella C, McFarlane A, Lamprou DA (2015) Influence of cholesterol on liposome stability and on in vitro drug release. Drug Deliv Transl Res 5(3):231–242. https://doi.org/10.1007/s13346-015-0220-8

Campaña JM, Arias M (2020) Nanofibers as a delivery system for Arbuscular Mycorrhizal fungi. ACS App Polym Mat 2(11):5033–5038. https://doi.org/10.1021/acsapm.0c00874

Castañeda LMF, Genro C, Roggia I, Bender SS, Bender RJ, Pereira CN (2014) Innovative Rice seed coating (*Oryza Sativa*) with polymer nanofibres and microparticles using the electrospinning method. J Res Updates Polym Sci 3:33–39

Chanchaichaovivat A, Ruenwongsa P, Panijpan B (2007) Screening and identification of yeast strains from fruits and vegetables: potential for biological control of postharvest chilli anthracnose (*Colletotrichum capsici*). Biol Control 42:326–335

Chandel S, Allan EJ, Woodward S (2010) Biological control of *Fusarium oxysporum* f.sp. *lycopersici* on tomato by *Brevibacillus brevis*. J Phytopathol 158 (7-8):470-478:470. https://doi.org/10.1111/j.1439-0434.2009.01635.x

Chauhan S, Wadhwa K, Vasudeva M, Narula N (2012) Potential of *Azotobacter* spp. as biocontrol agents against *Rhizoctonia solani* and *Fusarium oxysporum* in cotton (*Gossypium hirsutum*), guar (*Cyamopsis tetragonoloba*) and tomato (*Lycopersicum esculentum*). Arch of Agron Soil Sci 58:1365–1385

Chen L, Wang X, Ma Q, Bian L, Liu X, Xu Y, Zhang H, Shao J, Liu Y (2020) *Bacillus velezensis* CLA178-induced systemic resistance of Rosa multiflora against crown gall disease. Front Microbiol 11

Cook RJ (1993) Making greater use of introduced microorganisms for biological control of plant pathogens. Annu Rev Phytopathol 31:53–80. https://doi.org/10.1146/annurev.py.31.090193.000413

Dal Bello G, Mónaco C, Rollan MC, Lampugnani G, Arteta N, Abramoff C, Ronco L, Stocco M (2008) Biocontrol of postharvest grey mould on tomato by yeasts. J Phytopathol 156(5):257–263. https://doi.org/10.1111/j.1439-0434.2007.01351.x

De Gregorio PR, Michavila G, Ricciardi Muller L, de Souza BC, Pomares MF, Saccol de Sá EL, Pereira C, Vincent PA (2017) Beneficial rhizobacteria immobilized in nanofibers for potential application as soybean seed bioinoculants. PLoS One 12(5):e0176930. https://doi.org/10.1371/journal.pone.0176930

De La Fuente L, Quagliotto L, Bajsa N, Fabiano E, Altier N, Arias A (2002) Inoculation with *Pseudomonas fluorescens* biocontrol strains does not affect the symbiosis between rhizobia and forage legumes. Soil Biol Biochem 34(4):545–548. https://doi.org/10.1016/S0038-0717(01)00194-8

de Souza JT, Arnould C, Deulvot C, Lemanceau P, Gianinazzi-Pearson V, Raaijmakers JM (2003) Effect of 2, 4-diacetylphloroglucinol on Pythium: cellular responses and variation in sensitivity among propagules and species. Phytopathology 93(8):966–975

De Vleesschauwer D, Höfte M (2009) Rhizobacteria-induced systemic resistance. Adv Bot Res 51:223–281

Deshpande MV (2019) Nanobiopesticide perspectives for protection and nutrition of plants. In: Koul O (ed) Nano-biopesticides today and future perspectives. Academic Press, pp 47–68. https://doi.org/10.1016/B978-0-12-815829-6.00003-6

Dhir B (2017) Biofertilizers and biopesticides: eco-friendly biological agents. In: Advances in environmental biotechnology. Springer, pp 167–188

Donath E, Moya S, Neu B, Sukhorukov GB, Georgieva R, Voigt A, Bäumler H, Kiesewetter H, Möhwald H (2002) Hollow polymer shells from biological templates: fabrication and potential applications. Chem Europ J 8(23):5481–5485

Duffy B (1995) Use of *Gaeumannomyces graminis* var. *graminis*alone and in combination with fluorescent *Pseudomonas* spp. to suppress take-all of wheat. Plant Dis 79:907. https://doi.org/10.1094/PD-79-0907

Duhan JS, Kumar R, Kumar N, Kaur P, Nehra K, Duhan S (2017) Nanotechnology: the new perspective in precision agriculture. Biotechnol Rep (Amst) 15:11–23. https://doi.org/10.1016/j.btre.2017.03.002

El-Ghamry A, Mosa AA, Alshaal T, El-Ramady H (2018) Nanofertilizers vs. biofertilizers: new insights. Environ Biodiv Soil Secur 2(2018):51–72

Essghaier B, Bejji M, Jijakli H, Boudabous A, Sadfi-Zouaoui N (2009) High salt-tolerant protease from a potential biocontrol agent*Bacillus pumilus* M3-16. Ann Microbiol 59(3):553–558

Fakhrullin RF, Lvov YM (2012) "Face-lifting" and "make-up" for microorganisms: layer-by-layer polyelectrolyte nanocoating. ACS Nano 6(6):4557–4564

Farias B, Pirzada T, Mathew R, Sit TL, Opperman C, Khan SA (2019) Electrospun polymer nanofbers as seed coatings for crop protection. ACS Sustain Chem Eng 7:19848–19856

Fernando WGD, Nakkeeran S, Zhang Y (2005) PGPR: biocontrol and biofertilization. Springer Netherlands, Berlin, pp 67–109

Fialho MB, Toffano L, Pedroso MP, Augusto F, Pascholati SF (2010) Volatile organic compounds produced by *Saccharomyces cerevisiae* inhibit the in vitro development of *Guignardia citricarpa*, the causal agent of citrus black spot. W J Microbiol Biotechnol 26:925–932

Forgiarini A, Esquena J, González C, Solans C (2000) Studies of the relation between phase behavior and emulsification methods with nanoemulsion formation. Trends Collaid Interface Sci XIV:36–39

Forgiarini A, Esquena J, González C, Solans C (2001) Formation and stability of nano-emulsions in mixed nonionic surfactant systems. Trends Collaid Interface Sci XV:184–189

Franz B, Balkundi SS, Dahl C, Lvov YM, Prange A (2010) Layer-by-layer nano-encapsulation of microbes: controlled cell surface modification and investigation of substrate uptake in bacteria. Macromolec Biosci 10(2):164–172. https://doi.org/10.1002/mabi.200900142

Fryd MM, Mason TG (2012) Advanced nanoemulsions. Ann Rev Phys Chem 63(1):493–518. https://doi.org/10.1146/annurev-physchem-032210-103436

Gatahi D, Wanyika H, Kihurani A, Ateka E, Kavoo A (2015) Use of bio-nanocomposites in enhancing bacterial wilt plant resistance, and water conservation in greenhouse farming. In: The 2015 JKUAT scientific conference. agricultural sciences, technologies and global networking, p 52

Glandorf DCM, Verheggen P, Jansen T, Jorritsma JW, Smit E, Leeflang P, Wernars K, Thomashow LS, Laureijs E, Thomas-Oates JE, Bakker PAHM, Van Loon LC (2001) Effect of genetically modified *Pseudomonas putida* WCS358r on the fungal rhizosphere microflora of field-grown wheat. Appl Environ Microbiol 67(8):3371–3378. https://doi.org/10.1128/AEM.67.8.3371-3378.2001

Gordon TC, Pfender WF (2012) Effects of the mycoparasite *Sphaerellopsis filum* on overwintering survival of stem rust in perennial ryegrass. Plant Dis 96(10):1471–1481. https://doi.org/10.1094/PDIS-10-11-0837-RE

Gouda S, Kerry RG, Das G, Paramithiotis S, Shin H-S, Patra JK (2018) Revitalization of plant growth promoting rhizobacteria for sustainable development in agriculture. Microbiol Res 206: 131–140

Guerin J, Petit J, Burgain J, Borges F, Bhandari B, Perroud C, Desobry S, Scher J, Gaiani C (2017) *Lactobacillus rhamnosus* GG encapsulation by spray-drying: Milk proteins clotting control to produce innovative matrices. J Food Eng 193:10–19. https://doi.org/10.1016/j.jfoodeng.2016.08.008

Gupta A, Eral HB, Hatton TA, Doyle PS (2016) Nanoemulsions: formation, properties and applications. Soft Matter 12(11):2826–2841. https://doi.org/10.1039/c5sm02958a

Haas D, Défago G (2005) Biological control of soil-borne pathogens by fluorescent pseudomonads. Nat Rev Microbiol 3(4):307–319. https://doi.org/10.1038/nrmicro1129

Hafeez A, Razzaq A, Mahmood T, Jhanzab H (2015) Potential of copper nanoparticles to increase growth and yield of wheat. J Nanosci Adv Technol 1(1):1–6

Handelsman J, Stabb EV (1996, 1855) Biocontrol of soilborne plant pathogens. Plant Cell 8(10)

Hershkovitz V, Sela N, Taha-Salaime L, Liu J, Rafael G, Kessler C, Aly R, Levy M, Wisniewski M, Droby S (2013) De-novo assembly and characterization of the transcriptome of *Metschnikowia fructicola* reveals differences in gene expression following interaction with *Penicillium digitatum* and grapefruit peel. BMC Genomics 14(1):168. https://doi.org/10.1186/1471-2164-14-168

Hu X, Roberts D, Xie L, Qin L, Li Y, Liao X, Han P, Yu C, Xing L (2019) Seed treatment containing *Bacillus subtilis* BY-2 in combination with other *Bacillus* isolates for control of *Sclerotinia sclerotiorum* on oilseed rape. Biol Control 133:50. https://doi.org/10.1016/j.biocontrol.2019.03.006

Hung R, Lee S, Bennett JW (2013) *Arabidopsis thaliana* as a model system for testing the effect of *Trichoderma* volatile organic compounds. Fungal Ecol 6(1):19–26

Itelima J, Bang W, Onyimba I, Sila M, Egbere O (2018) Bio-fertilizers as key player in enhancing soil fertility and crop productivity: a review. J Microbiol Biotechnol Rep 2(1):22–28

Jafari SM (2017) Chapter 1 - An overview of nanoencapsulation techniques and their classification. In: Jafari SM (ed) Nanoencapsulation technologies for the food and nutraceutical industries. Academic Press, pp 1–34. https://doi.org/10.1016/B978-0-12-809436-5.00001-X

Jantzen M, Göpel A, Beermann C (2013) Direct spray drying and microencapsulation of probiotic *Lactobacillus reuteri* from slurry fermentation with whey. J Appl Microbiol 115(4):1029–1036. https://doi.org/10.1111/jam.12293

Jones E, Rabeendran, Stewart (2014) Biocontrol of *Sclerotinia sclerotiorum* infection of cabbage by *Coniothyrium minitans* and *Trichoderma* spp. Biocont Sci Technol 24:1363. https://doi.org/10.1080/09583157.2014.940847

Junaid JM, Dar NA, Bhat TA, Bhat AH, Bhat MA (2013) Commercial biocontrol agents and their mechanism of action in the management of plant pathogens. Int J Mod Plant Ani Sci 1(2):39–57

Kageyama K, Nelson EB (2003) Differential inactivation of seed exudate stimulation of *Pythium ultimum* sporangium germination by *Enterobacter* cloacae influences biological control efficacy on different plant species. Appl Environ Microbiol 69(2):1114–1120. https://doi.org/10.1128/aem.69.2.1114-1120.2003

Kanchiswamy CN, Malnoy M, Maffei ME (2015) Chemical diversity of microbial volatiles and their potential for plant growth and productivity. Front Plant Sci 6:151

Karthikeyan V, Sankaralingam A, Sevugapperumal N (2006) Biological control of groundnut stem rot caused by *Sclerotium rolfsii* (Sacc.). Arch Phytopathol Plant Protec 39:239–246. https://doi.org/10.1080/03235400500094688

Keswani C, Mishra S, Sarma BK, Singh SP, Singh HB (2014) Unraveling the efficient applications of secondary metabolites of various *Trichoderma* spp. Appl Microbiol Biotechnol 98(2):533–544. https://doi.org/10.1007/s00253-013-5344-5

Keswani C, Bisen K, Chitara MK, Sarma BK, Singh HB (2017) Exploring the role of secondary metabolites of Trichoderma in tripartite interaction with plant and pathogens. In: Singh JS, Seneviratne G (eds) Agro-environmental sustainability: volume 1: managing crop health. Springer International Publishing, Cham, pp 63–79. https://doi.org/10.1007/978-3-319-49724-2_4

Khalili E, Sadravi M, Naeimi S, Khosravi V (2012) Biological control of rice brown spot with native isolates of three *Trichoderma* species. Braz J Microbiol 43(1):297–305. https://doi.org/10.1590/s1517-83822012000100035

Khatun A, Farhana T, Sabir AA, Islam SMN, West HM, Rahman M, Islam T (2018) *Pseudomonas* and *Burkholderia* inhibit growth and asexual development of *Phytophthora capsici*. Z Naturforsch C 73(3–4):123–135. https://doi.org/10.1515/znc-2017-0065

Khodakovskaya MV, de Silva K, Biris AS, Dervishi E, Villagarcia H (2012) Carbon nanotubes induce growth enhancement of tobacco cells. ACS Nano 6(3):2128–2135. https://doi.org/10.1021/nn204643g

Kiss L (2003) A review of fungal antagonists of powdery mildews and their potential as biocontrol agents. Pest Manag Sci 59:475–483. https://doi.org/10.1002/ps.689

Kloepper JW, Ryu CM, Zhang S (2004) Induced systemic resistance and promotion of plant growth by *Bacillus* spp. Phytopathology 94(11):1259–1266. https://doi.org/10.1094/phyto.2004.94.11.1259

Konappa N, Dhamodaran N, Shanbhag SS, Sampangi MA, Krishnamurthy S, Arakere UC, Chowdappa S, Jogaiah S (2022) *Trichoderma*: a potential biopesticide for sustainable management of wilt disease of crops. In: Biopesticides. Elsevier, pp 261–275

Kottb M, Gigolashvili T, Großkinsky DK, Piechulla B (2015) *Trichoderma* volatiles effecting *Arabidopsis*: from inhibition to protection against phytopathogenic fungi. Front Microbiol 6:995

Koumoutsi A, Chen XH, Henne A, Liesegang H, Hitzeroth G, Franke P, Vater J, Borriss R (2004) Structural and functional characterization of gene clusters directing nonribosomal synthesis of bioactive cyclic lipopeptides in *Bacillus amyloliquefaciens* strain FZB42. J Bacteriol 186(4):1084–1096. https://doi.org/10.1128/jb.186.4.1084-1096.2004

Krell V, Jakobs-Schoenwandt D, Vidal S, Patel AV (2018) Encapsulation of *Metarhizium brunneum* enhances endophytism in tomato plants. Biol Control 116:62–73. https://doi.org/10.1016/j.biocontrol.2017.05.004

Krishnamoorthy V, Elumalai G, Rajiv S (2016) Environment friendly synthesis of polyvinylpyrrolidone nanofbers and their potential use as seed coats. New J Chem 40:3268–3276

Lassois L, Bellaire LL, Jijakli MH (2008) Biological control of crown rot of bananas with *Pichia anomala* strain K and *Candida oleophila* strain. Biol Control 45:410–418

Leena MM, Yoha KS, Moses JA (2021) Anandharamakrishnan C Nanofibers in food applications. In: Reference module in food science. Elsevier Inc., pp 1–9

Li S, Zhang N, Zhang Z, Luo J, Shen B, Zhang R, Shen Q (2013) Antagonist *Bacillus subtilis* HJ5 controls *Verticillium* wilt of cotton by root colonization and biofilm formation. Biol Fert Soil 49(3):295–303. https://doi.org/10.1007/s00374-012-0718-x

Liu W, Ye A, Liu W, Liu C, Han J, Singh H (2015) Behaviour of liposomes loaded with bovine serum albumin during *in vitro* digestion. Food Chem 175:16–24. https://doi.org/10.1016/j.foodchem.2014.11.108

Lopes MR, Klein MN, Ferraz LP, da Silva AC, Kupper KC (2015) *Saccharomyces cerevisiae*: a novel and efficient biological control agent for *Colletotrichum acutatum* during pre-harvest. Microbiol Res 175:93–99. https://doi.org/10.1016/j.micres.2015.04.003

Lorito M, Woo SL, Garcia I, Colucci G, Harman GE, Pintor-Toro JA, Filippone E, Muccifora S, Lawrence CB, Zoina A, Tuzun S, Scala F (1998) Genes from mycoparasitic fungi as a source for improving plant resistance to fungal pathogens. PNAS (USA) 95(14):7860–7865. https://doi.org/10.1073/pnas.95.14.7860

Maffei ME, Gertsch J, Appendino G (2011) Plant volatiles: production, function and pharmacology. Nat Prod Rep 28(8):1359–1380

Majumdar S (2017) Optimization of protease production from plant growth promoting *Bacillus amyloliquefaciens* showing antagonistic activity against phytopathogens. Int J Pharma Bio Sci 8:635–642. https://doi.org/10.22376/ijpbs.2017.8.2.b635-642

Maksimov IV, Abizgil'dina RR, Pusenkova LI (2011) Plant growth promoting rhizobacteria as alternative to chemical crop protectors from pathogens (review). App Biochem Microbiol 47(4): 333–345. https://doi.org/10.1134/S0003683811040090

Mari M, Martini C, Spadoni A, Rouissi W, Bertolini P (2012) Biocontrol of apple postharvest decay by *Aureobasidium pullulans*. Postharvest Biol Technol 73:56–62. https://doi.org/10.1016/j.postharvbio.2012.05.014

Mason TG, Wilking JN, Meleson K, Chang CB, Graves SM (2006) Nanoemulsions: formation, structure, and physical properties. J Phys Condens Matter 18(41):R635–R666. https://doi.org/10.1088/0953-8984/18/41/r01

Mavrodi DV, Mavrodi Olga V, Parejko James A, Bonsall Robert F, Kwak Y-S, Paulitz Timothy C, Thomashow Linda S, Weller David M (2012) Accumulation of the antibiotic phenazine-1-carboxylic acid in the rhizosphere of dryland cereals. App Environ Microbiol 78(3):804–812. https://doi.org/10.1128/AEM.06784-11

Meziane H, Van Der Sluis I, Van Loon LC, Höfte M, Bakker PAHM (2005) Determinants of *Pseudomonas putida* WCS358 involved in inducing systemic resistance in plants. Mol Plant Pathol 6(2):177–185. https://doi.org/10.1111/j.1364-3703.2005.00276.x

Migheli Q, González-Candelas L, Dealessi L, Camponogara A, Ramón D (1998) Transformants of *Trichoderma longibrachiatum* overexpressing the β-1,4-endoglucanase gene egl1 show enhanced biocontrol of *Pythium ultimum* on cucumber. Phytopathology 88:673–677. https://doi.org/10.1094/PHYTO.1998.88.7.673

Milgroom MG, Cortesi P (2004) Biological control of chestnut blight with hypovirulence: a critical analysis. Annu Rev Phytopathol 42:311–338. https://doi.org/10.1146/annurev.phyto.42.040803.140325

Mishra S, Arora NK (2012) Evaluation of rhizospheric *Pseudomonas* and *Bacillus* as biocontrol tool for *Xanthomonas campestris* pv *campestris*. W J Microbiol Biotechnol 28(2):693–702. https://doi.org/10.1007/s11274-011-0865-5

Mishra S, Keswani C, Abhilash P, Fraceto LF, Singh HB (2017) Integrated approach of Agri-nanotechnology: challenges and future trends. Front Plant Sci 8:471

Mishra P, Mishra J, Dwivedi SK, Arora NK (2020) Microbial enzymes in biocontrol of phytopathogens. In: Arora NK, Mishra J, Mishra V (eds) Microbial enzymes: roles and applications in industries. Springer, Singapore, pp 259–285. https://doi.org/10.1007/978-981-15-1710-5_10

Moradi Pour M, Saberi Riseh R, Skorik YA (2022) Sodium alginate-gelatin nanoformulations for encapsulation of *Bacillus velezensis* and their use for biological control of Pistachio Gummosis. Materials 15(6). https://doi.org/10.3390/ma15062114

Mosquera M, Giménez B, da Silva IM, Boelter JF, Montero P, Gómez-Guillén MC, Brandelli A (2014) Nanoencapsulation of an active peptidic fraction from sea bream scales collagen. Food Chem 156:144–150. https://doi.org/10.1016/j.foodchem.2014.02.011

Moyne AL, Shelby R, Cleveland TE, Tuzun S (2001) Bacillomycin D: an iturin with antifungal activity against *Aspergillus flavus*. J Appl Microbiol 90(4):622–629. https://doi.org/10.1046/j.1365-2672.2001.01290.x

Mukherjee PK, Horwitz BA, Herrera-Estrella A, Schmoll M, Kenerley CM (2013) *Trichoderma* research in the genome era. Annu Rev Phytopathol 51(1):105–129. https://doi.org/10.1146/annurev-phyto-082712-102353

Mukhopadhyay R, De N (2014) Nano clay polymer composite: synthesis, characterization, properties and application in rainfed agriculture. Global J Bio Biotechnol 3:133–138

Mukhtar T (2018) Management of root-knot nematode, *Meloidogyne incognita*, in tomato with two *Trichoderma* species. Pak J Zoology 50. https://doi.org/10.17582/journal.pjz/2018.50.4.sc15

Nedovic V, Kalusevic A, Manojlovic V, Levic S, Bugarski B (2011) An overview of encapsulation technologies for food applications. Proc Food Sci 1:1806–1815. https://doi.org/10.1016/j.profoo.2011.09.265

Nelson EB, Harman GE, Nash GT (1988) Enhancement of *Trichoderma* -induced biological control of pythium seed rot and pre-emergence damping-off of peas. Soil Biol Biochem 20(2):145–150. https://doi.org/10.1016/0038-0717(88)90030-2

Ordentlich A, Elad Y, Chet I (1988) The role of chitinase of *Serratia marcescens* in biocontrol of *Sclerotium rolfsii*. Phytopathology 78:84–87

Pal KK, Gardener BMS (2006) Biological control of plant pathogens. Plant Health Instruct 2:1117–1142. https://doi.org/10.1094/PHI-A-2006-1117-02

Palanisamy S, Thankappan S, Gandhi K, Uthandi S (2020) Comprehensive profiling of the VOCs of *Trichoderma longibrachiatum* EF5 while interacting with *Sclerotium rolfsii* and *Macrophomina phaseolina*. Microbiol Res 236:126436. https://doi.org/10.1016/j.micres.2020.126436

Pathak A, Sharma A, Johri B, Sharma A (2004) *Pseudomonas* strain GRP3 induces systemic resistance to sheath blight in rice. Int Rice Res Notes 29:35–36

Paulitz TC, Bélanger RR (2001) Biological control in greenhouse systems. Annu Rev Phytopathol 39:103–133. https://doi.org/10.1146/annurev.phyto.39.1.103

Pieterse CM, Zamioudis C, Berendsen RL, Weller DM, Van Wees SC, Bakker PA (2014) Induced systemic resistance by beneficial microbes. Annu Rev Phytopathol 52:347–375. https://doi.org/10.1146/annurev-phyto-082712-102340

Pour MM, Saberi-Riseh R, Mohammadinejad R, Hosseini A (2019) Investigating the formulation of alginate-gelatin encapsulated *Pseudomonas fluorescens* (VUPF5 and T17-4 strains) for controlling *Fusarium solani* on potato. Int J Biol Macromolec 133:603–613

Pour M, Saberi Riseh R, Skorik Y (2022) Sodium Alginate–Gelatin nanoformulations for encapsulation of *Bacillus velezensis* and their use for biological control of *Pistachio Gummosis*. Materials 15:2114. https://doi.org/10.3390/ma15062114

Prajapati S, Kumar N, Kumar S, Scholar LL, Maurya S (2020) Biological control a sustainable approach for plant diseases management: a review. J Pharmacogn Phytochem 9:1514–1523

Raguchander T, Saravanakumar D, Balasurramanian P (2011) Molecular approaches to improvement of biocontrol agents of plant diseases. J Biol Control 25(2):71–84

Ram RM, Keswani C, Bisen K, Tripathi R, Singh SP, Singh HB (2018) Biocontrol technology: eco-friendly approaches for sustainable agriculture. In: Omics technologies and bio-engineering. Elsevier, pp 177–190

Rivera-Méndez W, Obregón M, Morán-Diez ME, Hermosa R, Monte E (2020) *Trichoderma asperellum* biocontrol activity and induction of systemic defenses against *Sclerotium cepivorum* in onion plants under tropical climate conditions. Biol Control 141:104145. https://doi.org/10.1016/j.biocontrol.2019.104145

Rizvi R, Mahmood I, Tiyagi SA, Khan Z (2012) Conjoint effect of oil-seed cakes and *Pseudomonas fluorescens* on the growth of chickpea in relation to the management of plant-parasitic nematodes. Braz Arch Biol Tech 6:801–808

Robiglio A, Sosa MC, Lutz MC, Lopes CA, Sangorrín MP (2011) Yeast biocontrol of fungal spoilage of pears stored at low temperature. Int J Food Microbiol 147(3):211–216. https://doi.org/10.1016/j.ijfoodmicro.2011.04.007

Rojas EC, Jensen B, Jørgensen HJL, Latz MAC, Esteban P, Ding Y, Collinge DB (2020) Selection of fungal endophytes with biocontrol potential against *Fusarium*head blight in wheat. Biol Control 144:104222

Ryu C-M, Hu CH, Reddy M, Kloepper JW (2003) Different signaling pathways of induced resistance by rhizobacteria in *Arabidopsis thaliana* against two pathovars of *Pseudomonas syringae*. New Phytol 160(2):413–420

Ryu MC, Farag MA, Hu CH, Reddy MS, Kloepper JW, Paré PW (2004) Bacterial volatiles induce systemic resistance in *Arabidopsis*. Plant Physiol 134(3):1017–1026. https://doi.org/10.1104/pp.103.026583

Saberi Riseh R, Moradipour M (2021) A novel encapsulation of *Streptomyces fulvissimus* Uts22 by spray drying and its biocontrol efficiency against *Gaeumannomyces graminis*, the causal agent of take-all disease in wheat. Pest Manag Sci 77:4357. https://doi.org/10.1002/ps.6469

Saberi-Riseh R, Pour MM (2019) The effect of *Bacillus subtilis* Vru1 encapsulated in alginate - bentonite coating enriched with titanium nanoparticles against *Rhizoctonia solani* on bean. Int J Biol Macromol 150:1089–1097. https://doi.org/10.1016/j.ijbiomac.2019.10.197

Savchuk S, Dilantha Fernando WG (2004) Effect of timing of application and population dynamics on the degree of biological control of *Sclerotinia sclerotiorum* by bacterial antagonists. FEMS Microbiol Ecol 49(3):379–388. https://doi.org/10.1016/j.femsec.2004.04.014

Shanmugam H (2022) An insight on developing nanoformulations suitable for delivering plant beneficial microorganisms to crops under abiotic stresses. In: Mitigation of plant abiotic stress by microorganisms. Elsevier, pp 273–297

Shanmugam H, Acharya P, Jayaprakasha GK, Patil BS (2017) Nanoformulation and characterization of nomilin with different poly (lactic-co-glycolic acid) resomers and surfactants for the enhanced inhibition of α-amylase and angiotensin-converting-enzyme. J Func Foods 37:564–573

Shanmugam H, Rengarajan C, Nataraj S, Sharma A (2022) Interactions of plant food bioactives-loaded nano delivery systems at the nano-bio interface and its pharmacokinetics: an overview. Food Front 3(2):256–275

Sharma G (1996) Studies on the integrated management of banded leaf and sheath blight of maize caused by *Rhizoctonia solani*. Govind Ballabh Pant University of Agriculture and Technology, Pantnagar, India

Sharma S, Chen C, Navathe S (2019) A halotolerant growth promoting rhizobacteria triggers induced systemic resistance in plants and defends against fungal infection. Sci Rep 9(1):4054. https://doi.org/10.1038/s41598-019-40930-x

Silva HSA, Romeiro RS, Macagnan D, Halfeld-Vieira BA, Pereira MCB, Mounteer A (2004) Rhizobacterial induction of systemic resistance in tomato plants: non-specific protection and increase in enzyme activities. Biol Control 29(2):288–295. https://doi.org/10.1016/S1049-9644(03)00163-4

Singh HB, Mishra S, Fraceto LF, De Lima R (2018) Emerging trends in agri-nanotechnology: fundamental and applied aspects. CABI, USA, p 199

Siozios S, Tosi L, Ferrarini A, Ferrari A, Tononi P, Bellin D, Maurhofer M, Gessler C, Delledonne M, Pertot I (2015) Transcriptional reprogramming of the mycoparasitic fungus

*Ampelomyces quisqualis*during the powdery mildew host-induced germination. Phytopathology 105(2):199–209. https://doi.org/10.1094/phyto-01-14-0013-r

Slininger PJ, Behle RW, Jackson MA, Schisler DA (2003) Discovery and development of biological agents to control crop pests. Neotrop Entomol 32:183–195

Sood M, Kapoor D, Kumar V, Sheteiwy MS, Ramakrishnan M (2020) *Trichoderma*: the "secrets" of a multitalented biocontrol agent. Plants (Basel) 9(6):762. https://doi.org/10.3390/plants9060762

Szczech M, Shoda M (2004) Biocontrol of *Rhizoctonia*damping-off of tomato by *Bacillus subtilis*combined with *Burkholderia cepacia*. J Phytopathol 152(10):549–556. https://doi.org/10.1111/j.1439-0434.2004.00894.x

Tian B, Yang J, Lian L, Wang C, Li N, Zhang K-Q (2007) Role of an extracellular neutral protease in infection against nematodes by *Brevibacillus laterosporus* strain G4. App Microbiol Biotechnol 74(2):372–380. https://doi.org/10.1007/s00253-006-0690-1

Tortora ML, Díaz-Ricci JC, Pedraza RO (2011) *Azospirillum brasilense* siderophores with antifungal activity against *Colletotrichum acutatum*. Arch Microbiol 193(4):275–286. https://doi.org/10.1007/s00203-010-0672-7

Van Dijk K, Nelson EB (2000) Fatty acid competition as a mechanism by which *Enterobacter cloacae* suppresses *Pythium ultimum* sporangium germination and damping-off. App Environ Microbiol 66(12):5340–5347

Van Loon L, Bakker P (2007) Root-associated bacteria inducing systemic resistance. Plant-associated bacteria. Springer, In, pp 269–316

Van Wees SC, Van der Ent S, Pieterse CM (2008) Plant immune responses triggered by beneficial microbes. Curr Opin Plant Biol 11(4):443–448

Verhagen BW, Trotel-Aziz P, Couderchet M, Höfte M, Aziz A (2010) *Pseudomonas* spp.-induced systemic resistance to *Botrytis cinerea* is associated with induction and priming of defence responses in grapevine. J Exp Bot 61(1):249–260. https://doi.org/10.1093/jxb/erp295

Vinothini K, Daisy ERAC, Rajan M (2020) Mechanism of loading and release in nanocontainers. In: Smart Nanocontainers, pp 67–87. https://doi.org/10.1016/b978-0-12-816770-0.00005-8

Wallace RL, Hirkala DL, Nelson LM (2018) Efficacy of *Pseudomonas fluorescens* for control of *Mucor* rot of apple during commercial storage and potential modes of action. Can J Microbiol 64(6):420–431. https://doi.org/10.1139/cjm-2017-0776

Wang LW, Chi ZM, Chi Z, Li J, Wang XH (2009) Siderophore production by the marine-derived *Aureobasidium pullulans* and its antimicrobial activity. Bioresour Technol 100(9):2639–2641. https://doi.org/10.1016/j.biortech.2008.12.010

Wang M, Geng L, Sun X, Shu C, Song F, Zhang J (2020) Screening of *Bacillus thuringiensis* strains to identify new potential biocontrol agents against *Sclerotinia sclerotiorum* and *Plutella xylostella* in *Brassica campestris* L. Biol Control 145:104262. https://doi.org/10.1016/j.biocontrol.2020.104262

Wei F, Zhang Y (2019) Evaluation of the biocontrol potential of endophytic fungus *Fusarium solani* CEF559 against *Verticillium dahliae* in cotton plant. BioMed Res Int 2019:3187943. https://doi.org/10.1155/2019/3187943

Wei Y, Mao S, Tu K (2014) Effect of preharvest spraying *Cryptococcus laurentii* on postharvest decay and quality of strawberry. Biol Control 73:68–74. https://doi.org/10.1016/j.biocontrol.2014.02.016

Weller DM (2007) *Pseudomonas* biocontrol agents of soilborne pathogens: looking back over 30 years. Phytopathology 97(2):250–256. https://doi.org/10.1094/PHYTO-97-2-0250

Wilhite SE, Lumsden RD, Straney DC (2001) Peptide synthetase gene in *Trichoderma virens*.App. Environ Microbiol 67(3–12):5055–5062. https://doi.org/10.1128/aem.67.11.5055-5062.2001

Yaakov N, Karthik MK, Felfbaum R, Lahat M, Costa N, Belausov E, Ment D, Mechrez G (2018) Single cell encapsulation via pickering emulsion for biopesticide applications. ACS Omega 3(10):14294–14301

Yandigeri MS, Malviya N, Solanki MK, Shrivastava P, Sivakumar G (2015) Chitinolytic *Streptomyces vinaceusdrappus* S5MW2 isolated from Chilika lake, India enhances plant growth and biocontrol efficacy through chitin supplementation against *Rhizoctonia solani*. W J Microbiol Biotechnol 31(8):1217–1225

Zaghloul R, Hanafy E, Neweigy N, Khalifa N (2007) Application of biofertilization and biological control for tomato production. Paper presented at the 12th conference of microbiology, Cairo, Egypt, pp 198–212

Zambrano C, Moreno G, Sequeda-Castañeda L, García-Caicedo A, Albarracín D, Barreto-Charry L (2014) Determining the effectiveness of *Candida guilliermondii* in the biological control of *Rhizopus stolonifer* in postharvest tomatoes. Univ Sci 19:51–62. https://doi.org/10.11144/Javeriana.SC19-1.debc

Zaytseva O, Neumann G (2016) Carbon nanomaterials: production, impact on plant development, agricultural and environmental applications. Chem Biol Technol Agric 3. https://doi.org/10.1186/s40538-016-0070-8

Zhang S, Moyne A-L, Reddy M, Kloepper JW (2002) The role of salicylic acid in induced systemic resistance elicited by plant growth-promoting rhizobacteria against blue mold of tobacco. BiolControl 25(3):288–296

Zhao Y, Tu K, Shao X, Jing W, Su Z (2008) Effects of the yeast *Pichia guilliermondii* against *Rhizopus nigricans* on tomato fruit. Postharvest Biol Technol 49(1):113–120. https://doi.org/10.1016/j.postharvbio.2008.01.001

Zhao Q, Shen Q, Ran W, Xiao T, Xu D, Xu Y (2011) Inoculation of soil by *Bacillus subtilis* Y-IVI improves plant growth and colonization of the rhizosphere and interior tissues of muskmelon (*Cucumis melo* L.). Biol Fert Soils 47(5):507–514. https://doi.org/10.1007/s00374-011-0558-0

Zhou K, Yamagishi M, Osaki M (2008) *Paenibacillus* BRF-1 has biocontrol ability against *Phialophora gregata* disease and promotes soybean growth. Soil Sci Plant Nut 54(6): 870–875. https://doi.org/10.1111/j.1747-0765.2008.00308.x

Zuidam NJ, Shimoni E (2010) Overview of microencapsulates for use in food products or processes and methods to make them. In: Encapsulation technologies for active food ingredients and food processing. Springer, pp 3–29

Chapter 9
The Exploitation of Recombinant DNA Technology to Induce Biologics Directed to Biocontrol

Ömür Baysal and Kubilay Kurtuluş Baştaş

Abstract Plant production losses caused by insects, pests and pathogens are major problems in agricultural production. Over-use of pesticides impacts not only the environment but also human health, which grounds preparation for the negative sides of global warming and climate change. In addition, the overcome of resistance genes in plants by pathogens limits the crop management. As an alternative measurement, biologics can be used for all types of integrated pest management strategies. The profound studies providing high-output data involving omics technologies on biologics have revealed the chemical structure of antibiotics, enzymes and inhibitory compounds remaining unclear, which plays a major role in inhibition of the pathogen growth. They are also a genetically rich-gene pool that can be used depending on the characteristic property of the redesigned genetic profile of any microorganism, which is not harmful to the environment and health. To date, we are able to do genetic manipulation on the genomics data of any microorganism once we have an available sequence, which gives rise to expand of the capacity peculiar to the microorganisms based on expression and protein level. This updated chapter briefly focuses on the updated knowledge regarding biological agents related to disease and pest management programs.

Keywords Crop protection · Plant growth · Promoting bacteria · recombinant DNA technology

Ö. Baysal (✉)
Molecular Microbiology Unit, Department of Molecular Biology and Genetics, Faculty of Science, Muğla Sıtkı Koçman University, Muğla, Turkey
e-mail: omurbaysal@mu.edu.tr

K. K. Baştaş
Department of Plant Protection, Faculty of Agriculture, Selçuk University, Konya, Turkey
e-mail: kbastas@selcuk.edu.tr

K. K. Bastas et al. (eds.), *Microbial Biocontrol: Molecular Perspective in Plant Disease Management*, Microorganisms for Sustainability 49,
https://doi.org/10.1007/978-981-99-3947-3_9

9.1 Introduction

Global warming and food shortage and security are the problems that continuously threaten the whole world due to rapid population growth, and the harmful effect of plant parasites. The crop losses caused by biotic stresses are estimated at 30% in the World (Oerke 2006; Flood 2010) with biotic stress playing a major role in crop production in the last 50 years (Chakraborty and Newton 2011; Kumar 2022a, b). The resistant isolates of devastating pathogens belonging to bacteria and fungi are still controlled by fungicides though their hazardous and residue problems are a public concern (Nicot et al. 2016; Kumari et al. 2022).

Plants are the wellspring of supplements for some creatures (Cardinale et al. 2011) and they have root and flying frameworks which are presented to attacking of various organisms like microbes and growths. A portion of these microorganisms with plant pathogenic property causes disease influencing plant health (Dodds and Rathjen 2010; Pathak et al. 2022). However, some portion of the microorganisms have a place with rhizobacteria and arbuscular mycorrhizal parasites species showing harmonious ways of behaving advance plant growth (Pérez-de-Luque et al. 2017). Biological agents could be normal and safe in terms of their effect on humans and the environment (Selosse et al. 2004). Biocontrol of plant infection could be the way for diminishing the disease utilizing organic entity since fumigation does not become powerful and causes residue problems besides the development of resistance to chemicals when the microbe has been once again introduced (Weller et al. 2002; Singh et al. 2017, 2020). Infection concealment by presenting nonpathogenic microorganisms all around adjusted to growth on plant roots may likewise create an antimicrobial situation to repress the development of the microbe (Baysal et al. 2008, Baysal et al. 2013). There are three fundamental methods of activity that microorganisms could restrict the development of one /more microorganisms, which include antibiosis, mycoparasitism and competition for nutrients. Lately, unique control estimations have been created to improve the outcome of biocontrol in the wake of acquainting the living microorganisms with make a suppressive impact on the microbes to diminish the harm brought about by microorganisms (Compant et al. 2010; Singh et al. 2017).

Advances in molecular biological techniques make construction in the gene available for analysis and help for comprehension of pathogenic conduct in the microbial local area (Liu et al. 2010; Yadav et al. 2022).

9.2 Novel Control Strategies Using Agents

Pathogenic microorganisms have different methods of activity, which propose a limit with respect to transformation to new, which guarantees the connection between explicitness and organic control. Their genomes appear to be quality boxes on which control of their properties utilizing omics information and hereditary designing is more straightforward, which could be clear for the development of microbial specialists showing promising property (Palumbo et al. 2005). Besides, the development of microbes affects the degrees of pathogenic harmfulness, which produce explicit atoms and additionally trigger the initiation of different instruments as biocontrol specialists. All-natural control methodologies still need to be assessed under in vitro or nursery conditions. The appraisal of total information including transcriptomics, proteomics and metabolomics studies would give new understanding into ways of behaving of organisms in the rhizosphere and their environmental effect. In addition, microbe explicit phages have likewise encouraged the potential for biocontrol. Yet, it needs testing of their viability for phage arrangements relying upon the qualities and way of life of the phages (Ackermann et al. 2004).

Various plant illnesses are not being controlled with these synthetics; however, significant expense applications in the fields contrasted with the normal pay. Although crop loss caused by pests and pathogens depending on various conditions is getting an increase, the frequent use of pesticides in modern agriculture could be the major risks if they are inconveniently used by the growers. The risk of residue in/or on treated fruits, vegetables and grains besides in soil due to many chemicals used for the purpose of pesticide application is possible (EEA 2005; Kumar et al. 2021; Patel et al. 2022).

As opposed to the customarily improvement of microbial strains effectively utilized in the medication and food-drink production (Biot-Pelletier and Martin 2014), the economically accessible ones, which were hereditarily further developed ones, are considerably less than tried strains because of the enduring system of fostering these organic entities. Replacement of the hereditary construction utilizing genome rearranging (GR) called 'shuffling' is one of the most encouraging technologies for the screening of aggregate to the progress of microbial strains that permit us to consolidate the genotypes of parental strains conveying the ideal aggregates with the antifungal property as well as resistance to natural burdens without recombinant DNA methods (Magocha et al. 2018). Nevertheless, this innovation has been not disclosed. The bacterial BCAs having a place with *Bacillus subtilis* and *Streptomyces strains* have been done with GR method for the effective control of *Fusarium oxysporum* f. sp. melonis, *Phytophthora infestans* and *Fusarium oxysporum* f. sp. cucumerinum, separately (Zhao et al. 2014).

9.3 Genome Assembly

Collecting probability utilizing recombinant DNA innovation on a genome has extended our adaptability for producing new information on the hereditary design of any biocontrol quality. A reference genome is utilized for various reasons for works with zero in on quality articulation examinations, designated quality altering and marker-quality choice. In this stage, a great genome (high inclusion, scarcely any holes) depicting the properties of biocontrol specialists is a fundamental objective (Faino and Thomma 2014), for an assortment of quality to be gathered, and clarified expected for explicit reason (Çetinkaya et al. 2022). By cutting-edge methods utilizing NGS shotgun sequencing, we have recognized entire qualities liable for various enzymatic properties and arranged which can be related to the properties of any microorganism conveying organic control properties (Çetinkaya et al. 2022). In certain conditions, the applications acknowledged with a fragmented genome obtained from chosen molecular markers could be utilized, for example, microsatellites (Kumar 2022a, b) for a linkage map development (Beukeboom et al. 2010; Niehuis et al. 2010) and hereditary variety (Paspati et al. 2019). Genome gathering has different stages from the start of the sequencing from a solitary individual then followed by adjusting the successions into a get together until commenting on the gathering considering protein-coding data (Ekblom and Wolf 2014).

9.4 Gene Discovery

Mapping genes are used for quantitative characteristic loci studies (QTL) rather than a long practice in rearing and work to comprehend characters in promotion lines. High-throughput sequencing and genome-wide scale reads up for planning screens (Schlötterer et al. 2014) are utilized to recognize loci with various allele frequencies in different populace as far as surveying the aggregates of the objective characters (Bastıde et al. 2013). Today, with the open doors given by high-goal grouping information, it is presently conceivable to look at an extensive variety of genomic information so that we can recognize the proteins encoded by unambiguous qualities and we foresee the potential metabolites to be created.

We can also understand the pathways involved in these metabolic processes. We identify the individuals genetically close to this strain and relied on the database revealing the genomic differences by means of multi-locus sequence analyses to characterize the biological agent at the sub/species level. As a matter of fact, in a recent study, we have conducted, we have been able to reveal the potential of hormones and enzymes produced as a result of genome mapping on a bacterium that is able to degrade poly aromatic hydrocarbons besides its plant regulator effect to be used for enhancing a/biotic stress resistance in agricultural production (Çetinkaya et al. 2022). Gene expression is used in another strategy connected to the genetic structure of biocontrol features. It has recently become much simpler to use omics technologies (transcriptomes and proteomes, metabolomics), which permit entire RNA and protein expression profiles and enable extensive examination of hypothesized metabolic processes of a microorganism. The identification and quantification of patterns of gene expression based on microbial behaviour under biotic and abiotic environments can be aided by transcriptomic and proteomic data gathered under diverse situations (Wang et al. 2009). Studies on the gene-expression profiles of biocontrol agents can aid in understanding how they adapt to their environment and how the host responds in terms of defence (Félix and Barkoulas 2015).

9.5 Recombinant DNA Technology

Molecular techniques have given us the likelihood to disengage quality for one living being and express it in an alternate organic entity. The huge achievement has likewise been accomplished in recombinant DNA on viral disease (Maloy 2005). Regarding the studies on recombinant DNA technology, inserting the genes encoding the envelope protein of tobacco alfa mosaic virus and tobacco mosaic virus (TMV) into the genome of the tobacco plant restrained the improvement of the illness (Beachy et al. 1990).

Likewise, qualities encoding insecticidal proteins for *Bacillus thuringiensis* microscopic organisms were cloned into the genome of the host plant for lepidopteran and coleopteran pest to eliminate the adverse consequences of pesticides, which are against agrarian nuisances on the climate. Along these lines, the quality encoding the insecticidal protein in T plasmid vectors of *Agrobacterium tumefaciens* was additionally moved into tobacco, potato, tomato, rice and maize plants. Subsequently, a critical speed increase was gotten in bother control. Recombinant DNA innovation includes control of hereditary material got from an unfamiliar living being to get wanted qualities in chosen life forms. This innovation depends on the inclusion of DNA parts from a hereditary source conveying a helpful quality succession through a helpful vector (Berk and Zipursky 2000). Control in a living being's genome consolidates with one or a few new qualities and administrative components or obstructs the statement of the qualities, which gives benefits to expanding the articulation profile of the wanted genome (Bazan-Peregrino et al. 2013). Limitation catalysts applied to get different DNA sections are the particular objective arrangement, which is utilized to fix the ideal quality multiplicated in cloning and afterward move into articulation vector including marker qualities. The quality section could be brought into a host genome, which is developed to deliver different duplicates of DNA parts, lastly, the clones containing DNA fragments can be chosen and reaped (Venter 2007).

9.6 Gene Editing

Clustered interspaced short palindromic repeats (CRISPR) are another improvement of recombinant DNA innovation. This strategy has become the answer for a few issues in various species connected with enactment, concealment and erasure of qualities in harvests, yeast, and furthermore substantial cells and so on. The technique depends on the capacity of the Cas operon encoding Cas3 nucleases and different Cas proteins. crRNAs are ready for altering involving CRISPR for quality designing on a strain (Wang et al. 2016; Yadav et al. 2022). Acknowledgment and cleavage of unfamiliar DNA/RNA are a succession explicit interaction. Data about the unfamiliar hereditary material put away by the host framework consolidates with the CRISPR framework (Shmakov et al. 2015). *Cas*9t as a gene editing tool representing DNA endonucleases uses RNA molecules for specific targets (Gasiunas and Siksnys 2013). The Class 2 CRISPRCas system comprises a single protein effector used for genome editing operations (Mohanraju et al. 2016). The targeting of genes involved in the isolation process of homozygous gene knockouts by CRISPR-induced mutations can be used for 'potential antifungal targets' researches (Vyas et al. 2015). The studies suggest to the exploitation used to generate strains of the Native CRISPR-cas immune could provide resistance to different types of destructive viruses (Hynes et al. 2016) and it has an enormous role in stability and immunity enhancement in advanced biological systems. In fact, the CRISPR works to protect bacteria itself from viral attacks, which can be described as 'trap sequence' for virus invasion. As new spacers, DNA from a virus composed of short segments is inserted into the CRISPR sequence. Then production of CRISPR repeats and spacers in the bacterial DNA result in transcription, the formed RNA chain has short sequences after cutting into small fragments called CRISPR RNAs. These sequences are the guide data for bacteria to recognize and destroy viral material, which exactly matches the viral genome for further invasion of the virus (Pennisi 2013). Another editing tool called zinc finger nucleases (ZFNs) and transcription activator-like effector nucleases (TALENs) with chimeric nucleases are sequence-specific DNA binding modules, which give rise to cleavage on DNA.

9.7 RNA-Interference Technique

The method of RNA silencing opened the door for short non-coding RNAs to be used in RNA-mediated gene silencing. In both plants and animals, RNAi is a trigger mechanism that targets homologous mRNAs to prevent their transcription or destruction after translation. The primary result of this process is the RNA double-strands (dsRNA). Combining the synthesis of short RNAs (maximum 26 nucleotides (nt)) with the assistance of the argonaute protein results in sequence-specific control of gene expression at the transcriptional or translational level (RNA-based inhibition) (Issac 1992) (Mehrotra and Aggarwal 2003). In this situation, complementary messenger RNAs (mRNAs) are targeted for degradation by siRNAs within the RISC (Balasubramanian et al. 2009). In plants, RNA silencing is a mobile signal that is produced and spread throughout the plant. The target gene can also be silenced throughout the entire plant. One of the practical difficulties in RNAi research is the transformation of genetic fragments that produce active dsRNA or siRNA into various plant cells and tissues using vectors. These genetic fragments produce dsRNA for the gene(s) chosen by modified Agrobacterium (Balasubramanian et al. 2009).

One of the effective methods for developing plant tolerance to disease pathogens that cause losses in vital crops is RNAi technology. To silence genes involved in the virulence of many diseases, RNAi has been tested in a number of host-pathogen systems. Certain plant pathogens produce secondary metabolites and effector molecules involved in the case related to virulence. The molecular targets of these signaling pathways in the crop can be engineered, which results in disease resistance (Mazzoni et al. 2016). Genetic modification on the host to target the pathogen's virulence factors increases host resistance.

9.8 Superior Strains and Recombinant DNA Cloning

Today, at the point where recombinant DNA technology has come, a pathogen is a pest in an animal or an agricultural product can no longer is an obstacle to the ability to control another pest or pathogen in the agricultural field by expressing the genes by competent cells. With recombinant DNA technology, it is possible to use the desired genes from microorganisms with genome maps and use them by combining the enzymes they produce against pathogens with nanotechnology. All collected strains showing promising properties incorporated with genomic and proteomic data suggest various gene-rich pools to improve smart biologics (Baysal et al. 2013; Can 2022; Korkut 2022). Several microorganisms with environment-friendly properties are grown to utilize their required enzymes altered by genetic engineering and they are optimized under specific growth conditions related to process involving metabolic parameters for microbial technology (Ullah et al. 2015a, 2016a,b). The engineered microorganism belonging to the cyanobacteria genus produce hydrogen and other biofuels (Lomedico 1982; Ullah et al. 2015b).

Various strains of cyanobacteria are capable of converting CO_2 into reduced fuel molecules that transform into safe carbon energy sources. The modified microbial cells that are frequently utilized in the selection of superior strains and their mass manufacturing provide undesirable barriers in the production of some functional proteins. But different therapies are used in the cellular systems to address these problems. These challenges, which we mostly saw, are overcome by post-translational level therapies, cell stress response activation, or by boosting proteolytic activity. Resistance development also occurs in batch culture. The recombinant DNA technique uses competent *Escherichia coli* strains as a biological framework, allowing us to operate in controlled ways to manufacture the necessary molecules (Can 2022).

By cloning the genes utilizing a range of selectable marker systems, recombinant DNA technology has offered promising chances to also comprehend yeast biology. Site-directed mutagenesis was used to modify the yeast multiple cloning site sequence in order to assess the potential interaction between two proteins based on transcriptional level and expression (Anonymous 2008). These developments have been integrated with traditional genetic analysis and manipulation in yeast, the model organism for eukaryotic cells. Recombinant DNA technology's solutions to biological issues have increased our understanding of the structure and organization of specific genes (DeJong et al. 2006; Walker 2009). On the other hand, Actinomycetes used in pharmaceutical manufacturing in the health sciences and the manipulation of biosynthetic pathways could be the way for the creation of novel medications including recombinant designing. A DNA vaccination is another strategy for preventing many diseases that allows DNA to deliver the genes encoding dangerous proteins. The host immune system may be activated by certain virus particles transmitted by recombinant phage vectors rather than the entire genome (Roldao et al. 2010).

9.9 Recombinant DNA Technology in Biological Control of Insects

Due to their greater effectiveness and lower risk of causing the development of insect resistance, genetically engineered microbial insecticides in particular have many more benefits than natural microbial insecticides. Therefore, the creation of new microbial pesticides to control insect pests offers a tremendous potential for the utilization of recombinant technology. We can manipulate DNA using this technique to arrive at useful solutions. They might be the cause of the decline in the use of pest control agents. Commercially produced microbial insecticides made from specific insect pathogens, such as entomopathogenic fungi, baculoviruses, microsporidia and entomopathogenic nematodes and entomopathogenic bacteria belonging to *Bacillus thuringiensis*, are available for controlling insect pests in the agricultural and horticultural sectors (Wakefield 2018). Only a few of the 100 bacterial families identified as exo- and endopathogen microorganisms of insects of insects are now employed in commercial insect pest control (Chattopadhya et al. 2017). Sprays, powders, liquid concentrates, wettable powders and granules are some of their available forms. Currently, the commercially useful strains of bacteria include *Paenibacillus popilliae, Lysinibacillus sphaericus*, *Bacillus thuringiensis, Pseudomonas alcaligenes, Pseudomonas aureofaciens, Saccharopolyspora spinosa, Serratia entomophila* and *Streptomyces avermitilis* (Chattopadhya et al. 2017).

9.10 Genetically Engineered Entomopathogenic (GEE) Bacteria

Recombinant DNA strategies render hereditary changes conceivable on the most usually utilized bacterial strains used to examine individuals from the bacterial genera including Bacillus, Pseudomonas, Rhizobium and Cyanobacteria controlled to higher overexpressing of the poison qualities to give a wonderful insecticidal action (Jurat-Fuentes and Jackson 2012). Microbes showing entomopathogen properties against agrarian pest vermin and illness vectors increment the efficiency of the insect poisons, which creates the degrees of insecticidal proteins (Federici et al. 2006).

Through research on the quality articulation and guidelines for Bt poison, hereditary designing examinations gained importance. Additionally, Vip3 protein articulation was used to design location-specific recombination frameworks in an attempt to increase the insecticidal activity of Bt by combining Cry1Fa with Cry1Ac protoxin (Baum et al. 1999). Additionally, the distinct evidence of the zwittermicin-encoding quality has allowed researchers to alter the cry quality blend, which proposed the development of safe plants constructed using cry qualities from Bt. These evaluations have taken place in conjunction with the evaluation of protection against various annoyances and plant parasitic/pathogenic nematodes (Jouzani et al. 2017). In addition to anticancer treatments, the latest evaluations of Bt have confirmed that it has real limits in terms of promoting plant growth and the bioremediation of heavy metals. It also has adverse effects on pathogenic microorganisms (Jouzani et al. 2017). The property of E hereditarily encodes the insecticidal toxic substance protein complex (TccC) from *Pseudomonas taiwanensis.* Coli overexpressed and had an insecticidal effect on Drosophila melanogaster hatchlings, resulting in basic mortality (Liu et al. 2010).

Due to its pesticidal and fungicidal effects on pests and phytopathogenic life forms, Bt chitinase characteristics unquestionably set it apart from the experts in the field. To further the pesticidal development of Cry harms for the managing strategy of the pests, various chitinase features and Bt chitinases have been applied. Different chitinase (chi) properties from various microorganisms included into Bt have similarly demonstrated increased insecticidal activity. Thus, attempts at using Bacillus species against pests were tested using genetically reconstituted Bt strains. When utilized against *Plutella xylostella, B. velezensis* strain with Bt cry features also demonstrated insecticidal effectiveness (Yul et al. 2009) as well as *B. subtilis* strains. When tested against Tuta absoluta hatchlings, licheniformis strains carrying the Bt cry quality (cry1Ab) showed noticeably greater insecticidal activity than the wild-type Bt strain LM-466 (Theoduloz et al. 2003). Chitinase quality (chiB) cloned from *Serratia marcescens* was put into Bt, and recombinant Bt strains holding chi features were proven to move in an insecticidal manner frequently (Karabörklü et al. 2018). Similar to B., other Bacillus species *B. subtilis, B. velezensis,* and in recombinant assessments, licheniformis have been employed to enhance the pest-inhibitory power of biocontrol-trained experts. The superior chiABC chitinase from S. E was injected with marcescens. Chitinase activity, which was cloned into *E. coli*, produced results. *Helicoverpa armigera* and *Malacosoma neustria* (Danışmazoğlu et al. 2015). The chitinase gene chiA was found to be present in a new master's thesis that was finished in 2022, and its flanking regions were discovered using whole genome sequence data from our lab that is available in NCBI (Accession No. SAMN23480194) from the original Serratia strain inserted into *E. coli* led to increased chitinase activity when a cloned piece was put into an expression vector to fight *Myzus persicae* larvae (Can 2022). Another recent master's thesis demonstrated that the *Bacillus amyloliquefaciens* strain EU07 NCBI (Accession Nr. GCF 019997305.1) had high 1,3 beta endoglucanase activity, which led to a notable reduction in *Phytophthora citrophthora*'s mycelium growth. These investigations showed that using recombinant DNA technology and competent microorganisms aimed at specific pest and disease, the original gene pools of various bacterial pathogens might be leveraged to increase the biocontrol agent's potential.

9.11 GEE Fungi

Entomopathogenic organisms (EPF) are significant for the microbial control of pests because of their possibility connected with high conceptive limits, target-specific molding, short item times, and ability to deliver, which permit them to endure longer (Sinha et al. 2016). EPF causes diseases in various species belonging to the orders Lepidoptera, Orthoptera, Homoptera, Coleoptera and Diptera, in contrast to other regular control specialists. EPFs *Lecanicillium muscarium*, *Lecanicillium lecanii*, *Hirsutella thompsonii*, *Metarhizium anisopliae*, *Isaria fumosorosea* (previously *Paecilomyces fumosoroseus*) and *Nomuraea rileyi* are used extensively worldwide to control irritants (Wakefield 2018). Nearly 1000 parasitic species that are harmful to pests have been successfully created and formed (Wakefield 2018). The results of antitoxins, proteins, chemicals and other physiological regulators of the growths have been the focus of recombinant research. Proteases and chitinases are two substances that have been studied to improve the effects of contagious entomopathogenic specialists. Proteases, glyceraldehyde-3-phosphate dehydrogenase (hero), chitinases, corrosive trehalase, sterol transporters, esterases and benzoquinone oxidoreductase are examples of qualities that deliver subtilisin and have been tested against bothersome pests. A critical increase in the lethal dosages (LC50) of corrosive trehalase (ATM1) against *Locusta migratoria* was also observed upon overexpression in a recombinant Metarhizium acridum strain. Recombinant studies on parasitic entomopathogens that depended on chemicals and other physiological regulators that were used against pests were evaluated using RNA obstruction (RNAi) methods that expanded the declaration of diuretic chemical (MSDH) and physiological regulators like defenceless subsidiary flagging pathway hindrances, trypsin adjusting oostatic factors, serine proteinase obstacles, and pyrokinin-neuropeptide against *M. acridum*. Additionally, a recombinant strain of *Beauveria bassiana* that blocks an invulnerable-related flagging pathway demonstrated disinfectant activity toward *Galleria mellonella* and *Persicae*. Additionally, the mortality on the sprites of *Bemisia tabaci* has increased due to an *I. fumosorosea* strain created to communicate a dsRNA focusing on the quality encoding the defenceless partnered protein TLR7.

9.12 GEE Viruses

Recombinant DNA techniques are being employed to increase the effectiveness of entomopathogenic microorganisms, which are undoubtedly distinctive enough to be identified due to their pathogenicity toward insects (Karabörklü et al. 2018; Hails 2001). Recombinant studies have mainly concentrated on the death rates and pets of baculoviruses that reduce the pest control limits as well as various exogenous qualities encoding proteins that have an impact on the secretion of substances with insecticidal effort reason for toxin, chemical and compound properties (Slack et al.

2009). Baculoviruses have been modified to produce neurotoxins in the same ways as many species, including sea anemones, pests, scorpions, small insects and wasps. This is why the biocontrol effort has been added (Harrison and Hoover 2012). *Helicoverpa armigera*, *Heliothis virescens*, *Trichoplusia ni* and *Spodoptera exigua* have all been demonstrated to be susceptible to the recombinant baculovirus created by the inclusion of the AaIT quality into *Autographa californica*, which is in the various nucleopolyhedrovirus (AcMNPV). Entomopathogenic microorganisms (EPV) are powerful, ubiquitous microbes that have the potential to be used as biocontrol agents (Prasad and Srivastava 2016). Approximately 16 viral genes encoding specific protein families that are associated with RNA diseases, such as spoviruses, dicistroviruses, nodaviruses and tetraviruses, as well as DNA virus families, such as densoviruses, entomopoxviruses, ascoviruses, iridoviruses, nudiviruses and baculoviruses, have been tested in significant EPV gatherings other. Additionally, quality promoters and signal peptides that direct advancement have an impact on the insecticidal activity of baculoviruses when adjusted for relevant variables (Lacey et al. 2001; Prasad and Srivastava 2016).

9.13 GEE Microsporidia

The microsporidia are devoted intracellular parasites belonging to the genus Microspora (Solter et al. 2012). Microsporidia that are entomopathogenic are very pathogenic to a variety of earthy and greenish-blue vector pests (Corradi and Keeling 2009). However, because of their vast genome's complexity and problems with genomic structure, recombinant specialists are not allowed to create them (Mishra 2009). Several studies have been done on the roles that certain microsporidian proteins play in spore germination and illness. Particularly the *Nosema bombycis* ricin-B-lectin (RBL) quality was identified and the RBL quality extracted from *N. bombycis* was amplified and transmitted in recombinant *E. coli* BL21. Further research on Microspora must be incorporated in order to replace them in field chores.

9.14 GEE Nematodes

Steinernema and *Heterorhabditis*' two entomopathogenic nematodes have been used as effective biocontrol agents against a variety of pests. The studies on these EPNs' molecular hereditary characteristics have expanded our understanding of *Photorhabdus* and *Xenorhabdus* (Burnell 2002). However, certain studies on the congruity, infectivity and storage facility robustness of these nematodes produced encouraging results that could increase their viability in pathogenic activities. The *Caenorhabditis elegans* hsp70 heat-shock protein was injected into *Heterorhabditis bacteriophora* to increase their ability to cope with stress under in vitro settings (Gaugler et al. 1997). *Steinernema feltia* was also given overexpression of a trehalose phosphate synthase quality that had been discovered in *C. elegans*, which increased the control of nematodes (Vellai et al. 1999). Recombinant studies on entomopathogenic nematodes need to be continued and expanded because of the fascinating research on this subject.

9.15 Concluding Remarks

The world's agricultural enterprises and authorities have noticed to necessity of minimizing the fungicide use and its relief trouble with further sustainable styles for the future. Thus, the backing for natural control exploration and its perpetration are of significance. We believe that coming studies will pave for new ideas for the enhancement of sustainable strategies involving smart biologics directed with recombinant DNA technology. Several new genomic approaches involving omics wisdom technologies will help in assessing the bear proper styles for getting success grounded on microsatellite marker-grounded ways besides integrating whole knowledge on the inheritable armature in order to develop further effective pest operation programs. The rapid-fire development of genomic sequencing ways on gene sources results in clear data with lower cost (Wetterstrand 2019) regarding natural fields and operations (Supple and Shapıro 2018). They will also enable us to understand the genomic scale that we use by data mining on gene-rich pools for the purpose of opting for the promising genes to be used in practice at field conditions.

Knowledge about inheritable variation to design artificial chromosomes, either by traditional picky parentage (Zhang et al. 2018) or genomic selection (Xıa et al. 2020) with experimental elaboration (Lirakis and Magalhães 2019) inheritable enrichment stemming from gene pool sources will be helpful for developing effective control strategies against pathogens and pests. But we have to also give attention to genomic data to be attained from organisms living with us at the microscale. Multitudinous openings with genetically engineered microorganisms considered for natural control of factory pests, declination of poisonous wastes, mineralization of rare essence and other processes could be precious sources for improvement present installation incorporating with crop civilization. A better understanding of the genetics, biology and physiology of microorganisms in biotechnological ways will allow us to develop smart microorganisms in natural terrain.

We will be able to slow down the financial losses and damages brought on by crop losses in agricultural products, impairment in nature life in timbers, and other conditions thanks to the rapid advancements in recombinant DNA technology that have led to the development of new strategies against pests and conditions.

We would like to emphasize that the use of recombinant DNA technology in pest control is a promising research area with genetically modified microbes carrying many benefits in terms of long-term continuity and advanced effectiveness. The discovery of new microbial strains and their incredible gene-rich pools remaining for discovery with the assistance of advances in microbial technology will be dependent on the development of much more effective strategies involving gene pools and their combining with recombinant DNA technology to control pests and complaints.

References

Ackermann HW, Tremblay D, Moineau S (2004) Long-term bacteriophage preservation. World Fed Cult Collection Newslett 38:35–40

Anonymous (2008) https://doi.org/10.2144/btn-2018-0007

Balasubramanian S, Zheng D, Liu YJ et al (2009) Comparative analysis of processed ribosomal protein pseudogenes in four mammalian genomes. Genome Biol 10(1):1–10

Bastıde H, Betancourt A, Nolte V, Tobler R, Stöbe P, Futschık A, Schlötterer C (2013) A genome-wide, fine-scale map of natural pigmentation variation in Drosophila melanogaster. PLoS Genet 9:e1003534

Baum JA, Johnson TB, Carlton BC (1999) *Bacillus thuringiensis* natural and recombinant bioinsecticide products. In: Hall FR, Menn JJ (eds) Methods in biotechnology. Humana Press, Totowa, pp 189–210

Baysal Ö, Çalışkan M, Yeşilova Ö (2008) An inhibitory effect of a new *Bacillus subtilis* strain (EU07) against *fusarium oxysporum* f. sp. *radicis-lycopersici*. Physiol Mol Plant Pathol 73(1–3):25–32

Bazan-Peregrino M, Sainson RCA, Carlisle RC et al (2013) Combining virotherapy and angiotherapy for the treatment of breast cancer. Cancer Gene Ther 20(8):461–468

Beachy RN, Loesch-Fries S, Tumer NE (1990) Coat protein-mediated resistance against virus infection. Annu Rev Phytopathol 28(1):451–472

Berk A, Zipursky SL (2000) Molecular cell biology, vol 4. WHFreeman, New York, NY, USA

Beukeboom LW, Niehuis O, Pannebakker BA et al (2010) A comparison of recombination frequencies in intraspecific versus interspecific mapping populations of Nasonia. Heredity 104(3):302–309

Biot-Pelletier D, Martin VJ (2014) Evolutionary engineering by genome shuffling. Appl Microbiol Biotechnol 98(9):3877–3887

Burnell A (2002) Genetics and genetic improvement. In: Gaugler R (ed) Entomopathogenic nematology. CABI International, Wallingford, pp 241–263

Baysal Ö, Lai D, Xu HH, Siragusa M, Çalışkan M et al (2013) A Proteomic Approach Provides New Insights into the Control of Soil-Borne Plant Pathogens by *Bacillus* Species. PLoS One 8 (1):e53182. https://doi.org/10.1371/journal.pone.0053182

Can A (2022) Chitinase a gene region of *Serratia marcescens* GBS19 isolate cloning, expression, purification and investigation of insecticidal effect on agricultural Pest *Myzus persicae*, M.Sc. Thesis, Supervisor Prof. Dr. Ö. Baysal Muğla Sıtkı Koçman University, Mugla, Turkey

Cardinale BJ, Matulich KL, Hooper DU et al (2011) The functional role of producer diversity in ecosystems. Am J Bot 98(3):572–592

Çetinkaya A, Baysal Ö, Silme RS, Azim KM, Saleem F (2022) Genomic characterization and phytostimulative effect of a novel *Serratia* species. Genetika 54(1):341–367

Chakraborty S, Newton AC (2011) Climate change, plant diseases and food security: an overview. Plant Pathol 60(1):2–14

Chattopadhya P, Banerjee G, Mukherjee S (2017) Recent trends of modern bacterial insecticides for pest control practice in integrated crop management system. 3. Biotech 7(1):1–11

Compant S, Clément C, Sessitsch A (2010) Plant growth-promoting bacteria in the rhizo-and endosphere of plants: their role, colonization, mechanisms involved and prospects for utilization. Soil Biol Biochem 42(5):669–678

Corradi N, Keeling PJ (2009) Microsporidia: a journey through radical taxonomical revisions. Fungal Biol Rev 23(1–2):1–8

Danışmazoğlu M, Demir I, Sezen K, Muratoğlu H, Nalçacioğlu R (2015) Cloning and expression of chitinase A, B, and C (chiA, chiB, chiC) genes from *Serratia marcescens* originating from Helicoverpa armigera and determining their activities. Turk J Biol 39:78–87

DeJong JM, Liu Y, Bollon AP, Long RM, Jennewein S, Williams D, Croteau RB (2006) Genetic engineering of taxol biosynthetic genes in Saccharomyces cerevisiae. Biotechnol Bioeng 93(2): 212–224

Dodds PN, Rathjen JP (2010) Plant immunity: towards an integrated view of plant–pathogen interactions. Nat Rev Genet 11(8):539–548

EEA (2005). https://www.eea.europa.eu

Ekblom R, Wolf JB (2014) A field guide to whole-genome sequencing, assembly and annotation. Evol Appl 7(9):1026–1042

Faino L, Thomma BP (2014) Get your high-quality low-cost genome sequence. Trends Plant Sci 19(5):288–291

Federici BA, Park HW, Sakano Y (2006) Insecticidal protein crystals of *Bacillus thuringiensis*. In: Inclusions in prokaryotes. Springer, Berlin, Heidelberg, pp 195–236

Félix MA, Barkoulas M (2015) Pervasive robustness in biological systems. Nat Rev Genet 16(8): 483–496

Flood J (2010) The importance of plant health to food security. Food security 2(3):215–231

Gasiunas G, Siksnys V (2013) RNA-dependent DNA endonuclease Cas9 of the CRISPR system: holy grail of genome editing? Trends Microbiol 21(11):562–567

Gaugler R, Lewis E, Stuart RJ (1997) Ecology in the service of biological control: the case of entomopathogenic nematodes. Oecologia 109(4):483–489

Hails R (2001) Natural and genetically modified baculoviruses: environmentally friendly pest control or an ecological threat? Outlook on Agriculture 30(3):171–178

Harrison R, Hoover K (2012) Baculoviruses and other occluded insect viruses. Insect pathology:73–131

Hynes AP, Labrie SJ, Moineau S (2016) Programming native CRISPR arrays for the generation of targeted immunity. MBio 7(3):e00202–e00216

Issac S (1992) Fungal-plant interactions. Chapman & Hall

Jouzani GS, Valijanian E, Sharafi R (2017) *Bacillus thuringiensis*: a successful insecticide with new environmental features and tidings. Appl Microbiol Biotechnol 101(7):2691–2711

Jurat-Fuentes JL, Jackson TA (2012) Bacterial entomopathogens. Insect pathology 2:265–349

Karabörklü S, Azizoglu U, Azizoglu ZB (2018) Recombinant entomopathogenic agents: a review of biotechnological approaches to pest insect control. World J Microbiol Biotechnol 34(1):1–12

Korkut A (2022) Molecular characterization and antagonistic activity of *Bacillus amyloliquefaciens* strain EU07 and its mode of action against citrus Brown rot and stem gum disease pathogen *Phytophthora citrophthora*, M.Sc. Thesis, Supervisor Prof. Dr. Ö. Baysal Muğla Sıtkı Koçman University, Muğla, Turkey

Kumar A (2022a) Microbial biocontrol: sustainable agriculture and Phytopathogen management, vol 1. Springer Nature Chem, Cham, Switzerland, pp 1–369. https://doi.org/10.1007/978-3-030-87512-1

Kumar A (2022b) Microbial biocontrol: food security and post harvest management, vol 2. Springer Nature Chem, Cham, Switzerland, pp 1–334. https://doi.org/10.1007/978-3-030-87289-2

Kumar A, Zhimo Y, Biasi A, Salim S, Feygenberg O, Wisniewski M, Droby S (2021) Endophytic microbiome in the Carposphere and its importance in fruit physiology and pathology. In: Spadaro D, Droby S, Gullino ML (eds) Postharvest pathology. Plant pathology in the 21st century, vol 11. Springer, Cham, pp 73–88. https://doi.org/10.1007/978-3-030-56530-5_5

Kumari M, Qureshi KA, Jaremko M, White JF, Singh SK, Sharma VK, Singh KK, Santoyo G, Puopolo G, Kumar A (2022) Deciphering the role of endophytic microbiome in postharvest diseases management of fruits: opportunity areas in commercial up-scale production. Front Plant Sci 13:1026575. https://doi.org/10.3389/fpls.2022.1026575

Lacey LA, Frutos R, Kaya HK, Vail P (2001) Insect pathogens as biological control agents: do they have a future? Biol Control 21(3):230–248

Lirakis M, Magalhães S (2019) Does experimental evolution produce better biological control agents? A critical review of the evidence. Entomol Exp Appl 167(7):584–597
Liu JR, Lin YD, Chang ST, Zeng YF, Wang SL (2010) Molecular cloning and characterization of an insecticidal toxin from *Pseudomonas taiwanensis*. J Agric Food Chem 58:12343–12349
Lomedico PT (1982) Use of recombinant DNA technology to program eukaryotic cells to synthesize rat proinsulin: a rapid expression assay for cloned genes. Proc Natl Acad Sci 79(19): 5798–5802
Magocha TA, Zabed H, Yang M, Yun J, Zhang H, Qi X (2018) Improvement of industrially important microbial strains by genome shuffling: current status and future prospects. Bioresour Technol 257:281–289
Maloy OC (2005) Plant disease management. Plant Instr. https://doi.org/10.1094/PHI-I-2005-0202-01
Mazzoni L, Perez-Lopez P, Giampieri F, Alvarez-Suarez JM, Gasparrini M, Forbes-Hernandez TY, Battino M (2016) The genetic aspects of berries: from field to health. J Sci Food Agric 96(2): 365–371
Mehrotra RS, Aggarwal A (2003) Plant pathology, 2nd edn. Tata McGraw-Hill Publishing Company Limited, pp 411–423
Mishra SR (2009) Understanding forest biology. Discovery Publishing House PVT, New Delhi
Mohanraju P, Makarova KS, Zetsche B, Zhang F, Koonin EV, Van der Oost J (2016) Diverse evolutionary roots and mechanistic variations of the CRISPR-Cas systems. Science 353(6299): aad5147
Nicot PC, Stewart A, Bardin M, Elad Y (2016) Biological control and biopesticide suppression of botrytis-incited diseases. In: Botrytis–the fungus, the pathogen and its management in agricultural systems. Springer, Cham, pp 165–187
Niehuis O, Gibson JD, Rosenberg MS et al (2010) Recombination and its impact on the genome of the haplodiploid parasitoid wasp Nasonia. PLoS One 5(1):e8597
Oerke EC (2006) Crop losses to pests. J Agric Sci 144(1):31–43
Palumbo JD, Yuen GY, Jochum CC, Tatum K, Kobayashi DY (2005) Mutagenesis of β-1, 3-glucanase genes in *Lysobacter* enzymogenes strain C3 results in reduced biological control activity toward Bipolaris leaf spot of tall fescue and Pythium damping-off of sugar beet. Phytopathology 95(6):701–707
Paspati A, Ferguson KB, Verhulst EC, Urbaneja A, González-Cabrera J, Pannebakker BA (2019) Effect of mass rearing on the genetic diversity of the predatory mite Amblyseius swirskii. Entomol Exp Appl 167(7):670–681
Patel P, Kumar S, Kumar A (2022) Plant microbiota: a prospect to edge off postharvest loss. In: Biocontrol mechanisms of endophytic microorganisms. Academic Press, Elsevier, USA, pp 261–284. https://doi.org/10.1016/B978-0-323-88478-5.00006-7
Pathak P, Rai VK, Can H, Singh SK, Kumar D, Bhardwaj N, Roychowdhury R, de Azevedo LCB, Kaushalendra VH, Kumar A (2022) Plant-endophyte interaction during biotic stress management. Plan Theory 11(17):2203. https://doi.org/10.3390/plants11172203
Pennisi E (2013) The CRISPR craze. Science 341(6148):833–836
Pérez-de-Luque A, Tille S, Johnson I, Pascual-Pardo D, Ton J, Cameron DD (2017) The interactive effects of arbuscular mycorrhiza and plant growth-promoting rhizobacteria synergistically enhance host plant defences against pathogens. Sci Rep 7(1):1–10
Prasad V, Srivastava S (2016) Insect viruses. In: Ecofriendly pest management for food security. Academic Press, pp 411–442
Roldao A, Mellado MC, Castilho LR, Carrondo MJ, Alves PM (2010) Virus-like particles in vaccine development. Expert Rev Vaccines 9(10):1149–1176
Schlötterer C, Tobler R, Kofler R, Nolte V (2014) Sequencing pools of individuals individuals—mining genome-wide polymorphism data without big funding. Nat Rev Genet 15:749–763
Selosse MA, Baudoin E, Vandenkoornhuyse P (2004) Symbiotic microorganisms, a key for ecological success and protection of plants. C R Biol 327(7):639–648

Shmakov S, Abudayyeh OO, Makarova KS et al (2015) Discovery and functional characterization of diverse class 2 CRISPR-Cas systems. Mol Cell 60(3):385–397

Singh VK, Singh AK, Kumar A (2017) Disease management of tomato through PGPB: current trends and future perspective. 3 Biotech 7(4):255

Singh M, Srivastava M, Kumar A, Singh AK, Pandey KD (2020) Endophytic bacteria in plant disease management. In: Kumar A, Singh KV (eds) Microbial endophytes: prospects for sustainable agriculture. Woodhead Publication, Cambridge USA, pp 61–89. https://doi.org/10.1016/B978-0-12-818734-0.00004-8

Sinha KK, Choudhary AK, Kumari P (2016) Entomopathogenic fungi. In: Ecofriendly pest management for food security. Academic Press, pp 475–505

Slack J, Li Z, Escasa S, Doucet D, Ladd T et al (2009) Genomics of entomopathogenic viruses. In: Stock SP, Vandenberg J, Glazer I, Boemare N (eds) Insect pathogens: molecular approaches and techniques. CABI, Wallingford, pp 329–342

Solter LF, Becnel JJ, Oi DH (2012) Microsporidian entomopathogens. In: Vega FE, Kaya HK (eds) Insect pathology. Academic, London, pp 222–263

Supple MA, Shapıro B (2018) Conservation of biodiversity in the genomics era. Genome Biol 19: 131

Theoduloz C, Vega A, Salazar M, Gonzalez E, Meza-Basso L (2003) Expression of a *Bacillus thuringiensisδ*-endotoxin cry1Ab gene in *Bacillus subtilis* and *Bacillus licheniformis* strains that naturally colonize the phylloplane of tomato plants (*Lycopersicon esculentum*, Mills). J Appl Microbiol 94(3):375–381

Ullah MW, Ul-Islam M, Khan S, Kim Y, Park JK (2015a) Innovative production of bio-cellulose using a cell-free system derived from a single cell line. Carbohydr Polym 132:286–294

Ullah MW, Khattak WA, Ul-Islam M, Khan S, Park JK (2015b) Encapsulated yeast cell-free system: a strategy for cost-effective and sustainable production of bio-ethanol in consecutive batches. Biotechnol Bioprocess Eng 20(3):561–575

Ullah MW, Ul-Islam M, Khan S, Kim Y, Park JK (2016a) Structural and physico-mechanical characterization of bio-cellulose produced by a cell-free system. Carbohydr Polym 136:908–916

Ullah MW, Khattak WA, Ul-Islam M, Khan S, Park JK (2016b) Metabolic engineering of synthetic cell-free systems: strategies and applications. Biochem Eng J 105:391–405

Vellai T, Molnár A, Lakatos L, Bánfalvi Z, Fodor A, Sáringer G (1999) Transgenic nematodes carrying a cloned stress resistance gene from yeast. In: Glazer P, Richardson P, Boemare N, Coudert F (eds) Survival of entomopathogenic nematodes. Offce for Offcial Publications of the European Communities, Luxembourg, pp 105–119

Venter M (2007) Synthetic promoters: genetic control through cis engineering. Trends Plant Sci 12(3):118–124

Vyas VK, Barrasa MI, Fink GR (2015) A Candida albicans CRISPR system permits genetic engineering of essential genes and gene families. Sci Adv 1(3):e1500248

Wakefield ME (2018) Microbial biopesticides. In: Recent advances in stored product protection. Springer, Berlin, Heidelberg, pp 143–168

Walker GM (2009) Yeasts. In: Desk encyclopedia of microbiology. Academic Press/Elsevier, pp 1174–1187

Wang Z, Mark G, Snyder M, Gerstein M, Snyder M (2009) RNA-Seq: a revolutionary tool for transcriptomics. Nat Rev Genet 10:57–63

Wang R, Li M, Gong L, Hu S, Xiang H (2016) DNA motifs determining the accuracy of repeat duplication during CRISPR adaptation in Haloarcula hispanica. Nucleic Acids Res 44(9): 4266–4277

Weller DM, Raaijmakers JM, Gardener BBM, Thomashow LS (2002) Microbial populations responsible for specific soil suppressiveness to plant pathogens. Annu Rev Phytopathol 40(1): 309–348

Wetterstrand KA (2019) DNA Sequencing Costs: Data from the NHGRI GenomeSequencing Program (GSP). https://www.genome.gov/about-genomics/factsheets/DNA-Sequencing-Costs-Data. Accessed 10 May 2019

Xıa S, Pannebakker BA, Groenen MAM, Zwaan BJ, Bıjma P (2020) Quantitative genetics of wing morphology in the parasitoid wasp Nasonia vitripennis: hosts increase sibling similarity. Heredity 125:40–49

Yadav MK, Patel P, Kumar A, Modi A, Babu PJ, Chhakchhuak L, Singh G, Mishra AK, Longchar B, Singh PK (2022) Tailoring disease resilience crops through CRISPR/Cas.. Tailoring disease resilience crops through CRISPR/Cas. In: Kumar A (ed) Microbial biocontrol: food security and post harvest management. Springer, Cham. https://doi.org/10.1007/978-3-030-87289-2_7

Yul RJ, Qin Liu O, Choi JY, Wang Y, Shim HJ et al (2009) Construction of a recombinant Bacillus velezensis strain as an integrated control agent against plant diseases and insect pests. J Microbiol Biotechnol 19:1223–1229

Zhang G, Lı Y, Tıan C, Xu Y, Zhou H, Huang J, Wang J, Lıu H (2018) Intraspecific variations on thermal susceptibility in the predatory mite Neoseiulus barkeri Hughes (Acari: Phytoseiidae): responding to long-term heat acclimations and frequent heat hardenings. Biol Control 121:208–215

Zhao S, Tian CY, Du CM (2014) Production of hybrids of Streptomyces bikiniensis strain HD-087 by genome shuffling and enhancement of its bio-control activity against *Fusarium oxysporum* f. sp. *cucumerinum*. J Hortic Sci Biotechnol 89(2):147–152. https://journals.plos.org/plosone/article?id=10.1371/journal.pone.0053182

Chapter 10
Biocontrol of Cucurbit Bacterial Diseases

Sumer Horuz and Yesim Aysan

Abstract Cucurbit crops are widely produced and consumed worldwide. Plenty of plant pathogenic microorganisms cause a serious threat to the cucurbit production stability and quality of food and feed. Various control strategies are adopted for years. The integration of chemical fungicides in intensive agriculture has been the predominant control measurement for combatting plant diseases caused by phytopathogenic microorganisms. However, they are disrupting the quantity of beneficial microorganisms in the soil and the efficiency of soil fertility. Biological control can offer an environmentally friendly alternative to the use of synthetic chemicals for controlling various plant diseases. Certain bacterial, fungal, viral, or yeasts have the capability to prevent plants from various diseases in the natural environment and to replace chemicals. This chapter outlined the role of biological control agents against cucurbit bacterial diseases and understanding the mechanisms of those agents such as the production of antibiotics, siderophores, competition for nutrition, and stimulating plant growth.

Keywords Cucurbitaceae · Plant disease · Biocontrol · *Acidovorax*

10.1 Introduction

The cucurbits are a very diverse group of plant species including squash, pumpkin, melon, muskmelon, watermelon, and cantaloupe and they form in the plant family Cucurbitaceae. There are over 200 cucurbit diseases affecting all parts of the plant. Bacterial seedling blight and fruit blotch (BFB) disease caused by *Acidovorax citrulli* (*Ac*) is a devastating bacterial disease in cucurbits worldwide. However,

S. Horuz (✉)
Department of Plant Protection, Faculty of Agriculture, Erciyes University, Kayseri, Turkey
e-mail: shoruz@erciyes.edu.tr

Y. Aysan
Department of Plant Protection, Faculty of Agriculture, Cukurova University, Adana, Turkey

K. K. Bastas et al. (eds.), *Microbial Biocontrol: Molecular Perspective in Plant Disease Management*, Microorganisms for Sustainability 49,
https://doi.org/10.1007/978-981-99-3947-3_10

disease control is a challenge since it causes great losses in harvesting fruit. The effective disease management requires the best decisions to reduce the risks of great loss. Management strategies are mainly based on disease prevention and reduce disease spread in the growing area. There are several control options to prevent crops from bacterial diseases like BFB. Organic farming can be a solution among these options and has been shown to be profitable. However, the quantity of harvested products in organic farming is reduced in value than that obtained under conventional farming systems. Moreover, these quantities vary from crops to crops, even in countries and regions. The use of bacterial disease-resistant cultivars has significantly improved disease control. To our knowledge, there are no *Acidovorax citrulli*-resistant cultivars among cucurbit crops yet. Seed health testing technologies and several seed treatments are both intensively used management aids for controlling *Ac* infestations (Dutta et al. 2012). Seed treatments like heat therapy (Hopkins et al. 2003), fermentation (Hopkins and Thompson 2002), treatment with chlorine gas (Stephens et al. 2008), and some chemical and biological treatments (Hopkins et al. 2003; Horuz and Aysan 2018) have proven to be effective to reduce the pathogen transmission, but these treatments are not 100% effective in eradication the pathogen from the seeds.

In fact, the preventive effects of chemical fungicides in plant diseases have a longstanding date back over centuries. The chemicals are constituent to the crop growing in several countries around the world, resulting in high quantities of yields and high farmer profits. Economic input studies have shown conclusively that the production of some crops would be a big challenge in most parts of the world without the application of any fungicides against plant pathogens in plant protection. Chemical control of plant bacterial diseases is provided by copper-containing bactericides or fungicides. Copper-based fungicides are recommended to be applied before blooming, thus, at least two to three copper sprays are required for the plant foliage (Traore 2014). Antibiotics are also used as protective against bacterial infections but, they are quite inefficient since the pathogens have the ability to develop resistance in years.

Biological control strategies in agriculture have been established to be very efficient in many modes of action, particularly in the control of several plant pathogens by applying beneficial microorganisms called biological chemicals like bacteria, fungi, viruses, yeasts, or a cocktail of these to the plant organs and/or the soil. Treatments of biological control agents (BCAs) against plant pathogens might be very effective under the right conditions. The basic success of a BCA is that it is especially specific for a phytopathogen and thus, BCAs are described as safe to out of target organisms in the environment. However, challenges in the biocontrol of microbial pathogens in major agricultural crops include issues with environmental suitability, restrictions in production contents, requirements like refrigeration, and lack of reliability in control of the target pathogen. That is why, so far, biocontrol of plant diseases with the help of BCAs in agriculture has only met with restricted success.

This chapter is designed to summarize the biological control strategies used by several researchers against the most devastating cucurbit bacterial pathogens including *A. citrulli*, *Xanthomonas cucurbitae*, and *Pseudomonas syringae* pv. *lachrymans* worldwide. All the recent papers were analyzed and the results were discussed in detail.

10.2 Biocontrol of Bacterial Seedling Blight and Fruit Blotch (*Acidovorax citrulli*) in Cucurbits

The Gram-negative bacterium *A. citrulli* causes bacterial seedling blight and fruit blotch in cucurbits around the world (Schaad et al. 2008). The pathogenic bacterium is actually seed transmitted and infects all parts of cucurbit plants except roots at all growth stages. Cause of the seed infections, extended to seedling contaminations, moreover, cotyledons develop water-soaked lesions that are irregularly shaped, and seedling blights or rot on fruit (Popović and Ivanović 2015). Lesions can expand and turn into brown to reddish-brown color and become necrotic within time. Hypocotyls can also bear lesions and following collapsed tissues and die. Watermelon fruits are the most common susceptible plant organs to the impact of the disease. At first, the upper surface of the fruit develops small irregularly shaped lesions resulting in extension into blotches. These blotches can surround most of the fruit within a few days after infection. Afterward, the center of lesions can bear cracks and amber-colored exudates can come from the central blotch area. Those symptoms limit the marketable fruit (Horuz and Aysan 2018). For decades, due to limitations in controlling BFB, biocontrol strategy using biological microorganisms has gained a large scale of importance as an alternative control strategy during time (Chowdhury et al. 2015). Some beneficial bacteria like *Bacillus* spp., *Burkholderia* sp., *Curtobacterium* sp., *Microbacterium* sp., nonpathogenic bacterial species, *Pantoea* sp., *Pseudomonas* spp., *Streptomyces* spp., yeasts like *Pichia* spp. *Rhodotorula* spp., and endophytic fungus, *Neocosmospora* sp., have been isolated and tested against cucurbit bacterial pathogens. Bacillus and Pseudomonas genera are the most prominent rhizosphere and endophytic bacteria that are effective as BCAs.

10.2.1 Essential Bacterial Species Used in Control of A. citrulli

Several modes of action of *Bacillus* and *Pseudomonas* varieties were found to donate to the biocontrol, including antibiotic production, competition for food or place, triggering host systemic resistance, and promoting plant growth (Chowdhury et al. 2015). The bacterial antagonists are tested in vitro to check their antibiosis or competition ability and applied to plants or soil under in vivo conditions. Medeiros et al. (2009) treated healthy melon seeds with BCAs isolated from different crops including *Bacillus* sp. RAB9 (radish), *Bacillus* sp. MEN19 (melon), *Paenibacillus* lentimorbus MEN2 (melon), not identified R9 (radish), *Pseudomonas chlororaphis* JM339 (cotton), and *Enterobacter cloacae* PEP91 (cucumber), treated seeds sown and kept under greenhouse conditions. Among all, the bacterial antagonist *Bacillus* sp. RAB9 diminished the area under the disease progress curve (AUDPC) up to 47% and raised the incubation period by 35%, respectively, when applied to seeds. When sprayed on melon seedlings, the bacterium *Paenibacillus lentimorbus* code MEN2 inhibited disease development and minimized the AUDPC and disease index over 80%, and disease incidence by 77%, respectively. In another study conducted by Chun-Hao et al. (2015), around 200 putative antagonists were tested for their potential as biocontrol agents (BCA) against *Ac* in vitro by dual culture tests and in vivo plant treatments under greenhouse conditions. The calculation of the diameter of inhibition zones showed that 50% of biocontrol strains had antagonism to the *Ac* strain. Since the limitations of testing almost all strains under greenhouse conditions, scientists preferred to choose the most effective five antagonists (54, YF3, H13, 1JN5, and Y2) for further experiments. Among the tested antagonists, greenhouse experiments showed that the biocontrol efficacy of five antagonists ranged from 45.72 to 56.59%, in which, the bacterium *Bacillus amyloliquefaciens* 54 showed the highest antagonism. Additionally, bacterial strain exhibited increases in shoot length, and much darker green leaves as a result of increases in contents of leaf chlorophylls by 21.9%, 19.8%, and 21.4%, respectively, when compared with the control group. Twelve antagonistic bacterial strains from different soils in Nigeria, nine strains from rockwool, five strains from the soil in the greenhouse, and four strains from pepper fields in Korea were isolated using the serial dilution method and tested for their ability to inhibit *Ac* growth in Petri dishes. Among the tested 30 strains, 15 were characterized as different species of *Bacillus*, and the remaining belonged to other different genera. Two bacterial strains, *Paenibacillus polymyxa* and *Sinomonas atrocyanea* diminished BFB development in watermelon. Additionally, these strains showed the highest plant growth promoting activities (Adhikari et al. 2017). These works exhibited the multiple modes of action of bacterial strains.

Successful colonization of BCAs on plants is important for biocontrol efficacy. The production of surfactin, iturin, and fengycin, nonribosomally synthesized antibiotics, acts as an important factor in inhibiting plant microbes (Guo et al. 2014). Many antagonistic bacteria or fungi produce multiple antibiotics to eliminate other bacteria or fungi. Researchers demonstrated that surfactin showed a highly inhibitory effect against phytopathogenic microorganisms (Vollenbroich et al. 1997; Hwang et al. 2008; Wen et al. 2011; Sabate and Audisio 2013), and, also acts significant roles in biofilm formation, motility, colonization on host tissues, and trigger plant resistance against plant pathogens (Ghelardi et al. 2012; Luo et al. 2015; Rahman et al. 2015). Many *Bacillus* spp. can produce these antibiotics. Like *Bacillus subtilis* code 9407 isolated from an apple, produced fengycin as an antifungal compound (Fan et al. 2017a) and surfactin as a primary antibacterial compound to combat *Ac* in vitro and in vivo, respectively (Fan et al. 2017b).

Many BCAs are very susceptible to unstableness of the environmental conditions. Their effectiveness in vitro cannot always transfer to in vivo trials. For instance, many *Pseudomonas* species exhibit high efficacy in laboratory trials, but this cannot transfer into stability. Fessehaie and Walcott (2005) evaluated the efficacy of *Pseudomonas fluorescens* code A506, *A. a.* subsp. *avenae* and a bacterium code WS-1, chosen from a collection of BCAs and nonphytopathogenic *Acidovorax* sp. for the control of BFB in watermelon. In the in vitro assays, the strains showed transparent inhibition zones that were over 10 mm in diameter. Naturally, *Ac*-contaminated watermelon seeds were treated with the bacterial strains and results indicated that the strains postponed the attack of BFB seedling transmission. Watermelon seed treatment with *A. avenae* subsp. *avenae* code AAA 99-2 showed excellent suppression of BFB transmission under chamber room; however, these results could not reach under greenhouse trials due to unstable environmental conditions. Likewise, Horuz and Aysan (2018) isolated 322 candidate antagonists by sampling healthy leaves or blossoms and soils from 37 melon and watermelon crop fields in Turkey. Results from in vitro tests showed that 54 individuals were antagonistic strains to *Ac* growth indicating the form of transparent areas between 2.30 and 27.0 mm in diameter. The scientists have chosen 14 strains to characterize via morphological and molecular tools. The strains were in different genera including *Pseudomonas, Bacillus, Paenibacillus, Proteus, Curtobacterium,* and *Microbacterium.* When the watermelon seeds treated with bacterial antagonists, seedling blight disease severity and disease incidence on cotyledons were ranged from 4.57 to 36.53% and from 14.06 to 79.47, respectively. Among the antagonists, *Pseudomonas oryzihabitans*, *Microbacterium oxydans*, *Curtobacterium flaccumfaciens,* and *Pseudomonas fluorescens* highly reduced the *Ac* symptoms in seedlings. Horuz (2021) tested plant epiphytic bacterial strains to screen their ability to combat BFB on melon as seed treatment and in vivo greenhouse experiments, respectively. All the 14 individual bacterial antagonists which were isolated from a previous work by Horuz and Aysan (2018) were highly effective in controlling the disease incidence (DI) and disease severity (DS) in both experiments, reaching up to 94% and 92% reduction in DI, and 99% and 93% reduction in DS, respectively. The bacterial antagonist *P. oryzihabitans* code Ant-12 resulted in the lowest DI (up to 7%) and DS (up to 3%) when applied to the melon seeds. From the results of seed treatments, the most potential seven antagonistic bacteria were interpreted for resisting the blight damages on melon under greenhouse conditions. The bacterial strain *P. oryzihabitans* code Ant-12 reduced the DS and the AUDPC over 50% when compared to control plants treated with only the pathogen. These results are highly correlated with the knowledge that BCAs are very vulnerable to unstable factors that affect their colonization.

Nonpathogenic strains of a pathogen can also act as a BCA in disease control (Sneh 1998). A mutant of *A. citrulli* was tested for its efficacy to colonize on seeds as a biocontrol treatment. The mutant strain diminished disease seedling transmission by 82%, additionally, female watermelon blossoms resulted in 8% BFB seedling transmission when sprayed with that strain (Johnson et al. 2011).

Phosphorus is an essential nutrient required for the growth of plants. Some bacterial strains can solubilize the phosphate and make it useful for the plant and help in disease management. Zhang et al. (2021) treated melon seeds with the antagonistic *Burkholderia* sp. code "N3" and *Acinetobacter* sp. code "M01" to analyze the effect of the strains to control BFB and expression of genes involved in the BFB damage. Results showed that the DI of fruit blotch was diminished by 57% with "N3." Additionally, the plant height (29% and 20%), dry weight (28% and 14%), leaf area (36% and 27%), and nutrient uptakes (N, P, K) of melon plants increased after inoculation with two antagonists. The results indicated that 129 and 33 upregulated and downregulated genes were involved in the resistance of melon.

Yeasts are present in all environmental conditions and have been described as potential antagonists in managing various plant pathogens. Many unicellular fungi have been considered for biocontrol applications due to their antagonistic ability. Wang et al. (2009) from China isolated 463 putative yeast strains among 283 plant samplings. These yeast strains were isolated from apricot, celery, coriander, cucumber, lettuce, soybean, tea, and watermelon plants, respectively. Dual culture tests showed that 24 antagonists diminished *Ac* growth in petri dishes as exhibited by the appearance of transparent areas around the bacterial pathogen. The diameter of transparent zones was 8.0 mm and 19.7 mm for strain 074101-2 and strain 0732-1, respectively. Regarding the results obtained from petri tests, 11 antagonistic yeasts were assessed for biocontrol of BFB on hami melon leaves. The strain *Pichia anomala* reduced the DI up to 49% and the disease index to 19, compared to the control group in both experiments. The remaining yeast strains were ineffective, slightly effective, or intermediate in reducing disease development leaves of hami melon. In order to reduce the seed infestation by the bacterium, the scientists treated the seeds with the potential yeast *P. anomala* 0732-1 under greenhouse conditions. Treatment of melon seeds with the cultural filtrates of the strain *P. anomala* code 0732-1 resulted in mild symptoms (3 and 7%). Similarly, the efficacy of the yeasts *Rhodotorula aurantiaca*, *Rhodotorula glutinis*, and *P. anomala* code CC-2 separately or in combination with silicon was evaluated to enhance the synergistic effect in relation to the control of BFB in seedlings and melon plants. The combination of yeasts with Si decreased the DI and AUDPC and effectively defended the seedlings and plants from *Ac* infection. However, no statistically significant additive efficacies were observed in combination treatments compared with the treatments using the yeasts separately (Conceição et al. 2014).

Some saprophytic and endophytic fungi protect against *Ac*. Gao et al. (2017) tested the antimicrobial capacity of the cultural filtrate of the saprophytic fungus *Aspergillus niger*, citric and oxalic acid (4 and 60 mmol L^{-1}) to reduce the seedborne infection by *Ac*. The results in Petri dishes, it showed that all treatments inhibited the growth of *Ac*. The DI values were lower when the watermelon seeds were treated with *A. niger* (88%), oxalic acid (9%), and HCI (6%) in pot treatments. The study suggested using *A. niger*, citric and oxalic acid as disinfectants for the elimination of seedborne *Ac*. Many microbial BCAs produce plant growth regulatory hormones and volatiles that stimulate plant growth. From a study focused on searching the antimicrobial properties of endophytes against *Ac*, four endophytes showed potential in controlling the pathogen growth on disc diffusion assays by the appearance of transparent zones between 12 and 15 mm. The most promising strain, *Neocosmospora* sp., reduced the DS on seedlings up to 80% when the conidial suspension was inoculated into the soil. Structure analysis revealed that the fungus produced three secondary metabolites exhibiting antagonistic activity against *Ac*. The chemical profile in the planta experiment showed a desired colonization of endophytic microbe in both host plants when analyzed. The study showed that secondary metabolites of the fungus provide significant growth inhibition of *Ac* in planta trials (Klomchit et al. 2021).

Bacteria infecting viruses called bacteriophages have a potential to biocontrol of phytopathogenic bacteria. The phage therapy for the management of plant microbes has been reviewed years ago (Jones et al. 2007). In Korea, samples from soil, leaves, and stems of Solanaceous plants including cucumber, pumpkin, and watermelon were collected from a wide range of sites to isolate phages against *Ac*. After the plaque assays, 46 bacteriophages were isolated and these phages were tested against 42 pathogenic *Ac* strains and 40 other bacterial strains to set forth their host ranges. When the watermelon seeds were coated with the broad host range bacteriophage before inoculation with *Ac* in soil, coated seeds resulted in 96% germination and survival rate of seeds. The experiment on the protection ability of the phage against artificially inoculated seeds has shown an 88% germination rate in phage-coated seeds. As a result, seedlings coated with phages were completely able to survive at a rate of 100%. All these results suggested that the phage may have a potential effect on biocontrol against BFB (Rahimi-Midani et al. 2020).

10.3 Biocontrol of Angular Leaf Spot (*Pseudomonas syringae* pv. *lachrymans*) in Cucurbits

Angular leaf spot (ALS) recognizes as irregular water-soaked lesions on cucurbit green leaves. The lesions rapidly extend until the larger secondary veins, appear as angular lesions. At high relative humidities, a milky exudate bears on those parts. The lesions become dry, turn brown, and later, may drop. ALS can be minimized by using pathogen-free seeds, disease-resistant cultivars, various seed treatments, cultivation of the soil, reducing the relative humidity, and spraying of copper. However, there is few biocontrol studies until now. García-Gutiérrez et al. (2012) applied PGPRs for controlling powdery mildew induced by *Podosphaera fusca* and ALS disease in cucurbits. PGPR *Bacillus* and *Pseudomonas* strains isolated from the roots of melon plants were tested against the disease. Antimicrobial performance, production of siderophores and auxin, swimming and swarming motilities, and biofilm formation were evaluated. Among the used strains, two *B. subtilis* and one *B. cereus* strains, and two *P. fluorescens* strains were selected to apply the melon seedlings. The selected strains increased fresh weight up to 30%. Furthermore, all tested beneficial strains provided high disease control up to 60%. These results suggested that the PGPRs had a potential in integrated control of cucurbit diseases.

Akbaba and Ozaktan (2018) isolated individual endophytic bacteria from healthy cucumber root, stem, leaves, and flowers to test their potential for biocontrol of

P. lachrymans. Among them, 24 antagonists were tested in vitro for the indole-3-acetic acid and siderophore production, phosphate solubilization, and inhibition growth of the pathogenic bacterium. Two strains named *Ochrobactrum pseudintermedium* and *Pantoea agglomerans* were selected as the most promising BCAs, thus, were tested in pots. Both strains reduced the DI by >50% when compared to the control group. Two strains colonized seeds at the rate of 10^6–10^7 CFU/g a day after the seed treatments with bacteria. However, the colonization of strains in cucumber tissues was about 10^4 CFU/g in roots and 10^4–10^5 CFU/g in shoots, respectively when calculated 45 days after bacterization. In another study, Akköprü and Ozaktan (2018) investigated the effect of the acibenzolar-S-methyl (ASM) and the rhizobacterium *Pseudomonas putida* code AA11/1 for biocontrol of cucumber ALS disease under chamber room and soilless growing systems, respectively. The treatments significantly reduced the DI by 69% and 34% in the susceptible and by 92% and 21% in the tolerant cucumber cultivars, respectively. DI was limited by both treatments, whereas bacterial strain significantly increased cucumber yield by 70 and 30% in the tested cucumber cultivars in the soilless system.

Akköprü et al. (2021) monitored the long-term population dynamics of the two endophytic bacteria (EB), *Ochrobactrum* spp. and *Pantoea agglomerans* in cucumber plants. Also, the scientists evaluated the potential efficacy of the strains in managing the ALS disease of cucumber and its efficacy on plant growth parameters in the greenhouse conditions. Both strains survived in plant organs until the end of the trials, in which, population densities diminished from 10^5 to 10^3 CFU g plant $^{-1}$ with time. The EB applications increased the total yield by 22% and 21%, respectively. The strain *P. agglomerans* significantly reduced the DS by 41%. This strain might contribute into protect the yield losses and disease infections in soilless growing systems.

10.4 Biocontrol of Bacterial Spot (*Xanthomonas cucurbitae*) in Cucurbits

Bacterial spot of cucurbits is sporadic in winter and squash, pumpkin, cucumber, and gourds are in humid and warm areas. Symptoms are similar to those of the angular leaf spot. Initially, water-soaked lesions first appear on leaves and later become round or angular surrounding with a wide yellow halo. The studies on biocontrol attempts of bacterial spot of cucurbits are very rare. Sulley et al. (2021) collected a total of 271 antagonistic bacterial isolates from the Midwestern states of the US, which inhibited *X. cucurbitae* growth in petri dish assays. The identification of the antagonistic strains was completed based on 16S rRNA primers and multilocus sequence typing and they were mostly classified in the genera *Pseudomonas*, *Pantoea,* and *Serratia*. The study found that *B. amyloliquefaciens*, *B. velezensis*, *Pantoea agglomerans*, and *P. putida* had the potential to protect pumpkin plants from bacterial spot development both in greenhouse and field experiments. Two antagonistic bacteria, *B. amyloliquefaciens* and *P. putida* increased pumpkin growth. The results showed the suitability of the biocontrol agents *Bacillus* and *Pseudomonas* species and thus, needs more research.

10.5 Concluding Remarks

Plants and pathogens are in an interaction with a wide variety of organisms. These interactions can be direct or indirect and significantly affect plant health and growth in different ways. The direct antagonism detects by dual culture assays on agar medium. There are multiple and different mechanisms that BCAs use to protect plants from pathogens under different circumstances. Plenty of biocontrol works are available, but further development and adoption of those methods require a greater understanding of the interactions among plants, people, non-target organisms, and the environment. This chapter is presented and described the current status of research and application of biological products in cucurbit bacterial diseases.

For decades, high quantities of crop yield have been achieved by chemical applications called chemical pesticides, resistance in plant cultivars, and other approaches. However, the extensive advantage of these measures has been lacking owing to the ecological influences or long-term impacts (He et al. 2021). Among those alternatives, biocontrol appears the most reliable and promising method for ecofriendly and sustainable agriculture. Nevertheless, nowadays, it is inadequate to conclude that the application of BCAs could effectively improve the fitness of disease management and sustainability in agricultural areas.

Since a lot of potential antagonists for biocontrol have been identified and used against many plant pathogens, only a few biopesticides are in use globally. Like this, there is no registered biopesticide against cucurbit bacterial diseases. To fill this and have a wide advantage of these BCAs in biocontrol, the researchers must rely on running more research and following the way from identification of the antagonistic strains to production, formulation, and releasing to the agricultural industry.

References

Adhikari M, Yadav DR, Kim SW, Um YH, Kim HS, Lee SC, Song JY, Kim HG, Lee YS (2017) Biological Control of Bacterial Fruit Blotch of Watermelon Pathogen (Acidovorax citrulli) with Rhizosphere Associated Bacteria. Plant Pathol J 33(2):170–183

Akbaba M, Ozaktan H (2018) Biocontrol of angular leaf spot disease and colonization of cucumber (*Cucumis sativus* L.) by endophytic bacteria. Egypt J Biol Pest Control 28:14

Akköprü A, Ozaktan H (2018) Identification of rhizobacteria that ıncrease yield and plant tolerance to angular leaf spot disease in cucumber. Plant Prot Sci 54:67–73

Akköprü A, Akat S, Ozaktan H, Gül A, Akbaba M (2021) The long-term colonization dynamics of endophytic bacteria in cucumber plants, and their effects on yield, fruit quality and angular leaf spot disease. Sci Hortic 282:110005

Chowdhury SP, Hartmann A, Gao X, Borriss R (2015) Biocontrol mechanism by root-associated *Bacillus amyloliquefaciens* FZB42-a review. Front Microbiol 6:780. https://doi.org/10.3389/fmicb.2015.00780

Chun-Hao J, Fang W, Zhen-Yun Y, Ping X, Hong-Jiao K, Hong-Wei L, Yi-Yang Y, Jian-Hua G (2015) Study on screening and antagonistic mechanisms of Bacillus amyloliquefaciens 54 against bacterial fruit blotch (BFB) caused by *Acidovorax avenae* subsp. *citrulli*. Microbiol Res 170:95–104

Conceição CS, Felix KCS, Mariano RLR, Medeiros EV, Souza EB (2014) Combined effect of yeast and silicon on the control of bacterial fruit blotch in melon. Sci Hortic 174:164–170

Dutta B, Scherm H, Gitaitis RD, Walcott RR (2012) *Acidovorax citrulli* seed inoculum load affects seedling transmission and spread of bacterial fruit blotch of watermelon under greenhouse conditions. Plant Dis 96:705–711

Fan H, Ru J, Zhang Y, Wang Q, Li Y (2017a) Fengycin produced by *Bacillus subtilis* 9407 plays a major role in the biocontrol of apple ring rot disease. Microbiol Res 199:89–97. https://doi.org/10.1016/j.micres.2017.03.004

Fan H, Zhang Z, Li Y, Zhang X, Duan Y, Wang Q (2017b) Biocontrol of bacterial fruit blotch by *Bacillus subtilis* 9407 via surfactin-mediated antibacterial activity and colonization. Front Microbiol 8:1973

Fessehaie A, Walcott RR (2005) Biological control to protect watermelon blossoms and seed from infection by *Acidovorax avenae* subsp. *citrulli*. Phytopathology 95:413–419

Gao T, Hao F, Yang D, Bie Z, Li G (2017) Oxalic acid produced by *Aspergillus Niger* Y-1 is effective for suppression of bacterial fruit blotch of watermelon seedlings. Biol Control 112:28–33

García-Gutiérrez L, Romero D, Zeriouh H, Cazorla FM, Torés JA, de Vicente A, Pérez-García A (2012) Isolation and selection of plant growth-promoting rhizobacteria as inducers of systemic resistance in melon. Plant Soil 358:201–212

Ghelardi E, Salvetti S, Ceragioli M, Gueye SA, Celandroni F, Senesi S (2012) Contribution of surfactin and SwrA to flagellin expression, swimming, and surface motility in Bacillus subtilis. Appl Environ Microbiol 78(18):6540–6544

Guo Q, Dong W, Li S, Lu X, Wang P, Zhang X, Wang Y, Ma P (2014) Fengycin produced by *Bacillus subtilis* NCD-2 plays a major role in biocontrol of cotton seedling damping-off disease. Microbiol Res 169:533–540. https://doi.org/10.1016/j.micres.2013.12.001

He DC, Burdon JJ, Xie LH, Zhan J (2021) Triple bottom-line consideration of sustainable plant disease management: from economic, sociological and ecological perspectives. J Integr Agric 20:2581–2591

Hopkins DL, Thompson CM (2002) Seed transmission of *Acidovorax avenae* subsp *citrulli* in cucurbits. HortScience 37:924–926

Hopkins DL, Thompson CM, Hilgren J, Lovic B (2003) Wet seed treatment with peroxyacetic acid for the control of bacterial fruit blotch and other seedborne diseases of watermelon. Plant Dis 87: 1495–1499

Horuz S (2021) *Pseudomonas oryzihabitans:* a potential bacterial antagonist for the management of bacterial fruit blotch *(Acidovorax citrulli)* of cucurbits. J Plant Pathol 103(3):751–758

Horuz S, Aysan Y (2018) Biological control of watermelon seedling blight caused by *Acidovorax citrulli* using antagonistic bacteria from the genera *Curtobacterium*, *Microbacterium* and *Pseudomonas*. Plant Prot Sci 54:138–146

Hwang MH, Kim MH, Gebru E, Jung BY, Lee SP, Park SC (2008) Killing rate curve and combination effects of surfactin C produced from *Bacillus subtilis* complex BC1212 against pathogenic *Mycoplasma hyopneumoniae*. World J Microbiol Biotechnol 24:2277–2282. https://doi.org/10.1007/s11274-008-9752-0

Johnson KL, Minsavage GV, Le T, Jones JB, Walcott RR (2011) Efficacy of a nonpathogenic *Acidovorax citrulli* strain as a biocontrol seed treatment for bacterial fruit blotch of cucurbits. Plant Dis 95:697–704

Jones JB, Jackson LE, Balogh B, Obradovic A, Iriarte FB, Momol MT (2007) Bacteriophages for plant disease control. Annu Rev Phytopathol 45:245–262

Klomchit A, Calderin JD, Jaidee W, Watla-iad K, Brooks S (2021) Napthoquinones from Neocosmospora sp.—antibiotic activity against *Acidovorax citrulli*, the causative agent of bacterial fruit blotch in watermelon and melon. J Fungi 7:370

Luo C, Liu X, Zhou H, Wang X, Chen Z (2015) Nonribosomal Peptide Synthase Gene Clusters for Lipopeptide Biosynthesis in Bacillus subtilis 916 and Their Phenotypic Functions. Environ Microbiol 81(1):422–431

Medeiros FHV, Moraes ISF, Da Silva Neto EB, Silveira EB, Mariano RDLR (2009) Management of melon bacterial blotch by plant beneficial bacteria. Phytoparasitica 37:453–460

Popović T, Ivanović Ž (2015) Occurrence of *Acidovorax citrulli* causing bacterial fruit blotch of watermelon in Serbia. Plant Dis 99:886–886

Rahman A, Uddin W, Wenner NG (2015) Induced systemic resistance responses in perennial ryegrass against Magnaporthe oryzae elicited by semi-purified surfactin lipopeptides and live cells of Bacillus amyloliquefaciens. Mol Plant Pathol 16(6):546–558

Rahimi-Midani A, Kim J-O, Kim JH, Lim J, Ryu J-G, Kim M-K, Choi T-J (2020) Potential use of newly isolated bacteriophage as a biocontrol against *Acidovorax citrulli*. Arch Microbiol 202: 377–389

Sabate DC, Audisio MC (2013) Inhibitory activity of surfactin, produced by different Bacillus subtilis subsp. subtilis strains, against listeria monocytogenes sensitive and bacteriocin-resistant strains. Microbiol Res 168:125–129. https://doi.org/10.1016/j.micres.2012.11.004

Schaad NW, Postnikova E, Sechler A, Claflin LE, Vidaver AK, Jones JB, Agarkova I, Ignatov A, Dickstein E, Ramundo BA (2008) Reclassification of subspecies of *Acidovorax avenae* as *A. avenae* (Manns 1905) emend., *A. cattleyae* (Pavarino, 1911) comb. nov., *A. citrulli* (Schaad et al., 1978) comb. nov., and proposal of *A. oryzae* sp. nov. Syst Appl Microbiol 31:434–446

Sneh B (1998) Use of non-pathogenic or hypovirulent fungal strains to protect plants against closely related fungal pathogens. Biotechnol Adv 16:1–32

Stephens DJ, Schneider RW, Walcott R, Johnson CE (2008) A procedure, based on exposure to chlorine gas, for disinfesting watermelon seeds. Phytopathology 98:150–151

Sulley S, Babadoost M, Hind Sarah R (2021) Biocontrol agents from cucurbit plants infected with *Xanthomonas cucurbitae* for managing bacterial spot of pumpkin. Biol Control 163:104757

Traore SM (2014) Characterization of type three effector genes of *A. citrulli*, the causal agent of bacterial fruit blotch of cucurbits [PhD thesis.] Blacksburg, Virginia Polytechnic Institute and State University, p 146

Vollenbroich D, Ozel M, Vater J, Kamp RM, Pauli G (1997) Mechanism of inactivation of enveloped viruses by the biosurfactant surfactin from *Bacillus subtilis*. Biologicals 25:289–297. https://doi.org/10.1006/biol.1997.0099

Wang X, Li G, Jiang D, Huand HJ (2009) Screening of plant epiphytic yeasts for biocontrol of bacterial fruit blotch (*Acidovorax avenae* subsp. *citrulli*) of hami melon. Biol Control 50:164–171

Wen CY, Yin ZG, Wang KX, Chen JG, Shen SS (2011) Purification and structural analysis of surfactin produced by endophytic *Bacillus subtilis* EBS05 and its antagonistic activity against *Rhizoctonia cerealis*. Plant Pathol J 27:342–348. https://doi.org/10.5423/ppj.2011.27.4.342

Zhang J, Wang P, Xiao Q, Chen J (2021) Effect of phosphate-solubilizing bacteria on the gene expression and inhibition of bacterial fruit blotch in melon. Sci Hortic 282:110018

Chapter 11
Environment-Friendly Management of Plant Diseases by *Bacillus* Through Molecular Pathways

Haris Butt and Kubilay Kurtulus Bastas

Abstract Numerous genetic and climatic factors constantly endanger the quality and quantity of crop products. Synthetic pesticides and fertilizers, which are dangerous to the environment, human health, and animals, are applied as part of conventional management procedures to lessen the effects of these elements. Additionally, excessive usage of these pesticides has had negative effects, including the deaths of animals and farmers from lethal diseases. These days, using microorganisms to support plant growth and provide biological defense against pathogens is given a lot of attention. It has been determined that many bacterial and fungal species are advantageous to the plant hosts. *Bacillus* has been given specific attention in this regard as they are capable of producing pathogen-inhibiting metabolites and indirectly enhancing plant growth. These metabolites include various cell wall degrading enzymes like protease, lipopeptides, chitosanase, hydrogen cyanide, glucanase, and cellulase, which inhibit different disease-causing bacteria, fungi, and viruses. *Bacillus* mimics the harmful effects of pathogens upon contact with the plant hosts activating the host defense mechanisms by triggering resistance genes, proteins, phytohormones, and metabolites. Additionally, the ability to produce spores resistant to harsh environments favors the *Bacillus* spp. as a biocontrol agent. However, the underlying mechanisms for such responses induced by *Bacillus* spp. are not well documented in the literature. This chapter gives the current knowledge available regarding the molecular mechanisms adapted by *Bacillus* spp. related to the biocontrol of different plant disease-causing agents.

Keywords Bacillus · Plant pathogens · Biocontrol agents · Plant–microbe interactions · Bacterial metabolites

---

H. Butt · K. K. Bastas (✉)
Department of Plant Protection, Faculty of Agriculture, Selcuk University, Konya, Turkey
e-mail: kbastas@selcuk.edu.tr

K. K. Bastas et al. (eds.), *Microbial Biocontrol: Molecular Perspective in Plant Disease Management*, Microorganisms for Sustainability 49,
https://doi.org/10.1007/978-981-99-3947-3_11

11.1 Introduction

The World's population is growing faster than ever, and feeding them with ample quality food, feed, and fiber is becoming a huge challenge for the growers. Different diseases of crops caused by bacteria, fungi, viruses, nematodes, and other agents are causing huge yield losses. On the other hand, new threats are faced by global food security due to emerging diseases (Wulff et al. 2011). According to Savary et al. (2012), plant pathogens are responsible for 20–40% of losses in crop produce, while 100% loss of some crops has been recorded (Butt and Bastas 2022). Globally, around US$220 billion and about US$70 billion are lost due to plant pathogenic diseases and invaders, such as insects, respectively. The population is thought to cross 9.7 bn in the year 2050 to 10.9 bn in 2100. Therefore, it is inevitable to manage the crop production losses caused by pests to fulfill the dietary needs of people by practicing different disease management strategies. Such strategies include the use of genetically modified organisms (GMO). These organisms are artificially introduced with genes, using different engineering techniques, that can help them in resisting specific stress that can be biotic or abiotic, but this practice is not successful at the commercial level (Butt and Bastas 2022). Another strategy is the use of synthetic chemicals that poses hazardous effects on the environment, resulting in high production costs and chemical residues on food, polluting soil, and groundwater, and is a proven grave crisis for animals, humans, and other organisms (Latz et al. 2016). But the farmers today still rely on these chemicals due to their operational ease, and quick and effective results. Moreover, pathogens have mutated in response to excessive and abusive use of chemical pesticides (Seong et al. 2017). These issues have demanded more environmentally safe disease control strategies to keep intact the natural ecosystem. The use of microorganisms or natural enemies to plant pathogens has attracted scientists in recent years and has since been increasing. Several strains of crop-friendly microorganisms are known today to provide protection against pathogens along with improving plant growth as well.

Biological control refers to the inhibition of an organism (plant pathogens in our case) by any other living organisms known as microbial biocontrol agents (MBCA) (Köhl et al. 2019). MBCAs subdue the phytopathogens by adopting mechanisms like parasitism, antibiotic secretions, contesting or fighting for nutrients and space, and initiating systemic acquired resistance (SAR) within the host (Koumoutsi et al. 2004; Bloemberg and Lugtenberg 2001). Some species of *Streptomyces*, *Pseudomonas fluorescens*, and *Bacillus* are among bacterial MBCA, while among fungal MBCA *Trichoderma* spp., and *Ulocladiumatrum* spp. are considered very important (Abbas et al. 2019). Bacteria helpful to plants, non-pathogenic, enhance plants' physiological vigor, and substantiate host vigor against various pathogenic and environmental stresses are referred to as plant growth-promoting rhizobacteria (PGPR). They are capable of surviving within the soil and colonize the plant's rhizosphere and can be endophytic and epiphytic (Butt and Bastas 2022; Van Loon 2007; Dimkpa et al. 2009; Baez-Rogelio et al. 2017; Lastochkina et al. 2017; Seifi Kalhor et al. 2017). Antibiosis or hyperparasitism is a direct approach through which beneficial bacteria can kill the harmful bacterial cells, spores, and sclerotium-like resting bodies of fungi (Ghorbanpour et al. 2018). They may not necessarily antagonize or kill the pathogen but indirectly can interact with the plant to activate or induce resistance in it against the pathogen and can increase the already existing one (Conrath et al. 2015; Pieterse et al. 2014). Indirectly, they can also contest or fight for food and room with pathogens (Spadaro and Droby 2016).

A plant's local and systemic defenses, which are activated in response to microbial contact, are made up of various categories. Plenty of well-signaling molecules, like jasmonic acid (JA), ethylene (ET), and salicylic acid (SA) are in charge of controlling these systems. Ef-TU proteins, peptidoglycans, lipopolysaccharides, and bacterial flagellin are some of the extracellular chemicals that plants make that are capable of identifying pathogens. They are known as molecular patterns, such as pathogen-associated (PAMPs) or microbe-associated (MAMPs). In addition to these, other molecules may also be taken into account, such as the intracellularly located effector proteins Avra10, Avr3a, and Avrk. The "zigzag model" of the plant immune system consists of four phases of operation. In phase 1, microbe-associated PAMPs bind to specific cell surface-bound PRRs, causing the plant to recognize the pathogen and activate PAMP-triggered immunity (PTI), which eventually prevents colonization and proliferation. Some infections produce effectors in phase 2 that might elicit effector-triggered susceptibility, interfering with PTI. Phase 3 is when certain effectors are detected by nucleotide binding-leucine-rich repeat receptor molecules. This recognition triggers effector-triggered immunity in the plant (ETI). This develops defenses in the host at the infection site, such as a hypersensitive reaction (HR). Pathogens are under pressure of selection in phase 4 to increase their virulence in order to thwart ETI. This is done by forgoing the production of their old effectors or creating new ones until plants produce new receptors (Butt and Bastas 2022; Thoms et al. 2021).

There have been reports of the possibility of using bacteria from the genera *Agrobacterium, Alcaligenes, Bacillus, Rhizobium,* and *Pseudomonas* as biological control agents. The most researched of these is the genus *Bacillus*. Gram-positive (G^{+ve}), spore-producing Bacillus spp. belongs to the class Firmicutes and is virtually ubiquitous in the environment (Felske 2004; Connor et al. 2010; Pignatelli et al. 2009). Beneduzi et al. (2012) claim that *Bacillus* spp. use strategies such as the synthesis of siderophores and antibiotics, the secretion of hydrolytic enzymes, competition (for nutrients and space), and/or the development of systemic resistance to thwart infections. Extreme climatic conditions such as UV radiation, desiccation, heat, and chemical solvents can all be tolerated by *Bacillus subtilis* endospores (Gao et al. 2016; Ongena and Jacques 2008; Nagórska et al. 2007). As a result, even in adverse environments, it can still help the plant by increasing its resistance. It facilitates the creation of bacillus-based bioformulations and painless stocking (Knox et al. 2000; Mannanov and Sattarova 2001; Leelasuphakul et al. 2006; Aouadhi et al. 2016). It is also simple to culture and isolates. These characteristics have drawn researchers' interest in using *Bacillus* spp. as a biological control agent. *B. subtilis* has been approved for use in food processing industry by the USFDA (Denner and Gillanders 1996).

At present, the beneficial traits of *Bacillus* spp. in guarding plants against abiotic stresses (Dimkpa et al. 2009; Turan et al. 2014; Lastochkina et al. 2017; Aloo et al. 2019), and biotic stresses (García-Gutiérrez et al. 2013; Egamberdieva et al. 2017) are well documented. *Bacillus*-based fertilizers can break down complex nutrients into readily available ones, trigger defense responses, and can inhibit pathogen growth (García-Fraile et al. 2015; Kang et al. 2015a). Here, we discuss the current knowledge of molecular pathways adopted by *B. subtilis* to suppress plant pathogens.

11.2 Biological Control of Plant Diseases

Heimpel and Mills (2017) described biological control as the inhibition of plant pathogens by natural enemies or other living organisms and/or microorganisms. Among such microorganisms, those that can grow on artificial media and are significantly effective against plant pathogens are selected. After selection, antagonists are mass-produced and during a growing season are applied once or more than once in high densities, referred as "augmentative biological control" (van Lenteren et al. 2018; Eilenberg et al. 2001; Heimpel and Mills 2017). For the commercial use of this strategy against plant diseases, farmers use registered MBCA-based products produced by biocontrol and pesticide companies. Sometimes, antimicrobial metabolites can also be found inside these products in addition to the MBCA from the selected biocontrol agent while sometimes, only active ingredients can be antimicrobial metabolites without the living cells of the agent (Glare et al. 2012). The United States, Australia, Japan, Canada, New Zealand, Brazil, and Europe were able to register 101 MBCAs during the year 2017 for their use in the biological control of plant diseases (van Lenteren et al. 2018). As discussed earlier, MBCAs limit the pathogen populations by adopting different modes of action. Effects of antagonists on pathogens and the characteristics of the selected MBCAs vary according to the exploited mode of action. Different and important MBCAs-related aspects like the possibility of pathogen developing resistance, the risk to humans and the environment, crop physiology, specificity to the pathogen, and reliance on climatic conditions may vary depending upon the mode of action intended. Screening of a new potential antagonist is also affected by the preferred use of a particular mode of action, which is required for its envisaged application (Köhl et al. 2011).

11.3 Application of *Bacillus* spp. in Managing Plant Pathogens Biologically

Numerous studies have been conducted and the genus *Bacillus* is used to control plant diseases biologically (Abbas et al. 2019). They settle on plant tissues and can cling to and colonize (populate) the roots of different plants (Fan et al. 2011, 2012). This capability manifests with the introduction of substances having antibacterial and host resistance-inducing characteristics (Anckaert et al. 2021). *Bacillus* species have demonstrated their capacity to suppress a variety of bacterial, viral, and fungal diseases by root colonization, as demonstrated by Fira et al. (2018) and Miljaković et al. (2020). The growth and function of many pathogenic viral, fungal, bacterial, and nematode pathogens are directly affected by the biologically active compounds produced by *Bacillus* spp. These compounds include bacteriocins, antibiotics, biosurfactants, and enzymes (Latz et al. 2016; Barák 2017). These unique enzymes can lyse the mycelium of plant pathogenic fungi *like Alternaria alternata* and *Fusarium culmorum* and break down the chitin and glucans found in the cell walls of fungal pathogens. These enzymes consist of cellulases, proteases, and glucanases. *Bacillus* species like *Bacillus subtilis*, *Bacillus cereus*, and *Bacillus licheniformis* secrete cellulolytic enzymes that are assisting in the worldwide recycling of enormous amounts of cellulose (Seong et al. 2017). Surfactants cause the cell membrane of microorganisms to malfunction, which leads to cell death. Biosurfactants generated by *B. subtilis* strains suppressed the growth of *Candida albicans, Listeria monocytogenes, Botrytis cinerea,* and *Legionella pneumophila* (Chowdhury et al. 2015a).

Bacteriocins are proteins with antimicrobial properties but are reported as active against a few species. These proteins can act as bactericidal agents depending upon the environmental conditions, the growth phase of the pathogen, and the dose used. Antibiotics produced by *Bacillus* spp. can interrupt the functions of cell membranes, disrupt cell wall structure, interfere with protein synthesis and inhibit respiratory chain enzymes (Muslim et al. 2003).

Bacillus amyloliquefaciens and *B. subtilis* inhibited the growth of disease-causing bacteria such as *Xanthomonas axonopodis* pv. *vesicatoria, Ralstonia solanacearum, Erwinia amylovora, Pseudomonas syringae* pv. *syringae* (Pss), and *Xanthomonas arbori* (De Cal et al. 2009). As in the instance of *Bacillus velezensis* LS69, which has demonstrated its antibacterial powers against *Ralstonia solanacearum* and *Erwinia carotovora*, a single strain can combat several bacterial diseases (Liu et al. 2017). *B. amyloliquefaciens* Ba01 (Lin et al. 2018) and strain S1 (Gautam et al. 2019) have been shown to have antibacterial activity against *Streptomyces scabies* and *Clavibacter michiganensis*, respectively. *Botrytis cinerea, Magnaporthe oryzae, Fusarium oxysporum, Blumeria graminis, Colletotrichum acutatum*, and *Zymoseptoria tritici* have all been successfully controlled by *Bacillus* species. (Elanchezhiyan et al. 2018; Jiang et al. 2018; Rahman et al. 2015; Kildea et al. 2008; Wang et al. 2020a; Matzen et al. 2019; Ntushelo et al. 2019). As demonstrated by *B. velezensis* Y6 and F7, which demonstrated in vitro suppression of both

F. oxysporum and *Colletotrichum gloeosporioides*, a single strain can also inhibit multiple fungal pathogens (Cao et al. 2018). According to Lee and Ryu (2016), *B. amyloliquefaciens* 5B6 has demonstrated biocontrol action against the tobacco and pepper plants' *Cucumber mosaic virus,* and *B. amyloliquefaciens* Ba13 has been reported to control tomato plants' *Yellow leaf curl virus* (Guo et al. 2019).

Only *Bacillus* spp. known to be toxic to insects is *B. thuringiensis*. It synthesizes protein crystalline toxins Cry and Cyt, which dissolve in the midgut of the insect and releases prototoxin which is responsible for the disturbance of osmotic level in the plasma membrane. This leads to cell lysis and ultimately intestines are paralyzed eventually killing the host insect (Ahvenniemi et al. 2009). *Bacillus* spp. can solubilize phosphorus and potassium and can produce siderophores and phytohormones, which made their inclusion in the PGPR group possible (Koumoutsi et al. 2004). These are helpful in increasing the photosynthetic activity of plants hence improving plant performance. Some nutrients are not readily available in the soil to be utilized by the plants for their maturation, such as potassium, phosphorus, and nitrogen. *Bacillus lexus* and *Bacillus megaterium* secrete phosphatases, while *Bacillus laevolacticus*, *B. subtilis*, and *B. licheniformis* secrete phytases and both of these compounds can convert phosphorus into phosphate, which can be used by plants (Benitez et al. 2010). *Bacillus coagulans*, *Bacillus edaphicus*, *B. subtilis*, *Bacillus mycoides*, *B. megaterium*, *Bacillus circulans*, *B. cereus*, *B. velezensis*, and *B. firmus* can solubilize potassium into simpler forms to be easily consumed by plants (Perry et al. 2010).

B. velezensis, *B. subtilis*, *and B. thuringiensis* are known to produce phytohormones like indole-3-acetic acid (IAA), gibberellic acid, and abscisic acid, which help in germination and growth of plants. They also support the plants under environmental stresses by enhancing resistance (Scholz et al. 2011). Siderophores are iron-chelating compounds, which can help plants take up iron ions from soil which is difficult for plants under natural conditions due to their low solubility in water. Moreover, siderophores can make iron ions inaccessible to pathogens limiting their growth (Liu et al. 2007). *Bacillus* spp., especially *B. subtilis*, induce signaling molecules like jasmonic acid and ethylene, which in turn agglomerate phenolic compounds at the infection site. Tissues adjacent to the infected ones become rich with flavonoid content and pathogenesis related (PR) proteins are transported to all plant parts or organs eventually initiating the induction of systemic resistance (ISR), and it is very important as it determines the status of basal immunity in plant organisms (Liu et al. 2007). Figure 11.1 shows the rhizospheric and phyllospheric direct secretions of *Bacillus* spp. under different stress conditions. Butt and Bastas (2022) have recently listed the bioformulations based on *Bacillus* spp.

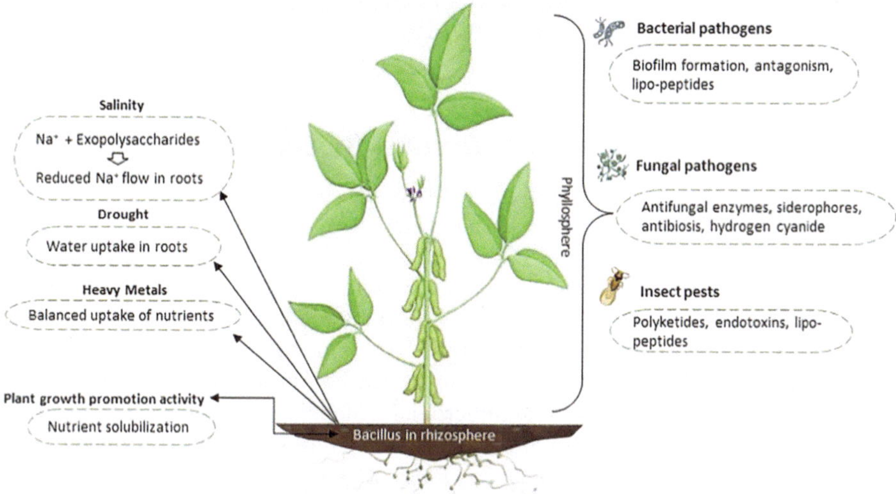

Fig. 11.1 Direct secretions of *Bacillus* spp. (Butt and Bastas 2022)

11.4 *Bacillus* spp. Secreted Metabolites

It is thought that the biocontrol activity of *Bacillus* species is related to their capacity to generate secondary metabolites (SMs) with antibacterial characteristics. The synthesis of these metabolites occupies up to 12% of the genome (Pandin et al. 2017; Liu et al. 2017; Molinatto et al. 2016; Chowdhury et al. 2015a). All organisms of *B. subtilis* group contain genes that are specifically devoted to the release of these compounds. Due to the existence of the majority of the genes encoding these metabolites, *B. velezensis* is the most effective and frequently utilized in biocontrol (Ye et al. 2018; Rabbee et al. 2019; Fan et al. 2018). *B. subtilis* and *B. amyloliquefaciens* are two other species that produce relatively fewer metabolites (Harwood et al. 2018; Andrić et al. 2020). The majority of commercially available strains have either the *Bacillus subtilis* or *Bacillus amyloliquefaciens* patent. However, whole-genome phylogenetic analysis showed that the majority of strains, including MBI600, GBO3, FZB42 or D747, and QST713, are descended from *B. velezensis* species (Dunlap et al. 2016; Fan et al. 2017a; Dunlap 2019).

According to Andersson and Hughes (2014), the molecules of these antimicrobial metabolites are extremely important since they have an impact on the microbial population and play a crucial ecological role. At low concentrations, they can initiate physiological processes like quorum sensing, resistance activation in the host plants, biofilm development, and antibiosis, which are non-lethal physiological responses like competition for nutrients and space (Xu et al. 2019). Peptide antibiotics are produced by the bacteria *B. subtilis*, *B. licheniformis*, *B. thuringiensis*, *B. amyloliquefaciens*, and *B. cereus* (Saxena et al. 2020). There are two distinct categories of peptide compounds: ribosomally synthesized peptides like bacteriocins and nonribosomally synthesized peptides like lipopeptides, bacitracin, gramicidin, glycopeptides, and tyrocidine (Fira et al. 2018).

11.5 Molecular Pathways of *Bacillus* Biocontrol

11.5.1 Antibiosis

Antibiosis refers to the antagonistic effect of one microorganism on another microorganism due to the activity of its toxic SMs. Bacterial genera reported to produce broad-spectrum antibiotics or antimicrobial compounds include *Pseudomonas, Bacillus*, *Pantoea, Streptomyces, Stenotrophomonas*, *Agrobacterium*, and *Serratia*. Because of its effective release of antimicrobial metabolites, antibiosis is recognized as the key defense mechanism of *Bacillus* species against numerous phytopathogens. This mechanism is adapted by *B. cereus, B. amyloliquefaciens, B. subtilis, B. licheniformis, B. mojavensis, B. pasteurii, B. pomilus, B. megaterium, B. mycoides,* and *B. sphaericus*. These metabolites are various lipopeptides or heterogeneous chemical complexes with small molecules. *Bacillus* spp. holds a special place in the biotechnology, agriculture, and pharmaceutical industries thanks to these lipopeptides with an antibacterial activity (Stein 2005; Nagórska et al. 2007; Maksimov et al. 2015).

Bacillus spp. can hinder the thriving of pathogens like *Ralstonia solanacearum*, *Pseudomonas savastanoi*, and *Xanthomonas axonopodis* in the soil and or in the plant tissues (Krid et al. 2012; Yi et al. 2013). The production of antibiotics suppresses and phytopathogenic fungi like *Botryosphaeria ribis*, *Fusarium oxysporum*, *Aspergillus flavus, Helminthosporium maydis*, *Alternaria solani*, *Colletotrichum gloeosporioides*, and *Phomopsis gossypii* and also gram-negative (G^{-ve}) and G^{+ve} bacteria (Duffy et al. 2003). The risk of resistance development by the pathogen due to selection pressure is minimized by using MCBAs as the antibiotics are only released when both come in close contact contrary to conventional pesticides (Butt and Bastas 2022).

11.5.2 Peptide Compounds

Peptide antibiotics predominate among the antibiotics produced by *Bacillus* species. Peptide antibiotics produced by *Bacillus* species are divided into ribosomally produced and nonribosomally produced peptides, taking into consideration the production pathways.

11.5.2.1 Ribosomally Synthesized Metabolites

According to Field et al. (2007) and Zou et al. (2018), bacteriocins peptide compounds are synthesized ribosomally most of which are smaller in size, amphiphilic, and heat stable, which can control many pathogenic and antibiotic-resistant bacteria by targeting their cell envelope. Usually, bacteriocins inhibit the growth and development of bacteria kindred to the producer, while other bacteriocins can act against bacteria not related to their own species (Field et al. 2007). According to Juturu and Wu (2018), these metabolites act by perforating the plasma membrane and by hindering the formation of the target pathogens' cell wall. Bacteriocins and similar substances like subtilosin A and B, subtilosin amylolysin, amysin, and thuricin are synthesized by different bacteria. These include B. *subtilis*, *B. amyloliquefaciens*, *Bacillus coagulans, B. cereus*, and *Bacillus thuringiensis* (Abriouel et al. 2011; Butt and Bastas 2022). Relatively uncomplicated in comparison to bacteriocins, dipeptide bacilysin is known to possess strong antibacterial traits against some pathogens (Chen et al. 2009). But much stronger antimicrobial properties are known to be possessed by the lipopeptides and peptides that are produced nonribosomally by *Bacillus* spp. (Fira et al. 2018).

11.5.2.2 Nonribosomally Synthesized Metabolites

Among the Bacillus synthesized peptide metabolites with antimicrobial properties, the major class is represented by cyclic lipopeptides (cLPs). These cLPs can differ in the fatty acid chain length, sequence of amino acids, and type of peptide cyclization (Ongena and Jacques 2008). Tyrocidine, bacitracin, and gramicidin are nonribosomal peptides (NRPs) produced by Bacillus whereas main lipopeptide families like iturin, surfactin, and fengycin synthesized by nonribosomal peptide synthetases (Cawoy et al. 2015; Dunlap et al. 2011). The chemical structures of iturin, surfactin, and fengycin are shown in Fig. 11.2. Lipopeptides (LPs) are known to damage the structure of the pathogen's cell membrane, and solubilize and perforate it (Fira et al. 2018). There are reports of an interaction between LPs and DNA (Zhang et al. 2013). LPs synthesized by *Bacillus* species play their part in plant defense activation (Han et al. 2015). Different types of peptide compounds produced nonribosomally by Bacillus were listed by Butt and Bastas (2022).

Surfactins are cyclic lipopeptides (cLPs) with extraordinary surfactant activity and are most thoroughly studied among lipopeptide families (Wang et al. 2018). Its amphiphilic nature, according to Ongena and Jacques 2008, enables it to integrate into the lipid layers of other organisms' cell membranes and cause disruption. Surfactin includes surfactin, bamilocyn, lichenysin, pumilacidin, and halobacilin and are reportedly produced by different species of *Bacillus* namely *Bacillus coagulans, B. subtilis, B. licheniformis,* and *Bacillus pumilus* (Butt and Bastas 2022). These are the first members to be identified and characterized among cLPs (Nair et al. 2016). They do their part in the motility and biocontrol properties of the

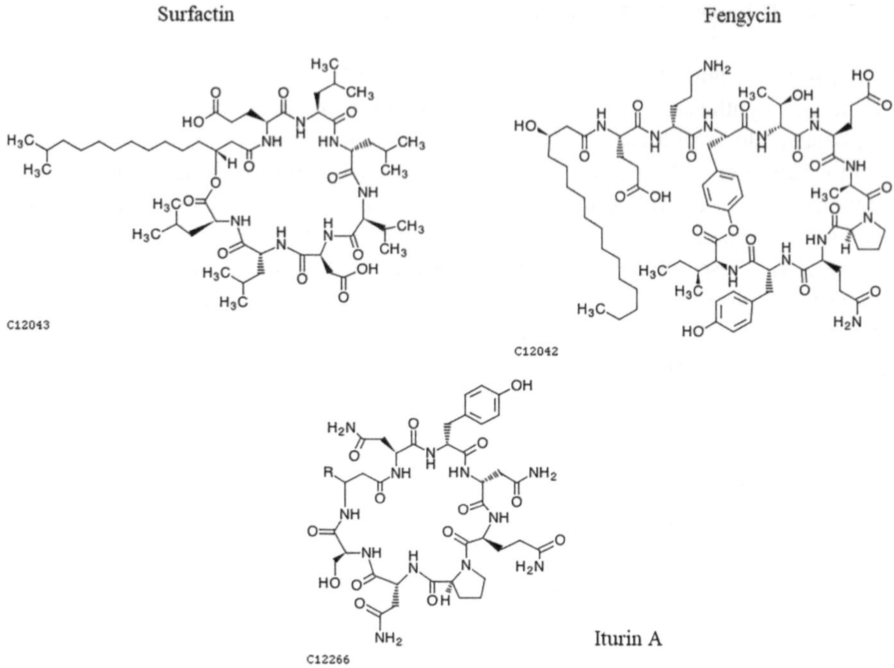

Fig. 11.2 Chemical structures obtained from KEGG COMPOUND Database (https://www.genome.jp/kegg/compound/)

Bacillus in the rhizosphere (Sabaté et al. 2020). Liu et al. (2015) stated that surfactins are stable under elevated saline and high environments, and heat stress. Surfactins were known as inhibitors of fibrin clots but recently their antibacterial, antiviral, and antimycoplasmic activities have been identified (Wang et al. 2018).

Surfactins have antibacterial capabilities but have little effect on plant pathogenic fungi, claim Meena and Kanwar (2015). However, some strains of *B. velezensis*, formerly *B. subtilis or B. amyloliquefaciens*, and *B. subtilis* (B9-5) have been reported to produce excessive amounts of surfactin when in contact with plant pathogenic fungi (Du et al. 2019; Harwood et al. 2018; Fan et al. 2017a; Dunlap et al. 2016; Torres Manno et al. 2019; Pandin et al. 2019; DeFilippi et al. 2018; Chowdhury et al. 2015a; Li et al. 2014). For instance, *B. velezensis* FZB42 overproduced surfactin when it detected the pathogen *R. solani* in the rhizosphere of the lettuce plant (Chowdhury et al. 2015b), and *B. velezensis* SQR9 strongly induced surfactin when *Phytophthora parasitica* and *Sclerotinia sclerotiorum* were present (Li et al. 2014).

On the contrary, Sarwar et al. (2018) reported the antifungal actions of purified surfactin A from surfactin producing strains NH-217 and NH-100 of *Bacillus* against *Fusarium moniliforme* in rice. In the light of these reports, it stays unclear why surfactin is overproduced in the presence of phytopathogenic fungi despite being a relatively poor antifungal compound. There are reports of surfactin being an active compound of *B. subtilis* in the biocontrol of different plant pathogens (Fan et al. 2017b). Surfactin producing *B. subtilis* 9407 has been highly effective against *Acidovorax citrulli* in vitro and has shown significant biocontrol on melon seedlings under greenhouse conditions. DsrfAB mutant that could not produce surfactin had lost both abilities showing the significance of surfactin to control pathogens biologically (Fan et al. 2017b). Surfactins have been reported to be essential in reducing the infections caused by *Pseudomonas syringae* on Arabidopsis plants, whereas surfactin-deficient mutant strain of *B. subtilis* failed to do so. Moreover, biologically relevant concentrations of surfactin when added to liquid culture suppressed the growth of *P. syringae* (Bais et al. 2004). These have also shown their antibacterial effects against human and plant bacterial pathogens such as *X. oryzae*, *R. solanacearum*, *Legionella pneumophila*, and *Listeria monocytogenes* (Luo et al. 2015; Loiseau et al. 2015; Sabaté and Audisio 2013; Yakimov et al. 1995; Naruse et al. 1990). Surfactin is an active antibiotic molecule in the range that is of higher concentration than the natural environment, i.e., 50–200 μM (Sarwar et al. 2018; Fan et al. 2017b; Liu et al. 2014; Debois et al. 2014; Raaijmakers et al. 2010; Jourdan et al. 2009). Surfactins' antimicrobial activity can be linked to the production of other cLPs. Fengycins A, B, and C are included in fengycin group, structurally are lipodecapeptides and are also known as plipastatins (Wang et al. 2015). According to Batool et al. (2011), five biosynthetic genes are responsible for the production of fengycins that are *ppsA*, *ppsB*, *ppsC*, *ppsD*, and *ppsE*. Fengycin A was found more abundantly in *Bacillus amyloliquefaciens* Q426 than B, while double the quantity of Fengycin B was found in *B. subtilis* F29-3 than A (Shu et al. 2002; Zhao et al. 2014). Fengycin, despite being less hemolytic as compared to iturin and surfactin, exhibits a valiant antifungal property and is known to suppress bacterial and fungal pathogens (Ongena and Jacques 2008). *B. amyloliquefaciens* and *B. subtilis* are reported to produce fengycin A and B. Fengycins are known to possess strong antifungal properties explicitly in resisting filamentous fungi (Stein 2005; Fira et al. 2018; Hu et al. 2007; Cawoy et al. 2015). In vitro management of black leaf streak in banana attributable to *Mycosphaerella fijiensis* was demonstrated by Villegas-Escobar et al. (2013) using EA-CB0015 strain of *B. subtilis* that produced a neoteric or unique isoform of Fengycin C. *B. subtilis* hindered the growth of *Monilinia* spp. minimizing the development of brown rot disease in peaches by producing fengycin-like LPs (Yánez-Mendizábal et al. 2011), while fengycin from *B. subtilis* CPA-8 strain reduced the development of *Monilinia fructicola* (Yánez-Mendizábal et al. 2012). Another brown rot disease agent *Botrytis cinerea* was reported to be effectively managed by fengycin-producing *B. subtilis* GA1 (Toure et al. 2004). Fengycin from *B. subtilis* 9407 was found effective in controlling *Botryosphaeria dothidea* in apple (Fan et al. 2017a). Iturins were first identified in Ituri, Democratic Republic of the Congo, in a *B. subtilis* strain. These are cyclic heptapeptides and include different types of bacillomycins, i.e., D, F, and L, iturins such as A, C, D, and E; mycosubtilin; and bacillopeptin (Mnif and Ghribi 2015; Ye et al. 2012; Maget-Dana and Peypoux 1994). Iturins are known to possess robust antifungal properties against different plant fungal pathogens but there are few reports of iturins exhibiting antibacterial properties (Wang et al. 2015; Maget-Dana and Peypoux 1994). *Bacillus* spp. known to produce iturins are *B. circulans*, *B. subtilis*, *Bacillus vallismortis*, *B. pumilus*, and *B. amyloliquefaciens* (Butt and Bastas 2022). LPs similar to iturin released by *B. subtilis* FZB24 were found effective against multiple plant pathogenic fungi (Krebs et al. 1996). *B. subtilis* hindered the growth of *Monilinia fructicola* by producing iturin (Gueldner et al. 1988). Investigation regarding the biocontrol of *Fusarium oxysporum* and *Rosellinia necatrix* revealed that iturin A was the main metabolite among other antibiotics produced by two different strains of *B. subtilis* (Cazorla et al. 2007). *B. subtilis* RB14 was found responsible to control *Rhizoctonia solani* on tomato seedlings through the production of iturin A. Yoshida et al. (2001) found iturin A2 responsible for in vitro growth inhibition *X. campestris* pv. *campestris* (*Xss*) produced by *B. amyloliquefaciens* RC-2. In case of tomato host, synergistic interaction between iturin A and surfactin of *B. subtilis* RB14 suppressed *Rhizoctonia solani* (Asaka and Shoda 1996).

The growth of *Penicillium digitatum* was hampered by fengycin and iturin A. (Waewthongrak et al. 2015). Iturin A and fengycin C, which were secreted by *B. subtilis*, were able to stop *Colletotrichum acutatum* from growing and developing symptoms of anthracnose in tamarillo fruits by 76%. (Arroyave-Toro et al. 2017). It is believed that *B. subtilis*'s simultaneous production of iturin A, fengycin, and bacillomycin is what causes *Podosphaera fusca*, a cucurbit pathogen, to be controlled (Romero et al. 2007). Similar to this, mycosubtilin and surfactin have demonstrated more significant biocontrol of *F. oxysporum* f. sp. *iridacearum* and *Botrytis cinerea* (Tanaka et al. 2015; Mihalache et al. 2018). It has been observed that fengycin and surfactin work well together to treat *Phytophthora infestans* (Wang et al. 2020b). *Bacillus licheniformis* N1 has been linked to the biocontrol of *B. cinerea* on tomato and strawberry plants through the production of iturin A and surfactin (Kong et al. 2010).

Pepper *Phytophthora* blight and cucumber *Fusarium* wilt were controlled as a result of *B. subtilis* producing iturin, bacilysin, and mersacidin simultaneously (Chung et al. 2008). On tomato plants, isomers of iturin A along with surfactin were effective against *P. syringae* after greenhouse application with *B. subtilis* (Hinarejos et al. 2016). *B. subtilis'* iturin prevented the growth of bacterial melon pathogens like *Xanthomonas campestris* pv. *cucurbitae* and *Pectobacterium carotovorum* subsp. *carotovorum.* Citrus canker-causing *Xanthomonas axonopodis* and *Xss* have deteriorated cell walls as a result of *B. subtilis* OG's production of surfactin and iturin. In mixture, application of different *Bacillus* spp. is thought to enhance the protection against a particular pathogen. This may be the result of the simultaneous production of different LPs from different individual biocontrol agents in a mixture (Butt and Bastas 2022).

Besides the aforementioned nonribosomally synthesized LPs, there are a few other types in this category such as bacitracins, kurstakins, tyrocidines, gramicidins, and polymyxins. Bacitracins mainly target Gram^{+ve} bacteria and are found in *B. subtilis*, *Bacillus sonorensis*, and *B. licheniformis* (Adimpong et al. 2012). According to Gélis-Jeanvoine et al. (2017), *B. thuringiensis and B. cereus* are recognized for producing the kurstakins, which work by rupturing the membranes of harmful bacteria and fungi. The main cause of *B. brevis*'s biocontrol action against Gram^{+ve} and Gram^{-ve} bacteria is the activity of tyrocidines and gramicidins it produces. Polymyxins made by *P. polymyxa* (*B. polymyxa*) damage the cells of Gram^{-ve} bacteria (Wan et al. 2018). Another class of nonribosomal metabolites known as polyketides is synthesized by enzyme complexes called polyketide synthases (PKs) (Bozhüyük et al. 2019; Winn et al. 2016; Dutta et al. 2014). Important polyketides from *Bacillus* are difficidins, macrolactins, and bacillaenes, and have far-ranging effects on bacterial pathogens like *Erwinia amylovora*, a serious threat to pome fruits production (Chen et al. 2009). PKs have the ability to interfere with the protein biosynthesis of plant pathogenic bacteria, while bacillaenes and macrolactins are antifungal to some extent (Caulier et al. 2019; Olishevska et al. 2019). Phosphono-oligopeptide rhizocticin produced by *B. subtilis* is unable to act against pathogenic bacteria but is reported to interfere with fungal and nematode growth (Borisova et al. 2010).

Different works have indicated the synthesis of iturin, surfactin, and gramicidin as the major attribute of *B. subtilis* efficacy against many postharvest pathogens/diseases (Ongena et al. 2005; Kong et al. 2010; Yánez-Mendizábal et al. 2011). A huge reported data in the literature points toward antibiosis as the main mechanism against postharvest diseases, but a number of researchers are demanding nonantibiotic producing beneficial microbes or antagonists for such control as well (Droby et al. 1992; Singh and Sharma 2007). However, it is clear that various LPs-producing bacteria strains play a direct role in the host's direct development of pathogen resistance. When compared to synthetic analogs, using bacterial strains that produce LPs is much safer because they are naturally biodegradable, safer for the environment, and able to survive in a wide range of pH and temperature. This approach is ideal for extensive agriculture and food production due to these desirable properties.

11.5.3 Quorum Quenching

Pathogen inhibition is not the only pathway adopted by *Bacillus* spp., to minimize disease severity, but species like *B. subtilis* can indirectly affect the virulence of the pathogen negatively (Pan et al. 2008). Quorum sensing (QS) is a communication process required by bacteria to develop biofilm, secrete SMs, and express different virulence factors to initiate a population onto the host. This is achieved by conducting various signals and regulators and interference with such signals can weaken the pathogen in its establishment. This mechanism of interference with QS is called quorum quenching (QQ) and might demonstrate itself as an important mechanism against phytopathogens (Helman and Chernin 2015). *N*-acyl–L-homoserine lactones (AHL) are well-known QS molecules. Enzymes such as lactonases are able to interfere with AHL and were found to exist in various *Bacillus* spp. and these are thought to be QQ enzymes (Dong et al. 2000; Kalia et al. 2011; Raafat et al. 2019). *P. syringae* was found to be inefficient in colonizing host roots due to the degradation of AHL through QQ by *B. cereus* INT1c leading to the pathogen motility

(Ananda et al. 2019). Several *Bacillus* spp. are known to degrade the *P. syringae* AHL which resulted in the reduced hypersensitivity response of tomato plants (Jose et al. 2019). Potato soft rot pathogen *Erwinia carotovora* depends upon autoinducers for its virulence. *B. subtilis* BS-1 has been reported to reduce potato soft rot symptoms due to the activity of an AiiA enzyme known to interfere with the QS signal. The role of the AiiA enzyme was further supported by the use of *Escherichia coli* supernatant that expressed the *aiiA* gene resulted in reduced the disease severity of *Erwinia carotovora* (Pan et al. 2008).

11.5.4 Induction of Systemic Resistance in Host Plant

Plants have evolved sophisticated defense mechanisms in response to pathogenic attacks to strengthen their resilience. In addition to inducing many defense mechanisms in the host plant, *Bacillus* spp. are well known for their direct attacks on pathogens via their metabolites. Systemic acquired resistance (SAR) and induced systemic resistance (ISR) are the results of these pathways (Pieterse et al. 2014). Several phytohormones, including SA, ABA, JA, and ethylene, modulate these pathways (García-Gutiérrez et al. 2013; Pieterse et al. 2012). These mechanisms are also known to be controlled by CLPs (De Vleesschauwer and Höfte 2009). Upon pathogenic attack, SAR is produced across the entire host plant organism, alerting the plant tissues that are not infected (Durrant and Dong 2004). SA is the mechanism that makes this strategy work. ISR, on the other hand, is a kind of resistance that the host plant instigates in response to interaction with advantageous microbes. For instance, ISR refers to the development of systemic resistance within the entire plant host following contact between beneficial bacteria and the roots of the plant (Shafi et al. 2017). This sort of resistance depends on signaling pathways that are mediated by JA and/or ethylene (Choudhary and Johri 2009). In the melon host, *B. subtilis* UMAF6639 triggered JA- and SA-dependent defense responses, which led to the host developing systemic resistance to powdery mildew (García-Gutiérrez et al. 2013). Today, it is understood that a variety of bacteria can produce auxins, cytokinins, gibberellins, ABA, JA, and SA (Dobbelaere et al. 2003; Dodd et al. 2010; Kudoyarova et al. 2014; Sivasakthi et al. 2013). Numerous bacterial genera and strains, including *Bacillus, Azospirillum, Brevibacterium, Pseudomonas*, and *Lysinibacillus*, are capable of synthesizing ABA and modifying the levels of the hormone in plants (Dodd et al. 2010; Belimov et al. 2014). Specifically, *Azospirillum lipoferum* USA59b treatment of maize plants under drought stress enhances ABA accumulation within the host, which in turn increases plant development (Cohen et al. 2009). Infected plants showed decreased levels of JA and ABA, while Bacillus-treated hosts showed increased levels of IAA, gibberellic acid (GA3) hormones, and SA (Kang et al. 2015b; Chowdappa et al. 2013). Pieterse et al. (2014) recently studied the mechanisms/pathways that MAMPs induce, leading to ISR in plants. From the rice rhizosphere, bacterial strains that can eliminate ABA while promoting growth in tomato hosts have been identified. This is accomplished via an

ABA-dependent pathway, highlighting the potential for a change in the host's internal hormonal balance in response to PGPB exposure (Kumar et al. 2012). Phytohormones are frequently used by pathogens to weaken the host's defenses, whereas PGPB controls the hormonal equilibrium of the host (Kumar et al. 2012). Additionally, through promoting the production of phytohormones and other chemicals, PGPB can cause several defense responses in plants (Maksimov et al. 2015). It is unclear how bacteria suppress viral infections in plants. ISR and improved plant growth against *Cucumber mosaic virus* were produced by *Bacillus* spp. in contact with a cucumber plant. A biofilm was created by *B. amyloliquefaciens* subsp. *plantarum* on the surfaces of the hosts, and it produced surfactin, which activated the hosts' ISR defenses against viral infections (Chowdhury et al. 2015a).

The creation of antifungal chemicals also activates defense systems in biologically regulated fruits and vegetables that have been harvested and kept. Plant ISR mechanisms were activated by *B. subtilis* iturin and fungicin synthesis via promoting the phenylpropanoid metabolism-related genes in plants (Falardeau et al. 2013). The lipoxygenase pathway enzymes in tomato and bean plants were activated by surfactin and fengycin from *B. subtilis* 168, increasing the plants' resistance to *B. cinerea* (Ongena et al. 2010). According to reports, mycosubtilin and surfactin trigger ISR in grape, tomato, melon, tobacco, and bean plants, whereas fengycin and surfactin are known to increase defense responses in potatoes (Farace et al. 2015). *Brassica napus* developed resistance to *Botrytis cinerea* after being exposed to a strain of *B. amyloliquefaciens* that produces surfactin (Sarosh et al. 2009). *Bacillus amyloliquefaciens* FZB42 also caused lettuce to develop resistance to *Rhizoctonia solani* by activating the JA/ethylene-dependent signaling protective pathway (Chowdhury et al. 2015b). Future research should primarily concentrate on gaining insights into the complex mechanisms by which LPs interact with plant cells (Butt and Bastas 2022). Some *Bacillus* strains can create volatile compounds (VOCs), which are substances that can easily travel over a long distance utilizing air diffusion and soil pores. These VOCs typically have low molecular weights of less than 300kD. These are known to trigger plant responses to biotic and abiotic stressors, resulting in the development of ISR (Pertry et al. 2009). VOCs generated by *B. subtilis* GB03, 2R, 3R Butanediol, and C13, respectively, can cause the induction of ISR (Pieterse et al. 2014). In order to reduce spot disease caused by *Xanthomonas axonopodis* pv. *vesicatoria*, VOCs from *B. amyloliquefaciens* induced ISR in the pepper (Choi et al. 2014). The interaction of phytopathogens with biological controls that cause resistance or with signaling molecules is indirect and does not put pathogens in danger from the pressure of natural selection to survive. As a result, microorganisms find it extremely challenging to develop resistance to the induction of resistance (Romanazzi et al. 2016), which is problematic when synthetic drugs are used.

11.6 Concluding Remarks

Phytopathogens pose a serious threat to crop production where scientists and farmers are working hard to produce enough food for the growing population. Chemical control creates environmental imbalance and has hazardous effects on human health. Biological control is an attractive approach to managing plant pathogens with natural antagonists conserving the environment. *Bacillus* strains are widely studied and used for this purpose and are popular among many scientists. These antagonists help minimize the chemical residues in fruits and vegetables during development and can alter their postharvest physiology thus enhancing the resistance to various postharvest pathogens. This prolongs the storage duration and preserves the freshness and nutritional values of products. Commercialization of these microbial antagonists is inevitable and requires further investigations regarding the mode of action of these microbial agents and underlying molecular mechanisms such as the detailed role of antibiotics individually and in combination under field and storage conditions. Nowadays, climate change is the top trend in all aspects of nature. Raising temperatures, drought, and elevated carbon dioxide levels can alter the characteristics of microbial communities and which will influence their activities. This means altered signals from the microbial community will receive an altered response from the plants. Hence, insufficient or short-term studies will not be enough as changes caused by climate change will have a long-term effect on plant–microbe interactions. Fluctuations in the beneficial microbes community will have a proportional effect on the pathogen's community. There is a need for extensive studies to develop and select cultivars that will adapt to the changing climate and aid in the adaptation of microorganisms within the same.

References

Abbas A, Khan SU, Khan WU, Saleh TA, Khan MHU, Ullah S, Ali A, Ikram M (2019) Antagonist effects of strains of *bacillus* spp. against *rhizoctonia solani* for their protection against several plant diseases: alternatives to chemical pesticides. Comp Ren Biol 342:124–135

Abriouel H, Franz CM, Omar NB, Gálvez A (2011) Diversity and applications of *Bacillus* bacteriocins. FEMS Microbiol Rev 35(1):201–232

Adimpong DB, Sørensen KI, Thorsen L, Stuer-Lauridsen B, Abdelgadir WS, Nielsen DS, Derkx PM, Jespersen L (2012) Antimicrobial susceptibility of *Bacillus* strains isolated from primary starters for African traditional bread production and characterization of the bacitracin operon and bacitracin biosynthesis. Appl Environ Microbiol 78(22):7903–7914

Ahvenniemi P, Wolf M, Lehtonen MJ, Wilson P, German-Kinnari M, Valkonen J (2009) Evolutionary diversification indicated by compensatory base changes in ITS2 secondary structures in a complex fungal species, Rhizoctonia solani. J Mol Evol 69(2):150–163

Aloo BN, Makumba BA, Mbega ER (2019) The potential of Bacilli rhizobacteria for sustainable crop production and environmental sustainability. Microbiol Res 219:26–39

Ananda M, Rusmana I, Akhdiya A (2019) Quorum quenching of *Bacillus cereus* INT1c against *pseudomonas syringae*. The 2nd international conference on mathematics, science and computer science, Indonesia, October 2019. J. Phys. Conf. Ser. 1277 012010. IOP Publishing

Anckaert A, Arias AA, Hoff G (2021) The use of *Bacillus* spp. as bacterial biocontrol agents to control plant. Burleigh Dodds Ser Agri Sci. https://doi.org/10.19103/AS.2021.0093.10
Andersson DI, Hughes D (2014) Microbiological effects of sublethal levels of antibiotics. Nat Rev Microbiol 12(7):465–478
Andrić S, Meyer T, Ongena M (2020) Bacillus responses to plant-associated fungal and bacterial communities. Front Microbiol 11:1350
Aouadhi C, Rouissi Z, Kmiha S, Mejri S, Maaroufi A (2016) Effect of sporulation conditions on the resistance of *Bacillus sporothermodurans* spores to nisin and heat. Food Microbiol 54:6–10
Arroyave-Toro JJ, Mosquera S, Villegas-Escobar V (2017) Biocontrol activity of *Bacillus subtilis* EA-CB0015 cells and lipopeptides against postharvest fungal pathogens. Biol Control 114:195–200
Asaka O, Shoda M (1996) Biocontrol of rhizoctonia solani damping-off of tomato with *Bacillus subtilis* RB14. Appl Environ Microbiol 62(11):4081–4085
Baez-Rogelio A, Morales-García YE, Quintero-Hernández V, Muñoz-Rojas J (2017) Next generation of microbial inoculants for agriculture and bioremediation. Microb Biotechnol 10(1): 19–21
Bais HP, Fall R, Vivanco JM (2004) Biocontrol of *Bacillus subtilis* against infection of Arabidopsis roots by *Pseudomonas syringae* is facilitated by biofilm formation and surfactin production. Plant Physiol 134(1):307–319
Barák I (2017) Spores and spore formers. Front Microbiol 8:1046
Batool M, Khalid MH, Hassan MN, Yusuf HF (2011) Homology modeling of an antifungal metabolite plipastatin synthase from the *Bacillus subtilis* 168. Bioinformation 7(8):384
Belimov AA, Dodd IC, Safronova VI, Dumova VA, Shaposhnikov AI, Ladatko AG, Davies WJ (2014) Abscisic acid metabolizing rhizobacteria decrease ABA concentrations in planta and alter plant growth. Plant Physiol Biochem 74:84–91
Beneduzi A, Ambrosini A, Passaglia LM (2012) Plant growth-promoting rhizobacteria (PGPR): their potential as antagonists and biocontrol agents. Genet Mol Biol 35:1044–1051
Benitez LB, Velho RV, Lisboa MP, da Costa Medina LF, Brandelli A (2010) Isolation and characterization of antifungal peptides produced by *Bacillus amyloliquefaciens* LBM5006. J Microbiol 48(6):791–797
Bloemberg GV, Lugtenberg BJ (2001) Molecular basis of plant growth promotion and biocontrol by rhizobacteria. Curr Opin Plant Biol 4(4):343–350
Borisova SA, Circello BT, Zhang JK, van der Donk WA, Metcalf WW (2010) Biosynthesis of rhizocticins, antifungal phosphonate oligopeptides produced by *Bacillus subtilis* ATCC6633. Chem Biol 17(1):28–37
Bozhüyük KA, Micklefield J, Wilkinson B (2019) Engineering enzymatic assembly lines to produce new antibiotics. Curr Opin Plant Biol 51:88–96
Butt H, Bastas KK (2022) Biochemical and molecular effectiveness of *Bacillus* spp. in disease suppression of horticultural crops. In: Seymen M (ed) Sustainable horticulture- microbial inoculants and stress interaction. Elsevier, Academic Press, pp 461–494
Cao Y, Pi H, Chandrangsu P, Li Y, Wang Y, Zhou H, Xiong H, Helmann JD, Cai Y (2018) Antagonism of two plant-growth promoting *Bacillus velezensis* isolates against *Ralstonia solanacearum* and *Fusarium oxysporum*. Sci Rep 8(1):1–4
Caulier S, Nannan C, Gillis A, Licciardi F, Bragard C, Mahillon J (2019) Overview of the antimicrobial compounds produced by members of the *Bacillus subtilis* group. Front Microbiol 10:302
Cawoy H, Debois D, Franzil L, De Pauw E, Thonart P, Ongena M (2015) Lipopeptides as main ingredients for inhibition of fungal phytopathogens by *Bacillus subtilis/amyloliquefaciens*. Microb Biotechnol 8(2):281–295
Cazorla FM, Romero D, Pérez-García A, Lugtenberg BJ, Vicente AD, Bloemberg G (2007) Isolation and characterization of antagonistic *Bacillus subtilis* strains from the avocado rhizoplane displaying biocontrol activity. J Appl Microbiol 103(5):1950–1959

Chen XH, Koumoutsi A, Scholz R, Schneider K, Vater J, Süssmuth R, Piel J, Borriss R (2009) Genome analysis of *Bacillus amyloliquefaciens* FZB42 reveals its potential for biocontrol of plant pathogens. J Biotechnol 140(1–2):27–37
Choi HK, Song GC, Yi HS, Ryu CM (2014) Field evaluation of the bacterial volatile derivative 3-pentanol in priming for induced resistance in pepper. J Chem Ecol 40(8):882–892
Chowdappa P, Kumar SM, Lakshmi MJ, Upreti KK (2013) Growth stimulation and induction of systemic resistance in tomato against early and late blight by *Bacillus subtilis* OTPB1 or *Trichoderma harzianum* OTPB3. Biol Control 65(1):109–117
Chowdhury SP, Hartmann A, Gao X, Borriss R (2015a) Biocontrol mechanism by root-associated *bacillus amyloliquefaciens* FZB42–a review. Front Microbiol 6:780
Chowdhury SP, Uhl J, Grosch R, Alquéres S, Pittroff S, Dietel K, Schmitt-Kopplin P, Borriss R, Hartmann A (2015b) Cyclic lipopeptides of *Bacillus amyloliquefaciens* subsp. *plantarum* colonizing the lettuce rhizosphere enhance plant defense responses toward the bottom rot pathogen *Rhizoctonia solani*. Mol Plant-Microbe Interact 28(9):984–995
Choudhary DK, Johri BN (2009) Interactions of Bacillus spp. and plants-with special reference to induced systemic resistance (ISR). Microbiol Res 164:493–513
Chung S, Kong H, Buyer JS, Lakshman DK, Lydon J, Kim SD, Roberts DP (2008) Isolation and partial characterization of *Bacillus subtilis* ME488 for suppression of soilborne pathogens of cucumber and pepper. Appl Microbiol Biotechnol 80(1):115–123
Cohen AC, Travaglia CN, Bottini R, Piccoli PN (2009) Participation of abscisic acid and gibberellins produced by endophytic *Azospirillum* in the alleviation of drought effects in maize. Botany 87(5):455–462
Connor N, Sikorski J, Rooney AP, Kopac S, Koeppel AF, Burger A, Cole SG, Perry EB, Krizanc D, Field NC, Slaton M (2010) Ecology of speciation in the genus Bacillus. Appl Environ Microbiol 76(5):1349–1358
Conrath U, Beckers GJ, Langenbach CJ, Jaskiewicz MR (2015) Priming for enhanced defense. Annu Rev Phytopathol 53(1):97–119
De Cal A, Sztejnberg A, Sabuquillo P, Melgarejo P (2009) Management *Fusarium* wilt on melon and watermelon by *Penicillium oxalicum*. Biol Control 51(3):480–486
De Vleesschauwer D, Höfte M (2009) Rhizobacteria-induced systemic resistance. Adv Bot Res 51: 223–281
Debois D, Jourdan E, Smargiasso N, Thonart P, De Pauw E, Ongena M (2014) Spatiotemporal monitoring of the antibiome secreted by Bacillus biofilms on plant roots using MALDI mass spectrometry imaging. Anal Chem 86(9):4431–4438
DeFilippi S, Groulx E, Megalla M, Mohamed R, Avis TJ (2018) Fungal competitors affect production of antimicrobial lipopeptides in *Bacillus subtilis* strain B9–5. J Chem Ecol 44(4): 374–383
Denner WH, Gillanders T (1996) The legislative aspects of the use of industrial enzymes in the manufacture of food and food ingredients. Stockton Press, New York, pp 397–412
Dimkpa CO, Merten D, Svatoš A, Büchel G, Kothe E (2009) Siderophores mediate reduced and increased uptake of cadmium by *Streptomyces tendae* F4 and sunflower (*Helianthus annuus*), respectively. J Appl Microbiol 107(5):1687–1696
Dobbelaere S, Vanderleyden J, Okon Y (2003) Plant growth-promoting effects of diazotrophs in the rhizosphere. Crit Rev Plant Sci 22(2):107–149
Dodd IC, Zinovkina NY, Safronova VI, Belimov AA (2010) Rhizobacterial mediation of plant hormone status. Ann Appl Biol 157(3):361–379
Dong YH, Xu JL, Li XZ, Zhang LH (2000) AiiA, an enzyme that inactivates the acylhomoserine lactone quorum-sensing signal and attenuates the virulence of *Erwinia carotovora*. PNAS 97(7): 3526–3531
Droby S, Chalutz E, Wilson CL, Wisniewski ME (1992) Biological control of postharvest diseases: a promising alternative to the use of synthetic fungicides. Phytoparasitica 20(1):149–153
Du Y, Ma J, Yin Z, Liu K, Yao G, Xu W, Fan L, Du B, Ding Y, Wang C (2019) Comparative genomic analysis of *Bacillus paralicheniformis* MDJK30 with its closely related species reveals

an evolutionary relationship between *B. paralicheniformis* and *B licheniformis*. BMC Genom 20(1):1–6

Duffy B, Schouten A, Raaijmakers JM (2003) Pathogen self-defense: mechanisms to counteract microbial antagonism. Annu Rev Phytopathol 41(1):501–538

Dunlap CA (2019) Taxonomy of registered *Bacillus* spp. strains used as plant pathogen antagonists. Biol Control 134:82–86

Dunlap CA, Schisler DA, Price NP, Vaughn SF (2011) Cyclic lipopeptide profile of three *Bacillus subtilis* strains; antagonists of *Fusarium* head blight. J Microbiol 49(4):603–609

Dunlap CA, Kim SJ, Kwon SW, Rooney AP (2016) *Bacillus velezensis* is not a later heterotypic synonym of *Bacillus amyloliquefaciens*; *Bacillus methylotrophicus*, *Bacillus amyloliquefaciens* subsp. *plantarum* and '*Bacillus oryzicola*' are later heterotypic synonyms of *Bacillus velezensis* based on phylogenomics. Int J Syst Evol Microbiol 66(3):1212–1217

Durrant WE, Dong X (2004) Systemic acquired resistance. Annu Rev Phytopathol 42(1):185–209

Dutta S, Whicher JR, Hansen DA, Hale WA, Chemler JA, Congdon GR, Narayan AR, Håkansson K, Sherman DH, Smith JL, Skiniotis G (2014) Structure of a modular polyketide synthase. Nature 510(7506):512–517

Egamberdieva D, Wirth SJ, Shurigin VV, Hashem A, Abd-Allah EF (2017) Endophytic bacteria improve plant growth, symbiotic performance of chickpea (*Cicer arietinum* L.) and induce suppression of root rot caused by *Fusarium solani* under salt stress. Front Microbiol 8:1887

Eilenberg J, Hajek A, Lomer C (2001) Suggestions for unifying the terminology in biological control. BioControl 46(4):387–400

Elanchezhiyan K, Keerthana U, Nagendran K, Prabhukarthikeyan SR, Prabakar K, Raguchander T, Karthikeyan G (2018) Multifaceted benefits of *Bacillus amyloliquefaciens* strain FBZ24 in the management of wilt disease in tomato caused by *Fusarium oxysporum* f. sp. *lycopersici*. Physiol Mol Plant Pathol 103:92–101

Falardeau J, Wise C, Novitsky L, Avis TJ (2013) Ecological and mechanistic insights into the direct and indirect antimicrobial properties of *Bacillus subtilis* lipopeptides on plant pathogens. J Chem Ecol 39(7):869–878

Fan B, Chen XH, Budiharjo A, Bleiss W, Vater J, Borriss R (2011) Efficient colonization of plant roots by the plant growth promoting bacterium *Bacillus amyloliquefaciens* FZB42, engineered to express green fluorescent protein. J Biotechnol 151(4):303–311

Fan B, Borriss R, Bleiss W, Wu X (2012) Gram-positive rhizobacterium *Bacillus amyloliquefaciens* FZB42 colonizes three types of plants in different patterns. J Microbiol 50(1):38–44

Fan H, Ru J, Zhang Y, Wang Q, Li Y (2017a) Fengycin produced by *Bacillus subtilis* 9407 plays a major role in the biocontrol of apple ring rot disease. Microbiol Res 199:89–97

Fan H, Zhang Z, Li Y, Zhang X, Duan Y, Wang Q (2017b) Biocontrol of bacterial fruit blotch by *Bacillus subtilis* 9407 via surfactin-mediated antibacterial activity and colonization. Front Microbiol 8:1973

Fan B, Wang C, Song X, Ding X, Wu L, Wu H, Gao X, Borriss R (2018) *Bacillus velezensis* FZB42 in 2018: the gram-positive model strain for plant growth promotion and biocontrol. Front Microbiol 9:2491

Farace G, Fernandez O, Jacquens L, Coutte F, Krier F, Jacques P, Clément C, Barka EA, Jacquard C, Dorey S (2015) Cyclic lipopeptides from *Bacillus subtilis* activate distinct patterns of defence responses in grapevine. Mol Plant Pathol 16(2):177–187

Felske AD (2004) Ecology of *Bacillus* species in soil. In: Ricca E (ed) Bacterial spore formers: probiotics and emerging applications, horizon bioscience, pp 35–44

Field D, Cotter P, Hill C, Ross RP (2007) Bacteriocin biosynthesis, structure, and function. In: Riley MA (ed) Research and applications in bacteriocins, vol 28. Horizon Bioscience, pp 5–41

Fira D, Dimkić I, Berić T, Lozo J, Stanković S (2018) Biological control of plant pathogens by *bacillus* species. J Biotechnol 285:44–55

Gao H, Xu X, Dai Y, He H (2016) Isolation, identification and characterization of *Bacillus subtilis* CF-3, a bacterium from fermented bean curd for controlling postharvest diseases of peach fruit. Food Sci Technol 22(3):377–385

García-Fraile P, Menéndez E, Rivas R (2015) Role of bacterial biofertilizers in agriculture and forestry. AIMS Bioeng 2(3):183–205

García-Gutiérrez L, Zeriouh H, Romero D, Cubero J, de Vicente A, Pérez-García A (2013) The antagonistic strain *Bacillus subtilis* UMAF 6639 also confers protection to melon plants against cucurbit powdery mildew by activation of jasmonate-and salicylic acid-dependent defence responses. Microb Biotechnol 6(3):264–274

Gautam S, Chauhan A, Sharma R, Sehgal R, Shirkot CK (2019) Potential of *Bacillus amyloliquefaciens* for biocontrol of bacterial canker of tomato incited by *Clavibacter michiganensis* ssp. *michiganensis*. Microb Pathog 130:196–203

Gélis-Jeanvoine S, Canette A, Gohar M, Caradec T, Lemy C, Gominet M, Jacques P, Lereclus D, Slamti L (2017) Genetic and functional analyses of krs, a locus encoding kurstakin, a lipopeptide produced by *Bacillus thuringiensis*. Res Microbiol 168(4):356–368

Ghorbanpour M, Omidvari M, Abbaszadeh-Dahaji P, Omidvar R, Kariman K (2018) Mechanisms underlying the protective effects of beneficial fungi against plant diseases. Biol Control 117: 147–157

Glare T, Caradus J, Gelernter W, Jackson T, Keyhani N, Köhl J, Marrone P, Morin L, Stewart A (2012) Have biopesticides come of age? Trends Biotechnol 30(5):250–258

Gueldner RC, Reilly CC, Pusey PL, Costello CE, Arrendale RF, Cox RH, Himmelsbach DS, Crumley FG, Cutler HG (1988) Isolation and identification of iturins as antifungal peptides in biological control of peach brown rot with *Bacillus subtilis*. J Agric Food Chem 36(2):366–370

Guo Q, Li Y, Lou Y, Shi M, Jiang Y, Zhou J, Sun Y, Xue Q, Lai H (2019) *Bacillus amyloliquefaciens* Ba13 induces plant systemic resistance and improves rhizosphere microecology against tomato yellow leaf curl virus disease. Appl Soil Ecol 137:154–166

Han Q, Wu F, Wang X, Qi H, Shi L, Ren A, Liu Q, Zhao M, Tang C (2015) The bacterial lipopeptide iturins induce *Verticillium dahliae* cell death by affecting fungal signalling pathways and mediate plant defence responses involved in pathogen-associated molecular pattern-triggered immunity. Environ Microbiol 17(4):1166–1188

Harwood CR, Mouillon JM, Pohl S, Arnau J (2018) Secondary metabolite production and the safety of industrially important members of the *Bacillus subtilis* group. FEMS Microbiol Rev 42(6): 721–738

Heimpel GE, Mills NJ (2017) Biological control. Cambridge University Press

Helman Y, Chernin L (2015) Silencing the mob: disrupting quorum sensing as a means to fight plant disease. Mol Plant Pathol 16(3):316–329

Hinarejos E, Castellano M, Rodrigo I, Bellés JM, Conejero V, López-Gresa MP, Lisón P (2016) *Bacillus subtilis* IAB/BS03 as a potential biological control agent. Eur J Plant Pathol 146(3): 597–608

Hu LB, Shi ZQ, Zhang T, Yang ZM (2007) Fengycin antibiotics isolated from B-FS01 culture inhibit the growth of *Fusarium moniliforme* Sheldon ATCC 38932. FEMS Microbiol Lett 272(1):91–98

Jiang CH, Liao MJ, Wang HK, Zheng MZ, Xu JJ, Guo JH (2018) *Bacillus velezensis*, a potential and efficient biocontrol agent in control of pepper gray mold caused by *Botrytis cinerea*. Biol Control 126:147–157

Jose PA, Krishnamoorthy R, Kwon SW, Janahiraman V, Senthilkumar M, Gopal NO, Kumutha K, Anandham R (2019) Interference in quorum sensing and virulence of the phytopathogen *Pseudomonas syringae* pv. *passiflorae* by *Bacillus* and *Variovorax* species. BioControl 64(4): 423–433

Jourdan E, Henry G, Duby F, Dommes J, Barthelemy JP, Thonart P, Ongena MA (2009) Insights into the defense-related events occurring in plant cells following perception of surfactin-type lipopeptide from *Bacillus subtilis*. Mol Plant-Microbe Interact 22(4):456–468

Juturu V, Wu JC (2018) Microbial production of bacteriocins: latest research development and applications. Biotechnol Adv 36(8):2187–2200

Kalia VC, Raju SC, Purohit HJ (2011) Genomic analysis reveals versatile organisms for quorum quenching enzymes: acyl-homoserine lactone-acylase and-lactonase. Open Microbiol J 5:1

Kang SM, Radhakrishnan R, Lee IJ (2015a) *Bacillus amyloliquefaciens* subsp. *plantarum* GR53, a potent biocontrol agent resists *Rhizoctonia* disease on Chinese cabbage through hormonal and antioxidants regulation. World J Microbiol Biotechnol 31(10):1517–1527

Kang SM, Radhakrishnan R, Lee KE, You YH, Ko JH, Kim JH, Lee IJ (2015b) Mechanism of plant growth promotion elicited by *Bacillus* sp. LKE15 in oriental melon. Acta Agric Scand B Soil Plant Sci 65(7):637–647

Kildea S, Ransbotyn V, Khan MR, Fagan B, Leonard G, Mullins E, Doohan FM (2008) *Bacillus megaterium* shows potential for the biocontrol of *Septoria tritici* blotch of wheat. Biol Control 47(1):37–45

Knox OG, Killham K, Leifert C (2000) Effects of increased nitrate availability on the control of plant pathogenic fungi by the soil bacterium *Bacillus subtilis*. Appl Soil Ecol 15(2):227–231

Köhl J, Postma J, Nicot P, Ruocco M, Blum B (2011) Stepwise screening of microorganisms for commercial use in biological control of plant-pathogenic fungi and bacteria. Biol Control 57(1): 1–2

Köhl J, Kolnaar R, Ravensberg WJ (2019) Mode of action of microbial biological control agents against plant diseases: relevance beyond efficacy. Front Plant Sci:845

Kong HG, Kim JC, Choi GJ, Lee KY, Kim HJ, Hwang EC, Moon BJ, Lee SW (2010) Production of surfactin and iturin by *Bacillus licheniformis* N1 responsible for plant disease control activity. Plant Pathol J 26(2):170–177

Koumoutsi A, Chen XH, Henne A, Liesegang H, Hitzeroth G, Franke P, Vater J, Borriss R (2004) Structural and functional characterization of gene clusters directing nonribosomal synthesis of bioactive cyclic lipopeptides in *Bacillus amyloliquefaciens* strain FZB42. J Bacteriol 186(4): 1084–1096

Krebs B, Ockhardt A, Hoeding B, Bendzko P, Maximov J, Etzel W (1996) Cyclic peptides from *Bacillus amyloliquefaciens* useful antimycotics, antivirals, fungicides, nematicides etc. DE19641213, p 26

Krid S, Triki MA, Gargouri A, Rhouma A (2012) Biocontrol of olive knot disease by *Bacillus subtilis* isolated from olive leaves. Ann Microbiol 62(1):149–154

Kudoyarova GR, Melentiev AI, Martynenko EV, Timergalina LN, Arkhipova TN, Shendel GV, Kuz'mina LY, Dodd IC, Veselov SY (2014) Cytokinin producing bacteria stimulate amino acid deposition by wheat roots. Plant Physiol Biochem 83:285–291

Kumar P, Dubey RC, Maheshwari DK (2012) *Bacillus* strains isolated from rhizosphere showed plant growth promoting and antagonistic activity against phytopathogens. Microbiol Res 167(8):493–499

Lastochkina O, Pusenkova L, Yuldashev R, Babaev M, Garipova S, Blagova DY, Khairullin R, Aliniaeifard S (2017) Effects of *Bacillus subtilis* on some physiological and biochemical parameters of *Triticum aestivum* L.(wheat) under salinity. Plant Physiol Biochem 121:80–88

Latz E, Eisenhauer N, Rall BC, Scheu S, Jousset A (2016) Unravelling linkages between plant community composition and the pathogen-suppressive potential of soils. Sci Rep 6(1):1–0

Lee GH, Ryu CM (2016) Spraying of leaf-colonizing *Bacillus amyloliquefaciens* protects pepper from *cucumber mosaic virus*. Plant Dis 100(10):2099–2105

Leelasuphakul W, Sivanunsakul P, Phongpaichit S (2006) Purification, characterization and synergistic activity of β-1, 3-glucanase and antibiotic extract from an antagonistic *Bacillus subtilis* NSRS 89-24 against rice blast and sheath blight. Enzym Microb Technol 38(7):990–997

Li B, Li Q, Xu Z, Zhang N, Shen Q, Zhang R (2014) Responses of beneficial *Bacillus amyloliquefaciens* SQR9 to different soilborne fungal pathogens through the alteration of antifungal compounds production. Front Microbiol 5:636

Lin C, Tsai CH, Chen PY, Wu CY, Chang YL, Yang YL, Chen YL (2018) Biological control of potato common scab by *Bacillus amyloliquefaciens* Ba01. PLoS One 13(4):e0196520. https://doi.org/10.1371/journal.pone.0196520

Liu Y, Chen Z, Ng TB, Zhang J, Zhou M, Song F, Lu F, Liu Y (2007) Bacisubin, an antifungal protein with ribonuclease and hemagglutinating activities from *Bacillus subtilis* strain B-916. Peptides 28(3):553–559

Liu J, Hagberg I, Novitsky L, Hadj-Moussa H, Avis TJ (2014) Interaction of antimicrobial cyclic lipopeptides from *Bacillus subtilis* influences their effect on spore germination and membrane permeability in fungal plant pathogens. Fungal Biol 118(11):855–861

Liu Q, Lin J, Wang W, Huang H, Li S (2015) Production of surfactin isoforms by *Bacillus subtilis* BS-37 and its applicability to enhanced oil recovery under laboratory conditions. Biochem Eng J 93:31–37

Liu G, Kong Y, Fan Y, Geng C, Peng D, Sun M (2017) Whole-genome sequencing of *Bacillus velezensis* LS69, a strain with a broad inhibitory spectrum against pathogenic bacteria. J Biotechnol 249:20–24

Loiseau C, Schlusselhuber M, Bigot R, Bertaux J, Berjeaud JM, Verdon J (2015) Surfactin from *Bacillus subtilis* displays an unexpected anti-legionella activity. Appl Microbiol Biotechnol 99(12):5083–5093

Luo C, Liu X, Zhou X, Guo J, Truong J, Wang X, Zhou H, Li X, Chen Z (2015) Unusual biosynthesis and structure of locillomycins from *Bacillus subtilis* 916. Appl Environ Microbiol 81(19):6601–6609

Maget-Dana R, Peypoux F (1994) Iturins, a special class of pore-forming lipopeptides: biological and physicochemical properties. Toxicology 87(1–3):151–174

Maksimov IV, Veselova SV, Nuzhnaya TV, Sarvarova ER, Khairullin RM (2015) Plant growth-promoting bacteria in regulation of plant resistance to stress factors. Russ J Plant Physiol 62(6): 715–726

Mannanov RN, Sattarova RK (2001) Antibiotics produced by Bacillus bacteria. Chem Nat Compd 37(2):117–123

Matzen N, Heick TM, Jørgensen LN (2019) Control of powdery mildew (*Blumeria graminis* spp.) in cereals by Serenade® ASO (*Bacillus amyloliquefaciens* (former subtilis) strain QST 713). Biol Control 139:104067

Meena KR, Kanwar SS (2015) Lipopeptides as the antifungal and antibacterial agents: applications in food safety and therapeutics. Biomed Res Int 2015:473050. https://doi.org/10.1155/2015/473050

Mihalache G, Balaes T, Gostin I, Stefan M, Coutte F, Krier F (2018) Lipopeptides produced by *Bacillus subtilis* as new biocontrol products against fusariosis in ornamental plants. Environ Sci Pollut Res 25(30):29784–29793

Miljaković D, Marinković J, Balešević-Tubić S (2020) The significance of *Bacillus* spp. in disease suppression and growth promotion of field and vegetable crops. Microorganisms 8(7):1037

Mnif I, Ghribi D (2015) Review lipopeptides biosurfactants: mean classes and new insights for industrial, biomedical, and environmental applications. J Pept Sci 104(3):129–147

Molinatto G, Puopolo G, Sonego P, Moretto M, Engelen K, Viti C, Ongena M, Pertot I (2016) Complete genome sequence of *Bacillus amyloliquefaciens* subsp. *plantarum* S499, a rhizobacterium that triggers plant defences and inhibits fungal phytopathogens. J Biotechnol 238:56–59

Muslim A, Horinouchi H, Hyakumachi M (2003) Biological control of *Fusarium* wilt of tomato with hypovirulent binucleate Rhizoctonia in greenhouse conditions. Mycoscience 44(2):77–84

Nagórska K, Bikowski M, Obuchowski M (2007) Multicellular behaviour and production of a wide variety of toxic substances support usage of *Bacillus subtilis* as a powerful biocontrol agent. Acta Biochim Pol 54(3):495–508

Nair D, Vanuopadath M, Nair BG, Pai JG, Nair SS (2016) Identification and characterization of a library of surfactins and fengycins from a marine endophytic *Bacillus* sp. J Basic Microbiol 56(11):1159–1172

Naruse N, Tenmyo O, Kobaru S, Kamei H, Miyaki T, Konishi M, Oki T (1990) Pumilacidin, a complex of new antiviral antibiotics production, isolation, chemical properties, structure and biological activity. J Antibiot 43(3):267–280

Ntushelo K, Ledwaba LK, Rauwane ME, Adebo OA, Njobeh PB (2019) The mode of action of *Bacillus* species against *Fusarium graminearum*, tools for investigation, and future prospects. Toxins 11(10):606

Olishevska S, Nickzad A, Déziel E (2019) Bacillus and Paenibacillus secreted polyketides and peptides involved in controlling human and plant pathogens. Appl Microbiol Biotechnol 103(3): 1189–1215

Ongena M, Jacques P (2008) Bacillus lipopeptides: versatile weapons for plant disease biocontrol. Trends Microbiol 16(3):115–125

Ongena M, Jacques P, Touré Y, Destain J, Jabrane A, Thonart P (2005) Involvement of fengycin-type lipopeptides in the multifaceted biocontrol potential of *Bacillus subtilis*. Appl Microbiol Biotechnol 69(1):29–38

Ongena M, Henry G, Thonart P (2010) The roles of cyclic lipopeptides in the biocontrol activity of *Bacillus subtilis*. In: Recent developments in management of plant diseases. Springer, Dordrecht, pp 59–69

Pan J, Huang T, Yao F, Huang Z, Powell CA, Qiu S, Guan X (2008) Expression and characterization of aiiA gene from *Bacillus subtilis* BS-1. Microbiol Res 163(6):711–716

Pandin C, Le Coq D, Canette A, Aymerich S, Briandet R (2017) Should the biofilm mode of life be taken into consideration for microbial biocontrol agents? Microb Biotechnol 10(4):719–734

Pandin C, Darsonval M, Mayeur C, Le Coq D, Aymerich S, Briandet R (2019) Biofilm formation and synthesis of antimicrobial compounds by the biocontrol agent *Bacillus velezensis* QST713 in an Agaricus bisporus compost micromodel. Appl Environ Microbiol 85(12):e00327–e00319. https://doi.org/10.1128/AEM.00327-19

Perry JD, Asir K, Halimi D, Orenga S, Dale J, Payne M, Carlton R, Evans J, Gould FK (2010) Evaluation of a chromogenic culture medium for isolation of *Clostridium difficile* within 24 hours. J Clin Microbiol 48(11):3852–3858

Pertry I, Václavíková K, Depuydt S, Galuszka P, Spíchal L, Temmerman W, Stes E, Schmülling T, Kakimoto T, Van Montagu MC, Strnad M (2009) Identification of *Rhodococcus fascians* cytokinins and their modus operandi to reshape the plant. Proc Natl Acad Sci 106(3):929–934

Pieterse CM, Van der Does D, Zamioudis C, Leon-Reyes A, Van Wees SC (2012) Hormonal modulation of plant immunity. Annu Rev Cell Dev Biol 28:489–521

Pieterse CM, Zamioudis C, Berendsen RL, Weller DM, Van Wees SC, Bakker PA (2014) Induced systemic resistance by beneficial microbes. Annu Rev Phytopathol 52:347–375

Pignatelli M, Moya A, Tamames J (2009) EnvDB, a database for describing the environmental distribution of prokaryotic taxa. Environ Microbiol Rep 1(3):191–197

Raafat MM, Ali-Tammam M, Ali AE (2019) Quorum quenching activity of *Bacillus cereus* isolate 30b confers antipathogenic effects in *Pseudomonas aeruginosa*. Infect Drug Resist 12:1583

Raaijmakers JM, De Bruijn I, Nybroe O, Ongena M (2010) Natural functions of lipopeptides from Bacillus and Pseudomonas: more than surfactants and antibiotics. FEMS Microbiol Rev 34(6): 1037–1062

Rabbee MF, Ali MS, Choi J, Hwang BS, Jeong SC, Baek KH (2019) *Bacillus velezensis*: a valuable member of bioactive molecules within plant microbiomes. Molecules 24(6):1046

Rahman A, Uddin W, Wenner NG (2015) Induced systemic resistance responses in perennial ryegrass against *Magnaporthe oryzae* elicited by semi-purified surfactin lipopeptides and live cells of *Bacillus amyloliquefaciens*. Mol Plant Pathol 16(6):546–558

Romanazzi G, Sanzani SM, Bi Y, Tian S, Martínez PG, Alkan N (2016) Induced resistance to control postharvest decay of fruit and vegetables. Postharvest Biol Technol 122:82–94

Romero D, De Vicente A, Rakotoaly RH, Dufour SE, Veening JW, Arrebola E, Cazorla FM, Kuipers OP, Paquot M, Pérez-García A (2007) The iturin and fengycin families of lipopeptides are key factors in antagonism of *Bacillus subtilis* toward *Podosphaera fusca*. Mol Plant-Microbe Interact 20(4):430–440

Sabaté DC, Audisio MC (2013) Inhibitory activity of surfactin, produced by different *Bacillus subtilis* subsp. *subtilis* strains, against *Listeria monocytogenes* sensitive and bacteriocin-resistant strains. Microbiol Res 168(3):125–129

Sabaté DC, Petroselli G, Erra-Balsells R, Audisio MC, Brandan CP (2020) Beneficial effect of *Bacillus* sp. P12 on soil biological activities and pathogen control in common bean. Biol Control 141:104131

Sarosh BR, Danielsson J, Meijer J (2009) Transcript profiling of oilseed rape (*Brassica napus*) primed for biocontrol differentiate genes involved in microbial interactions with beneficial *Bacillus amyloliquefaciens* from pathogenic *Botrytis cinerea*. Plant Mol Biol 70(1):31–45

Sarwar A, Hassan MN, Imran M, Iqbal M, Majeed S, Brader G, Sessitsch A, Hafeez FY (2018) Biocontrol activity of surfactin A purified from *Bacillus* NH-100 and NH-217 against rice bakanae disease. Microbiol Res 209:1–13

Savary S, Ficke A, Aubertot JN, Hollier C (2012) Crop losses due to diseases and their implications for global food production losses and food security. Food Secur 4(4):519–537

Saxena AK, Kumar M, Chakdar H, Anuroopa N, Bagyaraj DJ (2020) *Bacillus* species in soil as a natural resource for plant health and nutrition. J Appl Microbiol 128(6):1583–1594

Scholz R, Molohon KJ, Nachtigall J, Vater J, Markley AL, Süssmuth RD, Mitchell DA, Borriss R (2011) Plantazolicin, a novel microcin B17/streptolysin S-like natural product from *Bacillus amyloliquefaciens* FZB42. J Bacteriol 193(1):215–224

Seifi Kalhor M, Aliniaeifard S, Seif M, Javadi E, Bernard F, Li T, Lastochkina O (2017) Rhizobacterium *Bacillus subtilis* reduces toxic effects of high electrical conductivity in soilless culture of lettuce. In: International symposium on new technologies for environment control, energy-saving and crop production in greenhouse and plant, vol. 1227, pp 471–478

Seong Y, Yong J, Hak J, Yong K (2017) Isolation and identification of Nâ butylâ tetrahydroâ 5â oxofuranâ 2â carboxamide produced by *Bacillus* sp. L60 and its antifungal activity. J Basic Microbiol 57(3):283–288

Shafi J, Tian H, Ji M (2017) *Bacillus* species as versatile weapons for plant pathogens: a review. Biotechnol Biotechnol Equip 31(3):446–459

Shu HY, Lin GH, Wu YC, Tschen JS, Liu ST (2002) Amino acids activated by fengycin synthetase FenE. Biochem Biophys Res Commun 292(4):789–793

Singh D, Sharma RR (2007) Sustainable pest management

Sivasakthi S, Kanchana D, Usharani G, Saranraj P (2013) Production of plant growth promoting substance by *Pseudomonas fluorescens* and *Bacillus subtilis* isolates from paddy rhizosphere soil of Cuddalore District, Tamil Nadu, India. Int J Microbiol Res 4(3):227–233

Spadaro D, Droby S (2016) Development of biocontrol products for postharvest diseases of fruit: the importance of elucidating the mechanisms of action of yeast antagonists. Trends Food Sci Technol 47:39–49

Stein T (2005) *Bacillus subtilis* antibiotics: structures, syntheses and specific functions. Mol Microbiol 56(4):845–857

Tanaka K, Amaki Y, Ishihara A, Nakajima H (2015) Synergistic effects of [Ile7] surfactin homologues with bacillomycin D in suppression of gray mold disease by *Bacillus amyloliquefaciens* biocontrol strain SD-32. J Agric Food Chem 63(22):5344–5353

Thoms D, Liang Y, Haney CH (2021) Maintaining symbiotic homeostasis: how do plants engage with beneficial microorganisms while at the same time restricting pathogens? Mol Plant-Microbe Interact 34(5):462–469

Torres Manno MA, Pizarro MD, Prunello M, Magni C, Daurelio LD, Espariz M (2019) GeM-Pro: a tool for genome functional mining and microbial profiling. Appl Microbiol Biotechnol 103(7): 3123–3134

Toure Y, Ongena MA, Jacques P, Guiro A, Thonart P (2004) Role of lipopeptides produced by *Bacillus subtilis* GA1 in the reduction of grey mould disease caused by *Botrytis cinerea* on apple. J Appl Microbiol 96(5):1151–1160

Turan M, Ekinci M, Yildirim E, Güneş A, Karagöz K, Kotan R, Dursun A (2014) Plant growth-promoting rhizobacteria improved growth, nutrient, and hormone content of cabbage (*Brassica oleracea*) seedlings. Turk J Agric For 38(3):327–333

van Lenteren JC, Bolckmans K, Köhl J, Ravensberg WJ, Urbaneja A (2018) Biological control using invertebrates and microorganisms: plenty of new opportunities. BioControl 63(1):39–59

Van Loon LC (2007) Plant responses to plant growth-promoting rhizobacteria. Eur J Plant Pathol 119:243–254. https://doi.org/10.1007/s10658-007-9165-1

Villegas-Escobar V, Ceballos I, Mira JJ, Argel LE, Orduz Peralta S, Romero-Tabarez M (2013) Fengycin C produced by Bacillus subtilis EA-CB0015. J Nat Prod 76(4):503–509

Waewthongrak W, Pisuchpen S, Leelasuphakul W (2015) Effect of *Bacillus subtilis* and chitosan applications on green mold (*Penicilium digitatum* Sacc.) decay in citrus fruit. Postharvest Biol Technol 99:44–49

Wan Y, Stanovych A, Gori D, Zirah S, Kouklovsky C, Alezra V (2018) β, γ-diamino acids as building blocks for new analogues of Gramicidin S: synthesis and biological activity. Eur J Med Chem 149:122–128

Wang T, Liang Y, Wu M, Chen Z, Lin J, Yang L (2015) Natural products from *Bacillus subtilis* with antimicrobial properties. Chin J Chem Eng 23(4):744–754

Wang XQ, Zhao DL, Shen LL, Jing CL, Zhang CS (2018) Application and mechanisms of *Bacillus subtilis* in biological control of plant disease. In: Role of rhizospheric microbes in soil. Springer, Singapore, pp 225–250

Wang X, Yuan Z, Shi Y, Cai F, Zhao J, Wang J, Wang Y (2020a) *Bacillus amyloliquefaciens* HG01 induces resistance in loquats against anthracnose rot caused by *Colletotrichum acutatum*. Postharvest Biol Technol 160:111034

Wang Y, Zhang C, Liang J, Wang L, Gao W, Jiang J, Chang R (2020b) Surfactin and fengycin B extracted from *Bacillus pumilus* W-7 provide protection against potato late blight via distinct and synergistic mechanisms. Appl Microbiol Biotechnol 104(17):7467–7481

Winn M, Fyans JK, Zhuo Y, Micklefield J (2016) Recent advances in engineering nonribosomal peptide assembly lines. Nat Prod Rep 33(2):317–347

Wulff BB, Horvath DM, Ward ER (2011) Improving immunity in crops: new tactics in an old game. Curr Opin Plant Biol 14(4):468–476

Xu Z, Mandic-Mulec I, Zhang H, Liu Y, Sun X, Feng H, Xun W, Zhang N, Shen Q, Zhang R (2019) Antibiotic bacillomycin D affects iron acquisition and biofilm formation in *Bacillus velezensis* through a Btr-mediated FeuABC-dependent pathway. Cell Rep 29(5):1192–1202

Yakimov MM, Timmis KN, Wray V, Fredrickson HL (1995) Characterization of a new lipopeptide surfactant produced by thermotolerant and halotolerant subsurface *Bacillus licheniformis* BAS50. Appl Environ Microbiol 61(5):1706–1713

Yánez-Mendizábal V, Usall J, Viñas I, Casals C, Marín S, Solsona C, Teixidó N (2011) Potential of a new strain of *Bacillus subtilis* CPA-8 to control the major postharvest diseases of fruit. Biocontrol Sci Tech 21(4):409–426

Yánez-Mendizábal V, Zeriouh H, Viñas I, Torres R, Usall J, de Vicente A, Pérez-García A, Teixidó N (2012) Biological control of peach brown rot (*Monilinia* spp.) by *Bacillus subtilis* CPA-8 is based on production of fengycin-like lipopeptides. Eur J Plant Pathol 132(4):609–619

Ye YF, Li QQ, Gang FU, Yuan GQ, Miao JH, Wei LI (2012) Identification of antifungal substance (Iturin A2) produced by *Bacillus subtilis* B47 and its effect on southern corn leaf blight. J Integr Agric 11(1):90–99

Ye M, Tang X, Yang R, Zhang H, Li F, Tao F, Li F, Wang Z (2018) Characteristics and application of a novel species of *Bacillus*: *Bacillus velezensis*. ACS Chem Biol 13(3):500–505

Yi HS, Yang JW, Ryu CM (2013) ISR meets SAR outside: additive action of the endophyte *Bacillus pumilus* INR7 and the chemical inducer, benzothiadiazole, on induced resistance against bacterial spot in field-grown pepper. Front Plant Sci 4:122

Yoshida S, Hiradate S, Tsukamoto T, Hatakeda K, Shirata A (2001) Antimicrobial activity of culture filtrate of *Bacillus amyloliquefaciens* RC-2 isolated from mulberry leaves. Phytopathology 91(2):181–187

Zhang B, Dong C, Shang Q, Han Y, Li P (2013) New insights into membrane-active action in plasma membrane of fungal hyphae by the lipopeptide antibiotic bacillomycin L. Biochim Biophys Acta Biomembr 1828(9):2230–2237

Zhao P, Quan C, Wang Y, Wang J, Fan S (2014) *Bacillus amyloliquefaciens* Q-426 as a potential biocontrol agent against *Fusarium oxysporum* f. sp. *spinaciae*. J Basic Microbiol 54(5):448–456

Zou J, Jiang H, Cheng H, Fang J, Huang G (2018) Strategies for screening, purification and characterization of bacteriocins. Int J Biol Macromol 117:781–789

Chapter 12
Using QS in Biological Control as an Alternative Method

Mustafa Mirik and Cansu Oksel

Abstract The microorganisms are provided by the Quorum sensing (QS) mechanism to communicate between inter-species. The signaling molecules known as autoinducers are essential to this QS communication mechanism. The most typical autoinducer is known as *N*-acyl-homoserine lactones (AHLs). QS enables bacteria to cooperate, live, compete, endure in the environment, or colonize the host. The proliferation of nearby bacterial cells is the most crucial factor to activate for QA. So, a wide of behaviors in bacteria are provided by QS including bioluminescence, swarming, biofilm, motility, stress survival, and virulence factors. However, some studies indicate that the QS-disputing mechanism reduces adequately population density and virulence accordingly. The QS manipulation methods such as AHL-degradation and mimic act give promising approaches to controlling plant pathogen bacteria. So, it can lead to an alternative way of supporting biological control. This review focuses on the QS mechanism in plant bacteria and the different disrupting mechanisms. It demonstrates a novel method in biocontrol and the major outcomes of plant protection.

Keywords Plant bacteria · Quorum sensing · Signal molecules · Biological control · AHL

12.1 Introduction

The loss of agricultural productivity during harvest or postharvest storage conditions are a major challenge for the food security. These losses are generally led by b various pathogenic bacteria, fungi, nematodes (Kumar et al. 2021; Kumari et al. 2022). The management of bacterial diseases includes several methods such as sanitation, host resistance, cultural practices, and biological and chemical control;

M. Mirik (✉) · C. Oksel
Department of Plant Protection, Agricultural Engineering Faculty, Tekirdag Namik Kemal University, Tekirdag, Turkey
e-mail: mmiri@nku.edu.tr

K. K. Bastas et al. (eds.), *Microbial Biocontrol: Molecular Perspective in Plant Disease Management*, Microorganisms for Sustainability 49,
https://doi.org/10.1007/978-981-99-3947-3_12

however, finding an effective management method for bacterial diseases cannot be easy. Recently studies reported that most bacteria already have resistant chemical compounds like copper and antibiotics. Also, they are considered microbial toxins for the environment and human health. Nowadays some studies focused on how aminoglycosides inhibited the Quorum sensing (QS) system of *Chromobacterium violaceum* and make think about using QS to control bacterial disease (Zhang et al. 2011; Vasudevan et al. 2018).

QS is like a language based on cell–cell communication among bacteria (Miller and Bassler 2001). According to QS, the bacteria need to utilize the autoinducers like *N*-acyl-homoserine lactone (AHL) to provide communication between bacteria. This kind of communication needs a level of signal required for gene expression. However, the population of bacteria needs to reach the amount of density for occurring QS. The first basal level of AHL is produced in a low population density to activate the AHL synthase. Then in time AHL based on cell density increases to respond to bacterial activity. So, AHL molecules release into cells to turn to activate or suppress the coordinated expression of genes. It causes to create favorable conditions for the bacterial environment (Fuqua et al. 2001). Each bacterium that has different properties such as biofilm, motility, swarming, bioluminescence, and pathogenicity utilizes the QS mechanism (Meighen 1991; Meighen 1994; Zhu et al. 2000; Luo and Farrand 2001). The QS mechanism can be gained as an alternative to improve innovative approaches to disease control (Alt-Morbe et al. 1996; Loh et al. 2002a, b). Some mechanisms such as AHL-degradation and AHL-mimics are linked as QS-disrupting mechanisms (Liu et al. 2002; Teplitski et al. 2000).

The first study of QS-disrupting was presented from *Bacillus* sp. based on by aiiA gene from transgenic tobacco and potato plants. Then the following study about the expression of this gene showed that increasing the plant resistance against the phytopathogen bacteria namely *Pectobacterium carotovorum* (Dong et al. 2001; Mae et al. 2001). Recent findings indicate that blocking the AHL mechanism is possibly helping AHL-degrading enzymes and AHL mimics. Basically, blocking the AHL signal provides an option for biocontrol activity on plant pathogen bacteria (Steidle et al. 2002; Pierson and Weller 1994; Raupach and Kloepper 1998). This review focuses to shed new light on QS using an alternative and eco-friendly strategy in biocontrol.

12.2 What Is the Quorum Sensing?

Bacteria are highly interactive and display several collective social behaviors. They act together in the ecosystem. So, they can keep in touch with each other. QS is like a communication way of bacteria and conveys the concept of work when bacteria are crowded together (Bassler and Losick 2006). QS provides the accumulation of self-generated diffusible signaling molecules from each bacterial cell in the surrounding environment until the bacteria population can make a coordinated response. Therefore, the bacteria can survive to utilize the group behavior in the ecosystem (Diggle et al. 2007). The bacteria need signaling molecules called autoinducers in the QS mechanism to communicate. The signaling molecules are secreted by particular genes. The first OS system was declared in *Vibrio fischeri* which is marine bioluminescent bacteria (Nealson and Hastings 1979; Stevens and Greenberg 1997).

In the first phase of QS, LuxI synthases *N*-(3-oxo-hexanoyl)-L-homoserine lactone (3-oxoC6HL), also known as autoinducer (Al), from S-adenosyl methionine and 3-oxo-hexanoyl-acyl carrier protein. As a result, it enables Al to travel readily across bacterial membranes (Parsek et al. 1999; Watson et al. 2002; More et al. 1996). Then, 3-oxoC6HL begins to interact with its LuxR receptor. As a result, the luxI and luxR genes' promoter regions contain a palindromic lux box element, which the QS system is triggered to bind to with enhanced affinity. As autoregulators, these genes support a dual positive feedback system (reviewed in Stevens and Greenberg 1997). The lux regulon's remaining genes encode luminescence-related enzymes (reviewed in Bassler 2002; Fuqua and Greenberg 2002; Visick and McFall-Ngai 2000). Based on that, since the bacteria grew in liquid cultures, it is observed that only their cell population produced light sufficient.

12.3 Quorum Sensing Mechanism

The bacteria need an autoinducer signal molecule to communicate and coordinate the response. OS systems have revealed various autoinducer systems. The signal molecule can provide communication between species. Acyl-homoserine lactone (AHL), originally recognized as a signal molecule in 1980 (Pal and Gardener 2006; Mansfield et al. 2012). QS may be in responsible of virulence factors and infection processes in plant pathogenic bacteria, including lipopolysaccharides, plant cell wall degrading enzymes (PCWDES), type II secretion system (T3SS), and type IV secreted effectors (Hosseinzadeh et al. 2013; Deryabin et al. 2018). *N*-acyl-homoserine lactones (AHLs) and oligopeptides are utilized in the QS mechanism by Gram-negative and -positive bacteria, respectively (Kleerebezem et al. 1997; Fuqua et al. 2001; Fuqua and Greenberg 2002), and autoinducers-2 (Al-2) are present in both (Chen et al. 2002). In Gram-negative bacteria, AHLs can passively diffuse through a thin cell wall. For Gram-positive bacteria, the procedure is challenging in contrast to this. The ATP-binding cassette (ABC) must be actively carried by their autoinducers via a peptidoglycan cell wall, which must be comprised of peptides (Winsdor 2020).

Based on two components, LuxR-type (R) regulator and Luxl-type (I) protein, which act as a signal receptor and an AHL synthesis, respectively, AHLs are the most prevalent autoinducers. When there is a sufficient density or quorum for the creation of signals, QS occurs (Miller and Bassler 2001). Bacteria begin to create AHL signals at low densities until they approach the threshold concentration, at which point signaling molecules engage with the R protein to regulate the expression of the target genes (Fuqua et al. 1996; Teplitski et al. 2000; Zhang et al. 2002). That cutoff allows intracellular autoinducers to continue egressing from the cell, increasing their intracellular concentration. Autoinducers bind to their receptors when intracellular concentration rises, initiating signaling cascades that enable transcription factor activity and, consequently, gene expression (Winsdor 2020). The communication system between species or between species and non-species is provided by these signals (Schauder et al. 2001; Ganin et al. 2009; Kimura et al. 2009; Verma and Miyashiro 2013). The main contribution of QS to bacterial behavior is significant; Release from biofilms and virulence gene expression (Gill et al. 2015).

12.4 Quorum Sensing in Plant Bacteria

In order to support their communication system for infection or population on their host and the environment, many bacteria have used several signal molecules in QS (Helman and Chernın 2015). This communication method has some benefits. As an illustration, *Pseudomonas syringae* pv. *tabaci* causes wildfire on tobacco and is capable of acquiring iron (Shaw et al. 1997; Taguchi et al. 2006); *P. s.* pv. *syringae* causes bacterial canker on stone fruits and is capable of withstanding oxidative stress (Dumenyo et al. 1998; Quinones et al. 2004, 2005). Additionally, *Rhizobium radiobacter* transfers Ti plasmid genes to host plants to cause crown gall disease, and the bacterial wilt pathogen *Pantoea stewartii*, also known as the bacterial wilt of maize, establishes a biofilm on the host (Beck von Bodman and Farrand 1995; Koutsoudis et al. 2006; Zhang et al. 1993; Farrand et al. 2002). AHLs are regarded as the most prevalent kind of QS. Proteobacterial species number above 50 (Fuqua et al. 2001; Whitehead et al. 2001). Several bacteria, including *P. syringae pv. tabaci*, *Ralstonia solanacearum*, *R. radiobacter* (*Agrobacterium tumefaciens*), *Erwinia amylovora*, *Pectobacterium* sp., and *Dickeya* sp., have AHLs that are made up of various signal molecules, including *N*-3-produce oxohexano (Mansfield et al. 2012).

It is necessary for the pathogenicity, secreted virulence factor, type III secreted harpin, and antibiotics of some *Pectobacterium* species. Additionally, post-transcriptional regulation is linked to the effect of OHL (*N*-(3-oxohexanoyl)-L-homoserine lactone), which is one of the various forms of AHLs in *P. carotovorum* pv. *carotovorum*. Since RsmA mutants are OHL non-producing, they prevent OHL-deficient bacteria from generating exoproteins (Byers et al. 2002; Barnard et al. 2007; Barnard and Salmond 2007).

A notable example of QS is the *Pseudomonas* genus. The mechanism is crucial for plant survival and colony morphology (Loh et al. 2001). In QS, *Pseudomonas* contains unique R and I proteins. According to the earlier research, *Pseudomonas aeruginosa* can either respond specifically to LasR or to both LasR and RhlR, which are known to regulate the production of virulence genes. It implies that they are able to control particular genes to establish AHL signals. Consequently, it alters the number of microorganisms and their availability (Whitely and Greenberg 2001; Jimenez et al. 2012). *P. syringae* B728a has utilized the AHlI and AHlR components of the AHLs-QS system. Tobacco and soybean plants get wildfire disease from *P. s.* pv. *tabaci* 1528. Based on the expression of the LuxI-LuxR genes, the strain possesses swarming movement, flagellum formation, pili assembly, biofilm, and chemotaxis (Cheng et al. 2016, 2018). Some important plant pathogenic bacteria like *R. solanacearum* and *Xanthomonas campestris* can have different QS signals such as fatty acid and butyrolactone derivatives respectively. So, they can produce their own pathogenicity factors on the host (Barber et al. 1997; Chun et al. 1997; Flavier et al. 1997).

In *R. solanacearum* has Phc regulatory system. This system is responsible for host contact, microcolony, biofilm, and pathogenicity, depending on population density. PhcA which is a LysR-type regulator is activated by 3-OH palmitic acid methyl ester (3-OH PAME). The suitable level of 3-OH PAME leads to an increase in EPS, exoenzymes, motility, and siderophore production (Flavier et al. 1997). Moreover, *R. solanacearum* has LuxI-LuxR system called SolI-SolR. SolI provides the C6-HSL and C8-HSL molecules and is responsible for the expression of aidA gene which supports the pathogen to adapt to the temperature (Chen et al. 2009; Meng et al. 2015).

Besides, Rhizobium-legume symbiosis belongs to signal molecules between Rhizobia and plant. The symbiosis process is subjected to quorum regulation. *Rhizobium* species lead to nodulation on the root based on the several kinds of AHLs that lead to nodulation on the root. The quorum signal shows direct control of nodulation (Lithgow et al. 2000; Loh et al. 2001; Daniels et al. 2002).

R. radiobacter leads to producing opine hormone associated with plant growth and causes crown gall (Gelvin 2012; Bourras et al. 2015). So, AHLs have a role to transfer the signal of opine synthesis. Opine is associated with a type of Ti plasmids. LuxR provides the segregation system of the Ti plasmid (Su et al. 2008). Shortly, QS provides the conjugative transfer and replication of T' plasmid (Cho et al. 2007). *Erwinia amylovora* causes fire-blight to have OHHL and OHL signal molecules and EamRI based on LuxRI homologs. It produces amylovoran and levan as extracellular polysaccharides supporting AHLs (Rezzonico and Duffy 2007). *Dickeya dadantii* utilizes AHL signal molecule namely OHHL based on ExpI and ExpR. However, the previous findings didn't show any connection between pathogen and AHL deficiency (Nasser et al. 1998; Reverchon et al. 1998; Castang et al. 2006).

In *P. s.* subsp. *stewartii* has LuxI and LuxR associated with EsaI and EsaR in QS. EsaR-AHL increases the expression of resA and it leads to enhancing the production of stewartan and exopolysaccharide under the high cell density of the pathogen (von Bodman et al. 1998).

12.5 Using QS in Biological Control as an Alternative Method

Previous studies mentioned when the bacterial cell density is reduced, the pathogens can lose the ability or severity of pathogenicity. Based on the results, it is possible blocking the QS mechanism called QS-disrupting. The principle of the QS-disrupting mechanism is based on preventing bacterial communication. When the pathogen couldn't reach a sufficient amount of density, the signal molecules couldn't be transferred between cells. It means that they don't respond to the other signal. So, the pathogens aren't shown their own properties such as biofilm, swarming, stress survival, motility, and virulence factor (Fig. 12.1). The first study of QS-disrupting reported discovering the aiiA on *Bacillus* sp. that is obtained from transgenic tobacco and potato plants. This gene that is associated with AHL-lactonase was genetically modified to destroy the production of AHL based on QS systems of the phytopathogenic bacterium (Allen et al. 2016). So, some outcomes obtained increased plant disease resistance against *P. carotovorum* (Cui et al. 1995). With the blocking of bacterial signal molecules belonging to transgenic plants, the pathogen didn't reach enough population for signal molecules. It led to insufficient infection and gave hope to a more applicable strategy in the future which would be the application of microorganisms with a natural ability to degrade AHLs.

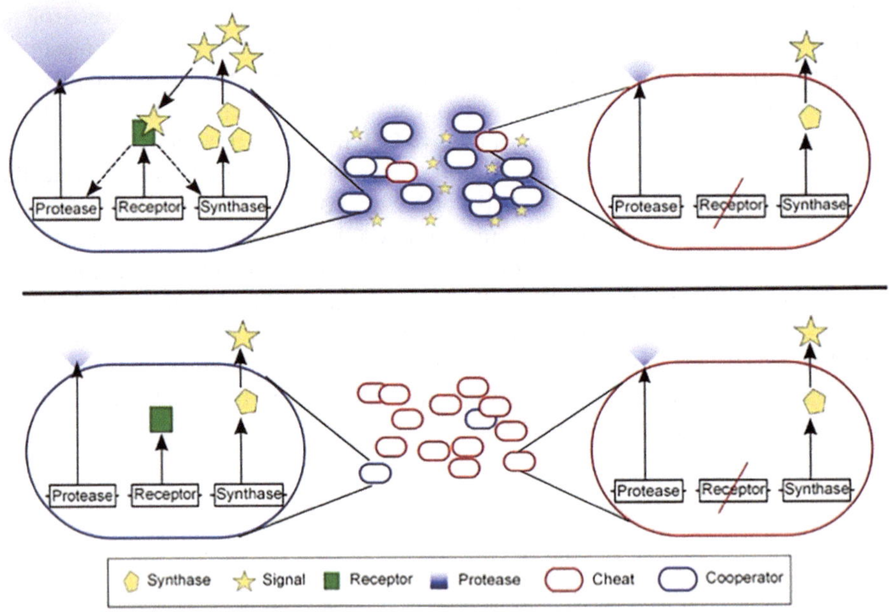

Fig. 12.1 The principle of the QS-disrupting mechanism. (Figure was reviewed in Allen et al. 2016)

QS-disrupting occurs the blocking AHL signals and provides more possibilities for biological control (de Kievit and Iglewski 2000; Molina et al. 2003). QS-disrupting consists of AHL degradation and mimicking of AHL.

12.5.1 AHL Degradation

AHL-degradation which acylase degrades AHL by removing bonds from the fatty acid chain lateral to the AHL ring was found in *R. solanacearum* GMT10000. This mechanism blocks to accumulation of AHL signals. Similarly, *P. syringae* has different AHL degradation signs to reduce bacterial cell density to make the biofilm formation lower (Chen et al. 2009; Shepherd and Lindow 2008) (Baltenneck et al. 2021). The previous studies indicated that some enzymes could inactivate AHL signals between bacteria. For example, aiiA (namely Al inactivation) encoded by *Bacillus* sp. offers the blocking.

The hydrolysis of the lactone ring produces AHL signals (Dong et al. 2000; Dong et al. 2001). However, the AHL signals in the bacterial agent *P. carotovorum*, which causes soft rot illnesses on potatoes, eggplant, carrots, and celery, were reduced as a result of the aiiA gene. There is evidence that the infection lessens the hosts' symptoms (Emmert and Handelsman 1999; Dong et al. 2000; Dong et al. 2001; Dong et al. 2002). Additionally, *Variavorax paradoxus* and Ralstonia strain XJ12 produced the enzyme aiiD, which was identified to slow down the hydrolysis of the AHL amide and the matching fatty acid (Leadbetter and Greenberg 2000; Lin et al. 2003). This outcome prompted transgenic plants to start expressing the QS-disrupting gene. One theory holds that the bacterium's breakdown of AHL signals has an impact on the virulence genes.

Therefore, the defense mechanism of plants would be blocked from infection and the spread of pathogens (Loh et al. 2002a, b). Based on laboratory studies, transgenic plants belonging to bacterial AHL degrading enzymes are stronger against pathogen infections.

Recent studies on *P. syringae* pv. *syringae* B728a indicated that it has an important role of QS in epiphytes. The non-AHL mutants of B728a showed decreased virulence in beans. The pathogen wasn't exhibited water-soaking symptoms (Elasri et al. 2001). Besides, GacA and GacS are mentioned that control QS signaling in *P. syringae* (Dumenyo et al. 1998; Chatterjee et al. 2003). A study has exhibited that pigB is controlled pigments xanthomonadin in *Xanthomonas campestris* pv. *campestris* strain B24. So, non-pigmented pigB mutants cannot synthesize EPS. It means the pathogen causes a few symptoms in cabbage. Following findings for *Xanthomonas oryzae* pv. *oryzae* show that rpfF mutants are responsible for iron-uptake systems caused by reducing the virulence of *X. oryzae* pv. *oryzae* (Chatterjee and Sonti 2002).

In *Dickeya zeae*, AHL promotes greater mobility and inhibits the growth of multicellular aggregates. Additionally, AHLs named expI were reduced but did not have the same impact on rice seed pathogenicity as AHLs (Hussain et al.

2007). Additionally, recent research reported that in *D. dadantii*, the QS ExpR-ExpI system supported the maceration capability of the plant cell wall degrading enzymes (PCWDES) (Ham et al. 2004; Nasser et al. 2012). These findings demonstrate the ability of ExpR-ExpI systems to limit the pathogenicity of an organism (Nasser et al. 1998; Hussain et al. 2007; Feng et al. 2019).

As apparent from the previous studies, most bacteria have AHL degrading enzymes. These enzymes which probably use QS disrupting could be implicated in good alternative control of bacterial diseases.

12.5.2 AHL-Mimics

Manefield et al. (2000) reported that some molecules called mimic molecules can be used to bind the AHL receptor. The mimic molecules were obtained from pea, soybean, rice, etc. The secretes of the plants have the ability to affect bacterial AHL signaling. For example, garlic includes some compounds using mimic molecules. Bjarnsholt et al. (2005) reported that garlic extracts have a strong QS-inhibiting activity on *P. aeruginosa*. Especially, one compound of garlic extracts namely 4,5,9-trithiadadeca-1,6,11-triene-9-oxide was found a significantly important to active QS-disrupting compound (Jakobsen et al. 2012). Besides, some studies recommend that plants respond to AHL molecules (Von Rad et al. 2008; Zhao et al. 2016). Some treatments of AHL molecules like oxo-C6 HSL can lead to increasing auxin and cytokinin. These properties may support plant development like root elongation and growth rate.

Some bacteria, like *P. carotovorum* and *P. syringae,* are known for their ability to create swimming motility and biofilms when AHL is produced. Salicylic acid, however, was found to decrease the swimming ability of these bacteria and to prevent the development of biofilms, according to earlier investigations (Lemos et al. 2014; Lagonenko et al. 2013). These results showed that the effects of salicylic acid and LuxR ligand on the virulence factors of phytopathogenic bacteria are identical (Lagonenko et al. 2013). Additionally, QS, carvacrol, and eugenol had a negative effect on *P. carotovorum* subsp. *brasiliense* both decreased PCWDES gene expression and biofilm formation (Joshi et al. 2016).

Additionally, the AHL demonstration may support the defense mechanism (Zarkani et al. 2013; Shrestha et al. 2019). Arabidopsis application with AHL priming indicated bacterial pathogen resistance. Similar to this, AHL priming with oxo-C14 HSL increased Arabidopsis' resistance to *P. syringae*. These findings assume that AHL-priming may be different for using mimic molecules and it can be helpful to induce resistant mechanism growth.

12.6 Concluding Remark

Plant–pathogen interactions directly affect plant health by utilizing and regulating QS that mediates interactions between plant microorganisms. Previous studies indicated that bacterial communication occurs in plants based on bacterial density. The bacteria in QA demonstrated that AHL includes signal producers. Manipulate these AHL signals using AHL-degrading enzymes and AHL-mimics could be helpful to reduce or control disease and maintain plant health. Using the QS-disrupting mechanism for plant protection gives a control strategy like eco-friendly and has no health risks. QS-disrupting can be a key biological control point that can manipulate the communication between plant–bacteria interactions.

References

Allen RC, McNally L, Popat R, Brown SP (2016) Quorum sensing protects bacterial co-operation from exploitation by cheats. Int Soc Microbial Ecol 10:1706–1716

Alt-Morbe J, Stryker JL, Fuqua C, Li PL, Farrand SK, Winans SC (1996) The conjugal transfer system of *Agrobacterium tumefaciens* octopine-type Ti plasmids is closely related to the transfer system of an IncP plasmid and distantly related to Ti plasmid vir genes. J Bacteriol 178:4248–4257

Baltenneck J, Reverchom S, Hommais F (2021) Quorum sensing regulation in phytopathogenic bacteria. Microorganisms 9:239

Barber CE, Tang JL, Feng JX, Pan MQ, Wilson TJ (1997) A novel regulatory system required for pathogenicity of *Xanthomonas campestris* is mediated by a small diffusible signal molecule. Mol Microbiol 24:555–566

Barnard AML, Salmond GPC (2007) Quorum sensing in Erwinia species. Anal Bioanal Chem 387: 415–423

Barnard AML, Bowden SD, Burr T, Coulthurst SJ, Monson RE, Salmond GPC (2007) Quorum sensing, virulence and secondary metabolite production in plant soft-rotting bacteria. Philos Trans R Soc B 362:1165–1183

Bassler BL (2002) Small talk. Cell-to-cell communication in bacteria. Cell 109:421–424

Bassler BL, Losick R (2006) Bacterially speaking. Cell 125:237–246

Beck von Bodman S, Farrand SK (1995) Capsular polysaccharide biosynthesis and pathogenicity in Erwinia stewartii require induction by an N-acylhomoserine lactone autoinducer. J Bacteriol 177:5000–5008

Bjarnsholt T, Jensen PO, Rasmussen TB, Christophersen L, Calum H, Hentzer M, Hougen HP, Rygaard J, Moser C, Eberl L, Høiby N, Givskov M (2005) Garlic blocks quorum sensing and promotes rapid clearing of pulmonary *Pseudomonas aeruginosa* infections. Microbiology 151: 3873–3880

Bourras S, Rouxel T, Meyer M (2015) Agrobacterium tumefaciens gene transfer: how a plant pathogen hacks the nuclei of plant and non plant organisms. Phytopathology 105:1288–1301

Byers JT, Lucas C, Salmond GP, Welch M (2002) Nonenzymatic turnover of an *Erwinia carotovora* quorum-sensing signaling molecule. J Bacteriol 184:1163–1171

Castang S, Reverchon S, Gouet P, Nasser W (2006) Direct evidence for the modulation of the activity of the *Erwinia chrysanthemi* quorum-sensing regulator ExpR by acylhomoserine lactone pheromone. J Biol Chem 281:29972–29987

Chatterjee S, Sonti RV (2002) rpfFmutants of *Xanthomonas oryzae* pv. *oryzae* are deficient for virulence and growth under low iron conditions. Mol Plant-Microbe Interact 15:463–471

Chatterjee A, Cui Y, Yang H, Collmer A, Alfano JR, Chatterjee AK (2003) GacA, the response regulator of a two-component system, acts as a master regulator in *Pseudomonas syringae* pv. *tomato* DC3000 by controlling regulatory RNA, transcriptional activators, and alternate sigma factors. Mol Plant Microbe Interact 16:1106–1117

Chen X, Schauder S, Potier N, Van Dorsselaer A, Pelczer I, Bassler BL, Hughson FM (2002) Structural identification of a bacterial quorum-sensing signal containing boron. Nature 415:545–549

Chen CN, Chen CJ, Liao CT, Lee CY (2009) A probable aculeacin acylase from the *Ralstonia solanacearum* GMI1000 is N-acyl-homoserine lactone acylase with quorum-quenching activity. BMC Microbiol 9:89

Cheng F, Ma A, Zhuang X, He X, Zhuang G (2016) N-(3-oxo-hexanoyl)-homoserine lactone has a critical contribution to the quorum-sensing-dependent regulation in phytopathogen *Pseudomonas syringae* pv. *tabaci* 11528. FEMS Microbiol Lett 363:fnw265

Cheng F, Ma A, Luo J, Zhuang X, Zhuang G (2018) N-acylhomoserine lactone-regulation of genes mediating motility and pathogenicity in *Pseudomonas syringae* pathovar *tabaci* 11528. Microbiol Open 6:e00440

Chun W, Cui J, Poplawsky AR (1997) Purification, characterization and biological role of a pheromone produced by *Xanthomonas campestris* pv. *campestris*. Physiol Mol Plant Pathol 51:1–14

Corbett M, Virtue S, Bell K, Birch P, Burr T, Hyman L, Lilley K, Poock S, Toth I, Salmond G (2005) Identification of a new quorum-sensing-controlled virulence factor in Erwinia carotovora subsp. atroseptica secreted via the type II targeting pathway. Mol Plant Microbe Interact 18: 334–342

Crépin A, Beury-Cirou A, Barbey C, Farmer C, Hélias V, Burini JF, Faure D, Latour X (2012) N-acyl homoserine lactones in diverse Pectobacterium and Dickeya plant pathogens: diversity, abundance, and involvement in virulence. Sensors (Basel) 12:3484–3497

Cui YY, Chatterjee A, Liu Y, Dumenyo CK, Chatterjee AK (1995) Identification of a global repressor gene, rsmA, of *Erwinia carotovora* subsp. *carotovora* that controls extracellular enzymes, N-(3-oxohexanoyl)-L-homoserine lactone and pathogenicity of soft-rotting in Erwinia spp. J Bacteriol 177:5108–5115

de Kievit TR, Iglewski BH (2000) Bacterial quorum sensing in pathogenic relationships. Infect Immun 68:4839–4849

Diggle SP, Griffin AS, Campbell GS, West SA (2007) Cooperation and conflict in quorum-sensing bacterial populations. Nature 450:411–414

Dong YH, Xu JL, Li XZ, Zhang LH (2000) AiiA, an enzyme that inactivates the acyl homoserine lactone quorum-sensing signal and attenuates the virulence of *Erwinia carotovora*. Proc Natl Acad Sci U S A 97:3526–3531

Dong YH, Wang LH, Xu JL, Zhang HB, Zhang XF, Zhang LH (2001) Quenching quorum-sensing-dependent bacterial infection by an N-acyl homoserine lactonase. Nature 411:813–817

Dong YH, Gusti AR, Zhang Q, Xu JL, Zhang LH (2002) Identification of quorum-quenching N-acyl homoserine lactonases from Bacillus species. Appl Environ Microbiol 68:1754–1759

Dumenyo C, Mukherjee A, Chun W, Chatterjee AK (1998) Genetic and physiological evidence for the production of N-acyl homoserine lactones by *Pseudomonas syringae* pv. *syringae* and other fluorescent plant pathogenic Pseudomonas species. Eur J Plant Pathol 104:569–582

Elasri M, Delorme S, Lemanceau P, Stewart G, Laue B (2001) Acylhomoserine lactone production is more common among plant-associated Pseudomonas spp. than among soilborne Pseudomonas spp. Appl Environ Microbiol 67:1198–1209

Emmert EAB, Handelsman J (1999) Biocontrol of plant disease: a (gram-) positive perspective. FEMS Microbiol Lett 171:1–9

Farrand SK, Qin YP, Oger P (2002) Quorum-sensing system of Agrobacterium plasmids: analysis and utility. Method Enzymol 358:452–484

Feng L, Schaefer AL, Hu M, Chen R, Greenberg EP, Zhou J (2019) Virulence factor identification in the banana pathogen *Dickeya zeae* ms2. Appl Environ Microbiol 85:e01611

Flavier AB, Clough SJ, Schell MA, Denny TP (1997) Identification of 3- hydroxypalmitic acid methyl ester as a novel autoregulator controlling virulence in *Ralstonia solanacearum*. Mol Microbiol 26:251–259

Fuqua C, Greenberg EP (2002) Listening in on bacteria: acyl-homoserine lactone signaling. Nat Rev Mol Cell Biol 3:685–695

Fuqua WC, Winans SC, Greenberg EP (1996) Census and consensus in bacterial ecosystems: the LuxR-LuxI family of quorumsensing transcriptional regulators. Annu Rev Microbiol 50:727–751

Fuqua C, Parsek MR, Greenberg EP (2001) Regulation of gene expression by cell to cell communication: acyl-homoserine lactone quorum sensing. Annu Rev Genet 35:439–468

Ganin H, Tang X, Meijler MM (2009) Inhibition of *Pseudomonas aeruginosa* quorum sensing by AI-2 analogs. Bioorg Med Chem Lett 19:3941–3944

Gelvin SB (2012) Traversing the cell: agrobacterium T-DNA's journey to the host genome. Front Plant Sci 3:52. [

Gill EE, Franco OL, Hancock RE (2015) Antibiotic adjuvants: diverse strategies for controlling drug-resistant pathogens. Chem Biol Drug Des 85:56–78

Ham JH, Cui Y, Alfano JR, Rodríguez-Palenzuela P, Rojas CM (2004) Analysis of *Erwinia chrysanthemi* EC16 pelE::uidA, pelL::uida, and hrpN::uidA mutants reveals strain-specific atypical regulation of the Hrp type III secretion system. Mol Plant-Microbe Interact 17:184–194

Helman Y, Chernın L (2015) Silencing the mob: disrupting quorum sensing as a means to fight plant disease. Mol Plant Pathol 16(3):316–329

Hussain MBBM, Zhang HB, Xu JL, Liu Q, Jiang Z, Zhang H (2007) The acyl-homoserine lactone-type quorum-sensing system modulates cell motility and virulence of *Erwinia chrysanthemi* pv. *zeae*. J Bacteriol 190:1045–1053

Jakobsen TH, van Gennip M, Phipps RK, Shanmugham MS, Christensen LD, Alhede M, Skindersoe ME, Rasmussen TB, Friedrich K, Uthe F, Jensen PO, Moser C, Nielsen KF, Eberl L, Larsen TO, Tanner D, Hoiby N, Bjarnsholt T, Givskov M (2012) Ajoene, a sulfur-rich molecule from garlic, inhibits genes controlled by quorum sensing. Antimicrob Agents Chemother 56:2314–2325

Jimenez PN, Koch G, Thompson JA, Xaiver KB, Cool RH, Quax WJ (2012) The multiple signaling systems regulating virulence in *Pseudomonas aeruginosa*. Microbiol Mol Biol Rev 76:46–65

Kimura S, Tateda K, Ishii Y, Horikawa M, Miyairi S, Gotoh N, Ishiguro M, Yamaguchi K (2009) Pseudomonas aeruginosa las quorum sensing autoinducer suppresses growth and biofilm production in Legionella species. Microbiology 155:1934–1939

Kleerebezem M, Quadri LEN, Kuipers OP, de Vos WM (1997) Quorum sensing by peptide pheromones and two-component signal-transduction systems in gram-positive bacteria. Mol Microbiol 24:895–904

Koutsoudis MD, Tsaltas D, Minogue TD, von Bodman SB (2006) Quorum sensing regulation governs bacterial adhesion, biofilm development, and host colonization in *Pantoea stewartii* subsp.*stewartii*. Proc Natl Acad Sci U S A 103:5983–5988

Leadbetter JR, Greenberg EP (2000) Metabolism of acyl-homoserine lactone quorum-sensing signals by *Variovorax paradoxus*. J Bacteriol 182:6921–6926

Lin YH, Xu JL, Hu JY, Wang LH, Ong SL, Leadbetter JR, Zhang LH (2003) Acyl-homoserine lactone acylase from Ralstonia strain XJ12B represents a novel and potent class of quorum-quenching enzymes. Mol Microbiol 47:849–860

Lithgow JK, Wilkinson A, Hardman A, Rodelas B, Wisniewski-Dye F, Williams P, Downie JA (2000) The regulatory locus cinRI in *Rhizobium leguminosarum* controls a network of quorum-sensing loci. Mol Microbiol 37:81–97

Liu B, Fillgrove WY, Anderson VE (2002) Triclosan inhibits enoyl-reductase of type I fatty acid synthase in vitro and is cytotoxic to MCF-7 and SKBr-3 breast cancer cells. Cancer Chemother Pharmacol 49:187–193

Loh J, Yuen-Tsai JPY, Welborn A, Stacey G (2001) Population density-dependent regulation of the *Bradyrhizobium japonicum* nodulation genes. Mol Microbiol 42:37–46

Loh J, Carlson RW, York WS, Stacey G (2002a) Bradyoxetin, a unique chemical signal involved in symbiotic gene regulation. Proc Natl Acad Sci U S A 99:14446–14451

Loh J, Pierson EA, Pierson LS III, Stacey G, Chatterjee A (2002b) Quorum sensing in plant-associated bacteria. Curr Opin Plant Biol 5:1–6

Luo ZQ, Farrand SK (2001) The *Agrobacterium tumefaciens* rnd homolog is required for TraR-mediated quorum-dependent activation of Ti plasmid tra gene expression. J Bacteriol 183:3919–3930

Mae A, Montesano M, Koiv V, Palva ET (2001) Transgenic plants producing the bacterial pheromone N-acyl-homoserine lactone exhibit enhanced resistance to the bacterial phytopathogen *Erwinia carotovora*. Mol Plant-Microbe Interact 14:1035–1042

Manefield M, Harris L, Rice SA, de Nys R, Kjelleberg S (2000) Inhibition of luminescence and virulence in the black tiger prawn (*Penaeus monodon*) pathogen *Vibrio harveyi* by intercellular signal antagonists. Appl Environ Microbiol 66:2079–2084

Meighen EA (1991) Molecular biology of bacterial bioluminescence. Microbiol Rev 55:123–142

Meighen EA (1994) Genetics of bacterial bioluminescence. Annu Rev Genet 28:117–139

Meng F, Babujee L, Jacobs JM, Allen C (2015) Comparative transcriptome analysis reveals cool virulence factors of *Ralstonia solanacearum* race 3 biovar 2. PLoS One 10:e0139090

Miller MB, Bassler BL (2001) Quorum sensing in bacteria. Annu Rev Microbiol 55:165–199

Mohammadi M, Geider K (2007) Autoinducer-2 of the fire blight pathogen *Erwinia amylovora* and other plant-associated bacteria. FEMS Microbiol Lett 266:34–41

Molina L, Constantinescu F, Michel L, Reimmann C, Duffy B, Defago G (2003) Degradation of pathogen quorum-sensing molecules by soil bacteria: a preventive and curative biological control mechanism. FEMS Microbiol Ecol 45:71–81

More MI, Finger LD, Stryker JL, Fuqua C, Eberhard A, Winans SC (1996) Enzymatic synthesis of a quorum-sensing autoinducer through use of defined sub-strates. Science 272:1655–1658

Mukherjee A, Cui Y, Liu Y, Chatterjee AK (1997) Molecular characterization and expression of the Erwinia carotovora hrpNEcc gene, which encodes an elicitor of the hypersensitive reaction. Mol Plant–Microbe Interact 10:462–471

Mukherjee A, Cui Y, Ma WL, Liu Y, Chatterjee AK (2000) hexA of Erwinia carotovora ssp carotovora strain Ecc71 negatively regulates production of RpoS and rsmB RNA, a global regulator of extracellular proteins, plant virulence and the quorum-sensing signal, N-(3-oxohexanoyl)-L-homoserine lactone. Environ Microbiol 2:203–215

Nasser W, Bouillant ML, Salmond G, Reverchon S (1998) Characterization of the *Erwinia chrysanthemi* expI–expR locus directing the synthesis of two N-acyl-homoserine lactone signal molecules. Mol Microbiol 29:1391–1405

Nasser W, Dorel C, Wawrzyniak J, Van Gijsegem F, Groleau MC, Déziel E, Reverchon S (2012) A new quorum sensing system controls the virulence of *Dickeya dadantii*. Environ Microbiol 15: 865–880

Nealson KH, Hastings JW (1979) Microbiol Rev 43:469–518

Parsek MR, Val DL, Hanzelka BL, Cronan JE, Greenberg EP (1999) Acyl homoserine-lactone quorum-sensing signal generation. Proc Natl Acad Sci U S A 92:1490–1494

Pierson EA, Weller DM (1994) Use of mixtures of fluorescent pseudomonads to suppress take-all and improve the growth of wheat. Phytopathology 84:940–947

Quinones B, Pujol CJ, Lindow SE (2004) Regulation of AHL production and its contribution to epiphytic fitness in *Pseudomonas syringae*. Mol Plant Microbe Interact 17:521–531

Quinones B, Dulla G, Lindow SE (2005) Quorum sensing regulates exopolysaccharide production, motility, and virulence in *Pseudomonas syringae*. Mol Plant Microbe Interact 18:682–693

Raupach GS, Kloepper JW (1998) Mixtures of plant growth-promoting rhizobacteria enhance biological control of multiple cucumber pathogens. Phytopathology 88:1158–1164

Reverchon S, Bouillant ML, Salmond G, Nasser W (1998) Integration of the quorum-sensing system in the regulatory networks controlling virulence factor synthesis in *Erwinia chrysanthemi*. Mol Microbiol 29:1407–1418

Rezzonico F, Duffy B (2007) The role of LuxS in the fire blight pathogen *Erwinia amylovora* is limited to metabolism and does not involve quorum sensing. Mol Plant Microbe Interact 20: 1284–1297

Schauder S, Shokat K, Surette MG, Bassler BL (2001) The LuxS family of bacterial autoinducers: biosynthesis of a novel quorum-sensing signal molecule. Mol Microbiol 41:463–476

Schenk ST, Schikora A (2014) AHL-priming functions via oxylipin and salicylic acid. Front Plant Sci 5:784

Schenk ST, Stein E, Kogel KH, Schikora A (2012) Arabidopsis growth and defense are modulated by bacterial quorum sensing molecules. Plant Signal Behav 7:178–181

Shaw PD, Ping G, Daly SL, Cha C, Cronan JE, Rinehart KL, Farrand SK (1997) Detecting and characterizing N-acyl-homoserine lactone signal molecules by thin-layer chromatography. Proc Natl Acad Sci U S A 94:6036–6041

Shepherd RW, Lindow SE (2008) Two dissimilar N-acyl-homoserine lactone acylases of *Pseudomonas syringae* influence colony and biofilm morphology. Appl Environ Microbiol 75:45–53

Shrestha A, Elhady A, Adss S, Wehner G, Böttcher C, Heuer H (2019) Genetic differences in barley govern the responsiveness to N-acyl homoserine lactone. Phytob J 3:191–202

Steidle A, Allesen-Holm M, Riedel K, Berg G, Givskov M, Molin S, Eberl L (2002) Identification and characterization of an N-acyl-homoserine lactone-dependent quorum-sensing system in *pseudomonas putida* strain IsoF. Appl Environ Microbiol 68:6371–6382

Stevens AM, Greenberg EP (1997) Quorum sensing in *Vibrio fischeri*: essential elements for activation of the luminescence genes. J Bacteriol 179:557–562

Su S, Khan SR, Farrand SK (2008) Induction and loss of Ti plasmid conjugative competence in response to the acyl-homoserine lactone quorum-sensing signal. J Bacteriol 190:4398–4407

Taguchi F, Ogawa Y, Takeuchi K, Suzuki T, Toyoda K, Shiraishi T, Ichinose Y (2006) A homologue of the 3-oxoacyl-(acyl carrier protein) synthase III gene located in the glycosylation Island of *Pseudomonas syringae* pv. *tabaci* regulates virulence factors via N-acyl homoserine lactone and fatty acid synthesis. J Bacteriol 188:8376–8384

Teplitski M, Robinson JB, Bauer WD (2000) Plants secrete substances that mimic bacterial N-acyl homoserine lactone signal activities and affect population density-dependent behaviors in associated bacteria. Mol Plant-Microbe Interact 13:637–648

Verma SC, Miyashiro T (2013) Quorum sensing in the squid-vibrio symbiosis. Int J Mol Sci 14(8): 16368–16401

Visick KL, McFall-Ngai MJ (2000) An exclusive contract: specificity in the Vibrio fischeri-Euptrymna scolopes partnership. J Bacteriol 182:1779–1787

Von Bodman SB, Majerczak DR, Coplin DLA (1998) Negative regulator mediates quorum-sensing control of exopolysaccharide production in *Pantoea stewartii* subsp. *stewartii*. Proc Natl Acad Sci USA 95:7687–7692

von Rad U, Klein I, Dobrev PI, Kottova J, Zazimalova E, Fekete A, Hartmann A, Schmitt-Kopplin P, Durner J (2008) Response of *Arabidopsis thaliana* to N-hexanoyl-DL-homoserine-lactone, a bacterial quorum sensing molecule produced in the rhizosphere. Planta 229:73–85

Watson WT, Minoque TD, Val DL, von Bodman S, Churchill ME (2002) Structural basis and specificity of acyl-homoserine lactone signal production in bacterial quorum sensing. Mol Cell 9:685–694

Whitehead NA, Barnard AML, Slater H, Simpson NJ, Salmond GPC (2001) Quorum sensing in gram-negative bacteria. FEMS Microbiol Rev 25:365–404

Whitely M, Greenberg EP (2001) Promoter specificity elements in *Pseudomonas aeruginosa* quorum sensing-controlled genes. J Bacteriol 183:5529–5534

Winsdor WJ (2020) How quorum sensing works. https://asm.org/Articles/2020/June/How-QuorumSensingWorks#:~:text=Bacterial%20communication%20relies%20on%20versatile,these%20signals%20vary%20between%20species

Zarkani AA, Stein E, Rohrich CR, Schikora M, Evguenieva-Hackenberg E, Degenkolb T (2013) Homoserine lactones influence the reaction of plants to rhizobia. Int J Mol Sci 14:17122–17146

Zhang L, Murphy PJ, Kerr A, Tate ME (1993) Agrobacterium conjugation and gene regulation by N-acyl-L-homoserine lactones. Nature 362:446–448
Zhang HB, Wang LH, Zhang LH (2002) Genetic control of quorum-sensing signal turnover in *agrobacterium tumefaciens*. Proc Natl Acad Sci U S A 99:4638–4643
Zhao Q, Li M, Jia Z, Liu F, Ma H, Huang Y (2016) AtMYB44 positively regulates the enhanced elongation of primary roots induced by N-3-OxoHexanoyl-homoserine lactone in *Arabidopsis thaliana*. Mol Plant-Microbe Interact 29:774–785
Zhu J, Oger PM, Schrammeijer B, Hooykaas PJJ, Farrand SK, Winans SC (2000) The basis of crown gall tumorigenesis. J Bacteriol 182:3885–3895

Chapter 13
Omics Technologies in the Plant–Microbe Interactions

Kubilay Kurtulus Bastas and Ajay Kumar

Abstract The investigation and integration of these chemical interactions are aided by studies of the microbiomes connected to plants. Through cutting-edge genomic techniques like meta-omics and comparative investigations, key aspects of endophytism are clarified. A thorough understanding of the host infestation mechanisms and the functions of endophytes can be used to improve agricultural eco-management processes such as biological control, bioremediation, and plant growth promotion. Modern methods including metagenomics, microarray, genome sequencing, metatranscriptomics, comparative genomics, and next-generation sequencing can be used to understand the relationships between plants and endophytes. By using contemporary methods and procedures, it is possible to investigate the alleged functions of endophytes in the ecology of the host plant. Identification of genes, mRNA, and proteins of biological control agents (BCAs) involved in the biocontrol mechanisms; comparison between strains and mutated strains with specific biological control properties; and concurrent studies of bi- or tri-trophic interactions related to the transcriptome or proteome analysis of the host plant, the plant pathogen, and the microbial biological control agents (MBCAs) are all possible. By sequencing BCA's genome, plant disease control can be significantly improved. Omics-based technologies are anticipated to offer considerably deeper insights into the biocontrol processes in this series of events, which will be crucial for the future commercialization of biocontrol drugs.

Keywords Omics · Endophytism · Microbiome · Endophytes · Genomics · Metagenomics

K. K. Bastas (✉)
Department of Plant Protection, Faculty of Agriculture, Selcuk University, Konya, Turkey
e-mail: kbastas@selcuk.edu.tr

A. Kumar
Amity Institute of Biotechnology, Amity University, Noida, India

K. K. Bastas et al. (eds.), *Microbial Biocontrol: Molecular Perspective in Plant Disease Management*, Microorganisms for Sustainability 49,
https://doi.org/10.1007/978-981-99-3947-3_13

13.1 Introduction

Food security is one of the most serious challenges for the rising global population. Currently huge amount of agricultural productivity, including crops and fruits, losses occur each year due to numerous biotic and abiotic stressors (Kumar 2022a, b). Due to plant diseases carried on by several pathogenic bacteria, crop productivity might endure losses of up to 40% (Valenzuela-Ruiz et al. 2020). According to FAO (2018) reports, the use of different agricultural practices to minimize the losses of plant diseases has been in practice without compromising the biosafety and nutritional value, and most part of the world is still using chemical pesticides to manage the plant pathogens or diseases during harvest or postharvest storage conditions (Kumari et al. 2022). But the utilization of chemical pesticides adversely affects the texture and food quality due to their chemical residue or toxic nature. In this regard the scientific community has moved toward ecofriendly methods or utilization of microbial biocontrol agent to control the plant diseases (Patel et al. 2022; Verma et al. 2022).

Ecologically friendly management of phytopathogens depends on the proper use of biological control agents (BCAs) (Heimpel and Mills 2017; Singh et al. 2017). According to Conrath et al. (2015) and Spadaro and Droby (2016), BCAs have the ability to compete with pathogens for resources and space or to stimulate host resistance. They can also inhibit the growth of pathogens through the process of antibiosis (Ghorbanpour et al. 2018; Singh et al. 2020). BCAs can also directly prevent the spread of pathogens by secreting secondary metabolites (SMs) with antimicrobial characteristics (Raaijmakers and Mazzola 2012). Sustainable agriculture can benefit greatly from the characterization and application of microorganisms as various formulations that can reduce the need for toxic synthetic pesticides and fertilizers while still managing phytopathogens in a sustainable manner (Khokhar et al. 2012).

With phenomenal advancements in the latest next-generation technology, it is now possible to combine and relate data of genomic studies to various computational tools otherwise known as bioinformatics tools. This combination has remarkably helped in the prediction of plant–microbe interactions (Lima-Mendez et al. 2015; Frantzeskakis et al. 2020; Terauchi et al. 2020), which is very essential for the development of BCAs with enhanced capabilities. Various such studies as gene editing, genome engineering, and many other latest approaches have been used for understanding the aforementioned interactions (Yadav et al. 2022; Kumar 2022a). Omics technologies have revolutionized molecular studies up to the level of proteins and recent discoveries in these technologies have made possible the understanding of different interactions between BCA and pathogens as well as between these two and the host, the identification of new microbes, and detailed studies of BCA residing within different agroecosystems.

However, with the use of omics technologies, processes related to bacterial physiology, pathogenicity, stress mechanisms, and the mode of action of secondary metabolites with antimicrobial characteristics have been thoroughly searched after and studied (Roemer and Boone 2013; Tang 2015). Identification of novel microorganisms with biological control properties requires deep insights into the plant–microbiome functions. Advancements in omics technologies and high-throughput sequencing techniques are now helping greatly in achieving these goals. Increased phyto-tolerance against pathogenic stresses requires the understanding of unknown underlying mechanisms to develop new-generation products with plant protection properties, especially stress protection agents (SPAs) (Berg et al. 2013). As of now, the efficacy of already available biopesticides is low when compared to synthetic compounds. The reasons behind this are thought to be the different responses of BCAs to different environments and their short life span (Singh 2014). On the contrary, antagonism or triggering of host defense responses by BCAs are capable of managing soil-borne diseases effectively, which are otherwise difficult to manage through the use of synthetic pesticides.

13.2 Identification of Microbes

Microbes were used to be identified by their isolation and culturing using various types of growth media including selective types, following morphological and physiological characterization and nutritional requirements (Sanzani et al. 2014). The 16S rRNA gene is being used for the taxonomic studies of BCAs at the molecular level for prokaryote taxonomic affiliation as it exists as a multigenic family and has been found in a similar group of organisms; its evolutionary progress is considered as very slow as its function has not been changed over time, the length of this gene being 1500 bp, which is not very ideal for the bacterial identification at the species level but good enough for technical and bioinformatic purposes (Aguilar-Marcelino et al. 2020). Similarly, the internal transcribed spacer (ITS) regions (ITS1, ITS2, and 5.8S rRNA gene) are strictly conserved at the genus level in eukaryotes, while their identification and taxonomic affiliation at the species level are hindered by its length (650–1100 bp) and restricted genetic diversity. Inter Simple Sequence Repeats (ISSR), Amplified Fragment Length Polymorphism (AFLP), Random Amplified Polymorphic DNA (RAPD), and Restriction Fragment Length Polymorphism (RFLP) are molecular techniques being practiced for the taxonomic affiliation of MBCA, while for detection purposes marker genes are used, which can be Sequence Characterized Amplified Region (SCAR), 16S rRNA, 28S rRNA, elongation factor, β-tubulin, RPB2, and ITS in polymerase chain reaction (PCR) test (Kumari et al. 2022). Selected regions or genes, type of restriction enzymes, and suitable molecular techniques are important factors to be considered for the taxonomic identifications. However, the molecular technique is a highly sensitive process; the use of quality control methods can be very crucial for the proper identification as the poor-quality DNA reads can be eliminated in the following analysis (Xi et al. 2019). Trimmomatic v0.36 (Bolger et al. 2014), checkM and Quast (Gurevich et al. 2013; Parks et al. 2015), SPAdes (Antipov et al. 2016), and Canu are a few of the bioinformatics tools frequently used for the analysis (Koren et al. 2017). The DNA–DNA hybridization standard (DDH), which is used to determine the taxonomic affiliation of bacterial species, has lately been replaced by in silico tools. Overall genome relatedness indices (OGRI) and average nucleotide identity (ANI) are the most significant in silico tools (Paulsen et al. 2005; Knief 2014; Richter et al. 2016).

Many genera have species that are antagonistic to various plant pathogens but can be dangerous or pathogenic to plants, animals, and humans such as *Bacillus*. Therefore correct taxonomic affiliation is inevitable for the multiplication and commercial use of biosafe BCA.

The development of new molecular techniques and in silico technologies, such as various next-generation sequencing (NGS) strategies and platforms, along with various genomic, phylogenomic, and bioinformatic tools, has allowed for a thorough understanding of the interaction between the evolution of microorganisms and their environment (Jagadeesan et al. 2019). The ability to pinpoint the genes in charge of colonization and host adaptation as well as the mechanisms behind their activity has also been facilitated by these technologies. The development of the "DNA barcoding" approach for molecularly identifying bacteria may help distinguish between various species by sequencing small DNA fragments from particular sections of the genome (Hebert et al. 2003).

These short fragments are utilized to design universal PCR primers as they harbor conserved flanking sites and have genetic variability and divergence at the species level thus providing a wide taxonomic application (Kress and Erickson 2008). Nowadays, the characterization of microbial communities from compound environmental samples is possible due to the advancement in metagenomic studies including meta-barcoding technology and supporting analytical software.

13.3 Plant–Microbe Beneficial Interaction

Plants interact with numerous microbial communities including bacteria, fungi, actinomycetes, and nematodes and share a complex relationship with each other. This interaction may be positive or negative, which means either beneficial or harmful (Pathak et al. 2022; Kumar et al. 2021). The beneficial interactions are of great importance in agriculture systems. There are many reports of such interactions in the literature that include various organisms such as plant growth-promoting rhizobacteria (PGPR) and mycorrhizal fungi (Gupta et al. 2020). Factors like soil type, soil condition, climate, and plant species are considered important for microbes to support plant growth by aiding nutrient uptake such as copper, sulfur, iron, nitrogen, and phosphorous. They also activate the resistance mechanisms of plants by promoting hormone production against biotic stresses.

Efficient management of plant pathogens depends upon the engineering of superior BCA. To do so, they should be provided with an improved environment favorable for their growth and development, or it can be achieved by making necessary changes in their genome using mutagens and genetic engineering, and genetic manipulation can also increase their stress resistance. Similarly, the genes encoding undesirable traits that hinder the performance of BCA can be silenced or deleted from the genome for its better performance. Recent developments in genetic-level research have improved our comprehension of the mechanism of action of BCA. By using techniques like whole genome sequencing, targeted gene sequence from a known database, and differential display techniques, the underlying mechanism of an organism's mode of action can be better understood (Abdel-Salarn et al. 2007). Biocontrol properties of *Pseudomonas*, *Bacillus*, and *Trichoderma* have been enhanced through whole genome sequencing studies. Techniques used other than the aforementioned ones are the study of gene activation and their overexpression, RNA silencing, and PCR tilling.

13.4 Omics Sciences

Latest omics technologies have revolutionized gene expression studies. Massart et al. (2015) stated that the modern "-omics" research progress has provided us with better insights into various tropic interactions in rhizosphere systems at reduced costs. Omics approaches and strategies in plant–microbe interactions are summarized in Fig. 13.1.

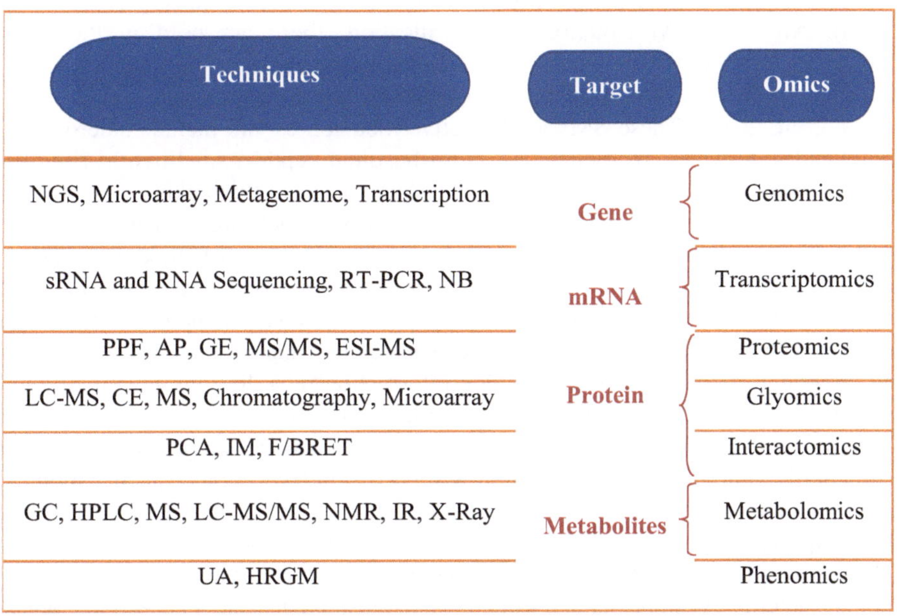

Fig. 13.1 Omics approaches and strategies in plant–microbe interactions. *AP* affinity purification, *CE* capillary electrophoresis, *F/BRET* fluorescence/bioluminescence resonance energy transfer, *GE* gel electrophoresis, *HRGM* high-resolution genome mapping, *IM* interactome mapping, *LC-MS* liquid chromatography-mass spectrometry, *MS* mass spectrometry, *NB* northern blot, *NGS* next-generation sequencing, *PCA* protein complementation assay, *PPF* protein/peptide fractionation, *UA* ultrastructural analysis. (Modified from Sharma et al. 2020)

13.5 The Omics Science in Microbial Biocontrol

For the effective and sustainable utilization of BCAs as alternative to chemical pesticides, it is necessary to optimize the viable doses and frequency of applications and also to identify suitable substrates that can provide them support under different stresses. This can be achieved by evaluating the regional microbiomes for their potential beneficial activities, and studying their mechanisms of action in different environments, as this can provide key information about their establishment and functions like antagonistic actions (Verma et al. 2022).

Through the identification of molecular patterns associated with microorganisms, diseases, or damage, BCAs can stimulate the host's systemic defense response against infections (Damage-Associated Molecular Patterns, DAMPs). An induced systemic reaction (ISR) is this kind of response. By identifying the pathogen and activating it in non-infected plant sections, this aids the plant in preventing its establishment and protects it against ongoing and subsequent phytopathogen attacks (Pathak et al. 2022). These defense responses are said to be activated or triggered by the accumulation of histones, DNA methylation alterations, and mitogen-activated kinase proteins (MAKPs) (Espinas et al. 2016).

Secondary metabolites (SMs) secreted by most of the BCAs are of great importance and are considered a green solution in the agriculture industry (Köhl et al. 2019). After the SM isolation and purification, these are used in the agro-biotechnology industry as they possess antagonistic abilities and can be used in the production of biofertilizers and biopesticides as carriers of chemical elements (Woo et al. 2014). These SMs are bioactive metabolites and include siderophore, polyketides, terpenes, lytic enzymes, non-ribosomal peptides, and hybrid peptide–peptide metabolites. The biosynthesis of SM is related to the shikimate acid pathway, sugar derivatives, and acetyl Co-A pathway, which are primary metabolic pathways (Pott et al. 2019), and an in-depth understanding of these pathways is crucial in product innovation for biocontrol through traditional or omics techniques.

The functions of different types of suppressive soils can now be studied or evaluated through advanced technologies such as metagenomics. Metagenomics is an integrated approach that can provide information regarding the microbial communities within the soil, which can prevent plant pathogens to cause disease or multiply within their desired host. Such soils are general suppressive soils that contain high microbial mass but with low suppression levels while specific suppressive soils contain one or more microbial species at high concentration levels with high suppressive activities (Mousa and Raizada 2016). Inoculation of chili pepper plant rhizosphere with *Bacillus amyloliquefaciens* elevated the diversity of stimulating bacteria (Acidobacteria, Firmicutes, Leptosphaeria, and Phaeosphaeriopsis) and also suppressed the *Fusarium* establishment (Shen et al. 2015). Many studies regarding the use of BCAs in plant disease management have been reported, but more extensive and integrated studies are required to develop affordable but still profitable, effective, and environmentally safe alternatives to establish a strong industry for the production of high-quality and safe bioproducts.

13.5.1 Genomics

The study of an organism's entire genome is referred to as genomics, which helps to incorporate components from genetics. It requires a number of procedures, such as combining multiple DNA sequencing techniques with recombinant DNA methods, then utilizing bioinformatics methodology to assemble and analyze the generated data to determine the structural and functional aspects of genomes.

Microbial habitats are a complex mixture of DNA whose real-time characterization is made easy by the metagenomics approach (Mitra 2019). This technique also provides the base for further RNA investigation such as metabolomics, metaproteomics, and metatranscriptomics, all three of which are capable to unveil the unknown metabolic capacity of a certain microbiota (Delmotte et al. 2009). The metabolic capacity of the bacterial community inside the rice plant roots was studied through metagenome analysis of microbes from the root endosphere after surface sterilization (Sessitsch et al. 2012). By combining the data obtained after transcriptional analysis of some genes through metagenomics, it was noticed that the bacterial microbiota was able to assist in the nitrogen cycle in plants (Sharma et al. 2020).

Through genomic analysis, characteristics of fungal germs like *Trichoderma* species can be revealed. There are some differences between the genomes of *Trichoderma virens* and *Trichoderma atroviride*, which have more metabolic genes than *Trichoderma reesei* (Mukherjee et al. 2012). Numerous genes in the *Trichoderma* biocontrol strains were discovered by the comparison of the genomic data of the non-mycoparasitic strain *T. reesei* and the mycoparasitic strain *Trichoderma harzianum* (Sharma et al., 2017). Genome mining of genes with unknown functions, comparative genomics between strains of the same species that may or may not possess antagonistic traits, or from related species, as well as transcription studies are all made possible by the availability of detailed genomic data, which creates the foundation for high-throughput analyses that accelerate the study of antagonistic traits.

The annotation process identifies genome features such as repetitive or mobile elements, gene clusters, and prophage integration for bacteria and plasmids. Secretion mechanisms and enzymes that break down fungal cell walls were found in the genome of a BCA that is hostile to fungi. In a BCA acting by antibiosis, genes associated in the synthesis of primary or secondary metabolites are targeted. Comparative genome analysis, which can reveal the genetic variations in their modes of living, can also be used to compare the genomes of pathogens and BCAs. A fascinating illustration of this kind of contrast is found in the *Pseudomonas fluorescens* group, which contains both disease strains and BCA. Since identical secondary metabolites have been found in both BCA and pathogen strains, more research is needed to understand the biocontrol capabilities of genome mining (Eyiwumi Olorunleke et al. 2015). Microbe sequencing is a well-established process specifically for bacteria that aids in the discovery of new natural products using genomics-guided approaches under genome mining. There are now several established methods for determining the metabolic byproducts of genes encoding biocontrol traits, and only a small number of products have been synthesized (Bachmann et al. 2014; Challis 2008). Through the use of genome mining techniques, the polyketide synthase and non-ribosomal peptide synthetase found in the *P. fluorescens* group led to the creation of new antibiotic compounds (Sharma et al. 2020). Eco-genomics has improved our understanding of the intricate connections between host, pathogen, and environment (Martin 2013). A greater understanding of a location's microbiota can be obtained using metagenomics and metatranscriptomics combined. Such knowledge can aid in the development of a variety of new biocontrol agents as well as the modulation of pathogen activity. Major plant pathogens' genomes have been sequenced thanks in large part to groundbreaking genomics discoveries, lower sequencing costs, and high-throughput sequencing. This has led to a thorough understanding of the pathogenesis process, potential imparted resistance, and the mechanism by which BCA inhibits pathogenesis. The genomes of pathogens, host plants, and helpful microorganisms can all be modified using several crucial genetic engineering tools, such as RNA interference (RNAi) and genome editing technologies (Klosterman et al. 2016; Collinge 2018). Given that both metabolomics and signalomics concentrate on omics techniques that target signaling molecules, they can work together to study how the native and imported microbiota interact (Mhlongo et al. 2018). Making important decisions, such as the ideal BCA dosage and application duration, can be made easier by being aware of these interactions. According to Mazzola and Freilich (2017), this method can help to build native microbial communities to strengthen the antagonistic capabilities of BCA and helper strains to increase the BCA's ability to survive and suppress pathogens (Massart et al. 2015). Furthermore, various microbiomes with various microorganisms and modes of action can offer better and more chances for managing plant diseases in a sustainable way (Gopal et al. 2013; Massart et al. 2015; Syed Ab Rahman et al. 2018).

13.5.1.1 Multigenomics

Comparative multigenome analysis can be used to comprehend the metabolic and genetic diversity of comparable or related microorganisms that interact with plants in diverse ways (Zuccaro et al. 2011). Comparing their genomes reveals the molecular markers that discriminate between closely related species of various microbiomes. Comparative genome research has shown the factors that influence habitat choice, such as variations in secretory, surface-attachment, metabolic, and transport proteins. Lipopolysaccharide and adhesins were discovered to be the molecules likely responsible for the disparate phenotypic behavior of related microorganisms (Monteiro et al. 2012).

Comparing the genomes of endophytic and non-endophytic isolates can reveal crucial characteristics required for the creation and maintenance of the connection between bacteria and host plants (Lopez-Fernandez et al. 2015). Comparing the genomes of closely related endophytes with various roles in the plants can also reveal mechanisms of adaptation and evolution (Schardl et al. 2013; Tezerji et al. 2015).

Through pan-genomic investigations, genetic factors affecting the endophytism of bacteria may now be closely examined. This indicates that this method may be useful for identifying unknown essential genes that are involved in a microbe's adaptability and evolution into an endophyte (Medlini et al. 2005; Mayer et al. 2014).

13.5.1.2 Metagenomics

Sequences of various microbes from different ecological communities can be analyzed through metagenomics and it also eliminates the need for isolation and cultivation of microbes. Understanding the metabolic and beneficial traits is inevitable in order to engineer the endophytes to incorporate necessary characteristics for better interaction with host plants. Metagenomic is greatly helpful in this regard as most of the endophytes known today are non-culturable and their endophytic microbial functions can be determined through this approach (Dinsdale et al. 2008). It is done by direct extraction of DNA from the whole population followed by gene content analysis. Metagenome analysis of rice-based root endophytes unveiled their presumptive functions like their various metabolic adaptations to their habitat (Sessitsch et al. 2012). This has suggested their high potential as antagonists against pathogens, plant-growth promoters, bioremediation, and as enhancers of plant stress resistance. Microbial communities from various habitats have maintained their functional diversity. Comparative metagenomics is useful for such study and has been used by Dinsdale et al. (2008) to describe the differences in functions of nine different microbes.

Next-generation sequencing (NGS) has accelerated and facilitated metagenomic research on novel characterization of bacteria (Akinsanya et al. 2015). It has given scientists a mechanistic approach for the quick and sensibly analysis of DNA sequences from an environmental sample (Jones 2010). Jumpponen et al. (2010) claim that 454 sequencing has facilitated genetic research on fungal microorganisms. The coexistence of mycorrhizal and endophytic fungi in a temperate forest in Japan has been proven using this sequencing method (Toju et al. 2013). The potential complex interactions involved in cohabitation can be further investigated using metaproteomics, metatranscriptomics, or metaproteogenomics techniques. The fact that there are no homologs in public databases for the large proportion of obtained sequences is one of the main limitations of the metagenomics technique. Before utilizing NGS technologies for experimental studies, such restrictions must be taken into account (Daniel et al. 2008; Jones 2010). The constraint would be partially resolved by genome sequencing analyses of the strains taken from the same environment.

13.5.2 Transcriptomics and Metatranscriptomics

The transcriptomics approach has made it possible to analyze microbial communities interacting with different plants and has also made it simple to characterize the potential transcripts responsible for distinct biological activities (Molina et al. 2012; Herrera-Estrella 2014; Sheibani-Tezerji et al. 2015). Modulating the expression of genes involved in the interaction with biotic and/or abiotic factors has become possible thanks to transcriptome analysis, which focuses on the study of gene expression, signaling cascades, controlled molecular mechanisms, and co-expression patterns (Aguilar-Marcelino et al. 2020). By contrasting the transcriptome analyses of the communities of interacting species, it is possible to get insight into how microbial communities respond to different environmental situations. In the same way that genome- and metagenome-based analyses may identify individual genes' presence or absence, the endophytic phenomena can be better understood by examining how certain genes express themselves in various microenvironments. A thorough understanding of the mechanism and character of mutualism between the host and symbiotic bacteria is given by dual RNA-seq transcriptional profiling. With the aid of dual RNA-seq technology, transcriptional profiling of the roots of the wheat plant revealed upregulation of genes related to nutrient uptake and cell cycle (Camilios-Neto et al. 2014). Through RNA-seq, it is possible to find genes with differential expression that cannot be found by microarray analysis alone. The microarray is still a popular tool for transcriptional profiling due to its low cost and simplicity in data storage and analysis. Comparative transcriptome analysis of plants free of endophytes and those infected by them can help to better understand the mechanisms underlying plant growth promotion properties and resistance to endophyte-mediated illness. On the other hand, by contrasting the distinct endophyte expression profiles inside and outside the host plant, it is possible to pinpoint the relationship-maintaining elements of interaction (Johnson et al. 2004; Dinkins et al. 2010; Ambrose and Belanger 2012). The determination of symbionts' endophytic lifestyles can be accelerated by combining genome and transcriptomics techniques since genomic investigations give a solid foundation for transcriptional analyses.

Omics technologies have greatly helped in understanding the relationship between soil suppressiveness (rhizosphere) and *Lysobacter* abundance. Genomics has unraveled that a large number of genes encoding antagonistic properties are shared between biocontrol *Lysobacter* strains, while transcriptomics has revealed the molecular mechanisms through which *Lysobacter* spp. tend to interact with other microorganisms and the environment (Puopolo et al. 2017).

Understanding the mechanisms of biocontrol during the plant–pathogen–BCA interaction has been the subject of numerous investigations in the literature. The pathogen's impact on expression profiles, the host plant's response to the pathogen, and the modulation of these profiles by the presence or absence of a BCA were all examined in these studies. For instance, the disease was reduced by 91% thanks to the transcriptional reprogramming in canola caused by the treatment with

Pseudomonas chlororaphis PA23. This was accomplished by 16 times downregulating the post-infection gene expression in canola that was upregulated (8000 genes) upon sole treatment with canola stem rot pathogen *Sclerotinia sclerotiorum*. Additionally, it was found that *P. chlororaphis* PA23 increased the host plant's defense gene expression, causing ISR and the formation of Reactive Oxygen Species (ROS) (Duke et al. 2017).

Transcriptomics has been used to examine the relationship between *Trichoderma*, the host plant, and the pathogen. Through transcriptome analysis, Reithner et al. (2011) investigated the interactions of *T. atroviride* at various phases with *Rhizoctonia solani*, *Phytophthora capsici*, and *Botrytis cinerea*. Contrary to the projected activation of the transcribed genes for *T. atroviride* based on genomic data, only a tiny portion of genes were discovered to be active during the interaction between *R. solani* and the host plant (Reithner et al. 2011). Transcriptome analysis studies are critical for setting standards for subsequent research such as identifying the genes causing mycoparasitism, common cellular pathways, and potential host advantages of *Trichoderma* strains. This study on the biosynthesis of trichodermin by *Trichoderma brevicompactum* presented a chance (Shentu et al. 2014). By analyzing the gene expression in the roots' outer layer after *T. virens* colonization, compromised proteins, transporters, and enzymes were found (Morán-Diez et al. 2015). Out of all the genes expressed, 888 differentially expressed genes (DEGs) were discovered through transcriptome research (Sun et al. 2016). The relationship between the DEGs and various processes that aid in encoding transcription factors, enzymes, and secondary metabolites is the primary function of the DEGs (Sun et al. 2016).

This methodology has been extensively applied to investigate the fungal BCA for genes encoding cell wall-degrading enzymes (CWDEs) such as proteases, glucanases, and chitinases (Massart and Jijakli 2007). In respect to bacterial biological control agents, the genes that are in charge of producing secondary metabolites have received the most attention. This method is thought to give a fairly in-depth understanding of the studied expressed genes, although it is unable to conduct a thorough analysis of the transcriptome. To get over this restriction, "finding driven" methods were created (Liu and Yang 2005; Massart and Jijakli 2006; Viterbo et al. 2004) to find genes that had varied transcription patterns under various circumstances without knowing them beforehand.

Through extensive transcriptome analysis of expressed genes of genetically modified BCA that have been created to integrate desirable features, gene functions and properties of biological control agents may be thoroughly examined (Montero-Barrientos et al. 2011; Trushina et al. 2013). Studies based on the transcriptome, which only include active mRNA species, are supported by the differential expression of genes and the appearance of unknown components during transcription (Meteignier et al. 2017). The results from transcriptome, transcriptome analysis, and proteome analysis can be used to further understand the molecular mechanisms of BCAs (Sharma et al. 2017; Lysøe et al. 2017; Jiang et al. 2019; Zhao et al. 2020). The information from ectomycorrhiza short-read next-generation transcriptomic sequencing was used to identify genes that encode particular metabolic pathways (Larsen et al. 2011).

With regard to a single partner, several researches on the patterns of gene expression of microbial symbionts have been conducted (Liao et al. 2016; Thomas et al. 2019; Oliveira et al. 2020), but only a few have covered the technique of transcriptional profiling (Ithal et al. 2007; Brotman et al. 2012; Breuillin et al. 2010). *Pseudomonas syringae* inoculation of *Arabidopsis* shoots led to noticeably changed transcript levels in the leaves, while *Pseudomonas fluorescens* WCS417r root treatment caused ISR (Verhagen et al. 2004). It was claimed that the transcript levels of *Arabidopsis* varied depending on the abilities of *P. fluorescens* (Sharma et al. 2020).

The shift in microbial activities induced by drought stress was determined using metatranscriptomics analysis of the rhizosphere. Many of these include catabolism of amino acids, carbohydrates, and secondary metabolites and feed root-associated microbes and rhizosphere secreted by plants through root exudates (Xu et al. 2021). Small RNA sequences were observed in the soybean plants through metatranscriptomic analysis that are not part of the plant's genome. The presence of different microbes that are symbiotic, free living, and pathogenic in nature was noticed after the comparative analysis of these sequences (Molina et al. 2012).

The sequencing of bacterial communities extracted from different habitats was done using metatranscriptomics, which helped in the understanding of the transcriptome of many microbes. During the development, differentially expressed rhizospheric bacterial genes were identified in *Arabidopsis* using metatranscriptomics (Chaparro et al. 2014; Chapelle et al. 2016). This approach allows the identification of microbial traits from the plant microbiome without culturing them. However, the disadvantages of this technique are the contamination of the plant microbiome with plant proteins and DNA, and the complex nature of this microbiome and inherited traits (Bulgarelli et al. 2015).

Transcriptomic analysis has an advantage over genomic analysis in that it aids in detection by controlling the actively transcribed genes. Even so, there are certain drawbacks to metatranscriptomics, such as the seldom ability to assign specific transcripts to the bacteria that make up high-quality reference genomes. Nano string technology based on hybridization can be utilized as an alternative to transcriptomic technology based on sequencing. In a sample of plant microbiota with mixed transcripts, our method can more effectively identify the bacterial transcripts. Therefore, the development of methods to enrich and detect bacterial transcripts will contribute to a better understanding of how bacteria interact with their host plants (Sharma et al. 2020).

13.5.3 Metabolomics

Molecular variations, detection, and quantification of small molecules at the time of plant–microbe interaction can be done using metabolomics (Sharma et al. 2020).

Responses of various microorganisms to different stresses (temperature, organic compounds, heavy metals) have been studied using this approach (Lankadurai et al. 2013). These instruments are crucial for identifying metabolic issues in bacteria treated with antibiotics (Aliferis and Jabaji 2011). This is accomplished by examining the metabolites that BCAs make and the alterations that they cause in the host tissues. Through a metabolomics technique, various kinds of secondary metabolites released by *Trichoderma* were found. These include those that are harmful to plants, those that cause ISR, and those that are harmful to pathogenic bacteria (Mukherjee et al. 2012; Kaliňák et al. 2014; Mazzei et al. 2016; Vinci et al. 2018). Metabolomics investigation shows that the grassroots' chemical exudation mechanism is what brings about the rhizospheric community (Zhalnina et al. 2018). Numerous researches using metabolomics to examine the relationship between plants and their diseases have been reported (Scandiani et al. 2015; Zhu et al. 2018). Negrel et al. (2018) evaluated the lipid indicators of *Plasmopara viticola* infection in grapevines, while a metabolomics approach was utilized to study *Phytophthora sojae* infection of soybean hypocotyls (Zhu et al. 2018).

The agriculture sector needs metabolomic techniques and other genomic approaches because they can aid in a better understanding of BCAs' mechanisms of action because these BCAs can produce a wide range of metabolites with significant agro-biotechnological value. For instance, using metabolomics, it was possible to identify the new metabolite isoharzianic acid produced by *Trichoderma harzianum*, and this metabolite was able to stop the growth of *Rhizoctonia solani* and *Sclerotinia sclerotiorum* (Vinale et al. 2014). After using isoharzianic acid, improvements in tomato seed germination and the induction of ISR in plants were also noted.

Similarly, metabolites such as giocladic acid, hetelidic acid, trichodermanone C, and bisorbicillinol were also reported to be produced by *Trichoderma* through metabolomics analysis Kang et al. 2011).

13.5.4 Proteomics

Proteomics is the systematic assessment of all proteins expressed throughout a specific time period in a given cell, tissue, biological fluid, or organism. The origin of protein creation and the site of action can be determined, and it can be used to detect alterations that occur after the translation processes (Yates et al. 2009). The proteomics method can be used to more effectively study the interactions between BCA and host plants to combat phytopathogens as well as between BCA and pathogen. The discovery of novel, naturally antagonistic proteins that may be isolated and utilized to combat phytopathogens to reduce the need for chemical pesticides is another benefit of this method. Using this method, it was possible to identify three unique proteins that *T. atroviride* produces in conjunction with *N*-acetyl-b-*D*-glucosaminidase and endochitinase to form fungal cell wall-degrading enzymes (CWDEs) in response to interaction with *Rhizoctonia solani* (Grinyer et al. 2005). Additionally, proteomic analysis can be used to identify ISR induction in plants treated with *Trichoderma*. Through this method, different effector proteins of both a positive and negative nature produced during the contact of *T. virens* maize root were discovered. These proteins had effectors that block the defense pathways necessary for ISR induction (Lamdan et al. 2015). In a study, it was revealed that an increase in membrane proteins such as chaperones and important enzymes identified to be involved in the regulation of membrane component manufacturing pathways was linked to a metabolic problem in *Pseudomonas aeruginosa* and *Serratia marcescens* after treatment with tea polyphenol (Yi et al. 2010, 2014). Besides understanding the fundamental biological reactions, identifying the proteins that are expressed differently during the biological control process, and finding proteins of biotechnological significance, proteomics has also been used to explore the characteristics of BCAs. For BCA fungi, strains of the *Trichoderma* genus were initially analyzed, and their proteomic fingerprinting was the first one to be available. For BCA bacteria, *Bacillus subtilis* IS58 was the first to be studied by the proteomics technique (Bernhardt et al. 1997; Schmid et al. 1997; Grinyer et al. 2004a, b: Lorito et al. 2010).

The biological regulation of phytopathogens can be studied specifically through the use of proteomics. An antifungal substance secreted by *Pseudozyma flocculosa*, glycolipid (flocculosin), for instance, assisted in the colonization of powdery mildew colonies. *P. flocculosa* flocculosin production was stimulated by certain stress or limiting circumstances, according to proteome analysis (Hammami et al. 2010). *B. amyloliquefaciens* strain (Buensanteai et al. 2008), *B. subtilis* (Baysal et al. 2013), *P. chlororaphis* (Kwasiborski et al. 2014), and *B. subtilis* strains KB-1122 have all had their proteomes compared (Zhang et al. 2009). Proteome analysis was used to compare two different strains: the MBA strain and a mutant strain designed to eliminate the antagonistic features, and a natural strain with low antagonistic activity and a BCA strain (Baysal et al. 2013; Zhang et al. 2009; Buensanteai et al. 2008; Kwasiborski et al. 2014). These studies have uncovered a large number of previously unknown proteins believed to be engaged in biological control as well as differentially expressed proteins connected to potential biological control functions (Kwasiborski et al. 2014). When *B. subtilis* and members of the *Trichoderma* genus interacted biotrophically, differential proteomics investigations were conducted to analyze the BCA proteome (Cheng et al. 2012; Yang et al. 2009).

The amount of CWDE protein was shown to significantly rise during the interaction between BCA and the pathogen, although this proteomic trend was directly tied to the type of the pathogen's cell walls. In response to *Trichoderma viride* parasitization, the plant pathogen *Schizophyllum commune* activated its defensive mechanisms as a consequence of the proteome investigation of the co-culture interaction between *B. subtilis* and *Magnaporthe grisea* (Ujor et al. 2012). Proteomics analysis and research can also be used to address difficulties with integrated applications, such as syntrophic relationships and compatibility, in microbial consortia with biocontrol capabilities—for instance, interactions between *T. atroviride* and *P. fluorescens* and *Fusarium oxysporum* (Faraji et al. 2013). A proteomics technique can be used to identify Microbe-Associated Molecular Patterns (MAMPs) or plant regulators involved in the interactions between BCA and plants.

MAMPs were discovered in BCA proteomes, including endopolygalacturonase (Morán-Diez et al. 2009), cyclophilin (Grinyer et al. 2004a, b), flagellin proteins (Kierul et al. 2015), and swollenin (Brotman et al. 2008). BCA-induced alterations in the proteome of rice (Kandasamy et al. 2009) have also been investigated.

Individual or combined colonization of host plants by BCA and the pathogen can have a different impact on protein profiles. It was observed that combined colonization had a greater transcriptional and translational profile post inoculation than the individual application, which indicated a larger protein expression profile caused by pathogen than the BCA (Kang et al. 2019). All proteins secreted by BCA can be analyzed through secretomics, a subset of proteomics. The 14-3-3 proteins, a protein family specific to the eukaryotes, were observed to interact with defense proteins resulting in the attack from pathogen effectors (Lozano-Durán and Robatzek 2015). Induction of plant defense against *Guignardia citricarpa* and *Fusarium solani* was found to be related to different proteins released as a result of *T. harzianum* mycoparasitism (Lima de Brida et al. 2017).

13.5.4.1 Metaproteomics

Through metaproteomics, a whole set of proteins from the microbial population of an environmental sample can be significantly described. Total proteins can be extracted from a microenvironment using direct or indirect lysis techniques (Maron et al. 2007). Total proteins can be recovered directly from plant endosphere using direct lysis method under varied stress and natural settings; consequently, the impact on endophytes' capacity to create metabolites can be assessed using protein fingerprint (Bhuyan et al. 2015). In both the presence and absence of endophytes, it is also possible to isolate the actual particular proteins from the overall host plant proteins that are engaged in their interaction. To examine the proteins made by PGPR strains in response to root exudates, the metaproteome of forest trees' phyllosphere can be evaluated by metaproteomics (Lambais et al. 2017). (Kierul et al. 2015). Proteomics is constrained by challenging microorganisms, poor protein value, reduced sensitivity to host proteins, and low protein consciousness. Similar restrictions apply to the identification of plant-related bacterial communities using proteomics, including the need for a comprehensive reference library, lower expression levels in complex samples followed by detection limits, and relatively low bacterial protein (Sharma et al. 2020).

13.5.4.2 Metaproteogenomics

Metaproteogenomics is capable of identifying more proteins than proteomics by combining the proteome of the environmental samples with its genome. It links the metaproteome to the metagenome of a single sample. Microbial communities found within rice rhizosphere and phyllosphere were studied using metaproteogenomic (Knief et al. 2012). The results indicated the expression of nifH genes only in the rhizosphere despite its presence in both microenvironments. Endospheric studies through this approach can reveal more significant information regarding the endophyte functionality. Functional and genetic diversity of microbial floras can be linked together through the data provided by the metaproteogenome characterization. Metaproteogenomics now targets functional studies of proteins that are involved in the interaction of plants with non-culturable endophytes. Interactions between plants and bacteria can be determined successfully by studying the secretion systems of bacteria associated with plants (Downie 2010). Using proteogenomics, specific characteristics of bacteria found in the phyllosphere have been identified. Studies involving proteogenomic analysis can be conducted using tools like the bacterial proteogenomic pipeline. The method provides insights into potential endophytic lifestyle strategies. Metagenomics and metaproteomics can be combined to overcome the limitation of the lack of closely related reference genomes in metaproteomics, which prevents the identification of many proteins.

13.6 Concluding Remarks

Detailed understanding of host endophyte interactions is necessary to comprehend plant probiotics. The adoption of a multidisciplinary approach is necessary to better comprehend the unavoidable variables involved in the creation and maintenance of symbiotic partnerships. These investigations also aid in the interpretation of how endophytes assist their host plants thrive and tolerate stress. Endophyte-mediated processes require predictive and explanatory models, which call for the use of contemporary "omics"-generated complementary data connected to other system biology methodologies. The chance to map out the intricate web of interactions between hosts and host-associated microorganisms and endophytes will be significant. This will enable more effective and sustainable exploitation of the biotechnological potential of various natural microbial populations and microbial communities linked with plants (Ciancio et al. 2019).

Another technique based on genomic technologies uses genome-wide mutant libraries for screening (Gray et al. 2015). In order to find new potential targets, mutations that exhibit resistance to a specific compound can be chosen through sequencing techniques, followed by genetic comparison (Brazas and Hancock 2005; Bachmann et al. 2014; Köser et al. 2014). This eliminates the challenges associated with finding new targets through mutant libraries or protein interaction assays (Nijman 2015). Studies of plant microbiomes can be used to develop strategies for sustainable agriculture, including as biological control, the use of BCA for the creation of biofertilizers, and the generation of stress-resistant crops. The development of plant microbiome investigations is essential for the future of plant breeding and biotechnology. To optimize the advantages of the complete microbiome, the extension of the plant microbiome as an integrated biomarker should be taken into consideration. Improved classification of plant microbiomes can stop plant disease epidemics and the introduction of human diseases into plants. Complex cellular processes during plant–microbe interactions can be examined in numerous ways using omics technology. Genomic analysis can be used to compare various microorganisms, hosts, and their interactions. Comparative and integrated genomics may examine how new strains have evolved and how they interact with host plants, and it can be very helpful in developing plans for the agriculture sector's long-term sustainability (Sharma et al. 2020).

Studies on biological control have intensified due to the development of omics technology. The development of bioinformatics and technology techniques has greatly increased our understanding of the biocontrol capabilities of BCA and the methods they employ. Thanks to NGS technologies, detailed comprehensions of the interactions between BCA and plant-associated microbial communities are now possible (Massart et al. 2015). Genome sequencing can be the first step in the characterization of biological control agents and is likely to serve as a solid foundation for all further research, regardless of the methodology. For the BCA to incorporate greater biological control features, suitable experimental designs and biological explanations in omics are essential. The omic techniques have so far been used singly or in succession. By combining metatranscriptomic and metabarcoding data, it is possible to gain more insight into the composition and operation of the microbiome in various contexts. By anticipating the function of each taxon that is found, bioinformatics technologies have the potential to provide information on the ecology of these bacteria.

The ability to alter the essential functions of plant hosts is made possible by the most modern genetic engineering technique, Clustered Regularly Interspaced Palindromic Repeats (CRISPR) (Roman-Reyna et al. 2020). The microbiome-based (Deng et al. 2021; Horton et al. 2014; Wallace et al. 2018) and similar approaches are used to recruit particular microorganisms through genetic loci, and these loci can be deconstructed by using mapping populations and vast plant germplasm collections. Besides evaluating the effects of these alterations on the plant microbiome as a whole, it is also possible to use the CRISPR system to eliminate certain species from a varied microbial community or even specific alleles within them. Integrative analyses for holo-omics require computational resources such as appropriate data storage facilities, data processing, data analysis, and workflows with appropriate modeling and quality control (Muñoz-Benavent et al. 2020). The Transkingdom network (TransNet) analytic tool, which can integrate and query holo-omics data, was recently developed for this purpose. Data types suitable for TransNet analysis include methylation data, gene expression, miRNAs, host and bacterial gene expression, proteins, and metabolites (Rodrigues et al. 2018).

Four essential technologies are required for system biology studies: genomics, transcriptomics, proteomics, and metabolomics. The use of these technologies in conjunction with meta-omics research is growing rapidly. Although each type of data is significant on its own, together they are considerably more valuable. Genetic information, as molecular machinery, reveals the potential capabilities of a microenvironment, but the precise expression and function are still unknown. On the other hand, excluding the regulation of protein levels, genome expression is explored by transcriptomics under diverse environmental situations. Proteomic investigations identify the functional proteins needed by microorganisms for survival. However, these researches are only finished once the predicted protein data and the actual data have been compared (Hettich et al. 2013). To analyze transcriptomic and proteomic data effectively, however, genome-based research is required. When metagenomics data are combined with metatranscriptomic and metaproteomic data, it would be entirely clear what an endophyte is doing and how it might develop. Each of the strategies mentioned depends on the others in that each produces data that complete the others (Kaul et al. 2016).

References

Abdel-Salarn MS, Abd El-Halim MM, El-Harnshary OIM (2007) Improvement of Pseudomonas antagonism against Fusarium oxysporum through protoplast fusion: I-fusants induction. Res J Cell Mol Biol 1:37–41

Aguilar-Marcelino L, Mendoza-de-Gives P, Al-Ani LKT, López-Arellano ME, Gómez-Rodríguez O, Villar-Luna E, Reyes-Guerrero DE (2020) Using molecular techniques applied to beneficial microorganisms as biotechnological tools for controlling agricultural plant pathogens and pest. In: Sharma V, Salwan R, Al-Ani LKT (eds) Molecular aspects of plant beneficial microbes in agriculture, pp 333–349. https://doi.org/10.1016/b978-0-12-818469-1.00027-4

Akinsanya MA, Goh JK, Lim SP, Ting AY (2015) Metagenomics study of endophytic bacteria in Aloe vera using next-generation technology. Genom Data 6:159–163. https://doi.org/10.1016/j.gdata.2015.09.004

Aliferis KA, Jabaji S (2011) Metabolomics – a robust bioanalytical approach for the discovery of the modes-of-action of pesticides: a review. Pestic Biochem Physiol 100:105–117. https://doi.org/10.1016/j.pestbp.2011.03.004

Ambrose KV, Belanger FC (2012) SOLiD-SAGE of endophyteinfected red fescue reveals numerous effects on host transcriptome and an abundance of highly expressed fungal secreted proteins. PLoS One 7:e53214. https://doi.org/10.1371/journal.pone.0053214

Antipov D, Hartwick N, Shen M, Raiko M, Lapidus A, Pevzner PA (2016) plasmidSPAdes: assembling plasmids from whole genome sequencing data. Bioinformatics 32(22):3380–3387

Bachmann BO, Van Lanen SG, Baltz RH (2014) Microbial genome mining for accelerated natural products discovery: is a renaissance in the making? J Ind Microbiol Biotechnol 41(2):175–184

Baysal Ö, Lai D, Xu HH et al (2013) A proteomic approach provides new insights into the control of soil-borne plant pathogens by Bacillus species. PLoS One 8(1):e53182

Berg G, Zachow C, Müller H, Philipps J, Tilcher R (2013) Next-generation bio-products sowing the seeds of success for sustainable agriculture. Agronomy 3(4):648–656

Bernhardt J, Volker U, Volker A, Antelmann H, Schmid R, Mach H, Hecker M (1997) Specific and general stress proteins in *Bacillus subtilis*—a two-dimensional protein electrophoresis study. Microbiology 143:999–1017

Bhuyan SK, Bandyopadhyay P, Yadava PK (2015) Extraction of proteins for two-dimensional gel electrophoresis and proteomic analysis from an endophytic fungus. Protoc Exch. https://doi.org/10.1038/protex.2015.084

Bolger AM, Lohse M, Usadel B (2014) Trimmomatic: A flexible trimmer for Illumina sequence data. Bioinformatics 30(15):2114–2120

Brazas MD, Hancock REW (2005) Using microarray gene signatures to elucidate mechanisms of antibiotic action and resistance. Drug Discov Today 10:1245–1252

Breuillin F, Schramm J, Hajirezaei M et al (2010) Phosphate systemically inhibits development of arbuscular mycorrhiza in Petunia hybrida and represses genes involved in mycorrhizal functioning. Plant J 64(6):1002–1017

Brotman Y, Briff E, Viterbo A, Chet I (2008) Role of swollenin, an expansin-like protein from Trichoderma, in plant root colonization. Plant Physiol 147(2):779–789

Brotman Y, Lisec J, Méret M, Chet I, Willmitzer L, Viterbo A (2012) Transcript and metabolite analysis of the *Trichoderma*-induced systemic resistance response to Pseudomonas syringae in Arabidopsis thaliana. Microbiology 158(1):139–146

Buensanteai N, Yuen GY, Prathuangwong S (2008) The biocontrol bacterium *Bacillus amyloliquefaciens* KPS46 produces auxin, surfactin and extracellular proteins for enhanced growth of soybean plant. Thai J Agric Sci 41(3–4):101–116

Bulgarelli D, Garrido-Oter R, Münch PC et al (2015) Structure and function of the bacterial root microbiota in wild and domesticated barley. Cell Host Microbe 17(3):392–403

Camilios-Neto D, Bonato P, Wassem R, Tadra-Sfeir MZ, BrusamarelloSantos LC, Valdameri G et al (2014) Dual RNA-seq transcriptional analysis of wheat roots colonized by Azospirillum brasilense reveals up-regulation of nutrient acquisition and cell cycle genes. BMC Genomics 15: 378. https://doi.org/10.1186/1471-2164-15-378

Challis GL (2008) Mining microbial genomes for new natural products and biosynthetic pathways. Microbiology 154:1555–1569

Chaparro JM, Badri DV, Vivanco JM (2014) Rhizosphere microbiome assemblage is affected by plant development. ISME J 8(4):790–803. https://doi.org/10.1038/ismej.2013.196. PMID: 24196324

Chapelle E, Mendes R, Bakker PAH, Raaijmakers JM (2016) Fungal invasion of the rhizosphere microbiome. ISME J 10(1):265–268

Cheng CH, Yang CA, Peng KC (2012) Antagonism of *Trichoderma harzianum* ETS 323 on Botrytis cinerea mycelium in culture conditions. Phytopathology 102:1

Ciancio A, Pieterse CM, Mercado-Blanco J (2019) Harnessing useful rhizosphere microorganisms for pathogen and pest biocontrol. Front Microbiol 10:1935

Collinge DB (2018) Transgenic crops and beyond: how can biotechnology contribute to the sustainable control of plant diseases? Eur J Plant Pathol 152(4):977–986

Conrath U, Beckers GJ, Langenbach CJ, Jaskiewicz MR (2015) Priming for enhanced defense. Annu Rev Phytopathol 53(1):97–119

Daniel GH, Cristopher PF, Annelise EB (2008) Advantages and limitations of next-generation sequencing technologies: a comparison of electrophoresis and non- electrophoresis method. Electrophoresis 29:4618–4626. https://doi.org/10.1002/elps.200800456

Delmotte NL, Kniefa C, Chaffronb S, Innerabnera G, Roschitzkic B, Schlapbachc R et al (2009) Community proteogenomics reveals insights into the physiology of phyllosphere bacteria. Proc Natl Acad Sci U S A 106:16428–16433. https://doi.org/10.1073/pnas.0905240106

Deng S, Caddell DF, Xu G, Dahlen L, Washington L, Yang J, Coleman-Derr D (2021) Genome wide association study reveals plant loci controlling heritability of the rhizosphere microbiome. ISME J 15(11):3181–3194

Dinkins RD, Barnes A, Waters W (2010) Microarray analysis of endophyte-infected and endophyte-free tall fescue. J Plant Physiol 167:1197–1203. https://doi.org/10.1016/j.jplph.2010.04.002

Dinsdale EA, Edwards RA, Hall D, Angly F, Breitbart M, Brulc JM et al (2008) Functional metagenomic profiling of nine biomes. Nature 452:629–632. https://doi.org/10.1038/nature06810

Downie JA (2010) The roles of extracellular proteins, polysaccharides and signals in the interactions of rhizobia with legume roots. FEMS Microbiol Rev 34:150–170. https://doi.org/10.1111/j.1574-6976.2009.00205.x

Duke KA, Becker MG, Girard IJ, Millar JL, Dilantha Fernando WG, Belmonte MF, de Kievit TR (2017) The biocontrol agent Pseudomonas chlororaphis PA23 primes Brassica napus defenses through distinct gene networks. BMC Genomics 18(1):1–16

Espinas NA, Saze H, Saijo Y (2016) Epigenetic control of defense signaling and priming in plants. Front Plant Sci 7:1201

Eyiwumi Olorunleke F, Phuong Kieu N, Höfte M (2015) Recent advances in pseudomonas biocontrol pp. bacteria-plant interactions: advanced research and future trends. Caister Academic Press, Wymondham, UK, pp 167–198

FAO (2018) El futuro de la alimentación y la agricultura: Vías alternativas hacia el 2050. Versión Resumida. Rome. 64 pp. http://www.fao.org/3/CA1553ES/ca1553es.pdf (Consulta, Mayo 2020)

Faraji M, Ahmadzadeh M, Behboudi K et al (2013) Study of proteome pattern of Pseudomonas fluorescens strain UTPF68 in interaction with Trichoderma atroviride strain P1 and tomato. Arch Adv Biosci 4(1)

Frantzeskakis L, Di Pietro A, Rep M, Schirawski J, Wu CH, Panstruga R (2020) Rapid evolution in plant-microbe interactions - a molecular genomics perspective. New Phytol 225(3):1134–1142. https://doi.org/10.1111/nph.15966. PMID: 31134629.

Ghorbanpour M, Omidvari M, Abbaszadeh-Dahaji P, Omidvar R, Kariman K (2018) Mechanisms underlying the protective effects of beneficial fungi against plant diseases. Biol Control 117: 147–157

Gopal M, Gupta A, Thomas GV (2013) Bespoke microbiome therapy to manage plant diseases. Front Microbiol 4:355

Gray AN, Koo B, Shiver AL, Peters JM, Osadnik H, Gross CA (2015) High-throughput bacterial functional genomics in the sequencing era. Curr Opin Microbiol 27:86–95. https://doi.org/10.1016/j.mib.2015.07.012

Grinyer J, McKay M, Herbert B, Nevalainen H (2004a) Fungal proteomics: mapping the mitochondrial proteins of a *Trichoderma harzianum* strain applied for biological control. Curr Genet 45(3):170–175

Grinyer J, McKay M, Nevalainen H, Herbert BR (2004b) Fungal proteomics: initial mapping of biological control strain *Trichoderma harzianum*. Curr Genet 45(3):163–169

Grinyer J, Hunt S, McKay M, Herbert BR, Nevalainen H (2005) Proteomic response of the biological control fungus *Trichoderma atroviride* to growth on the cell walls of Rhizoctonia solani. Curr Genet 47:381388. https://doi.org/10.1007/s00294-005-0575-3

Gupta P, Kumar V, Usmani Z, Rani R, Chandra A, Gupta VK (2020) Implications of plant growth promoting *Klebsiella* sp. CPSB4 and Enterobacter sp. CPSB49 in luxuriant growth of tomato plants under chromium stress. Chemosphere 240:124944

Gurevich A, Saveliev V, Vyahhi N, Tesler G (2013) QUAST: quality assessment tool for genome assemblies. Bioinformatics 29(8):1072–1075. https://doi.org/10.1093/bioinformatics/btt086

Hammami W, Chain F, Michaud D, Bélanger RR (2010) Proteomic analysis of the metabolic adaptation of the biocontrol agent Pseudozyma flocculosa leading to glycolipid production. Proteome Sci 8(1):1–9

Hebert PDN, Cywinska A, Ball SL, deWaard JR (2003) Biological identifications through DNA barcodes. Proc R Soc Lond B Biol Sci 270:313–321

Heimpel GE, Mills NJ (2017) Biological control. Cambridge University Press

Herrera-Estrella A (2014) Genome-wide approaches toward understanding mycotrophic Trichoderma species. In: Biotechnology and biology of Trichoderma. Elsevier, pp 455–464

Hettich RL, Pan C, Chourey K, Giannone RJ (2013) Metaproteomics: harnessing the power of high performance mass spectrometry to identify the suite of proteins that control metabolic activities in microbial communities. Anal Chem 85:4203–4214. https://doi.org/10.1021/ac303053e

Horton MW, Bodenhausen N, Beilsmith K et al (2014) Genome-wide association study of Arabidopsis thaliana leaf microbial community. Nat Commun 5(1):1–7

Ithal N, Recknor J, Nettleton D, Hearne L, Maier T, Baum TJ, Mitchum MG (2007) Parallel genome-wide expression profiling of host and pathogen during soybean cyst nematode infection of soybean. Mol Plant-Microbe Interact 20(3):293–305

Jagadeesan B, Gerner-Smidt P, Allard MW, Leuillet S, Winkler A, Xiao Y, Chaffron S, Vossen JVD, Tang S, Katase M, McClure P, Kimura B, Chai LC, Chapman J, Grant K (2019) The use of next generation sequencing for improving food safety: translation into practice. Food Microbiol 79:96–115. https://doi.org/10.1016/j.fm.2018.11.005

Jiang CH, Yao XF, Mi DD, Li ZJ, Yang BY, Zheng Y, Ql YJ, Guo JH (2019) Comparative transcriptome analysis reveals the biocontrol mechanism of Bacillus velezensis F21 against Fusarium wilt on watermelon. Front Microbiol. https://doi.org/10.3389/fmicb.2019.00652

Johnson LJ, Johnson RD, Schardl CL, Panaccione DG (2004) Identification of differentially expressed genes in the mutualistic association of tall fescue with Neotyphodium coenophialum. Physiol Mol Plant Pathol 63:305–317. https://doi.org/10.1016/j.pmpp.2004.04.001

Jones WJ (2010) High throughput sequencing and metagenomics. Estuaries Coast 33:944–952. https://doi.org/10.1007/s12237-009-9182-8

Jumpponen A, Jones KL, Mattox JD, Yaege C (2010) Massively parallel 454-sequencing of fungal communities in Quercus spp. ectomycorrhizas indicates seasonal dynamics in urban and rural sites. Mol Ecol 19:41–53. https://doi.org/10.1111/j.1365-294X.2009.04483.x

Kaliňák M, Šimkovič M, Žemla P, Maťaťa M, Molnár T, Liptaj T et al (2014) Changes in metabolome and in enzyme activities during germination of Trichoderma atroviride conidia. FEMS Microbiol Lett 357:201207. https://doi.org/10.1111/1574-6968.12516

Kandasamy S, Loganathan K, Muthuraj R et al (2009) Understanding the molecular basis of plant growth promotional effect of *Pseudomonas fluorescens* on rice through protein profiling. Proteome Sci 7(1):1–8

Kang DJ, Kim JY, Choi JN, Liu KH, Lee CH (2011) Chemotaxonomy of Trichoderma spp. using mass spectrometry-based metabolite profiling. J Microbiol Biotechnol 21(1):5–13

Kang X, Wang L, Guo Y et al (2019) A comparative transcriptomic and proteomic analysis of hexaploid wheat's responses to colonization by Bacillus velezensis and Gaeumannomyces graminis, both separately and combined. Mol Plant-Microbe Interact 32(10):1336–1347

Kaul S, Sharma T, Dhar K, M. (2016) "Omics" tools for better understanding the plant–endophyte interactions. Front Plant Sci 7:955

Khokhar MK, Renu G, Sharma R (2012) Biological control of plant pathogens using biotechnological aspects: a review. Epidemiology 1:277–289

Kierul K, Voigt B, Albrecht D, Chen XH, Carvalhais LC, Borriss R (2015) Influence of root exudates on the extracellular proteome of the plant growth-promoting bacterium Bacillus amyloliquefaciens FZB42. Microbiology 161(1):131–147

Klosterman SJ, Rollins JR, Sudarshana MR, Vinatzer BA (2016) Disease management in the genomics era—summaries of focus issue papers. Phytopathology 106(10):1068–1070

Knief C (2014) Analysis of plant microbe interactions in the era of next generation sequencing technologies. Front Plant Sci 5:216

Knief C, Delmotte N, Chaffron S, Stark M, Innerebner G, Wassmann R et al (2012) Metaproteogenomic analysis of microbial communities in the phyllosphere and rhizosphere of rice. ISME J 6:1378–1390. https://doi.org/10.1038/ismej.2011.192

Köhl J, Kolnaar R, Ravensberg WJ (2019) Mode of action of microbial biological control agents against plant diseases: relevance beyond efficacy. Front Plant Sci 10:845. https://doi.org/10.3389/fpls.2019.00845

Koren S, Walenz BP, Berlin K, Miller JR, Bergman NH, Phillipy AM (2017) Canu: scalable and accurate longread assembly via adaptive k-mer weighting and repeat separation. Genome Res 27:722–736. https://doi.org/10.1101/gr.215087.116

Köser CU, Ellington MJ, Peacock SJ (2014) Whole-genome sequencing to control antimicrobial resistance. Trends Genet 30:401–407. https://doi.org/10.1016/j.tig.2014.07.003

Kress WJ, Erickson DL (2008) DNA barcodes: genes, genomics, and bioinformatics. Proc Natl Acad Sci 105:2761–2762

Kumar A (2022a) Microbial biocontrol: sustainable agriculture and phytopathogen management, vol 1. Springer Nature Chem, Cham, Switzerland, pp 1–369. https://doi.org/10.1007/978-3-030-87512-1

Kumar A (2022b) Microbial biocontrol: food security and post harvest management, vol 2. Springer Nature Chem, Cham, Switzerland, pp 1–334. https://doi.org/10.1007/978-3-030-87289-2

Kumar A, Zhimo Y, Biasi A, Salim S, Feygenberg O, Wisniewski M, Droby S (2021) Endophytic microbiome in the Carposphere and its importance in fruit physiology and pathology. In:

Spadaro D, Droby S, Gullino ML (eds) Postharvest pathology. Plant pathology in the 21st century, vol 11. Springer, Cham, pp 73–88. https://doi.org/10.1007/978-3-030-56530-5_5

Kumari M, Qureshi KA, Jaremko M, White JF, Singh SK, Sharma VK, Singh KK, Santoyo G, Puopolo G, Kumar A (2022) Deciphering the role of endophytic microbiome in postharvest diseases management of fruits: opportunity areas in commercial up-scale production. Front Plant Sci 13:1026575. https://doi.org/10.3389/fpls.2022.1026575

Kwasiborski A, Bajji M, Renaut J, Delaplace P, Jijakli MH (2014) Identification of metabolic pathways expressed by Pichia anomala Kh6 in the presence of the pathogen Botrytis cinerea on apple: new possible targets for biocontrol improvement. PLoS One 9(3):e91434

Lambais MR, Barrera SE, Santos EC, Crowley DE, Jumpponen A (2017) Phyllosphere metaproteomes of trees from the Brazilian Atlantic forest show high levels of functional redundancy. Microb Ecol 73(1):123–134

Lamdan NL, Shalaby S, Ziv T, Kenerley CM, Horwitz BA (2015) Secretome of Trichoderma interacting with maize roots: role in induced systemic resistance. Mol Cell Proteomics 14(4): 10541063. https://doi.org/10.1074/mcp.M114.046607

Lankadurai BP, Nagato EG, Simpson MJ (2013) Environmental metabolomics: an emerging approach to study organism responses to environmental stressors. Environ Rev 21:180–205. https://doi.org/10.1139/er-2013-0011

Larsen PE, Sreedasyam A, Trivedi G, Podila GK, Cseke LJ, Collart FR (2011) Using next generation transcriptome sequencing to predict an ectomycorrhizal metabolome. BMC Syst Biol 5(1):1–14

Liao HL, Chen Y, Vilgalys R (2016) Metatranscriptomic study of common and host-specific patterns of gene expression between pines and their symbiotic ectomycorrhizal fungi in the Genus Suillus. PLoS Genet 12(10):e1006348. https://doi.org/10.1371/journal.pgen.1006348. PMID: 27736883

Lima de Brida A, Rosa JMO, de Oliveira CMG, de Castroe Castro BM, Serrão JE, Zanuncio JC et al (2017) Entomopathogenic nematodes in agricultural areas in Brazil. Sci Rep 7:45254

Lima-Mendez G, Faust K, Henry N et al (2015) Determinants of community structure in the global plankton interactome. Science 348(6237):1262073

Liu PG, Yang Q (2005) Identification of genes with a biocontrol function in *Trichoderma harzianum* mycelium using the expressed sequence tag approach. Res Microbiol 156:416–423

Lopez-Fernandez S, Sonego P, Moretto M, Pancher M, Engelen K, Pertot I et al (2015) Whole-genome comparative analysis of virulence genes unveils similarities and differences between endophytes and other symbiotic bacteria. Front Microbiol 6:419. https://doi.org/10.3389/fmicb.2015.00419

Lorito M, Woo SL, Harman GE, Monte E (2010) Translational research on Trichoderma: from'omics to the field. Annu Rev Phytopathol 48:395–417

Lozano-Durán R, Robatzek S (2015) 14-3-3 proteins in plant-pathogen interactions. Mol Plant-Microbe Interact 28:511518. https://doi.org/10.1094/MPMI-10-14-0322-CR

Lysøe E, Dees MW, Brurberg MB (2017) A three-way transcriptomic interaction study of a biocontrol agent (Clonostachys rosea), a *fungal pathogen (Helminthosporium solani), and a potato host (Solanum* tuberosum). Mol Plant-Microbe Interact 30(8):646–655

Maron P, Ranjard L, Mougel C, Lemanceau P (2007) Metaproteomics: a new approach for studying functional microbial ecology. Microb Ecol 53:486–493. https://doi.org/10.1007/s00248-006-9196-8

Martin F (2013) The ecological genomics of fungi. Wiley, Hoboken, p 38

Massart S, Jijakli MH (2006) Identification of differentially expressed genes by cDNA-amplified fragment length polymorphism in the biocontrol agent Pichia anomala (strain Kh5). Phytopathology 96(1):80–86

Massart S, Jijakli HM (2007) Use of molecular techniques to elu- cidate the mechanisms of action of fungal biocontrol agents: a review. Microbiol Methods 69:229–241

Massart S, Martinez-Medina M, Jijakli MH (2015) Biological control in the microbiome era: challenges and opportunities. Biol Control 89:98–108

Mayer PD, Chan WY, Rubagotti E, Venter SN, Toth IK, Birch P et al (2014) Analysis of the Pantoea ananatis pan-genome reveals factors underlying its ability to colonize and interact with plant, insect and vertebrate hosts. BMC Genomics 15:404. https://doi.org/10.1186/1471-2164-15-404

Mazzei P, Vinale F, Woo SL, Pascale A, Lorito M, Piccolo A (2016) Metabolomics by proton high-resolution magic-angle-spinning nuclear magnetic resonance of tomato plants treated with two secondary metabolites isolated from Trichoderma. J Agric Food Chem 64(18):35383545. https://doi.org/10.1021/acs.jafc.6b00801

Mazzola M, Freilich S (2017) Prospects for biological soilborne disease control: application of indigenous versus synthetic microbiomes. Phytopathology 107(3):256–263

Medlini D, Donati C, Tettelin H, Masignani V, Rappuoli R (2005) The microbial pan-genome. Curr Opin Genet Dev 15:589–594. https://doi.org/10.1016/j.gde.2005.09.006

Meteignier LV, El Oirdi M, Cohen M et al (2017) Translatome analysis of an NB-LRR immune response identifies important contributors to plant immunity in Arabidopsis. J Exp Bot 68(9): 2333–2344

Mhlongo MI, Piater LA, Madala NE, Labuschagne N, Dubery IA (2018) The chemistry of plant–microbe interactions in the rhizosphere and the potential for metabolomics to reveal signaling related to defense priming and induced systemic resistance. Front Plant Sci 9:112

Mitra S (2019) Multiple data analyses and statistical approaches for analyzing data from metagenomic studies and clinical trials. In: Evolutionary genomics, pp 605–634. https://doi.org/10.1007/978-1-4939-9074-0_20

Molina LG, da Fonseca GC, de Morais GL, de Oliveira LFV, Carvalho JB, Kulcheski FR et al (2012) Metatranscriptomic analysis of small RNAs present in soybean deep sequencing libraries. Genet Mol Biol 35:292–303. https://doi.org/10.1590/S1415-47572012000200010

Monteiro RA, Balsanelli E, Tuleski T, Faoro H, Cruz ML, Wassem R et al (2012) Genomic comparison of the endophyte Herbaspirillum seropedicae SmR1 and the phytopathogen *Herbaspirillum rubrisubalbicans* M1 by suppressive subtractive hybridization and partial genome sequencing. FEMS Microbiol Ecol 80:441–451. https://doi.org/10.1111/j.1574-6941.2012.01309.x

Montero-Barrientos M, Hermosa R, Cardoza RE, Gutiérrez S, Monte E (2011) Functional analysis of the Trichoderma harzianum nox1 gene, encoding an NADPH oxidase, relates production of reactive oxygen species to specific biocontrol activity against *Pythium ultimum*. Appl Environ Microbiol 77(9):3009–3016

Morán-Diez E, Hermosa R, Ambrosino P, Cardoza RE, Gutiérrez S, Lorito M, Monte E (2009) The ThPG1 endopolygalacturonase is required for the *Trichoderma harzianum*–plant beneficial interaction. Mol Plant-Microbe Interact 22(8):1021–1031

Morán-Diez ME, Trushina N, Lamdan NL, Rosenfelder L, Mukherjee PK, Kenerley CM et al (2015) Host-specific transcriptomic pattern of *Trichoderma virens* during interaction with maize or tomato roots. BMC Genomics 16(8). https://doi.org/10.1186/s12864-014-1208-3

Mousa WK, Raizada MN (2016) Natural disease control in cereal grains. In: Reference module in food science, pp 1–7. https://doi.org/10.1016/B978-0-08-100596-5.00206-7

Mukherjee PK, Horwitz BA, Kenerley CM (2012) Secondary metabolism in Trichodermaa genomic perspective. Microbiology 158:3545

Muñoz-Benavent M, Hartkopf F, Van Den Bossche T et al (2020) gNOMO: a multi-omics pipeline for integrated host and microbiome analysis of non-model organisms. NAR Genomics Bioinformat 2(3):lqaa058

Negrel L, Halter D, Wiedemann-Merdinoglu S, Rustenholz C, Merdinoglu D, Hugueney P, Baltenweck R (2018) Identification of lipid markers of *Plasmopara viticola* infection in grapevine using a non-targeted metabolomic approach. Front Plant Sci 9:360

Nijman SM (2015) Functional genomics to uncover drug mechanism of action. Nat Chem Biol 11: 942–948. https://doi.org/10.1038/nchembio.1963

Oliveira VH, Ullah I, Dunwell JM, Tibbett M (2020) Mycorrhizal symbiosis induces divergent patterns of transport and partitioning of Cd and Zn in Populus trichocarpa. Environ Exp Bot 171: 103925

Parks DH, Imelfort M, Skennerton CT, Hugenholtz P, Tyson GW (2015) CheckM: assessing the quality of microbial genomes recovered from isolates, single cells, and metagenomes. Genome Res 25:1043–1055. https://doi.org/10.1101/gr.186072.114

Patel P, Kumar S, Kumar A (2022) Plant microbiota: a prospect to edge off postharvest loss. In: Biocontrol mechanisms of endophytic microorganisms. Academic Press/Elsevier, pp 261–284. https://doi.org/10.1016/B978-0-323-88478-5.00006-7

Pathak P, Rai VK, Can H, Singh SK, Kumar D, Bhardwaj N, Roychowdhury R, de Azevedo LCB, Kaushalendra VH, Kumar A (2022) Plant-endophyte interaction during biotic stress management. Plan Theory 11(17):2203. https://doi.org/10.3390/plants11172203

Paulsen IT, Press CM, Ravel J et al (2005) Complete genome sequence of the plant commensal Pseudomonas fluorescens Pf-5. Nat Biotechnol 23(7):873–878

Pott DM, Osorio S, Vallarino JG (2019) From central to specialized metabolism: an overview of some secondary compounds derived from the primary metabolism for their role in conferring nutritional and organoleptic characteristics to fruit. Front Plant Sci 835

Puopolo G, Tomada S, Pertot S (2017) The impact of the omics era on the knowledge and use of Lysobacter species to control phytopathogenic micro- organisms. J Appl Microbiol 124:15–27

Raaijmakers JM, Mazzola M (2012) Diversity and natural functions of antibiotics produced by beneficial and plant pathogenic bacteria. Annu Rev Phytopathol 50:403–424

Reithner B, Ibarra-Laclette E, Mach RL, Herrera-Estrella A (2011) Identification of mycoparasitism-related genes in Trichoderma atroviride. Appl Environ Microbiol 77(13): 4361–4370

Richter M, Rosselló-Móra R, Oliver Glöckner F, Peplies J (2016) JSpeciesWS: a web server for prokaryotic species circumscription based on pairwise genome comparison. Bioinformatics 32(6):929–931

Rodrigues RR, Shulzhenko N, Morgun A (2018) Transkingdom networks: a systems biology approach to identify causal members of host–microbiota interactions. In: Microbiome analysis. Humana Press, New York, NY, pp 227–242

Roemer T, Boone C (2013) Systems-level antimicrobial drug and drug synergy discovery. Nat Chem Biol 9:222–231. https://doi.org/10.1038/nchembio.1205

Roman-Reyna V, Pinili D, Borja FN, Quibod IL, Groen SC, Alexandrov N et al (2020) Characterization of the leaf microbiome from whole-genome sequencing data of the 3000 rice genomes project. Rice 13(1):1–8

Sanzani SM, Li Destri Nicosia MG, Faedda R, Cacciola SO, Schena L (2014) Use of quantitative PCR detection methods to study biocontrol agents and phytopathogenic fungi and oomycetes in environmental samples. J Phytopathol 162(1):1–13

Scandiani MM, Luque AG, Razori MV et al (2015) Metabolic profiles of soybean roots during early stages of *Fusarium tucumaniae* infection. J Exp Bot 66(1):391–402

Schardl CL, Young CA, Hesse U, Amyotte SG, Andreeva K (2013) Plant-symbiotic fungi as chemical engineers: multi-genome analysis of the Clavicipitaceae reveals dynamics of alkaloid loci. PLoS Genet 9:e1003323. https://doi.org/10.1371/journal.pgen.1003323

Schmid R, Bernhardt J, Antelmann H, Völker A, Mach H, Völker U, Hecker M (1997) Identification of vegetative proteins for a two-dimensional protein index of *Bacillus subtilis*. Microbiology 143(3):991–998

Sessitsch A, Hardoim P, Döring J, Weilharter A, Krause A, Woyke T, Mitter B, Hauberg-Lotte L, Friedrich F, Rahalkar M, Hurek T, Sarkar A, Bodrossy L, van Overbeek L, Brar D, van Elsas JD, Reinhold-Hurek B (2012) Functional characteristics of an endophyte community colonizing rice roots as revealed by metagenomic analysis. Mol Plant-Microbe Interact 25(1):28–36

Sharma V, Salwan R, Sharma PN, Gulati A (2017) Integrated translatome and proteome: approach for accurate portraying of widespread multifunctional aspects of Trichoderma. Front Microbiol 8:1602. https://doi.org/10.3389/fmicb.2017.01602. Published 2017 Aug 29

Sharma M, Sudheer S, Usmani Z, Rani R, Gupta P (2020) Deciphering the omics of plant-microbe interaction: perspectives and new insights. Curr Genomics 21(5):343–362

Sheibani-Tezerji R, Rattei T, Sessitsch A, Trognitz F, Mitter B (2015) Transcriptome profiling of the endophyte *Burkholderia phytofirmans* PsJN indicates sensing of the plant environment and drought stress. MBio 6:e00621–e00615. https://doi.org/10.1128/mBio.0062115

Shen Z, Ruan Y, Chao X (2015) Rhizosphere microbial community manipulated by 2 years of consecutive biofertilizer application associated with banana *Fusarium* wilt disease suppression. Biol Fert Soils 51:553–562. https://doi.org/10.1007/s00374-015-1002-7

Shentu X-P, Liu W-P, Zhan X-H, Xu Y-P, Xu J-F, Yu X-P et al (2014) Transcriptome sequencing and gene expression analysis of *Trichoderma brevicompactum* under different culture conditions. PLoS One 9(4):e94203. https://doi.org/10.1371/journal.pone.0094203

Singh HB (2014) Management of plant pathogens with microorganisms. PINSA 80:443–454

Singh VK, Singh AK, Kumar A (2017) Disease management of tomato through PGPB: current trends and future perspective. 3 Biotech 7(4):255

Singh M, Srivastava M, Kumar A, Singh AK, Pandey KD (2020) Endophytic bacteria in plant disease management. In: Kumar A, Singh KV (eds) Microbial endophytes: prospects for sustainable agriculture. Woodhead Publication, Cambridge USA, pp 61–89. https://doi.org/10.1016/B978-0-12-818734-0.00004-8

Spadaro D, Droby S (2016) Development of biocontrol products for postharvest diseases of fruit: the importance of elucidating the mechanisms of action of yeast antagonists. Trends Food Sci Technol 47:39–49

Sun Q, Jiang X, Pang L, Wang L, Li M (2016) Functions of thga1 gene in Trichoderma harzianum based on transcriptome analysis. Biomed Res Int 8329513:19. https://doi.org/10.1155/2016/8329513

Syed Ab Rahman SF, Singh E, Pieterse CMJ, Schenk PM (2018) Emerging microbial biocontrol strategies for plant pathogens. Plant Sci 267:102–111

Tang Y (2015) Non-genomic omic techniques. In: Tang Y, Sussman M, Liu D, Poxton I, Schwartzman J (eds) Molecular medical microbiology. Academic Press, London, pp 399–406

Terauchi R, KokiFujisaki K, Shimizu M, Oikawa K, Takeda T, Takagi H, Abe A, Okuyama Y, Yoshida K, Saitoh H (2020) Using genomics tools to understand plant resistance against pathogens: A case study of Magnaporthe-rice interactions; applied plant biotechnology for improving resistance to biotic. Stress:181–188

Tezerji RS, Naveed M, Jehl MA, Sessitsch A, Rattei T, Mitte B et al (2015) The genomes of closely related Pantoea ananatis maize seed endophytes having different effects on the host plant differ in secretion system genes and mobile genetic elements. Front Microbiol 6:440. https://doi.org/10.3389/fmicb.2015.00440

Thomas J, Kim HR, Rahmatallah Y et al (2019) RNA-seq reveals differentially expressed genes in rice (Oryza sativa) roots during interactions with plant-growth promoting bacteria, Azospirillum brasilense. PLoS One 14(5):e0217309

Toju H, Yamamoto S, Sato H, Tanabe AS, Gilbert GS, Kadowaki K (2013) Community composition of root-associated fungi in a Quercusdominated temperate forest: "codominance" of mycorrhizal and rootendophytic fungi. Ecol Evol 3:1281–1293. https://doi.org/10.1002/ece3.546

Trushina N, Levin M, Mukherjee PK, Horwitz BA (2013) PacC and pH–dependent transcriptome of the mycotrophic fungus *Trichoderma virens*. BMC Genomics 14(1):1–21

Ujor VC, Peiris DG, Monti M, Kang AS, Clements MO, Hedger JN (2012) Quantitative proteomic analysis of the response of the wood-rot fungus, Schizophyllum commune, to the biocontrol fungus, Trichoderma viride. Lett Appl Microbiol 54(4):336–343

Valenzuela-Ruiz V, Gálvez-Gamboa GT, Villa-Rodríguez ED, Parra-Cota FI, Santoyo G, de los Santos-Villalobos S. (2020) Lipopeptides produced by biological control agents of the genus Bacillus: a review of analytical tools used for their study. Revista Mexicana de Ciencias Agrícolas 11(2):419–432. https://doi.org/10.29312/remexca.v11i2.2191

Verhagen BW, Glazebrook J, Zhu T, Chang HS, Van Loon LC, Pieterse CM (2004) The transcriptome of rhizobacteria-induced systemic resistance in Arabidopsis. Mol Plant-Microbe Interact 17(8):895–908

Verma H, Kant C, Singh SK, White JF, Kumar A, Droby S (2022) Biocontrol potential of microbial consortia: approaches in food security and disease management. In: Kumar A (ed) Microbial biocontrol: sustainable agriculture and phytopathogen management. Springer, Cham. https://doi.org/10.1007/978-3-030-87512-1_

Vinale F, Manganiello G, Nigro M, Mazzei P, Piccolo A, Pascale A et al (2014) A novel fungal metabolite with beneficial properties for agricultural applications. Molecules 19(7):9760–9772

Vinci G, Cozzolino V, Mazzei P, Monda H, Spaccini R, Piccolo A (2018) An alternative to mineral phosphorus fertilizers: the combined effects of Trichoderma harzianum and compost on Zea mays, as revealed by 1H NMR and GCMS metabolomics. PLoS One 13(12):e0209664. https://doi.org/10.1371/journal.pone.0209664

Viterbo A, Harel M, Chet I (2004) Isolation of two aspartyl proteases from Trichoderma asperellum expressed during colonization of cucumber roots. FEMS Microbiol Lett 238(1):151–158

Wallace JG, Kremling KA, Kovar LL, Buckler ES (2018) Quantitative genetics of the maize leaf microbiome. Phytobiomes J 2:208–224

Woo SL, Ruocco M, Vinale F, Nigro M, Marra R, Lombardi N, Pascale A, Lanzuise S, Manganiello G, Lorito M (2014) Trichoderma-based products and their widespread use in agriculture. Open Mycol J 8:71–126

Xi W, Gao Y, Cheng Z, Chen C, Han M, Yang P, Xiong G, Ning K (2019) Using QC-blind for quality control and contamination screening of bacteria DNA sequencing data without reference genome. Front Microbiol 10:1–15. https://doi.org/10.3389/fmicb.2019.01560

Xu L, Pierroz G, Wipf HML, Gao C, Taylor JW, Lemaux PG, Coleman-Derr D (2021) Holo-omics for deciphering plant-microbiome interactions. Microbiome 9(1):1–11

Yadav MK, Patel P, Kumar A, Modi A, Babu PJ, Chhakchhuak L, Singh G, Mishra AK, Longchar B, Singh PK (2022) Tailoring disease resilience crops through CRISPR/Cas. In: Kumar A (ed) Microbial biocontrol: food security and post harvest management. Springer, Cham. https://doi.org/10.1007/978-3-030-87289-2_7

Yang HH, Yang SL, Peng KC, Lo CT, Liu SY (2009) Induced proteome of Trichoderma harzianum by Botrytis cinerea. Mycol Res 113(9):924–932

Yates JR, Ruse CI, Nakorchevsky A (2009) Proteomics by mass spectrometry: approaches, advances, and applications. Annu Rev Biomed Eng 11:49–79. https://doi.org/10.1146/annurev-bioeng-061008-124934

Yi SM, Zhu JL, Fu LL, Li JR (2010) Tea polyphenols inhibit Pseudomonas aeruginosa through damage to the cell membrane. Int J Food Microbiol 144:111–117. https://doi.org/10.1016/j.ijfoodmicro.2010.09.005

Yi S, Wang W, Bai F, Zhu J, Li J, Li X et al (2014) Antimicrobial effect and membrane-active mechanism of tea polyphenols against *Serratia marcescens*. World J Microb Biotechnol 30: 451–460. https://doi.org/10.1007/s11274-013-1464-4

Zhalnina K, Louie KB, Hao Z, Mansoori N, da Rocha UN, Shi S et al (2018) Dynamic root exudate chemistry and microbial substrate preferences drive patterns in rhizosphere microbial community assembly. Nat Microbiol 3(4):470–480

Zhang CX, Zhao X, Han F, Yang MF, Chen H, Chida T, Shen SH (2009) Comparative proteome analysis of two antagonist *Bacillus subtilis* strains. J Microbiol Biotechnol 19(4):351–357

Zhao H, Zhou T, Xie J, Cheng J, Chen T, Jiang D, Fu Y (2020) Mycoparasitism illuminated by genome and transcriptome sequencing of Coniothyrium minitans, an important biocontrol fungus of the plant pathogen sclerotinia sclerotiorum. Microbial Genomics 6(3)

Zhu L, Zhou Y, Li X, Zhao J, Guo N, Xing H (2018) Metabolomics analysis of soybean hypocotyls in response to *Phytophthora sojae* infection. Front Plant Sci 9:1530

Zuccaro A, Lahrmann U, Guldener U, Langen G, Pfiffi S (2011) Endophytic life strategies decoded by genome and transcriptome analyses of the mutualistic root symbiont *Piriformospora indica*. PLoS Pathog 7:e1002290. https://doi.org/10.1371/journal.ppat.1002290

Chapter 14
Plant Disease Management Using Anti-quorum Sensing Cues with an Emphasis on *Pseudomonas syringae* Pathovars

A. Manikandan, R. Anandham, P. Arul Jose, R. Krishnamoorthy, M. Senthilkumar, I. Johnson, R. Raghu, and N. O. Gopal

Abstract Plant microbiomes encompass different communities of microorganisms either as epiphytic regions such as phyllosphere, or rhizosphere regions and as

---

A. Manikandan
Department of Agricultural Microbiology, Tamil Nadu Agricultural University (TNAU), Coimbatore, Tamil Nadu, India

Present Address: Institute of Ecology and Earth Sciences, University of Tartu, Tartu, Estonia
e-mail: manikandanariyan@gmail.com

R. Anandham (✉) · N. O. Gopal
Department of Agricultural Microbiology, Tamil Nadu Agricultural University (TNAU), Coimbatore, Tamil Nadu, India
e-mail: anandham@tnau.ac.in; nogopal@tnau.ac.in

P. Arul Jose
Department of Agricultural Microbiology, TNAU, Madurai, Tamil Nadu, India

R. Krishnamoorthy
Department of Agricultural Microbiology, TNAU, Madurai, Tamil Nadu, India

Present Address: Department of Crop Management, Vanavarayar Institute of Agriculture, Pollachi, Tamil Nadu, India
e-mail: moorthy.micro@gmail.com

M. Senthilkumar
Agricultural College and Research Institute, TNAU, Eachangkottai, Tamil Nadu, India
e-mail: msenthilkumar@tnau.ac.in

I. Johnson
Department of Plant Pathology, TNAU, Coimbatore, Tamil Nadu, India
e-mail: johnson.i@tnau.ac.in

R. Raghu
Department of Agricultural Microbiology, Tamil Nadu Agricultural University (TNAU), Coimbatore, Tamil Nadu, India

Department of Plant Biotechnology, TNAU, Coimbatore, India
e-mail: raghu.r@tnau.ac.in

K. K. Bastas et al. (eds.), *Microbial Biocontrol: Molecular Perspective in Plant Disease Management*, Microorganisms for Sustainability 49,
https://doi.org/10.1007/978-981-99-3947-3_14

endophytes in different plant tissues, which showed saprophytic, symbiotic, or pathogenic nature. Quorum sensing (QS) is a gene regulatory mechanism that controls crucial bacterial biological processes, including a variety of virulence characteristics. The concentration of chemical signal molecules known as autoinducers, which are produced and released by quorum-sensing bacteria, rises with cell density. A change in gene expression results from the observation of a minimal threshold stimulatory concentration of an autoinducer. Bacteria of both the Gram-positive and Gram-negative types use quorum-sensing communication circuits to control a wide range of physiological processes. Symbiosis, pathogenicity, competence, conjugation, generation of antibiotics, motility, sporulation, and biofilm development are some of these processes. This chapter covered numerous QS molecules' methods for detection as well as diverse disease control strategies.

Keywords Quorum sensing (QS) · Plant microbiomes · *Pseudomonas syringae* pathovars · Gene regulations

14.1 Introduction

Plant microbiomes encompass different communities of microorganisms either as epiphytic regions such as phyllosphere, or rhizosphere regions and as endophytes in different plant tissues, which showed saprophytic, symbiotic, or pathogenic nature. Among the pathogenic bacteria, many diseases are predominantly caused by Gram-negative bacteria. Quorum sensing (QS) is a gene regulatory mechanism that controls crucial bacterial biological processes, including a variety of virulence characteristics. QS according to cell density, signal molecules are synthesized and kept at a constant concentration. Through transcription activators, signal molecules can boost their synthesis at low doses. When the concentration of these signal molecules reaches a specific level, gene expression takes place (Pereira et al. 2013). There are three groups of signal molecules and receptors: (1) LuxI proteins in Gram-negative bacteria produce *N*-acyl homoserine lactones (AHLs), and monitoring their length and oxidation state can reveal information about their population density; (2) Oligopeptides or autoinducing peptides, containing 5–34 amino acid residues, generated by gram-negative bacteria for intercellular communication. Numerous of these peptides are exported using specialized procedures, where they then undergo a variety of post-translational changes before being recognized by nearby cells using membrane-based receptors in two-component regulatory systems. (3) The communication molecule Autoinducer-2 (AI-2) is generated in Gram-positive and Gram-negative bacteria. LuxS family of proteins generates a furanosyl borate diester, which has been chemically characterized.

When this signaling pathway or bacterial QS activity is interrupted, microbial pathogenicity is decreased. There are numerous methods for disrupting QS systems, and they will be helpful in the treatment of bacterial diseases that depend on QS. This prompted research into using QS inhibitors to prevent QS which is termed as quorum quenching (QQ) mechanism. Several methods can be used for this, including creating antibodies that are specific to QS signal molecules, degrading QS signal molecules enzymatically, or using medications that block QS signal molecules. These methods prevent the spread of antibiotic resistance by reducing the physical contact of the cells and are useful to study pathogenic microorganisms without a reduction in growth rate. The identification of *Pseudomonas syringae* pathovars' QS signal molecules and various QS inhibition techniques are covered in detail in this chapter.

14.1.1 Quorum Sensing

QS refers to a well-known interaction mechanism in bacteria through cell-to-cell contact and expresses the idea that some characteristics are only shown by grouped cells (Miller and Bassler 2001). This enables them to cooperate and supports the idea that for the survival of a specific bacterium in nature or to colonize a specific host, it benefits from cooperative collective behavior.

AHLs are the diffusible signal molecules generated by certain Gram-negative bacteria that are employed for pathogen control based on the cell numbers (Fuqua and Winans 1996). The LuxI-protein family synthase typically produces signal molecule AHLs, and this depends upon the population density, when a sufficient concentration is achieved (thus the term "quorum") then they bind to the specific receptors. This activates the expression of the target gene (Fuqua and Greenberg 2002). QS-based AHLs production has been found to be closely connected to the regulation of nearly all virulence factors in *Pantoea stewartii* ssp. *stewartii* such as the production of exopolysaccharides (EPS) (Beck von Bodman and Farrand 1995).

Generally, AHLs are released by numerous species of phytopathogenic pseudomonads such as *P. syringae and P. corrugata* (Cha et al. 1998). The homologs of LuxI and LuxR have been discovered in several of these bacteria (Dumenyo et al. 1998; Elasri et al. 2001). In *P. syringae* pv. *syringae* strain B3A, inactivating the ahlI, AHL synthase gene reduced epiphytic vitality, eliminated AHL synthesis, and changed colony morphology. Hernandez and Lindow (2021) provided a more comprehensive analysis of QS's importance in the different microbial strains. In the previously published report it has been reported that in some strains homologous genes ahlI and ahlR are transcribed convergently, then further AHL signal autoinducers ahlI synthesis (Elasri et al. 2001). 3-oxo-C6HL is the primary product of the AhlI synthase. At least two levels control the ahlI. However, authors reported two-component GacA/GacS signal transduction system, generally reported in *Pseudomonas*, primarily controls ahlI expression. Second, the AefR transcription factor, of TetR family of transcription factors, controls it as well. Because ahlI expression is significantly decreased but not eliminated in the aefR mutant strain, it is possible to conclude that, at low population density, AhlI is positively regulated by aefR. The route is effectively blocked by aefR expression, which is suppressed as the high population is denser in an AHL-dependent manner. By use of this mechanism, *P. syringae* pv. *syringae* B728a regulates QS behavior.

Evidence is also reported for the role of AefR in epiphytic fitness by up-regulation (initial colonization) and down-regulation (later stages of colonization) of some fitness genes in strain B728a devoid of swarming motility at lower cell density. But mutant AefR and ahlR strains of B728a, characterized by the production of less AHL, exhibit higher motility under similar conditions. Hence this strain B728 is motile until a favorable microhabitat is identified, but becomes non-motile when sedentary aggregates are formed. Exo Poly Saccharide (EPS) production is lowered in aefR and ahlR mutants rendering them hypersensitive to oxidative stress and decreasing the survival of these mutants on the leaf surface (Von Bodman et al. 2003).

Moreover, mutants lacking AHL in B728a show reduced virulence at later stages of infection in beans. It could be inferred that the AHL-mediated regulation plays a pivotal role in pathogen invasion by adopting different strategies. QS is initiated only after the colonization of bacteria which is achieved in a couple of days. This was proved by expression analysis of *gfp* reporter gene fused with *ahlI* in B728a cells on bean leaves. It was shown that until 2 days after inoculation, the signal synthase gene was expressed in a small number of bacteria. Thereafter, expression increased with time and exhibited a strong correlation (Hernandez and Lindow 2021).

14.1.2 Pseudomonas syringae *Pathovars*

A pathogenic bacterium *Pseudomonas syringae*, which affects more than 200 plant species, is economically significant because it causes spots, canker, and blights (Agrios 2005). The characteristic symptoms of *P. syringae* pv. *coriandricola* infections are necrosis, leaf spots, and water-soaked lesions on leaves, with a reduced number of fruits. During cold and rainy conditions, these initial infection foci may spread quickly (Toben and Rudoph 1996). Cucumber pathogen *P. syringae* pv. *lachrymans* which causes angular leaf spots of the leaves and fruit rot and it's also considered one of the 50 pathovars of *P. syringae* species (Young et al. 1996). However, different symptoms like water-soaked lesions, distinguished angular leaf spots on leaves, and water-soaked lesions on fruits were reported in the malformed fruits in cucumber caused by the pathogen *P. syringae*. The details of pathogens and their causal diseases are presented in Table 14.1.

14.1.3 QS in Pseudomonas syringae *Pathovars*

In general most of the Gram-positive bacteria and *P. syringae* use AHL as a signal molecule during quorum sensing. However, the signal molecules are controlled by the AhlI AHL synthase gene and AhlR AHL regulator gene. The cellular precursors catalyze AhlI's synthesis of 3-oxo-C6-HSL. AhlR triggers AhlI transcription through a positive feedback mechanism that results in the production of a stable

Table 14.1 List of *Pseudomonas syringae* pathovar strains

Species	Host	Disease	References
P. syringae pv. *tomato* DC30000	Tomato	Brown-black leaf spots	Preston (2000)
P. syringae pv. *maculicola* ES4326	Crucifers	Bacterial leaf spot	Hernandez and Lindow (2021)
P. syringae pv. *aesculi*	Horse chestnut	Bleeding canker	Green et al. (2009)
P. syringae pv. *coriandricola* GSPB 1965	Coriander	Bacterial umbel blight and seed decay	Toben and Rudoph (1996)
P. syringae pv. *actinidiae*	Kiwifruit	Bacterial canker	Vanneste et al. (2010)
P. syringae pv. *lachrymans*	Cucumber	Angular leaf spot	Olczak-Woltman et al. (2007)
P. syringae pv. *syringae*	Olives	Bacterial canker	Ashorpour et al. (2008)
P. syringae pv. *alisalensis*	Broccoli	Bacterial blight	Cintas et al. (2006)
P. syringae pv. *tomato*	Tomato	Bacterial speck	Bashan and de Bashan (2002)
P. syringae pv. *avii*	Wild cherry	Bacterial canker	Ménard et al. (2003)
P. syringae pv. *coryli*	Hazelnut	Twig dieback	Scortichini et al. (2005)
P. syringae pv. *morsprunorum*	Stone fruit	Bacterial canker	Latorre and Jones (1979)
P. syringae pv. *papulans*	Apple	Blister spot	Bedford et al. (1988)
P. syringae pv. *persicae*	Peach	Bacterial canker	Barzic and Guittet (1996)
P. syringae pv. *syringae*	Mango	Apical necrosis	Cazorla et al. (1998)
P. syringae pv. *phaseolicola*	Common bean	Halo blight	Fernández-Sanz et al. (2016)
P. syringae pv. *pisi*	Field pea	Bacterial blight	Hollaway and Bretag (1995)
P. syringae pv. *garcae* CFPB1634	Coffee	Bacterial blight	Belan et al. (2016)
P. syringae pv. *aceris*	Maple	Spotted bacterial disease	Takikawa et al. (1991)
P. syringae pv. *atrofaciens*	Wheat and barley	Basal glume rot	Toben et al. (1989)
P. syringae pv. *aptata*	Garden beet	Leaf spot	Takikawa et al. (1991), Arabi et al. (2006)
P. syringae pv. *glycinea*	Soybean	Bacterial blight	Hattermann and Ries (1989)
P. syringae pv. *japonica*	Barley and wheat	Bacterial black node	Fukuda et al. (1990)
P. syringae pv. *lapsa*	Wheat	Leaf streak and black chaff	Amanifar (2020)
P. syringae pv. *panici*	Rice	Bacterial brown stripe	Yaoita et al. (1984)
P. syringae pv. *persicae*	Peach	Decline syndrome	Young (1987)
P. syringae pv. *tabaci*	Coffee	Bacterial leaf spot	Destéfano et al. (2010)

(continued)

Table 14.1 (continued)

Species	Host	Disease	References
P. syringae pv. *berberidis*	Berberis	Leaf spot and leaf fall	Roberts and Preece (1984)
P. syringae pv. *glycinea* 2159	Soybean	Bacterial blight	Osman et al. (1986)
P. syringae pv. *helianthi*	Sunflower	Bacterial leaf spot	Arsenijević et al. (1994)
P. syringae pv. *passiflorae*	Passion fruit	Grease spot	Reid (1938)
P. syringae pv. *mori*	Mulberry	Bacterial blight	Krawczyk and Łochyńska (2020)
P. syringae pv. *oryzae*	Rice	Bacterial halo blight	Krawczyk and Łochyńska (2020)

molecule containing the AHL signal. As cell concentrations increase, more AHL is produced. The AhlI-AhlR quorum-sensing pathway is influenced by other regulatory proteins (Dumenyo et al. 1998; Fuqua and Greenberg 2002).

14.1.4 *Detection of QS Signal Molecule Production by* Pseudomonas syringae *Pathovars Under In Vitro Condition*

14.1.4.1 AHL Reporter Plate Bioassays Using Indicator Strains

The synthesis of AHL signal molecules is the initial step in recognizing if a bacterial strain has a LuxI/R QS system. Each AHL biosensor is based on a distinct LuxR family protein, indicating that it is specific to the corresponding AHL and, in some circumstances, closely related AHLs. Because many biosensors only detect a small spectrum of AHLs, it is critical to employ several biosensors when screening a bacterium for AHL synthesis. Every biosensor should respond to AHLs differently depending on its structural features. For QS signal detection, a variety of indicator strains are employed. As a reporter strain, *Agrobacterium tumefaciens* A136, without the Ti plasmid (pCF218, pCF372), is utilized to detect a variety of acyl HSLs (3-oxo-C4 to 3-oxo-C12-HSLs, C5-C10-HSLs). By expressing galactosidase activity, *A. tumefaciens* strain NT1 (pSVB33, pJM749), which lacks the Ti plasmid and is incapable of producing AHL with the atraG::lacZ fusion, is utilized to identify C6–C12-alkanoyl, 3-oxo-alkanoyl, and 3-hydroxy-alkanoyl side chain HSLs. In order to identify AHL-regulated violacein production, *Chromobacterium violaceum*, which lacks the ability to generate *N*-hexanoyl-L-homoserine lactone produced from the wild-type ATCC 31532, is employed. The use of *C. violaceum* CV026 and *A. tumefaciens* NT1 strains were used to identify the production of HSLs by various *Burkholderia* strains (Poonguzhali et al. 2007). Also, Shaw et al. (1997) used *A. tumefaciens* strain NT1 (pDCI41E33) for QS detection in *P. aeruginosa* PAO1 and

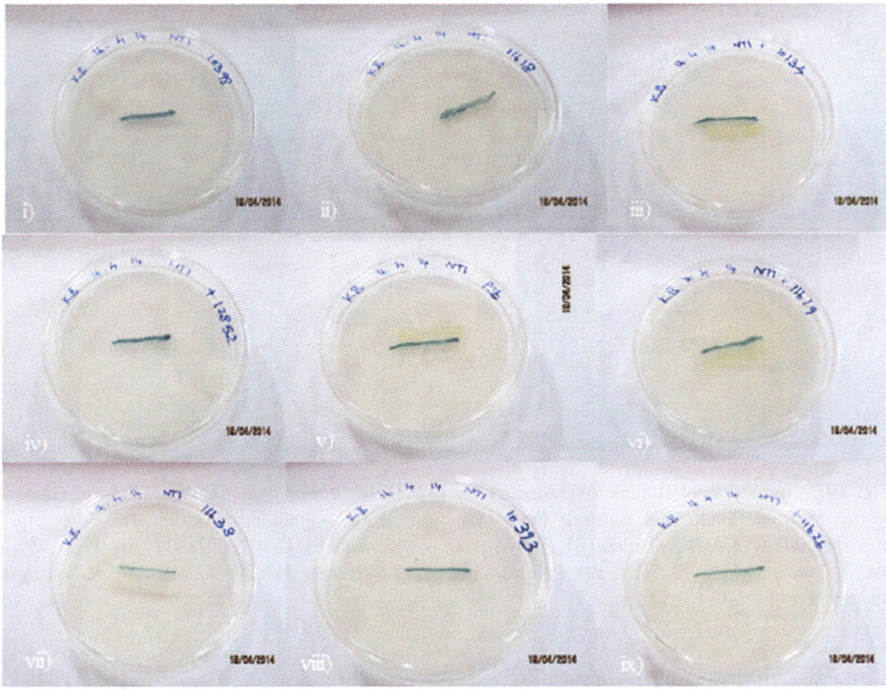

Fig. 14.1 Induction of β-galactosidase activity in *Agrobacterium tumefaciens* NT1 by *Pseudomonas syringae* pathovars through the production of QS signal compounds. (i) *P. syringae* pv. *garcae* KACC 10398, (ii) *P. syringae* pv. *helianthi* KACC 11618, (iii) *P. syringae* pv. *syringae* KACC 10134, (iv) *P. syringae* pv. *persicae* KACC 12852, (v) *P. syringae* pv. *tomato* MTCC1604, (vi) *P. syringae* pv. *panici* KACC 11619, (vii) *P. syringae* pv. *japonica* KACC 11638, (viii) *P. syringae* pv. *glycinea* KACC 10393, (ix) *P. syringae* pv. *atrofaciens* KACC 11626

P. syringae pv. *tabaci* 2024. Similarly, Jose et al. (2019) used *A. tumefaciens* strains A136 and NT1 for QS detection in *P. syringae* pv. *passiflorae*.

In dual culture assay with *A. tumefaciens* NT1, the following *P. syringae* pathovars such as *P. syringae* pv. *atrofaciens* KACC 11626, *P. syringae* pv. *berberidis* KACC 12841, *P. syringae* pv. *garcae* KACC 10398, *P. syringae* pv. *glycinea* KACC 10393, *P. syringae* pv. *helianthi* KACC 11618, *P. syringae* pv. *japonica* KACC 11638, *P. syringae* pv. *panici* KACC 11619, *P. syringae* pv. *passiflorae* KACC 12846, *P. syringae* pv. *persicae* KACC 12852, *P. syringae* pv. *syringae* KACC 10134, *P. syringae* pv. *tomato* MTCC1604 induced blue coloration (β-galactosidase activity) indicating that these *Pseudomonas* pathovars might have secreted C6-to-C12-alkanoyl, 3-oxo-alkanoyl, and 3-hydroxy-alkanoyl side chain HSLs (Fig. 14.1).

Similarly, except *P. syringae* pv. *berberidis* KACC 12841, *P. syringae* pv. *garcae* KACC 10398, *P. syringae* pv. *syringae* KACC 10134, *P. syringae* pv. *tomato* MTCC1604 all the *Pseudomonas syringae* pathovars listed above induced the blue coloration (β-galactosoidase activity) in *A. tumefaciens* A136

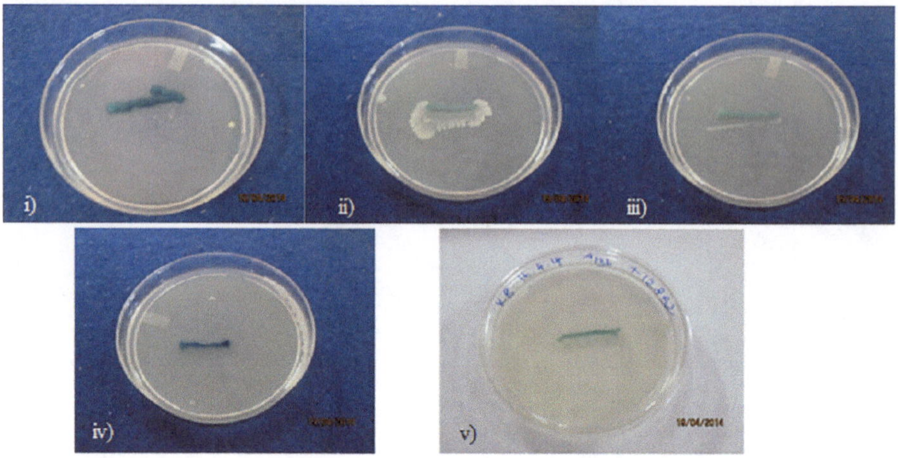

Fig. 14.2 Induction of β-galactosidase activity in *Agrobacterium tumefaciens* A136 by *Pseudomonas syringae* pathovars through the production of QS signal compounds. (i) *P. syringae* pv. *atrofaciens* KACC 11626, (ii) *P. syringae* pv. *berberidis* KACC 12841, (iii) *P. syringae* pv. *helianthi* KACC 11618, (iv) *P. syringae* pv. *passiflorae* KACC 12846, (v) *P. syringae* pv. *persicae* KACC 12852

indicating that these *Pseudomonas* pathovars might not have secreted acyl HSLs (3-oxo-C4 to 3-oxo-C12-HSLs, C5-C10-HSLs). Interestingly, none of the *P. syringae* pathovars did induce the violacein in *C. violaceum* CV026 indicating that it did not produce C4- and C6-HSLs (Fig. 14.2).

14.1.4.2 Extraction and Identification of QS Signal Molecules Using Reverse-Phase TLC

TLC separation in conjunction with bio-detection can be used to directly identify the acyl-HSL signal molecules accumulated in bacterial culture supernatants. On C18 reversed-phase plates, AHLs from spent culture supernatants of late exponential-stage cells are partially characterized by TLC (Shaw et al. 1997). This organic solvent significantly boosts the sensitivity of biosensors; AHLs frequently partition into the organic phase, and drying eliminates the solvent. The AHL biosensor strain's soft-agar solution is applied to TLC plates after the sample extracts and a number of standards are placed there (Schaefer et al. 2000). Using *A. tumefaciens* strain NT1(pDCI41E33) as a reporter stain in TLC technique, AHL was found in *P. aeruginosa* PAO1, *P. syringae* pv. *tabaci* 2024, and *Vibrio fischeri* MJ1 (Shaw et al. 1997). *Burkholderia* strains' production of HSL was verified in C18 reverse-phase TLC plates using *C. violaceum* CV026 as an indicator strain (Poonguzhali et al. 2007). The TLC and HPLC-MS/MS analyses were used to find *N*-butyryl-L-homoserine lactone (BHL) and *N*-hexanoyl-L-homoserine lactone (HHL) in *P. aeruginosa* CGMCC AHL compounds. AHL compounds were identified in the

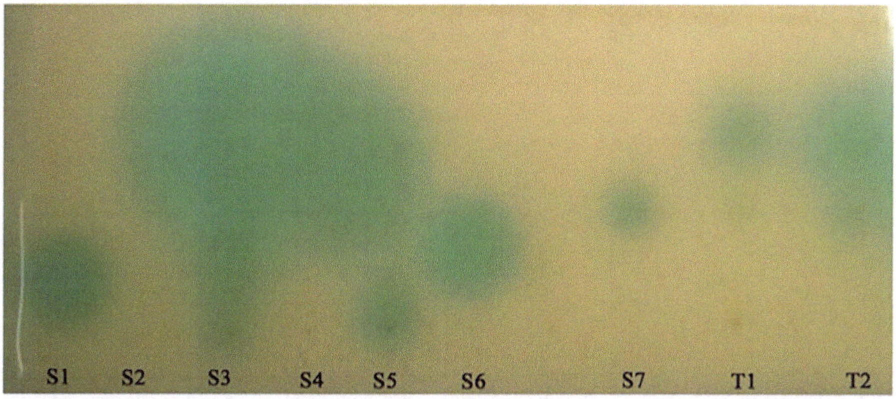

Fig. 14.3 Identification of QS signal molecules using reverse-phase TLC in comparison with standard AHL compounds. S1: *N*-Octonoyl-DL-Homoserine lactone; S2: L-Homoserine lactone hydrochloride; S3: *N*-(beta ketocaproli)-DL-homoserine lactone; S4: *N*-Tetradeconoyl-DL-homoserine lactone; S5: *N*-(3 oxododeconoyl)-L-homoserine lactone; S6: *N*-Heptanoyl-DL-homoserine lactone; S7: *N*-Hexanoyl-DL-homoserine lactone; T1: AHL compound produced by *P. syringae* pv. *passiflorae* KACC 12846; T2: AHL compound produced by *P. syringae* pv. *helianthi* KACC 11618

following TLC-based bioassay by contrasting them with reference AHL compounds. It was revealed that strains *P. syringae* pv. *passiflorae* KACC 12846 and *P. syringae* pv. *helianthi* KACC 11618 produced *N*-tetradeconoyl-DL-homoserine lactone (C14) and *N*-hexanoyl-DL-hemoserine lactone (C6) (Fig. 14.3).

14.1.4.3 Genes Responsible for AHL Production

The *luxR* and *aefR* genes from the consensus sequence of the respective genes in the whole genome data of different *P. syringae* pathovars encode the acyl-HSL synthase regulator and epiphytic fitness factor, respectively. The *luxR* and *aefR* genes were compared with GenBank databases using the BLASTx on the NCBI website (http://www.ncbi.nlm.nih.gov/). The LuxR homologs and AefR with already studied *P. syringae* pathovars were chosen for phylogenetic analysis. MEGA version 5.2 was used to create neighbor-joining phylogenetic trees for both proteins (Tamura et al. 2011), with 1000 bootstrap replicates.

The LuxR homologue from *P. syringae* pv. *passiflorae* forms a distinct clade from the cluster made up of *P. syringae* pv. *coriandricola* and *P. syringae* pv. *alisalensis*, according to a phylogenetic tree constructed from the deduced amino acid sequences of the LuxR homologue (Fig. 14.4). The *P. syringae* pv. *passiflorae* shared a high degree of homology and falls with *P. syringae* pv. *pisi* EGH43749 in the neighbor-joining phylogenetic tree (Fig. 14.5) made from determined amino acid sequences of AefR.

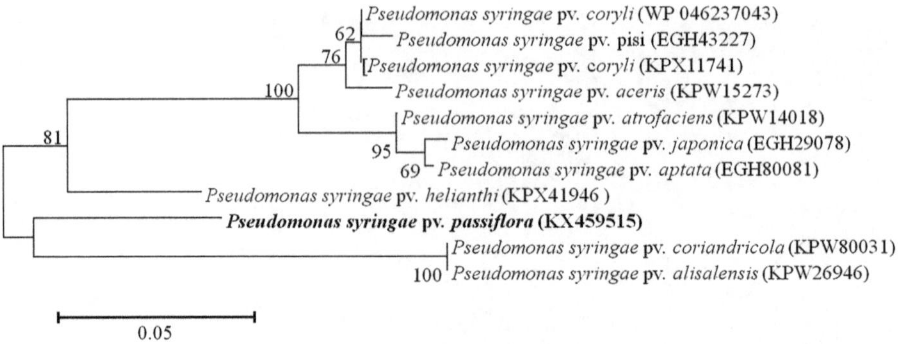

Fig. 14.4 Using the LuxR homologue of *P. syringae* pv. *passiflorae* LMG 2237s deduced amino acid sequences and those of other known *P. syringae* pv., a phylogenetic tree was constructed using neighbor-joining methods. Over 50 Bootstrap values were displayed. The scale bar shows 0.05 substitutions for each position of an amino acid

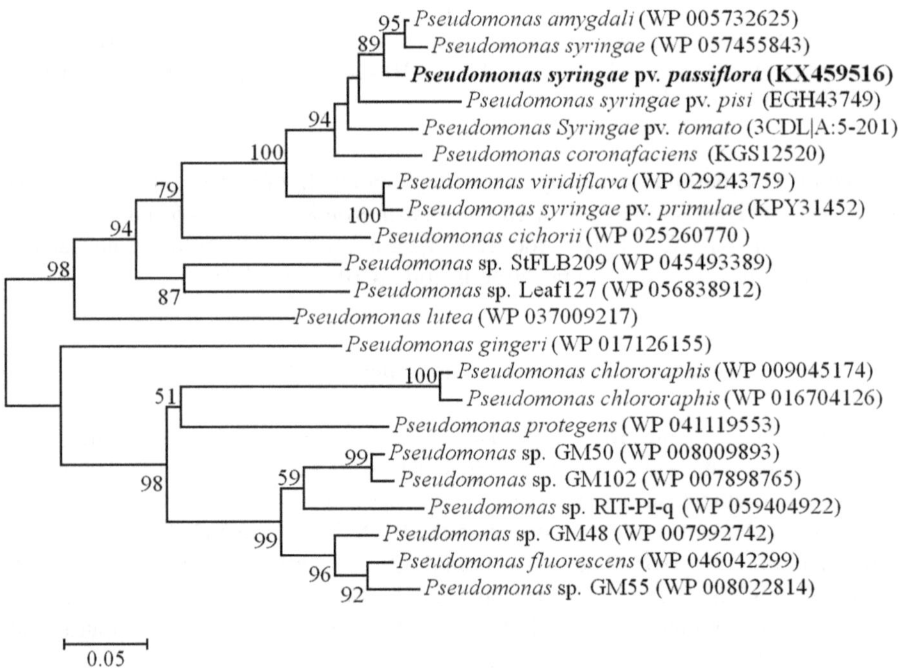

Fig. 14.5 Using the deduced amino acid sequences of *P. syringae* pv. *passiflorae*'s AefR and those of other known *P. syringae*, a phylogenetic tree was constructed using neighbor-joining methods. Bootstrap percentages over 50% were displayed. The scale bar shows 0.05 substitutions for each position of an amino acid

14.1.5 QS Signal Inhibition Strategies and Their Impact on Plant Health

Numerous strategies have been devised to stop the QS, lowering the incidence of plant diseases and enhancing plant health. Enzymes like lactonase and acylase are produced by different bacteria, especially *Bacillus* species, and they break down AHL through the QQ process. It has also been demonstrated that volatile organic compounds obstruct the QS signal. Dimethyl sulfide (DMS), dimethyl disulfide (DMDS), and dimethyl trisulfide (DMTS) were generated by *Serratia plymuthica* and strongly inhibited the growth of *Agrobacterium* (Dandurishvili et al. 2011). The QS is regulated by many plant molecules that release a variety of substances that can impede bacterial signals. The release of QS by plants can be suppressed or even enhanced by many plants' production of AHL mimics (Tannières et al. 2017; Teplitski et al. 2011).

14.1.5.1 QS AHL Degradation Through Enzymatic Interference

Another method for blocking QS has been discovered: QS inhibitory enzymes. The first AHL-degrading enzymes were discovered in a *Bacillus* species, and the aiiA gene was found to be the culprit (AI inactivation). AHL is inactivated by lactone ring hydrolysis, which is mediated by lactone ring hydrolysis. This enzyme is thought to be mostly used by bacteria to break down AHL instead of QQ (LaSarre and Federle 2013). Numerous other bacteria, including *Variovorax paradoxus* and *Ralstonia* strain, have been found to possess QQ abilities (Hanano et al. 2014). By rupturing the amide bond of AHL, the acylase enzyme (aiiD) produced by *Ralstonia* sp. causes AHL to degrade (Lin et al. 2003). The production of AHL by the transformed bacteria decreased when aiiD acylase was expressed in *P. aeruginosa*. It also decreased proteolytic enzyme synthesis and swarming. The details of AHL degrades enzymes from various microorganisms are presented in Table 14.2.

14.1.5.2 Disruption of QS by Volatile Organic Compounds (VOCs) of Bacteria

Volatiles are low molecular, diffusible compounds generated by plant-associated bacteria and have the ability for inter-kingdom communication (Mendes et al. 2013). Alcohols, ketones, sulfur compounds, organic acids, nitrogen compounds, dimethyl sulfide, and dimethyl trisulfide are the major chemical classes of microbial volatile organic compounds (mVOCs) (Schenkel et al. 2015; Schulz and Dickschat 2007). Approximately 2000 different types of VOCs with various synonyms have been reported from around 1000 bacterial species (Lemfack et al. 2018). Many of these volatile molecules are bestowed with antimicrobial activity against pathogens (Manikandan et al. 2022), induced systemic resistance in plants (Farag et al.

Table 14.2 AHL degrades enzymes from various microorganisms

Strains	Gene	Enzyme activity	References
Pseudomonas aeruginosa QSP01	*quiP* and *pvdO*	AHL acylase	Khalid et al. (2022)
Pseudomonas putida QQ3	*quiP* and *pvdO*	AHL acylase	Khalid et al. (2022)
Bacillus cereus QSP03	*aiiA*	AHL lactonase	Khalid et al. (2022)
Bacillus subtilis QSP10	*aiiA*	AHL lactonase	Khalid et al. (2022)
Bacillus cereus 29q	*aiiA*	AHL lactonase	Torabi Delshad et al. (2018)
Agrobacterium tumefaciens 41q	*aiiA*	AHL lactonase	Torabi Delshad et al. (2018)
Stenotrophomonas maltophilia 43q	*aiiA*	AHL lactonase	Torabi Delshad et al. (2018)
Lysinibacillus sp. Gs50	*aiiA*	AHL lactonase	Garge and Nerurkar (2016)
Arthrobacter sp. IBN110	*ahlD*	AHL lactonase	Roche et al. (2004)
Agrobacterium tumefaciens	*attM, aiiB*	AHL lactonase	Carlier et al. (2003)
Pseudomonas aeruginosa PAO1	*pvdQ*	AHL acylase	Huang et al. (2003)
Chryseobacterium sp.	*AidC*	AHL lactonase	Wang et al. (2012)
Streptomyces sp.	*ahlM*	AHL acylase	Sun-Yang Park et al. (2005)
Bacillus thuringiensis	*AiiA*	AHL lactonase	Su-Jin Park et al. (2008)
Sphingopyxis alaskensis DSM 13593	*qsdS*	AHL lactonase	Morohoshi et al. (2019)
Bacillus megaterium CYP102 A1	–	Oxidoreductase	Chowdhary et al. (2007)

2013), and have been linked to plant growth (Blom et al. 2011). Strains *P. fluorescens* and *S. plymuthica* also inhibited the pathogens *P. chlororaphis* 449, and *P. aeruginosa* PAO1 due to a severe reduction of AHL activity. Similarly, Dimethyl disulfide (DMDS) has been reported in *B. subtilis* (Manikandan et al. 2022). The mode of action of interruption of QS by VOCs was discovered to be reversible inhibition of AHL synthesis suggesting that VOCs and AHL can compete to activate the QS response in the same bacteria (Grandclément et al. 2016).

14.1.5.3 Phytochemicals as QS Inhibitors

Plant-derived compounds with QS inhibitory activity have been used in traditional medicine since ancient times. Plant-based compounds are mostly secondary metabolites, with phenols or oxygen substituted derivatives accounting for the vast majority of them. Secondary metabolites have numerous advantages, involving antimicrobial properties against pathogenic microbes (Choo et al. 2006). The major plant antimicrobial properties include phenolics, saponins, tannins, coumarins, terpenoids, phenolic acids, quinones, and alkaloids are principally responsible

Table 14.3 Various phytochemicals used for the QS inhibition

Source	Plant	Compound	References
Ginger oil	*Zingiber officinale* (ginger)	6-gingerol	Han-Shin Kim et al. (2015)
		Zingerone 2-butanone, 4-(4-hydroxy-3-methoxyphenyl)-	Ahmad et al. (2015)
Plant extract	*Pinus sylvestris* (pine)	a-Terpineol	Bose et al. (2020)
Bulb extracts	*Allium sativum* (garlic)	Ajoene	Jakobsen et al. (2012)
		Allicin	Lihua et al. (2013)
Plant extract	*Melaleuca alternifolia* (tea tree)	Terpinen-4-ol	Bukvički et al. (2016)
Extract	*Curcuma longa* (turmeric)	Curcumin	Rudrappa and Bais (2008)
		Eugenol	Zhou et al. (2013)
Bark extracts	*Cinnamomum camphora* (Cinnamomum)	Cinnamaldehyde	Niu and Gilbert (2004)
Extracts	*Origanum vulgare* (oregano)	Carvacrol	Ojo-Fakunle et al. (2013)
Roots	*Scutellaria baicalensis* (Chinese skullcap)	Baicalein	Chen et al. (2016)
Extracts	*Vitis vinifera* (grapes)	Naringenin	Vandeputte et al. (2011)
Extracts	*Coffea arabica* (coffee tree)	Caffeine	Norizan et al. (2013)
Essential oil	*Elettaria cardamomum* (cardamomum)	Cineol	Jaramillo-Colorado et al. (2012)

for plants (Teplitski et al. 2000). Because of differences in structure and chemical composition, the QS inhibitory action of these compounds varies.

The first anti-QS substances identified were halogenated furanones synthesized by the benthic marine macroalgae *Delisea pulchra*. They inhibited QS-regulated behaviors by binding to LuxR type proteins competitively. As a result, there is an increase in their proteolytic degradation rate without killing the bacteria (Choo et al. 2006). Dwivedi and Singh (2016) used the microtiter plate method to examine the impact of embelin and piperine, two natural substances, on the ability of *Streptococcus mutants* to generate biofilms. Malabaricone C, inhibited violacein pigment production by *C. violaceum* CV026 when grown in the presence of a cognate signaling molecule, N-3-oxohexanoyl-homoserine lactone. The compound malabaricone C is extracted from the bark of *Myristica cinnamomea*. Additionally, it prevented the development of biofilms and QS-controlled pyocyanin in *P. aeruginosa* PAO (Chong et al. 2011). The phytochemicals derived from various plants used for QS inhibition are mentioned in Table 14.3.

14.1.6 Strategies to Interfere with QS

In Gram-negative QS systems, three distinct targets could be targeted to disrupt the QS. They are the signal receptor, the signal molecule, and the signal generator. The strategy employed in plant protection is that the QS mechanism is hindered so that the "quorum" of the pathogens is not achieved to cause disease. Toward this the three plausible approaches to inhibit the QS are:

- Arrest the synthesis of signal molecules.
- Degradation or inactivation of signal molecules.
- Analogues of receptors that compete with signal molecules.

The simple method for inhibiting QS is to interfere with AHL synthesis. In simple terms, QS does not occur if no AHL is produced. The earlier studies on QS inhibitors focused on AHL receptor antagonists. Furthermore, there were few studies on AHL synthesis inhibitors, and very minimal data is available. Several analogues of S-adenosylmethionine (SAM), the second substrate for LuxI synthases, were discovered to prevent the LuxI reaction (Fig. 14.6) (Schaefer et al. 2000). Furthermore, there have been two significant investigations conducted. An analogue of C8-HSL that binds to AHL synthase and inhibits its enzymatic activity was found in the first investigation (Chung et al. 2011). The second investigation discovered substances that inhibit the action of the SAM recycling enzyme 5′-methylthioadenosine nucleosidase (MTAN, often known as "Pfs") (Gutierrez et al. 2009). AHL molecules can be broken down by enzymes. AHL compounds are broken down by three enzymes: (1) AHL lactonases (which hydrolyze the AHL lactone ring to produce the corresponding *N*-acyl homoserine), (2) AHL acylase (which breaks down the AHL amide bond to produce a free fatty acid and a lactone ring), and (3) AHL

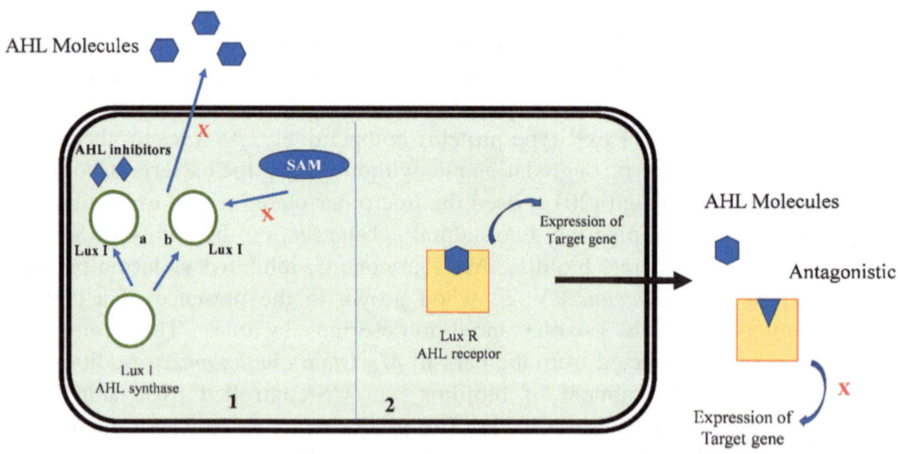

Fig. 14.6 Different mechanisms of QS inhibition in Gram-negative bacteria. (1) Inhibition of the biosynthesis of AHL molecules (*a* inhibiting LuxI; *b* blocking SAM biosynthesis). (2) Application of inductor antagonists

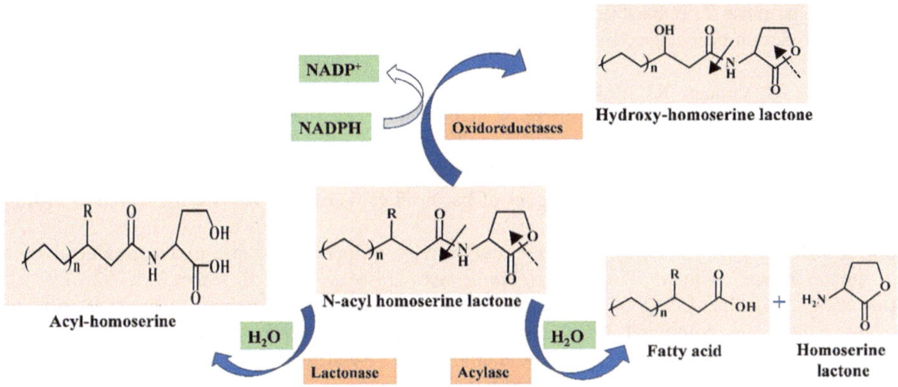

Fig. 14.7 Various mechanisms of AHL degrading enzymes

oxidoreductases (which modify the signal QS molecule by oxidizing it before it is eventually broken down). The list of these enzymes produced by diverse microbes is extensive and may be found in (Fig. 14.7 and Table 14.2). Utilizing inductor antagonists is another strategy for preventing bacterial spread (Bodede et al. 2018).

14.2 Conclusion

This chapter covered numerous QS molecules' methods for detection as well as diverse disease control strategies. A significant difficulty is posed by the large number of genetic regulatory mechanisms that have been identified in the synthesis of the diverse signal molecule, as well as by the complexity of bacterial pathogenicity and the environmental factors that affect plant growth. To prevent the QS-regulated pathogenicity of bacterial pathogens, the biocontrol agent applied in the plant–soil environment must be stable. In the context of integrated plant disease management, however QS-based strategy will develop an efficient method of disease management. Our knowledge of the regulatory mechanisms governing QS-regulated functions increases along with developments in molecular and biotechnological techniques.

References

Agrios GN (2005) Plant pathology. Elsevier

Ahmad, B., Rehman, M. U., Amin, I., Arif, A., Rasool, S., Bhat, S. A., . Bilal, S. (2015). A review on pharmacological properties of zingerone (4-(4-Hydroxy-3-methoxyphenyl)-2-butanone). Sci World J, 2015, 1

Amanifar N (2020) First report of *Pseudomonas syringae* pv. *apsa* causing leaf streak and black chaff of winter wheat in Southwest Iran. Plant Dis 104:276–276

Arabi F, Nikravesh Z, Babaeezad V, Rezaeian V, Rahimian H (2006) Occurrence of bacterial leaf spot of garden beet caused by *Pseudomonas syringae* pv. *aptata* in Iran. Iran J Plant Pathol 42: 655–671

Arsenijević M, Vennete RJ, Maširević S (1994) *Pseudomonas syringae* pv. *helianthi* (Kawamura 1934) dye, Wilkie et Young 1978, a pathogen of sunflower (*Helianthus annuus* L.). J Phytopathol 142:199–208

Ashorpour M, Niknejad Kazempour M, Ramezanie M (2008) Occurrence of *Pseudomonas syringae* pv. *syringae* the causal agent of bacterial canker on olives (*Olea europaea*) in Iran. Sci Asia 34:323–326

Barzic MR, Guittet E (1996) Structure and activity of persicomycins, toxins produced by a *pseudomonas syringae* pv. *Persicae/Prunus persica* isolate. Eur J Biochem 239:702–709

Bashan Y, de Bashan LE (2002) Reduction of bacterial speck (*Pseudomonas syringae* pv. *tomato*) of tomato by combined treatments of plant growth-promoting bacterium, *Azospirillum brasilense*, streptomycin sulfate, and chemo-thermal seed treatment. Eur J Plant Pathol 108: 821–829

Beck von Bodman S, Farrand SK (1995) Capsular polysaccharide biosynthesis and pathogenicity in *Erwinia stewartii* require induction by an N-acylhomoserine lactone autoinducer. J Bacteriol 177:5000–5008

Bedford KE, MacNeill BH, Bonn WG, Dirks VA (1988) Population dynamics of *Pseudomonas syringae* pv. *papulans* on Mutsu apple. Can J Plant Pathol 10:23–29

Belan LL, Pozza EA, de Oliveira Freitas ML, Raimundi MK, de Souza RM, da Cruz Machado J (2016) Occurrence of '*Pseudomonas syringae* pv. *garcae* in coffee seeds. Aust J Crop Sci 10: 1015–1021

Blom D, Fabbri C, Connor EC, Schiestl FP, Klauser DR, Boller T, Weisskopf L (2011) Production of plant growth modulating volatiles is widespread among rhizosphere bacteria and strongly depends on culture conditions. Environ Microbiol 13:3047–3058

Bodede O, Shaik S, Chenia H, Singh P, Moodley R (2018) Quorum sensing inhibitory potential and in silico molecular docking of flavonoids and novel terpenoids from *Senegalia nigrescens*. J Ethnopharmacol 216:134–146

Bose SK, Chauhan M, Dhingra N, Chhibber S, Harjai K (2020) Terpinen-4-ol attenuates quorum sensing regulated virulence factors and biofilm formation in *Pseudomonas aeruginosa*. Future Microbiol 15:127–142

Bukvički D, Cirić A, Soković M et al (2016) *Micromeria thymifolia* essential oil suppresses quorum-sensing signaling in *Pseudomonas aeruginosa*. Nat Prod Commun 11: 1934578X1601101232

Carlier A, Uroz S, Smadja B, Fray R, Latour X, Dessaux Y, Faure D (2003) The Ti plasmid of *Agrobacterium tumefaciens* harbors an attM-paralogous gene, aiiB, also encoding N-acyl homoserine lactonase activity. Appl Environ Microbiol 69:4989–4993

Cazorla FM, Torés JA, Olalla L, Pérez-García A, Farré JM, de Vicente A (1998) Bacterial apical necrosis of mango in southern Spain: a disease caused by *Pseudomonas syringae* pv. *syringae*. Phytopathology 88:614–620

Cha C, Gao P, Chen Y-C, Shaw PD, Farrand SK (1998) Production of acyl-homoserine lactone quorum-sensing signals by gram-negative plant-associated bacteria. Mol Plant-Microbe Interact 11:1119–1129

Chen Y, Liu T, Wang K et al (2016) Baicalein inhibits *Staphylococcus aureus* biofilm formation and the quorum sensing system in vitro. PLoS One 11:e0153468

Chong YM, Yin WF, Ho CY et al (2011) Malabaricone C from *Myristica cinnamomea* exhibits anti-quorum sensing activity. J Nat Prod 74:2261–2264

Choo JH, Rukayadi Y, Hwang JK (2006) Inhibition of bacterial quorum sensing by vanilla extract. Lett Appl Microbiol 42:637–641

Chowdhary PK, Keshavan N, Nguyen HQ, Peterson JA, González JE, Haines DC (2007) *Bacillus megaterium* CYP102A1 oxidation of acyl homoserine lactones and acyl homoserines. Biochemistry 46:14429–14437

Chung J, Goo E, Yu S et al (2011) Small-molecule inhibitor binding to an N-acyl-homoserine lactone synthase. Proc Natl Acad Sci 108:12089–12094

Cintas NA, Koike ST, Bunch RA, Bull CT (2006) Holdover inoculum of *Pseudomonas syringae* pv. *alisalensis* from broccoli raab causes disease in subsequent plantings. Plant Dis 90:1077–1084

Dandurishvili N, Toklikishvili N, Ovadis M et al (2011) Broad-range antagonistic rhizobacteria *Pseudomonas fluorescens* and *Serratia plymuthica* suppress Agrobacterium crown gall tumours on tomato plants. J Appl Microbiol 110:341–352

Destéfano SAL, Rodrigues LMR, Beriam LOS, Patrício FRA, Thomaziello RA, Rodrigues Neto J (2010) Bacterial leaf spot of coffee caused by *Pseudomonas syringae* pv *tabaci* in Brazil. New Dis Rep 59:22

Dumenyo CK, Mukherjee A, Chun W, Chatterjee AK (1998) Genetic and physiological evidence for the production of N-acyl homoserine lactones by *Pseudomonas syringae* pv. *syringae* and other fluorescent plant pathogenic *Pseudomonas* species. Eur J Plant Pathol 104:569–582

Dwivedi D, Singh V (2016) Effects of the natural compounds embelin and piperine on the biofilm-producing property of *Streptococcus* mutans. J Tradit Complement Med 6:57–61

Elasri M, Delorme S, Lemanceau P et al (2001) Acyl-homoserine lactone production is more common among plant-associated *Pseudomonas* spp. than among soilborne *Pseudomonas* spp. Appl Environ Microbiol 67:1198–1209

Farag MA, Zhang H, Ryu C-M (2013) Dynamic chemical communication between plants and bacteria through airborne signals: induced resistance by bacterial volatiles. J Chem Ecol 39: 1007–1018

Fernández-Sanz AM, Rodicio MR, González AJ (2016) *Pseudomonas syringae* pv. *phaseolicola* isolated from weeds in bean crop fields. Lett Appl Microbiol 62:344–348

Fukuda T, Azegami K, Tabei H (1990) Histological studies on bacterial black node of barley and wheat caused by *Pseudomonas syringae* pv. *japonica*. Jpns J Phytopathol 56:252–256

Fuqua C, Greenberg EP (2002) Listening in on bacteria: acyl-homoserine lactone signalling. Nat Rev Mol Cell Biol 3:685–695

Fuqua C, Winans SC (1996) Conserved cis-acting promoter elements are required for density-dependent transcription of *agrobacterium tumefaciens* conjugal transfer genes. J Bacteriol 178: 435–440

Garge SS, Nerurkar AS (2016) Attenuation of quorum sensing regulated virulence of *Pectobacterium carotovorum* subsp. *carotovorum* through an AHL lactonase produced by *Lysinibacillus* sp. Gs50. PLoS One 11(12):e0167344

Grandclément C, Tannières M, Moréra S, Dessaux Y, Faure D (2016) Quorum quenching: role in nature and applied developments. FEMS Microbiol Rev 40:86–116

Green S, Laue B, Fossdal CG, A'Hara SW, Cottrell JE (2009) Infection of horse chestnut (*Aesculus hippocastanum*) by *Pseudomonas syringae* pv. *aesculi* and its detection by quantitative real-time PCR. Plant Pathol 58:731–744

Groenhagen U, Baumgartner R, Bailly A, Gardiner A, Eberl L, Schulz S, Weisskopf L (2013) Erratum to: production of bioactive volatiles. J Chem Ecol 39:1343–1345

Gutierrez J, Crowder T, Rinaldo-Matthis A et al (2009) Transition state analogs of 5′-methylthioadenosine nucleosidase disrupt quorum sensing. Nat Chem Biol 5:251–257. https://doi.org/10.1038/nchembio.153

Hanano A, Harba M, Al-Ali M, Ammouneh H (2014) Silencing of *Erwinia amylovora* sy69 AHL-quorum sensing by a *Bacillus* simplex AHL-inducible aiiA gene encoding a zinc-dependent N-acyl-homoserine lactonase. Plant Pathol 63:773–783

Hattermann DR, Ries SM (1989) Motility of *Pseudomonas syringae* pv. *glycinea* and its role in infection. Phytopathology 79:284–289

Hernandez MN, Lindow SE (2021) Contact-dependent traits in *Pseudomonas syringae* B728a. PLoS One 16:e0241655

Hollaway GJ, Bretag TW (1995) The occurrence of *Pseudomonas syringae* pv. *pisi* in field pea (*Pisum sativum*) crops in the Wimmera region of Victoria, Australia. Austr Plant Pathol 24:133–136

Huang JJ, Han J-I, Zhang L-H, Leadbetter JR (2003) Utilization of acyl-homoserine lactone quorum signals for growth by a soil pseudomonad and *Pseudomonas aeruginosa* PAO1. Appl Environ Microbiol 69:5941–5949

Jakobsen TH, van Gennip M, Phipps RK et al (2012) Ajoene, a sulfur-rich molecule from garlic, inhibits genes controlled by quorum sensing. Antimicrob Agents Chemother 56:2314–2325

Jaramillo-Colorado B, Olivero-Verbel J, Stashenko EE, Wagner-Döbler I, Kunze B (2012) Anti-quorum sensing activity of essential oils from Colombian plants. Nat Prod Res 26:1075–1086

Jose PA, Krishnamoorthy R, Kwon S-W et al (2019) Interference in quorum sensing and virulence of the phytopathogen *Pseudomonas syringae* pv. *passiflorae* by *Bacillus* and *Variovorax* species. BioControl 64:423–433

Khalid SJ, Ain Q, Khan SJ et al (2022) Targeting Acyl Homoserine Lactones (AHLs) by the quorum quenching bacterial strains to control biofilm formation in *Pseudomonas aeruginosa*. Saudi J Biol Sci 29:1673–1682

Kim H-S, Lee S-H, Byun Y, Park H-D (2015) 6-Gingerol reduces *Pseudomonas aeruginosa* biofilm formation and virulence via quorum sensing inhibition. Sci Rep 5:1–11

Krawczyk K, Łochyńska M (2020) Identification and characterization of *PSEUDOMONAS syringae* pv. *mori* affecting white mulberry (*Morus alba*) in Poland. Eur J Plant Pathol 158: 281–291

LaSarre B, Federle MJ (2013) Exploiting quorum sensing to confuse bacterial pathogens. Microbiol Mol Biol Rev 77:73–111

Latorre BA, Jones AL (1979) *Pseudomonas morsprunorum*, the cause of bacterial canker of sour cherry in Michigan, and its epiphytic association with *P. syringae*. Phytopathology 69:335–339

Lemfack MC, Gohlke B-O, Toguem SMT, Preissner S, Piechulla B, Preissner R (2018) mVOC 2.0: a database of microbial volatiles. Nucleic Acids Res 46:D1261–D1265

Lihua L, Jianhui W, Jialin Y, Yayin L, Guanxin L (2013) Effects of allicin on the formation of *Pseudomonas aeruginosa* biofilm and the production of quorum-sensing controlled virulence factors. Pol J Microbiol 62:243

Lin YH, Xu JL, Hu J, Wang LH, Ong SL, Leadbetter JR, Zhang LH (2003) Acyl-homoserine lactone acylase from *Ralstonia* strain XJ12B represents a novel and potent class of quorum-quenching enzymes. Mol Microbiol 47:849–860

Manikandan A, Jaivel N, Johnson I et al (2022) Suppression of *Macrophomina* root rot, *Fusarium* wilt and growth promotion of some pulses by antagonistic rhizobacteria. Physiol Mol Plant Pathol 121:101876

Ménard M, Sutra L, Luisetti J, Prunier JP, Gardan L (2003) *Pseudomonas syringae* pv. avii (pv. nov.), the causal agent of bacterial canker of wild cherries (*Prunus avium*) in France. Eur J Plant Pathol 109:565–576

Mendes R, Garbeva P, Raaijmakers JM (2013) The rhizosphere microbiome: significance of plant beneficial, plant pathogenic, and human pathogenic microorganisms. FEMS Microbiol Rev 37: 634–663

Miller MB, Bassler BL (2001) Quorum sensing in bacteria. Annu Rev Microbiol 55:165–199

Morohoshi T, Kamimura Y, Sato N, Iizumi T (2019) Distribution and characterization of N-acylhomoserine lactone (AHL)-degrading activity and AHL lactonase gene (qsdS) in *Sphingopyxis*. J Biosci Bioeng 127:411–417

Niu C, Gilbert ES (2004) Colorimetric method for identifying plant essential oil components that affect biofilm formation and structure. Appl Environ Microbiol 70:6951–6956

Norizan SNM, Yin W-F, Chan K-G (2013) Caffeine as a potential quorum sensing inhibitor. Sensors 13:5117–5129

Ojo-Fakunle VT, Woertman J, Veldhuizen EJ, Burt SA (2013) Sub-lethal concentrations of carvacrol (from oregano) inhibit bacterial quorum sensing and formation of biofilms. Planta Med 79:SL51

Olczak-Woltman H, Masny A, Bartoszewski G, Płucienniczak A, Niemirowicz-Szczytt K (2007) Genetic diversity of *Pseudomonas syringae* pv. *lachrymans* strains isolated from cucumber leaves collected in Poland. Plant Pathol 56:373–382

Osman SF, Fett WF, Fishman ML (1986) Exopolysaccharides of the phytopathogen *Pseudomonas syringae* pv. *glycinea*. J Bacteriol 166(1):66–71

Park S-Y, Kang H-O, Jang H-S, Lee J-K, Koo B-T, Yum D-Y (2005) Identification of extracellular N-acylhomoserine lactone acylase from a *Streptomyces* sp. and its application to quorum quenching. Appl Environ Microbiol 71:2632–2641

Park S-J, Park S-Y, Ryu C-M, Park S-H, Lee J-K (2008) The role of AiiA, a quorum-quenching enzyme from *Bacillus thuringiensis*, on the rhizosphere competence. J Microbiol Biotechnol 18: 1518–1521

Pereira CS, Thompson JA, Xavier KB (2013) AI-2-mediated signalling in bacteria. FEMS Microbiol Rev 37:156–181

Poonguzhali S, Madhaiyan M, Sa T (2007) Quorum-sensing signals produced by plant-growth promoting *Burkholderia* strains under in vitro and in planta conditions. Res Microbiol 158:287–294

Preston GM (2000) *Pseudomonas syringae* pv. tomato: the right pathogen, of the right plant, at the right time. Mol Plant Pathol 1:263–275

Reid WD (1938) Grease spot of passion fruit. N Z J Sci Technol Sect A 20

Roberts SJ, Preece TF (1984) A note on *Pseudomonas syringae* pv. *berberidis* infections of Berberis: Aetiology of a leaf spot and leaf fall disease in England. J Appl Bacteriol 56:507–513

Roche DM, Byers JT, Smith DS, Glansdorp FG, Spring DR, Welch M (2004) Communications blackout? Do N-acylhomoserine-lactone-degrading enzymes have any role in quorum sensing? Microbiology 150:2023–2028

Rudrappa T, Bais HP (2008) Curcumin, a known phenolic from Curcuma longa, attenuates the virulence of *Pseudomonas aeruginosa* PAO1 in whole plant and animal pathogenicity models. J Agric Food Chem 56:1955–1962

Schaefer AL, Hanzelka BL, Parsek MR, Greenberg EP (2000) Detection, purification, and structural elucidation of the acylhomoserine lactone inducer of Vibrio fischeri luminescence and other related molecules, pp 288–301

Schenkel D, Lemfack MC, Piechulla B, Splivallo R (2015) A meta-analysis approach for assessing the diversity and specificity of belowground root and microbial volatiles. Front Plant Sci 6:707

Schulz S, Dickschat JS (2007) Bacterial volatiles: the smell of small organisms. Nat Prod Rep 24: 814–842

Scortichini M, Rossi MP, Loreti S et al (2005) *Pseudomonas syringae* pv. *coryli*, the causal agent of bacterial twig dieback of Corylus avellana. Phytopathology 95:1316–1324

Shaw PD, Ping G, Daly SL, Cha C, Cronan JE Jr, Rinehart KL, Farrand SK (1997) Detecting and characterizing N-acyl-homoserine lactone signal molecules by thin-layer chromatography. Proc Natl Acad Sci 94:6036–6041

Takikawa Y, Tsuyumu S, Goto M (1991) Occurrence of bacterial leaf spot of maple incited by *Pseudomonas syringae* pv. *aceris* in Japan. Ann Phytopathol Soc Jpn 57:724–728

Tamura K, Peterson D, Peterson N, Stecher G, Nei M, Kumar S (2011) MEGA5: molecular evolutionary genetics analysis using maximum likelihood, evolutionary distance, and maximum parsimony methods. Mol Biol Evol 28:2731–2739

Tannières M, Lang J, Barnier C, Shykoff JA, Faure D (2017) Quorum-quenching limits quorum-sensing exploitation by signal-negative invaders. Sci Rep 7:1–10

Teplitski M, Robinson JB, Bauer WD (2000) Plants secrete substances that mimic bacterial N-acyl homoserine lactone signal activities and affect population density-dependent behaviors in associated bacteria. Mol Plant-Microbe Interact 13:637–648

Teplitski M, Mathesius U, Rumbaugh KP (2011) Perception and degradation of N-acyl homoserine lactone quorum sensing signals by mammalian and plant cells. Chem Rev 111:100–116

Toben H, Rudoph K (1996) *Pseudomonas syringae* pv. *coriandricola*, incitant of bacterial umbel blight and seed decay of coriander (*Coriandrum sativum* L.) in Germany. J Phytopathol 144: 169–178

Toben H, Mavridis A, Rudolph KWE (1989) Basal glume rot (*Pseudomonas syringae* pv. *atrofaciens*) on wheat and barley in FRG and resistance screening of wheat 1. EPPO Bull 19:119–125

Torabi Delshad S, Soltanian S, Sharifiyazdi H, Haghkhah M, Bossier P (2018) Identification of N-acyl homoserine lactone-degrading bacteria isolated from rainbow trout (*Oncorhynchus mykiss*). J Appl Microbiol 125:356–369

Vandeputte OM, Kiendrebeogo M, Rasamiravaka T, Stevigny C, Duez P, Rajaonson S, El Jaziri M (2011) The flavanone naringenin reduces the production of quorum sensing-controlled virulence factors in *Pseudomonas aeruginosa* PAO1. Microbiology 157:2120–2132

Vanneste JL, Yu J, Cornish DA (2010) Molecular characterisations of *Pseudomonas syringae* pv. *actinidiae* strains isolated from the recent outbreak of bacterial canker on kiwifruit in Italy. N Z Plant Protect 63:7–14

Von Bodman SB, Bauer WD, Coplin DL (2003) Quorum sensing in plant-pathogenic bacteria. Annu Rev Phytopathol 41:455–482

Wang W-Z, Morohoshi T, Someya N, Ikeda T (2012) AidC, a novel N-acylhomoserine lactonase from the potato root-associated cytophaga-flavobacteria-bacteroides (CFB) group bacterium *Chryseobacterium* sp. strain StRB126. Appl Environ Microbiol 78:7985–7992

Yaoita T, Fujimaki Y, Abe T, Tsujimoto K (1984) Control of rice bacterial brown stripe disease [caused by *Pseudomonas syringae* pv. *panici*] multiplied in Hatomune auto-sproutor with Kasugamycin. J Bacteriol 194:5693–5694

Young JM (1987) New plant disease record in New Zealand: *Pseudomonas syringae* pv. *Persicae* from nectarine, peach, and Japanese plum. N Z J Agric Res 30:235–247

Young JM, Jones DS, Gillings M (1996) Relationships between populations of *Pseudomonas syringae* pv. *persicae* determined by restriction fragment analysis. Plant Pathol 45:350–357

Zhou L, Zheng H, Tang Y, Yu W, Gong Q (2013) Eugenol inhibits quorum sensing at sub-inhibitory concentrations. Biotechnol Lett 35:631–637

Chapter 15
Molecular Approaches on Biocontrol of Postharvest Fungal Plant Pathogens: Antagonistic Yeast Model

Pervin Kinay-Teksur

Abstract Postharvest decays limit storage life and cause significant losses on fruit and vegetables. For a long time, synthetic fungicides have been used successfully for controlling the postharvest pathogens. In recent years, due to their risks, studies on alternatives to the chemical fungicides on postharvest rots control have garnered high attention. In this concern, the uses of biological and physical methods have gained great importance. Yeasts are single-celled eukaryotic microorganisms that belong to Kingdom Fungi. As the epiphytic yeasts always exist on fruit and vegetable surfaces and compete with other microorganisms, they take attention on biological control. The management of postharvest pathogens with antagonistic yeasts, including pre- and postharvest uses, has gained great importance in the last 30 years. As a result of these studies some successful antagonist yeasts were found, and their formulations were carried out and commercialized. The use of antagonistic yeasts may differ with target pathogens and host; hence they can be applied at pre- or postharvest stages. After the 1990s, molecular methods have pioneered in providing very important developments in the biological control of fungi, as well as in all areas of plant pathology. To find a good antagonistic microorganism and deliver to conventional use is not easy and short process. These processes include isolation, identification, biology, ecology, physiology, biological efficacy-testing, commercial product development, and registration of yeasts for agricultural applications. Including all the steps, from the initial discovery to registration and product development, needs the work of specialist scientists from different disciplines together, including new technologies. In researches, genome sequencing is now commonly used for both identification and classification of pathogen and biocontrol agents for the control of the plant diseases. The interactions of the antagonist microorganisms with the pathogen and the host have been highly detectable and understandable by using omics technologies. This review supplies a summary of the usage of the molecular instruments for screening and identification of the antagonistic yeasts

P. Kinay-Teksur (✉)
Department of Plant Protection, Faculty of Agriculture, Ege University, Izmir, Turkiye
e-mail: pervin.kinay@ege.edu.tr

K. K. Bastas et al. (eds.), *Microbial Biocontrol: Molecular Perspective in Plant Disease Management*, Microorganisms for Sustainability 49,
https://doi.org/10.1007/978-981-99-3947-3_15

and comprehension of their mechanisms of action on postharvest biological control of fruit and vegetables.

Keywords Epiphytic yeast · Postharvest · Decay · Biological control · Omics · Molecular

15.1 Introduction

Fungi and fungi-like microorganisms are the main pathogens that cause devastating diseases and significant economic crop losses. Intensive fungicides are used in the chemical control of these diseases every year. The problems caused by synthetic chemical fungicides regarding public and environmental health have brought about the search for new and reliable alternative methods. Postharvest diseases and disorders cause very important *economic* losses in terms of crop quality and quantity of fresh fruit and vegetables, by 25–50%, depending on countries and technologies that are used during handling process. From harvest to consumer in different channels, many biological, pathological, physiological, and mechanical factors lead to deterioration of products. Inconvenient harvest and postharvest handling, transportation, packaging, and storage of crops influence the storage life and marketing acceptability, and cause important losses on fruit and vegetables (FAO 2011; Nunes 2012; Gustavsson et al. 2011). Many facultative parasite and saprophyte fungal pathogens cause deterioration on fresh fruit and vegetables. *Botrytis*, *Alternaria*, *Aspergillus*, *Geotrichum*, *Mucor*, *Monilinia*, *Penicillium*, and *Rhizopus* are main fungal genera that are responsible for postharvest decays of fresh produce. Meanwhile, *Pythium* and *Phytophthora*, fungal-like organisms that contaminate the product at preharvest stage, are also important sources of spoilage (Snowdon 1990; Barkai-Golan 2001).

Microorganisms that include bacteria, yeasts, and other filamentous fungi as a community commonly occupy the fruit and vegetable surface. Yeasts have different relationship with other communities such as symbiotic interactions: mutualistic, commensalistic, or parasitic (Starmer and Lachance 2011). The biological control system uses the advantage of these relationships between microorganisms in the direction of the human benefit through the understanding their interactions very well. The use of natural microbial antagonists on crop surface is an eco-friendly, sustainable, and improvable control measure compared to chemical fungicides.

Yeasts are microorganisms that have always existed in human history and been a part of human life throughout the ages. They are unicellular eukaryotic fungus multiplied by budding or fission in anamorphic stage. They are very important components of several complex mechanisms in ecosystems. The well-known model yeast in the world, *Saccharomyces cerevisiae*, is utilized in basic research, biotechnological applications, and food production. Natural habitat of *S. cerevisiae* is the surface of fruits, and humans have used it for thousands of years to produce alcoholic beverages and bread (Carlile et al. 2005; Kurtzman et al. 2011).

To control postharvest decays of some fruit such as citrus, cherry, banana, and apple, synthetic fungicides have been used widely in packinghouses. The extreme and continual usage of fungicides has led to several problems: fungicide resistance, environmental pollution, and the harmful effect on human health (Eckert et al. 1994; Smilanick et al. 2006; Kinay et al. 2007). Therefore, residue problems in crops and increasing awareness in health issues of public concerns have forced researchers to find alternative methods to minimize the postharvest losses of fruit and vegetables (Palou et al. 2008). Biological control serves as effective and safe alternative method to the chemicals, which is very popular since last 30 years (Wilson and Wisniewski 1989, 1994; Sharma et al. 2009; Wisniewski et al. 2016).

As biocontrol agent, yeasts were studied on a wide range of host, including newly harvested vegetables and fruits, and even grains and grass (Droby et al. 2009, 2016; Wisniewski et al. 2016; Schisler et al. 1995; Petersson and Schnürer 1995; Allen et al. 2004; Ray et al. 2011). From these studies, many antagonistic yeast species were identified and their semi-commercial and commercial tests were carried out. Some of them have even been provided for commercial use to control postharvest decays of fresh vegetables and fruits (Droby et al. 2016). In numerous studies, successful antagonistic yeast isolates were reported and they were also tested in large-scale commercial pilot packinghouses (Droby et al. 1993; Hofstein et al. 1994; Usall et al. 2001; Long et al. 2006). Since then, many antagonists have been identified and registered as commercial products to control postharvest diseases of different fruits and vegetables. *Aureobasidium pullulans* (Boniprotect, EU) on pome fruit at preharvest stage; *Candida oleophila* (Nexy, Belgium, EU) and *C. oleophila* (Aspire, United States) on citrus and pome fruit; *Candida sake* (Candifruit, Spain) on pome fruit; *Cryptococcus albidus* (Yield plus, South Africa) on pome fruit; and *Metschnikowia fructicola* (Shemer, Netherlands) on table grape, strawberry, and sweet potato are antagonistic yeasts and commercially available as biofungicides for the postharvest disease control (Spadaro and Droby 2016).

Primarily these studies have been carried out with classical morphological and biochemical diagnostic methods for many years. After new developments in molecular methods, the identification of fungi has been started to be carried out at a much more advanced level with sequence-based and genomic technologies for over two decades. The characterization of yeast antagonists with genomic-based methods is rapid, less expensive, and less inclined to technical mistakes. The aim of this paper is to give a brief overview about yeast and yeast-like microorganisms, which are an important component of postharvest biological control, and new molecular techniques used in these studies.

15.2 Epiphytic Yeasts as Postharvest Biocontrol Agents: Gene Isolation and Sequencing

Epiphytic yeasts naturally existing on aerial fruit and vegetable surfaces may combat with postharvest pathogens successfully. Over the last 30 years, a great effort in finding safe alternatives to chemicals for managing agricultural production systems and controlling plant diseases has been made (Wilson and Pusey 1985). There are two basic approaches to the use of microbial antagonists for controlling diseases of fruits and vegetables at postharvest stage. *First*, use of microorganisms that already exist on the crop surfaces that can be increased and handled: the use of antagonistic yeasts that are epiphytic and present on fruits and vegetable surfaces is convenient and safe in terms of adaptation of microorganisms. *Second*, use of microorganisms that can be artificially applied to postharvest pathogens: if the antagonist has a wide range of pathogens on different crops, it may give an opportunity to use antagonist on different commodities (Sharma et al. 2009).

Biological control with yeasts was reported to be a hopeful alternative against chemical fungicides for controlling postharvest diseases of fresh crops. The use of epiphytic yeasts for pathogen control on postharvest stage of fresh crops has been worked on many different products with yeasts and bacteria in different countries in the world (Chalutz et al. 1988; Wilson and Wisniewski 1989; Kinay et al. 1998; Kinay et al. 2002; Droby et al. 2016; Wisniewski et al. 2016). *C. oleophila* (Strain I-182) was developed as a commercial product. In these studies, identification of antagonistic yeasts and their mode of action has been traditionally carried out with morphological and biochemical tests.

Molecular methods give new perspectives on understanding of microbial environment at postharvest stage of fresh crops. In ribosomal DNA (rDNA), internal transcribed spacer (ITS) regions (ITS1, 5.8S, and ITS2) have been generally used on phylogenetic relationships for fungal barcoding. Since these regions easily get amplified and sequenced with universal primers, a consortium of mycologists has selected the ITS as the official barcode for fungi (Schoch et al. 2012). Nowadays, new-generation DNA-sequencing technologies provide the genome of microorganisms to be rapidly determined. The genetically rapid identification of antagonistic microorganisms by these technologies also has shortened the selection process of antagonistic microorganisms. The information obtained from sequencing provides the raw data for the field of bioinformatics. The first eukaryotic microorganism was *S. cerevisiae*, which was entirely sequenced with the genome of 12.5 Mb (Goffeau et al. 1996). Besides, it is a good model for genetic analysis (Cherry et al. 2012). Whole genome sequencing has become an increasingly routine technique in molecular biology and biological control studies. The genomics analyses speed up the biocontrol studies by identifying genome features of the potential antagonistic strains. In omics technologies, metabarcoding allows for the simultaneous identification of different microorganisms and provides helpful information for characterizing the microbiome diversity (Abdelfattah et al. 2018). In a study, a metagenomics technique based on ITS2 region was used to determine the fungal diversity on apple fruit. They found that fungi belonging to phylum Ascomycota were dominant in apple fruit (91.6%) and *Penicillium* was the most common genus in fungal populations in the peel and wounded tissue of apple. In their results, *Cryptococcus* (9.20%) was one of the most plentiful genera observed. Other potential yeast antagonists, *Metschnikowia* and *Wickerhamomyces*, were also found in peel and wound of apple fruit (Abdelfattah et al. 2016).

Yeasts play very important role as biocontrol agents inhibiting the growth of postharvest decay due to fungi. They also gain importance in the reduction of mycotoxin contamination in food. Detoxification capabilities of *Metschnikowia pulcherrima* strains have been tested on patulin mycotoxin and yeast cell walls did not absorb patulin, and they completely degraded the toxin (Reddy et al. 2011). In two studies, two major products of patulin, Z-/E-ascladiol and desoxypatulin acid, were identified after biodegradation by microorganisms (Zhu et al. 2015; Li et al. 2019). On apple treated with *Rhodotorula glutinis*, patulin accumulation was found lower than on non-treated fruit (Castoria et al. 2005). The detoxification of patulin by the yeast *Rhodotorula mucilaginosa* was determined up to 90% at 35°C (Li et al. 2019).

Next-generation sequencing is a comprehensive technology that can be applied to reveal the whole genome, transcriptome, and DNA–protein interaction. While genome sequencing provides reliable results in the isolation, selection, and assessment of antagonistic yeasts as biological control agents, it also supplies very important information on the host–pathogen–biocontrol agent relationships (Massart et al. 2015). Whole genome sequences of the yeasts *M. fructicola* (Piombo et al. 2018), *Rhodotorula graminis* WP1 (Firrincieli et al. 2015), and *Candida oleophila* (Sui et al. 2020) were identified using new sequencing technologies, carried out in different studies.

15.3 Factors Affecting Antagonistic Yeasts

The colonizing ability on the surface of fruit and vegetables, obtaining nutrients easily, and surviving under different environmental conditions are very important criteria to select antagonistic microorganisms. There is a strict interaction between microbial populations, host, and environment. To find and use an antagonist yeast as a biological control agent successfully, one needs to understand relationships between host, pathogen, and environment. The disease triangle, host–pathogen–environment, also directs influence on the success of antagonistic yeast (Fig. 15.1). Thus, the figure represents the modules in postharvest fungal plant diseases: that the interactions of a susceptible host, a high-virulent pathogen, and favorable environmental conditions for the disease progress have influence on effectiveness of biocontrol agents (Moore et al. 2011). On the contrary, postharvest disease development with antagonistic yeasts may be minimized or prevented upon elimination of unfavorable factors in these three causal components in an integrated control strategy perspective. Temperature, pH, and moisture have direct effect on antagonistic

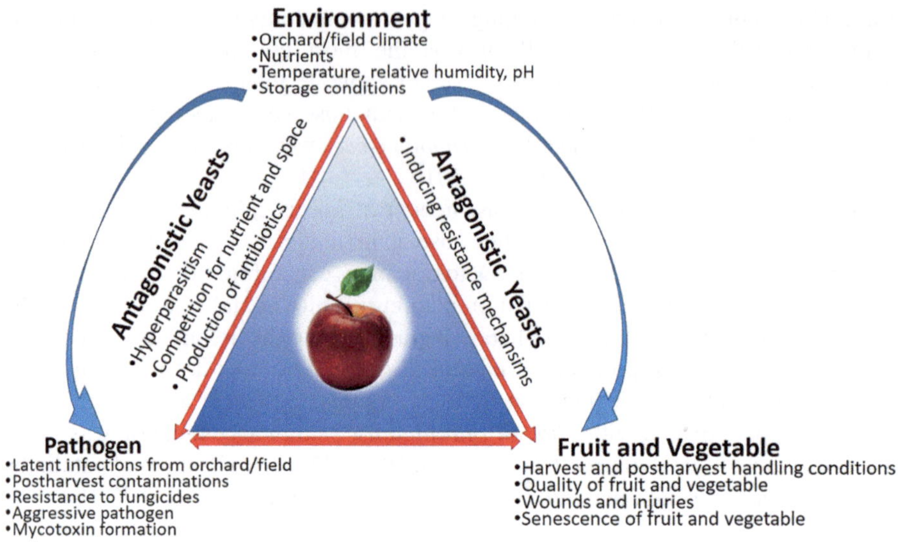

Fig. 15.1 Interactions between host, pathogen, environment, and antagonistic yeasts

yeasts' viability on surfaces and thus their efficacy (Teixidó et al. 1998; Liu et al. 2011, 2012). These factors and association between pathogen and biocontrol agents can be explained in more detail with the use of advanced molecular techniques, such as proteomics, transcriptomics, genomics, and metabolomics.

The preharvest environmental factors such as temperature, humidity, *nutrient availability*, pH, and usage of pesticides can influence species diversity and population of antagonistic yeasts. Pesticides usage during season in orchard/field, especially fungicides, may limit the species diversity. Yeasts always exist on the peel of fresh fruits and vegetables and epiphytic yeasts are a natural component of plant microbiota. Many species of yeasts are found on the superficial part of fresh crops epiphytically where they are able to find exudates including nutrients from natural openings. The exudates from fruit and vegetable surfaces such as simple sugars and organic acids are the main nutrient source for yeasts, which live epiphytically on peel of fresh products. The presence of water and nutrients for epiphytic yeasts varies in fruit and vegetables from different ecological and geographical groups. The yeasts are not just on the surface of fruits and vegetables available; they can stay alive throughout the year in other parts of the trees or plants. On the formation of epiphytic yeast communities, the exudates that are secreted from plant cells have a direct influence (Kurtzman et al. 2011). Nutrients exuded or leached by the host such as fruit peel exudates are important to feed both pathogens and biocontrol agents. The amount and ingredient of these exudates depends on maturity of fruit and vegetables. The nutritional content of immature and ripe crop varies from that of a mature fruit. Therefore, the application time of antagonist yeasts in terms of colonization gains importance in practice. The antagonistic yeasts as biological control agents can colonize the surface of fruit/vegetables for extended duration, even under improper conditions (Sharma et al. 2009). The leaves and flowers of 25 plants were examined by means of epiphytic yeast populations in a study and it was found that yeast population on the leaves changed regularly along the year. The dynamics of yeast were at the maximum level in autumn (Glushakova and Chernov 2007). The yeasts *Cryptococcus infirmominiatus*, *Cryptococcus laurentii*, and *R. glutinis*, applied on pear fruit before three weeks of harvest, maintained high population levels during the harvest (Benbow and Sugar 1999).

Nowadays, yeasts that exist naturally on the peel of fruits and vegetables have become an important actor in postharvest disease control as biological control agents. Therefore, there are some advantages to use biological control at postharvest stage. The antagonistic yeasts, which can remarkably minimize fruit rots, can be applied directly onto the intended area by dipping or spraying applications in different steps of packinghouses (Janisiewicz and Korsten 2002). The conditions on postharvest stage can be easily managed such as temperature and relative humidity that are very important for colonization of antagonistic yeasts. These factors have direct influence on the success of implementation and antagonistic yeasts. These advantages make yeasts the most studied microbial organisms as biological control agents (Wisniewski et al. 2016). In a study, the amplicon sequencing was utilized to observe the effect of washing and waxing applications on the microbial populations of apple after treatments and storage at 2°C for six months. They found that fungi and bacteria, particularly the epiphytic microflora of surface tissues, were impacted by sanitation practices (Abdelfattah et al. 2020).

Epiphytic yeasts are isolated easily from healthy fruit and vegetable surfaces and grow very fast in synthetic media. The yeasts are well adapted to resist many stress conditions and hence yeasts are common microorganisms on fruit and vegetable surfaces. They are generally tolerant to wide pH ranges, high salinity, and drought conditions (Carlile et al. 2005; Kurtzman et al. 2011; Sui et al. 2015). These are very important advantages for yeasts to live and colonize even in stress conditions. Common yeasts on fruit and vegetable surfaces belong to the genera *Candida*, *Cryptococcus*, *Rhodotorula*, *Sporobolomyces*, *Tilletiopsis*, *Pichia*, and *Torulopsis* (Carlile et al. 2005; Kurtzman et al. 2011). Most antagonistic yeasts belong to Ascomycota phyla such as *Candida oleophila*, *C. sake*, *Metschnikowia pulcherrima*, *M. fructicola*, *Pichia guilliermondii*, *Pichia membranifaciens*, and *P. anomala*; some genera belong to Basidiomycota such as *C. laurentii*, *C. albidus*, and *R. glutinis*; and some of them are called yeast-like organisms, such as *A. pullulans* (Kurtzman et al. 2011).

Microbial populations living on upper surfaces of fruit and vegetables have a great potential for competition with pathogens. The new trend on postharvest handling of fresh fruits and vegetables to control diseases after harvest is microbiome approaches (Kusstatscher et al. 2020; Wassermann et al. 2022). These approaches provide very important information for healthy and long storage life of the crop by detecting the microbiome on the surfaces of the fruit and vegetables.

15.4 Mode of Action of Antagonistic Yeasts and Molecular Tools

Yeasts as antagonist microorganisms show effect on postharvest pathogens using different *mechanisms* of action. The biocontrol activities or modes of action of antagonistic yeasts for control of pathogens include competition for nutrient and space between the pathogens, biofilm formation, antibiotic production, direct parasitism to pathogen, and induced resistance in crop (Sharma et al. 2009; Droby et al. 2009; Freimoser et al. 2019). To find a successful antagonistic and commercially available yeast for practical applications, the criteria are mostly related to their *mechanism* of action on pathogens. To identify the modes of action of antagonist is important for making convenient commercial formulation, and determining application methods and their efficacy (Zhang et al. 2011). The initial screening for identifying the *mechanism* of action of biological control agents has been carried out in in vivo and in vitro tests. Even multiple modes of action and their combinations can be tested with the help of well-prepared bioassays on nutrient media toward in vitro production of antimicrobial metabolites (Köhl et al. 2019). Nowadays, the genes and gene clusters, which encode enzymes and proteins, antimicrobial compounds, and secondary metabolites produced by antagonistic organisms, may be identified easily by using advanced molecular tools. In omics technologies, metabolomics, proteomics, and transcriptome methods are utilized to investigate and clarify the mechanisms of biological control (Massart et al. 2015; Apaliya et al. 2017, Apaliya et al. 2019; Zhao et al. 2020).

Induced resistance: The biocontrol agents may stimulate the biochemical or structural defense mechanisms that already exist in crop. The regional responses in the surrounding cells include accumulation of pathogenesis-related (PR) proteins; formation of phenolics, callose, lignin, and phytoalexins; and induction of reactive oxygen species (ROS), which inhibit the penetration of pathogens that occurs after the defense mechanism is induced with biotic or abiotic elicitors. The induced resistance can occur as systemic acquired resistance (SAR) or induced systemic resistance (ISR) in which SAR is triggered by pathogens while ISR is activated by beneficial microorganisms (Romera et al. 2019). If an antagonist is present at the infection site, some antimicrobial substances could be synthesized by the biological control agent or plant, which is triggered by antagonist. After emergence of the induced resistance on crop, a local protection layer can be formed because of the induced resistance (Prasannath 2017). In several studies, it was reported that the antagonistic yeasts are induced as defense mechanisms on fruit and vegetables (Spadaro and Droby 2016; Romanazzi et al. 2016). Antagonistic yeasts induce several biochemical defense responses in wound surface. Droby et al. (2002) found that *C. oleophila* application raised the ethylene biosynthesis, phenylalanine ammonia (PAL) lyase activity, and phytoalexin accumulation. Increase in β-1,3-glucanase activity and PAL was reported with different antagonistic yeast applications on fruit (Yao and Tian 2005; Tian et al. 2006; Xu et al. 2008). Further studies with improved molecular methods such as proteomics and transcriptome technologies on this subject continue to provide detailed scientific data. Transcriptome technology has been studied by different research teams. The advancing of biocontrol activity of yeast antagonists by biotic elicitor was examined in order to determine the molecular mechanism, and it was found that there was a substantial increase in peroxidase activity compared to the control (Apaliya et al. 2019). In in vitro test of *Botrytis cinerea* with antagonistic yeasts (*Pichia fermentans*, *Issatchenkia terricola*, and *Wickerhamomyces anomalus*), the authors identified with exometabolome analysis the phenylpropanoid trans-cinnamic acid and the alkaloid indole-3-carboxaldehyde metabolites that caused injury to the fungal membrane and raised ROS activity (Fernandez-San Millan et al. 2022).

Recent studies examined the enhancement efficacy of yeast antagonist with some stimulators, such as β-glucans and chitosan. In proteomics studies, the expression, structure, functions, interactions, and modifications of proteins are characterized by identifying gene products. PR proteins, induced by yeast *P. membranifaciens* and salicylic acid (SA), in peach fruit were examined with proteomic methods. It was found that both the yeast and SA increased the resistance and delayed the initiating of *Penicillium expansum* infection on peach fruit. Both treatments enhanced PR proteins and expression of the catalase gene (Chan et al. 2007). In the study in which was examined the effect of β-glucan of *Cryptococcus podzolicus* against blue mold using transcriptome technology, it was determined that β-glucan induced by *C. podzolicus* can raise the rate of polysaccharide usage and enhance the ability to adapt to oxidative stress by improving its antioxidant capacity. They found also β-glucan inhibits ripening and senescence of pears (Zhao et al. 2020).

The competition for nutrient and space: Microorganisms compete with other inhabitants for the space and nutrients in order to grow, feed, and survive on fruit and vegetable surfaces. The microorganism that first settles in that area dominates and takes advantage of this competition in an antagonistic relationship. The antagonistic microorganism covers the crop surface by rapid colonizing and prevents colonization of pathogens. One of the most desired mechanisms of action is the nutrient and space competition between the postharvest pathogen and microbial antagonist. The yeast antagonists mostly compete for space or use of some nutrients, especially consumption of carbohydrates, with the pathogen (Sharma et al. 2009; Freimoser et al. 2019).

Most of postharvest disease agents are "wound pathogens," hence the colonization on wound surface rapidly is a very important advantage for the antagonist. In the wound sites, *spores* of *fungal* pathogens germinate immediately and make infection very fast because of rich nutrients *leakage* from punctured cells, which stimulate the germination. The candidates of antagonistic yeasts have been selected primarily according to their ability of rapid colonization and growing in wounds surface and in addition the mode of action of competition with the pathogen for nutrients and space (Droby et al. 2009). Therefore, the effective biocontrol agents should have the capability of colonizing wounds rapidly, growing very fast, and using nutrients effectively in wound surfaces (Mercier and Wilson 1994). In the study, the results showed that endophytic yeast *Candida guilliermondii* is able to enter through the healthy cuticle of immature tomato fruits and colonize on plant tissues (Infante et al. 2012).

In many biological control studies especially on siderophore bacteria, the importance of iron competition between microorganisms has been shown. Iron is an essential element for all living organisms and it takes part in numerous cellular processes; therefore, iron deficiency affects the synthesis of DNA and cell cycle progress in microorganisms (Symeonidis and Marangos 2012). Biocontrol agent that is able to bind Fe from plants can successfully colonize on the host (Nevitt 2011). For antagonistic yeast, iron is one of the most important nutrient sought in competition with pathogen (Spadaro and Droby 2016) Iron is a critically important element for both growth of pathogen and biocontrol agent. Pretscher et al. (2018) confirmed iron depletion for strains of *M. pulcherrima* as an important biocontrol mechanism of action. The ability of competition for nutrients in wounds with other microorganisms is substantial in order to colonize rapidly, grow, and survive in also low concentrations of nutrients for antagonist agents.

Parasitism and lytic enzymes: In the mode of action called hyperparasitism/mycoparasitism, the antagonist directly attacks pathogen propagules, feeds with, and kills it. Antagonists produce cell wall degrading enzymes such as chitinases, cellulases, and β-1,3-glucanase for degradation of host cell walls to attack the target fungus (Sharma et al. 2013; Pretscher et al. 2018; Saravanakumar et al. 2008; Freimoser et al. 2019; Köhl et al. 2019). The same mechanisms are found in yeast antagonists having the mode of action of hyperparasitism. In many studies, it was demonstrated that the secretion of the exo-1,3-β-glucanase by antagonistic yeasts is one of the biocontrol modes of action against postharvest pathogens. It was detected that the antagonistic yeasts *P. guilliermondii*, *Pichia anomala*, *C. oleophila*, and *A. pullulans* secreted exo-b-1,3-glucanase and chitinase on the cell walls of *B. cinerea* (Wisniewski et al. 1991; Castoria et al. 2001; Grevesse et al. 2003; Bar-Shimon et al. 2004; Guerrero et al. 2014). In recent studies, gene expressions of this kind of enzyme activities have also been investigated. In a study, yeast exo-β-1,3-glucanase extract obtained from yeast transformed with the exg1 gene was evaluated as an inhibitory agent of *Colletotrichum lupini* and *B. cinerea* (Oelofse et al. 2009). The expression of glucanase genes of Wickerhamomyces anomalus, with Reverse Transcriptase Quantitative PCR (RT-qPCR), was determined by using specific primers (Parafati et al. 2017). In a study, it was found that the expression of the PR-8 gene was induced in apple fruit after *B. cinerea* infection and *C. oleophila* treatment (Liu et al. 2013).

Biofilm formation: Some yeast species may form a biofilm on fruit and vegetable surfaces as a part of biocontrol mechanism. Biofilms are created by the aggregate of microorganisms adhering to surfaces. Biofilm formation in medical mycology is defined very well by Donlan and Costerton (2002). The colonization on surfaces and biofilm formation in fungi were previously indicated for *Candida albicans* in human and bakers' yeast *S. cerevisiae* (Hawser and Douglas 1994; Reynolds and Fink 2001). In recent years, biofilm formation on the inside of wounds was demonstrated as a possible mechanism of biocontrol agents (Vila and Rozental 2016; Spadaro and Droby 2016). It was found that *S. cerevisiae* strain was able to form a biofilm and effective against *P. expansum* on apple fruit. The biofilm formation of the yeast antagonist *P. fermentans* secreted antimicrobial substances and formed a biofilm on the inner surface of wounds against *Monilinia fructicola* (Giobbe et al. 2007).

Volatile organic compounds and antibiotic production: Volatile organic compounds (VOCs) are secreted by antagonistic yeasts against pathogens after harvest period (Freimoser et al. 2019). Some yeast antagonists were tested for synthesizing antifungal VOCs. The VOC synthesis of antagonists has proven to minimize remarkably postharvest rots on artificially wounded tissues of strawberry and mandarin (Parafati et al. 2017). VOCs secreted by antagonistic yeasts may inhibit mycotoxin formation of pathogens by preventing spore germination and mycelial growth (Hua et al. 2014; Farbo et al. 2018; Fiori et al. 2014; Tilocca et al. 2019).

Secretion of antimicrobial compounds such as toxins may act as a mode of action for some yeast antagonists, called "killer" yeasts. Some yeast can secrete toxins or low-molecular glycoproteins against postharvest pathogens (Perez et al. 2016). The killer toxin produced by *P. membranifaciens* was observed against *B. cinerea* (Santos and Marquina 2004). Effective yeast isolates from tangerine against sour rot were identified as *Rhodotorula minuta*, *Candida azyma*, and *A. pullulans*. Their biocontrol modes of action were identified as killer activity and hydrolytic enzyme production (Ferraz et al. 2016).

Antibiotic production as a mode of action has been mostly studied and associated with bacterial antagonists. The screening for antibiotic production of antagonists usually is carried out in vitro in terms of formation of antibiotic zones with dual-culture tests. Since microorganisms have ability to gain resistance to the antibiotics quickly, this mode of action mostly is not desirable for yeast antagonists and eliminated at the beginning of the studies. The potential to gain resistance to antibiotics of postharvest pathogens and transmission risk to human pathogens through food chain are the main concerns for not preferring using these kinds of compounds on fruit and vegetables (Sharma et al. 2009). The inhibition mechanisms of yeasts as biological control microorganisms have not been completely figured out, since they are complex microorganisms and there are some synergistic relationships. Metabolomic analyses give support to explain antagonistic mechanisms essential to biocontrol strategies with yeast and allow to answer critical questions. Molecular studies on the mechanisms of action of biocontrol agents are relatively new and few in number. In the near future, it is thought that there will be an increase in studies using omics technologies to elucidate the mode of actions.

15.5 Formulations of Antagonistic Yeast

To process the discovery, identification, and improving of a biocontrol agent is a long-term and expensive undertaking (Droby et al. 2016). After finding the successful biocontrol agent, commercialization of this microorganism is critical for delivery and use in practice. Since the biological control agents are living organisms, their formulations should have a long shelf life where their effectiveness is retained for

Table 15.1 Some examples of antagonistic yeast isolates studied in different formulations

Yeast name	Formulation type	References
Metschnikowia pulcherrima	Freeze-drying	Spadaro et al. (2010)
M. pulcherrima *Pichia guilliermondii*	Granular formulation	Kinay and Yildiz (2008)
Candida sake CPA-1	Dry formulation	Carbo et al. (2018)
Candida sake CPA-1	Liquid formulation Freeze-drying	Abadias et al. (2001a, b, 2003a, b); Torres et al. (2003)
Cryptococcus laurentii	Freeze-drying	Li and Tian (2006)
C. laurentii	Liquid formulation	Liu et al. (2009)
Pichia anomala	Freeze-drying	Melin et al. (2011)
P. anomala	Dry formulation	Melin et al. (2007)

long periods. Besides, the formulation of biocontrol agents should be inexpensive for mass production, have sufficient colony forming units, and be simple to prepare and implement to target host.

The formulation of biocontrol products should consist of different sugars as carbon sources, and be carried in dried formulations, adhesives, emulsifiers and adjuvants, surfactants, and some anti-stress substances. The advantageous features of yeasts are: cheap and easy cultivation for mass production such as in molasses material, short generation times, and long life in an appropriate formulation and storage conditions. The advantage of development of the formulation of antagonistic yeasts is their ability to grow very fast in solutions containing high nutrients such as sugars. The storage of yeasts is the most well-known type of formulation that has been in life for a long time in human history. The yeast antagonists are formulated as the refrigerated liquid form, frozen products, and granular formulations (Abadias et al. 2003b; Janisiewicz and Korsten 2002; Kinay and Yildiz 2008). Formulating as granule form of product may preserve it from extreme environmental conditions and allow manageable delivery of the microorganism from the formulation (Shabana et al. 2003). The most well-known is the instant dry baker's yeast (*S. cerevisiae*), which is formulated in a granular form and has a shelf life of more than one year during cold storage conditions. Formulations of biological control agents and the storage conditions have very important influence on the efficacy and shelf life of the product. The shelf life of a commercial biological control product should be as long as possible. It is expected that the formulated bioproduct should be effective for at least six months and preferentially for two years during appropriate storage conditions (Pusey 1994). If the storage and shelf-life conditions for formulation are not suitable, the viable yeast populations in the product will decline in a short period (Janisiewicz and Korsten 2002).

As commercial biocontrol products, dry and liquid formulations have been prepared for antagonists (Table 15.1). Freeze-drying formulations of antagonist yeasts were studied for different species such as *C. sake*, *C. laurentii*, and *R. glutinis* (Li et al. 2008). *W. anomalus* and *Meyerozyma guilliermondii* were formulated with different materials, such as starch, gelatin, and carnauba wax

(Lima et al. 2013). *M. pulcherrima* and *P. guilliermondii* were formulated in granule form and found effective both to prolong yeast viability for up to six months and to retain yeast effectiveness to control decay of citrus fruit (Kinay and Yildiz 2008). Liquid formulations of yeast were also studied with *C. sake* (Abadias et al. 2003a), *Rhodotorula minuta* 77 (Patiño-Vera et al. 2005), and *Pichia anomala* J121 71, 78 (Melin et al. 2007). Abadias et al. (2003a) showed that a liquid formulation of the yeast isolate was effective and lengthened viability of cells, compared to the dried formulation. It was suggested that dry formulations provide longer shelf life and greater convenience during the product storage and distribution compered to liquid formulations (Sui et al. 2015; Melin et al. 2007).

Bio-encapsulation and microencapsulation technologies provide protection to biocontrol agents from unsuitable conditions, as well as improve their stability because of controlled release from formulations. For biological control agents, microencapsulation to provide living cells with a physical barrier against the external environment has been tested on bio-insecticides and antagonistic *Trichoderma* species (Chen et al. 2013; Liu and Liu 2009). Among microencapsulation methods, spray-drying is cost effective in different industries. There have been some studies in the brewery and bakery industries, but hardly any studies on microencapsulation formulations of antagonistic yeasts (Paramera et al. 2014). To obtain high-quality bioproduct with a good bioformulation process requires a good coordination from laboratory to industry. The new molecular tools also influence the bioformulation field.

15.6 Concluding Remarks

In the public concern, the major problem is the effects of chemical fungicides on human health and environment. Therefore, biopesticides have gained importance as an alternative to synthetic chemicals in the last two decades These developments in biological control in both pre- and postharvest stages in the world have pushed most of the important pesticide companies to be incorporated with small biopesticide firms. In the near future, the biopesticides market will grow very fast and take its place in the global markets.

The genome sequencing of biocontrol agents and omics technologies provide better insights for understanding their mode of actions and effective use of them in plant disease control strategies. In future, the omics-based technologies will play an important role in formulation and commercialization of biocontrol agents. Recent advanced molecular techniques have provided also new insights for explaining the mode of action of biocontrol agents. The development of new agricultural products based on yeasts requires interdisciplinary studies. These kinds of studies require specialists from different disciplines: specialists to identify and characterize the genetics; specialists for biological activity, field trial, and for culturing microorganisms on a large scale; formulation specialists to develop and improve efficacy and survival; biosafety experts to assess potential environmental and human health issues, and for the registration process; etc. It is obvious that the high-throughput technologies such as sequencing and other omics technologies will play very important role during all the steps of discovering efficient antagonistic microorganisms and commercializing them. Thus, these technologies will also facilitate and accelerate the studies by revealing unexplained relationships between host–pathogen–antagonist. In future, the biocontrol activity of antagonistic yeasts with the knowledge provided from omics technologies could be improved by manipulation of the environment, using mixtures of antagonistic organisms and integration of biocontrol with other control methods.

References

Abadias M, Benabarre A, Teixidó N, Usall J, Viñas I (2001a) Effect of freeze drying and protectants on viability of the biocontrol yeast Candida sake. Int J Food Microbiol 65:173–182

Abadias M, Teixidó N, Usall J, Benabarre A, Viñas I (2001b) Viability, efficacy, and storage stability of freeze-dried biocontrol agent *Candida sake* using different protective and rehydration media. J Food Prot 64:856–861

Abadias M, Teixidó N, Usall J, Viñas I (2003a) Optimization of growth conditions of the postharvest biocontrol agent *Candida sake* CPA-1 in a lab-scale fermenter. J Appl Microbiol 95:301–309

Abadias M, Usall J, Teixidó N, Viñas I (2003b) Liquid formulation of the postharvest biocontrol agent *Candida sake* CPA-1 in isotonic solutions. Phytopathology 93:436–442

Abdelfattah A, Wisniewski M, Droby S, Schena L (2016) Spatial and compositional variation in the fungal communities of organic and conventionally grown apple fruit at the consumer point-of-purchase. Hortic Res 3(1):16047

Abdelfattah A, Malacrinò A, Wisniewski M, Cacciola SO, Schena L (2018) Metabarcoding: a powerful tool to investigate microbial communities and shape future plant protection strategies. Biol Control 120:1–10

Abdelfattah A, Whitehead SR, Macarisin D, Liu J, Burchard E, Freilich S, Dardick C, Droby S, Wisniewski M (2020) Effect of washing, waxing and low-temperature storage on the postharvest microbiome of apple. Microorganisms 8(6):944

Allen TW, Quayyum HA, Burpee LL, Buck JW (2004) Effect of foliar disease on the epiphytic yeast communities of creeping bentgrass and tall fescue. Can J Microbiol 50(10):853–860

Apaliya MT, Yang Q, Zhang H, Zheng X, Zhao L, Zhang X, Kwaw E, Tchabo W (2019) Proteomics profile of *Hanseniaspora uvarum* enhanced with trehalose involved in the biocontrol efficacy of grape berry. Food Chem 274:907–914

Barkai-Golan R (2001) Postharvest diseases of fruits and vegetables. In: Development and Control. Elservier B. V, Amsterdam, Netherlands

Bar-Shimon M, Yehuda H, Cohen L, Weiss B, Kobeshnikov A, Daus A, Goldway M, Wisniewski M, Droby S (2004) Characterization of extracellular lytic enzymes produced by the yeast biocontrol agent *Candida oleophila*. Curr Genet 45:140–148

Benbow JM, Sugar D (1999) Fruit surface colonization and biological control of postharvest disease of pear by preharvest yeast application. Plant Dis 83:839–844

Carbo A, Torres R, Teixido N, Usall J, Medina A, Magan N (2018) Impact of climate change environmental conditions on the resilience of different formulations of the biocontrol agent Candida sake CPA-1 on grapes. Lett Appl Microbiol 67(1):2–8

Carlile MJ, Watkinson SC, Gooday GW (2005) The fungi, 2nd ed. Elsevier Academic Press, 588 pp

Castoria R, De Currtis F, Lima G, Caputo L, Pacifico S, De Cicco V (2001) Aureobasidium pullulans (LS-30) an antagonist of postharvest pathogens of fruits: study on its modes of action. Postharvest Biol Technol 22:7–17

Castoria R, Morena V, Caputo L, Panfili G, De Curtis F, De Cicco V (2005) Effect of the biocontrol yeast *Rhodotorula. glutinis* strain LS11 on patulin accumulation in stored apples. Phytopathology 95:1271–1278

Chalutz E, Ben-Arie R, Droby S, Cohen L, Weiss B, Wilson CL (1988) Yeasts as biocontrol agents of postharvest diseases of fruit. Phytoparasitica 16:69–75

Chan Z, Qin G, Xu X, Li B, Tian S (2007) Proteome approach to characterize proteins induced by antagonist yeast and salicylic acid in peach fruit. J Proteome Res 6(5):1677–1688

Chen KN, Chen CY, Lin YC, Chen MJ (2013) Formulation of a novel antagonistic bacterium based biopesticide for fungal disease control using microencapsulation techniques. J Agric Sci 5:153–163

Cherry JM, Hong EL, Amundsen C, Balakrishnan R, Binkley G, Chan ET, Wong ED (2012) *Saccharomyces* genome database: the genomics resource of budding yeast. Nucleic Acids Res 40:D700–D705

Donlan RM, Costerton JW (2002) Biofilms: survival mechanisms of clinically relevant microorganisms. Clin Microbiol Rev 15:167–193

Droby S, Hofstein R, Wilson CL, Wisniewski M, Fridlender B, Cohen L, Weiss B, Daus A, Timar D, Chalutz E (1993) Pilot testing of *Pichia guilliermondii*: a biocontrol agent for postharvest diseases of citrus fruit. Biol Control 3:47–52

Droby S, Vinokur V, Weiss B, Cohen L, Daus A, Goldschmidt EE, Porat R (2002) Induction of resistance to *Penicillium digitatum* in grapefruit by the yeast biocontrol agent *Candida oleophila*. Phytopathology 92:393–399

Droby S, Wisniewski M, Macarisin D, Wilson CL (2009) Twenty years of postharvest biocontrol research: is it time for a new paradigm? Postharvest Biol Technol 52:137–145

Droby S, Wisniewski M, Teixidó N, Spadaro D, Jijakli MH (2016) The science, development, and commercialization of postharvest biocontrol products. Postharvest Biol Technol 122:22–29

Eckert JW, Sievert JR, Ratnayake M (1994) Reduction of imazalil effectiveness against citrus green mold in California packinghouses by resistant biotypes of Penicillium digitatum. Plant Dis 78: 971–974

FAO Report (2011) Global food losses and food waste: extent, causes and prevention. FAO, Rome. Available at http://www.fao.org/docrep/014/mb060e/mb060e.pdf

Farbo MG, Urgeghe PP, Fiori S, Marcello A, Oggiano S, Balmas V, Hassan ZU, Jaoua S, Migheli Q (2018) Effect of yeast volatile organic compounds on ochratoxin A-producing aspergillus carbonarius and A. ochraceus. Int J Food Microbiol 284:1–10

Fernandez-San Millan A, Gamir J, Larraya L, Farran I, Veramendi J (2022) Towards understanding of fungal biocontrol mechanisms of different yeasts antagonistic to *Botrytis cinerea* through exometabolomic analysis. Biol Control 2022:105033

Ferraz LP, Cunha TD, da Silva AC, Kupper KC (2016) Biocontrol ability and putative mode of action of yeasts against Geotrichum citri-aurantii in citrus fruit. Microbiol Res 188–189:72–79

Fiori S, Urgeghe PP, Hammami W, Razzu S, Jaoua S, Migheli Q (2014) Biocontrol activity of four non- and low-fermenting yeast strains against *Aspergillus carbonarius* and their ability to remove ochratoxin A from grape juice. Int J Food Microbiol 189:45–50

Firrincieli A, Otillar R, Salamov A, Schmutz J, Khan Z, Redman RS, Fleck ND, Lindquist E, Grigoriev IV, Doty SL (2015) Genome sequence of the plant growth promoting endophytic yeast *Rhodotorula graminis* WP1. Front Microbiol 6:978

Freimoser FM, Rueda-Mejia MP, Tilocca B, Migheli Q (2019) Biocontrol yeasts: mechanisms and applications. World J Microbiol Biotechnol 35:154

Giobbe S, Marceddu S, Scherm B, Zara G, Mazzarello VL, Budroni M et al (2007) The strange case of a biofilm-forming strain of *Pichia fermentans*, which controls *Monilinia* brown rot on apple but is pathogenic on peach fruit. FEMS Yeast Res 7(8):1389–1398

Glushakova AM, Chernov IY (2007) Seasonal dynamic of the numbers of epiphytic yeasts. Microbiology 76(5):590–595

Goffeau A, Barrell BG, Bussey H, Davis RW, Dujon B, Feldmann H, Galibert F, Hoheisel JD, Jacq C, Johnston M, Louis EJ, Mewes HW, Murakami Y, Philippsen P, Tettelin H, Oliver SG (1996) Life with 6000 genes. Science 274(5287):546

Grevesse C, Lepoivre P, Jijakli MH (2003) Characterization of the exoglucanase-encoding gene PaEXG2 and study of its role in the biocontrol activity of Pichia anomala strain K. Phytopathology 93:1145–1152

Guerrero V, Guigon C, Berlanga D, Ojeda D (2014) Complete control of *Penicillium expansum* on apple fruit using a combination of antagonistic yeast *Candida oleophila*. Chil J Agric Res 74(4): 427–431

Gustavsson J, Cederberg C, Sonesson U, Otterdijk R, Meybeck A (2011) Global food losses and food waste: extent causes and Food and Agriculture Organization (FAO) of the United Nations

Hawser SP, Douglas LJ (1994) Biofilm formation by Candida species on the surface of catheter materials in vitro. Infect Immun 62:915–921

Hofstein R, Fridlender B, Chalutz E, Droby S (1994) Large scale production and pilot testing of biological control agents for postharvest diseases. In: Wilson CL, Wisniewski M (eds) Biological control of postharvest diseases – theory and practice. CRC Press, Boca Raton, pp 89–100

Hua SST, Beck JJ, Sarreal SBL, Gee W (2014) The major volatile compound 2-phenylethanol from the biocontrol yeast, *Pichia anomala*, inhibits growth and expression of aflatoxin biosynthetic genes of *Aspergillus flavus*. Mycotoxin Res 30:71–78

Infante EP, Marquinez X, Moreno G (2012) Tomato peel (*Solanum lycopersicum* L.) colonization by the endophyte yeast *Candida guilliermondii* (Castellani) Langeron et Guerra. Agron Colombiana 30(3):388–394

Janisiewicz WJ, Korsten L (2002) Biological control of postharvest diseases of fruits. Annu Rev Phytopathol 40:411–441

Kinay P, Yildiz M (2008) The shelflife and effectiveness of granular formulations of *Metschnikowia pulcherrima* and *Pichia guilliermondii* yeast isolates that control postharvest decay of citrus fruit. Biol Control 45(3):433–440

Kinay P, Yildiz M, Droby S, Yildiz F, Cohen L, Weiss B (1998) Evaluation of antagonistic activity of epiphytic yeasts against rot pathogens of mandarin Orange and grapefruit. In: Duffy BK, Rosenberger U, Défago G (eds) Molecular Approches in biological control IOBC/wprs bull, vol. 21(9), 291–296

Kinay P, Yildiz M, Droby S (2002) Biocontrol of *Penicillium* decays with epiphytic yeasts on Satsuma mandarins. In: Elad Y, Köhl J, Shtienberg D (eds) Proceedings of the seventh WG meeting influence of abiotic and biotic factors on biocontrol agents. Kusadasi, Turkey, May 22–25, 2002, IOBC WPRS Bull, pp 207–210

Kinay P, Mansour MF, Mlikota- Gabler F, Margosan DA, Smilanick JL (2007) Characterization of fungicide-resistant isolates of *Penicillium digitatum* collected in California. Crop Prot 26:647–656

Köhl J, Kolnaar R, Ravensberg WJ (2019) Mode of action of microbial biological control agents against plant diseases: relevance beyond efficacy. Front Plant Sci 10:845

Kurtzman CP, Fell JW, Boekhout T (2011) *The yeasts*: a taxonomic study, 5th edn. Elsevier, Amsterdam

Kusstatscher P, Cernava T, Abdelfattah A, Gokul J, Korsten L, Berg G (2020) Microbiome approaches provide the key to biologically control postharvest pathogens and storability of fruits and vegetables. FEMS Microbiol Ecol 96(7)

Li BQ, Tian SP (2006) Effects of trehalose on stress tolerance and biocontrol efficacy of *Cryptococcus laurentii*. J Appl Microbiol 100:854–861

Li BQ, Zhou ZW, Tian SP (2008) Combined effects of endo- and exogenous trehalose on stress tolerance and biocontrol efficacy of two antagonistic yeasts. Biol Control 46:187–193

Li X, Tang H, Yang C, Meng X, Liu B (2019) Detoxification of mycotoxin patulin by the yeast Rhodotorula mucilaginosa. Food Control 96:47–52

Lima JR, Gondime MF, Oliveira TA, Oliveira FSA, Gonçalves LRB, Viana FMP (2013) Use of killer yeast in the management of postharvest papaya anthracnose. Postharvest Biol Technol 83: 58–64

Liu CP, Liu SD (2009) Formulation and characterization of the microencapsulated entomopathogenic fungus *Metarhizium anisopliae* MA12. J Microencapsul 26:377–384

Liu J, Tian SP, Li BQ, Qin GZ (2009) Enhancing viability of two biocontrol yeasts in liquid formulation by applying sugar protectant combined with antioxidant. Biol Control 54:817–824

Liu J, Wisniewski M, Droby S, Norelli J, Hershkovitz V, Tian SP (2011) Increase in antioxidant gene transcripts, stress tolerance and biocontrol efficacy of *Candida oleophila* following sublethal oxidative stress exposure. FEMS Microbiol Ecol 80:578–590

Liu J, Wisniewski M, Droby S, Tian S, Hershkovitz V, Tworkoski T (2012) Effect of heat shock treatment on stress tolerance and biocontrol efficacy of *Metschnikowia fructicola*. FEMS Microbiol Ecol 76:145–155

Liu J, Wisniewski M, Artilip T, Sui Y, Droby S, Norelli J (2013) The potential role of *PR-8* gene of apple fruit in the mode of action of the yeast antagonist, *Candida oleophila*, in postharvest biocontrol of *Botrytis cinerea*. Postharvest Biol Technol 85:203–209

Long CA, Deng BX, Deng XX (2006) Pilot testing of *Kloeckera apiculata* for the biological control of postharvest diseases of citrus. Ann Microbiol 56(1):13–17

Massart S, Perazzolli M, Höfte M, Pertot I, Jijakli MH (2015) Impact of the omic technologies for understanding the modes of action of biological control agents against plant pathogens. BioControl 60:725–746. https://doi.org/10.1007/s10526-015-9686-z

Melin P, Håkansson S, Schnürer J (2007) Optimisation and comparison of liquid and dry formulations of the biocontrol yeast *Pichia anomala* J121. Appl Microbiol Biotechnol 73:1008–1016

Melin P, Schnürer J, Hakansson S (2011) Formulation and stabilisation of the biocontrol yeast Pichia anomala. Antonie Van Leeuwenhoek 99:107–112

Mercier J, Wilson CL (1994) Colonization of apple wounds by naturally occurring microflora and introduced *Candida oleophila* and their effect on infection by Botrytis cinerea during storage. Biol Control 4:138–144

Moore D, Robson GD, Trinci APJ (2011) 21st century guidebook to fungi, 2nd ed, Cambridge University Press, UK, 640 pages

Nevitt T (2011) War-Fe-re: iron at the core of fungal virulence and host immunity. Biometals 24(3): 547–558

Nunes C (2012) Biological control of postharvest diseases of fruit. Eur J Plant Pathol 133:181–196

Oelofse D, Dubery I, Berger DK (2009) Exo-ß-1,3-glucanase from yeast inhibits *Colletotrichum lupini* and *Botrytis cinereaspore* germination. J Phytopathol 157:1–6

Palou L, Smilanick JL, Droby S (2008) Alternatives to conventional fungicides for the control of citrus postharvest green and blue moulds. Stewart Posthar Rev 4:1–16

Parafati L, Vitale A, Restuccia C, Cirvilleri G (2017) Performance evaluation of volatile organic compounds by antagonistic yeasts immobilized on hydrogel spheres against gray, green and blue postharvest decays. Food Microbiol 63:191–198

Paramera EI, Karathanos VT, Konteles SJ (2014) Chapter 23: Yeast cells and yeast-based materials for microencapsulation. In: Microencapsulation in the food industry a practical implementation guide, 267–281

Patiño-Vera M, Jiménez B, Balderas K, Ortiz M, Allende R, Carrillo A, Galindo E (2005) Pilot-scale production and liquid formulation of *Rhodotorula minuta*, a potential biocontrol agent of mango anthracnose. J Appl Microbiol 99(3):540–550

Perez MF, Contreras L, Garnica NM, FernaÂndez-Zenoff MV, FarõÂas ME, Sepulveda M et al (2016) Native killer yeasts as biocontrol agents of postharvest fungal diseases in lemons. PLoS One 11(10):e0165590

Petersson S, Schnürer J (1995) Biocontrol of mold growth in high-moisture wheat stored under airtight conditions by *Pichia anomala, Pichia guilliermondii,* and *Saccharomyces cerevisiae*. Appl Environ Microbiol 61:1027–1032

Piombo E, Sela N, Wisniewski M, Hoffmann M, Gullino ML, Allard MW et al. (2018) Genome sequence, assembly and characterization of two *Metschnikowia fructicola* strains used as biocontrol agents of postharvest diseases

Prasannath K (2017) Plant defense-related enzymes against pathogens: a review. AGRIEAST J Agric Sci 11(1):38–48

Pretscher J, Fischkal T, Branscheidt S, Jäger L, Kahl S, Schlander M, Thines E, Claus H (2018) Yeasts from different habitats and their potential as biocontrol agents. Fermentation 4(31):1–17

Pusey PL (1994) Enhancement of biocontrol agents for postharvest diseases and their integration with other control strategies. In: Wilson CL, Wisniewski ME (eds) Biological control of postharvest diseases, theory and practice. CRC Press, Boca Raton, WV, USA, pp 77–88

Ray RC, Swain MR, Panda Lata SH (2011) Microbial control of postharvest diseases of fruits, vegetables, roots, and tubers. In: Parmar SA, N. Kuhad R. (eds) Bioaugmentation, biostimulation and biocontrol. Soil biology, vol 108. Springer, Berlin, Heidelberg, pp 311–355

Reddy KRN, Spadaro D, Gullino ML, Garibaldi A (2011) Potential of two *Metschnikowia pulcherrima* (yeast) strains for in vitro biodegradation of patulin. J Food Prot 74(1):154–156

Reynolds TB, Fink GR (2001) 'Bakers' yeast, a model for fungal biofilm formation. Science 291(5505):878–881

Romanazzi G, Sanzani SM, Bi Y, Tian S, Martínez PG, Alkan N (2016) Induced resistance to control postharvest decay of fruit and vegetables. Postharvest Biol Technol 122:82–94

Romera FJ, García MJ, Lucena C, Martínez-Medina A, Aparicio MA, Ramos J, Alcántara E, Angulo M, Pérez-Vicente R (2019) Induced systemic resistance (ISR) and Fe deficiency responses in dicot plants. Front Plant Sci 10:287

Santos A, Marquina D (2004) Killer toxin of *Pichia membranifaciens* and its possible use as a biocontrol agent against grey mould disease of grapevine. Microbiology (Reading) 150:2527–2534

Saravanakumar D, Ciavorella A, Spadaro D, Garibaldi A, Gullino ML (2008) *Metschnikowia pulcherrima* strain MACH-1 outcompetes *Botrytis cinerea, Alternaria alternata* and *Penicillium expansum* in apples through iron depletion. Postharvest Biol Technol 49(1):121–128

Schisler DA, Kurtzman CP, Bothast RJ, Slininger PJ (1995) Evaluation of yeasts for biological control of fusarium dry rot of potatoes. Am Potato J 72:339–353

Schoch CL, Seifert KA, Huhndorf S, Robert V, Spouge JL, Levesque CA et al (2012) Nuclear ribosomal internal transcribed spacer (ITS) region as a universal DNA barcode marker for fungi. Proc Natl Acad Sci USA 109

Shabana YM, Stover DM, Sauerborn J (2003) Granular Pesta formulation of *Fusarium oxysporum* f. sp. *orthoceras* for biological control of sunflower broomrape: efficacy and shelf-life. Biol Control 26(2):189–201

Sharma RR, Singh D, Singh R (2009) Biological control of postharvest diseases of fruits and vegetables by microbial antagonists: a review. Biol Control 50:205–221

Sharma AVD, Diwevidi S, Singh KK, Pawar M, Jerman LB, Singh S, Singh Srivastawa D (2013) Biological control and its important in agriculture. Int Jour of Biotech Bioeng Res 4(3):175–180

Smilanick JL, Mansour MF, Mlikota- Gabler F, Goodwine WR (2006) The effectiveness of pyrimethanil to inhibit germination of *Penicillium digitatum* and to control citrus green mold after harvest. Postharvest Biol Technol:75–85

Snowdon, A.L. 1990. A Color Atlas Postharvest Dieseases & Introduction & Fruits. Wolfe Scientific, 302 pages

Spadaro D, Droby S (2016) Development of biocontrol products for postharvest diseases of fruit: the importance of elucidating the mechanisms of action of yeast antagonists. Trends Food Sci Technol 47:39–49

Spadaro D, Ciavorella A, Lopez G, Garibaldi A, Gullino ML (2010) Effect of culture age, protectants, and initial cell concentration on viability of freeze-dried cells of *Metschnikowia pulcherrima*. Can J Microbiol 56:809–815

Starmer WT, Lachance M (2011) A yeast ecology. In: Kurtzman CP, Fell JW, Boekhout T (eds) The yeasts – a taxonomic study, 5th edn. Elsevier, Amsterdam, pp 65–86

Sui Y, Wisniewski M, Droby S, Liu L (2015) Responses of yeast biocontrol agents to environmental stress. Appl Environ Microbiol 81(9):2968–2975

Sui Y, Wisniewski M, Droby S, Piombo E, Wu X, Yue J (2020) Genome sequence, assembly, and characterization of the antagonistic yeast *Candida oleophila* used as a biocontrol agent against post-harvest diseases. Front Microbiol 11:295

Symeonidis A, Marangos M (2012) Iron and microbial growth. In: Roy PK (ed) Insight and control of infectious disease in global scenario. InTech, Rijeka, Croatia, pp 289–332

Teixidó N, Usall J, Gutiérrez O, Viñas I (1998) Effect of the antagonist Candida sake on apple surface microflora during cold and ambient (shelf life) storage. Eur J Plant Pathol 104:387–398

Tian SP, Wan YK, Qin GZ, Xu Y (2006) Induction of defense responses against Alternaria rot by different elicitors in harvested pear fruit. Appl Microbiol Biotechnol 70:729–734

Tilocca B, Balmas V, Hassan ZU, Jaoua S, Migheli Q (2019) A proteomic investigation of *Aspergillus carbonarius* exposed to yeast volatilome or to its major component 2-phenylethanol reveals major shifts in fungal metabolism. Int J Food Microbiol 306:108265

Torres R, Usall J, Teixidó N, Abadias M, Viñas I (2003) Liquid formulation of the biocontrol agent *Candida sake* by modifying water activity or adding protectants. J Appl Microbiol 94:330–339

Usall J, Teixido N, Torres R, Ochoa de Eribe X, Vinas I (2001) Pilot tests of *Candida sake* (CPA-1) applications to control postharvest blue mold on apple fruit. Postharvest Biol Technol 21(2): 147–156

Vila TVM, Rozental S (2016) Biofilm formation as a pathogenicity factor of medically important fungi. In: Sultan S (ed) Fungal pathogenicity. InTechOpen, London, UK, pp 1–24

Wassermann B, Abdelfattah A, Cernava T, Wicaksono W, Berg G (2022) Microbiome-based biotechnology for reducing food loss post harvest. Curr Opin Biotechnol 78:102808

Wilson CL, Pusey PL (1985) Potential for biological control of postharvest plant diseases. Plant Dis 69:375–378

Wilson CL, Wisniewski M (1989) Biological control of postharvest diseases of fruits and vegetables: an emerging technology. Annu Rev Phytopathol 27:425–441

Wilson CL, Wisniewski M (1994) Biological control of postharvest diseases: theory and practice. CRC Press, Boca Raton, FL, p 182

Wisniewski M, Wilson CL, Droby S, Chalutz E, El Ghaouth A, Stevens C (1991) Postharvest biocontrol: the discovery of new concepts and applications. In: Vincent C, Goettal M, Lazarovits G (eds) Biological control: international case studies, vol 440. CABI Publishing, UK, pp 262–273

Wisniewski M, Droby S, Norelli J, Liu J, Schena L (2016) Alternative management technologies for postharvest disease control: the journey from simplicity to complexity. Postharvest Biol Technol 122:3–10

Xu XB, Qin G, Tian SP (2008) Effect of microbial biocontrol agents on alleviating oxidative damage of peach fruit subjected to fungal pathogen. Int J Food Microbiol 126:153–158

Yao HJ, Tian SP (2005) Effects of a biocontrol agent and methyl jasmonate on postharvest diseases of peach fruit and the possible mechanisms involved. J Appl Microbiol 98:941–950

Zhang D, Spadaro D, Garibaldi A, Gullino ML (2011) Potential biocontrol activity of a strain of Pichia guilliermondii against grey mold of apples and its possible modes of action. Biol Control 57(3):193–201

Zhao L, Wang Y, Wang Y, Li B, Gu X, Zhang X, Boateng NAS, Zhang H (2020) Effect of β-glucan on the biocontrol efficacy of Cryptococcus podzolicus against postharvest decay of pears and the possible mechanisms involved. Postharvest Biol Technol 160:111057

Zhu R, Feussner K, Wu T, Yan F, Karlovsky P, Zheng X (2015) Detoxification of mycotoxin patulin by the yeast Rhodosporidium paludigenum. Food Chem 179:1–5

Chapter 16
The Potential Application of Entomopathogenic Fungi (EF) in Insect Pest Management

Manisha Mishra

Abstract Biological control based on entomopathogenic fungi (EF) is an environmentally safe and promising substitute for chemical pesticides in agriculture. EF range from extremely pathogenic forms to naturally occurring ones and are often present in the soil, plant rhizosphere, or sometimes as endophytes of plants. By the diverse niches they occupy, biochemistry of the insect cuticle, and enzymes and toxins they release, these EF-based contact bioinsecticides are known to control agricultural pests, mosquitoes, termites, etc. The EF attack the insect and the complexity of attacks and counterattacks depends upon the communication between the attacking EF and the insect host, which leads to a co-evolutionary arms race between the fungus and the insect. It has been well understood that lipids present in the epicuticle are important in the stress and virulence of EF. Further, to generate novel strains of EF with enhanced virulence and improved resistance to stress, genetic engineering is employed. Nowadays, with the percolation of nanotechnology in agriculture, EF-based myconanopesticides can be fabricated. EF are incorporated into formulations and marketed after getting registered as mycoinsecticides, and they can be used either alone or in combination with plant extracts to achieve synergistic pest control.

Keywords Entomopathogenic fungi · Biological control · Virulence · Genetic engineering · Formulations · Nanotechnology

Highlights

- The entomopathogenic fungi (EF) can be used for the biological control of insect pests.
- Understanding the molecular mechanism of biocontrol action of EF can help in enhancing virulence and stress tolerance.

M. Mishra (✉)
University Department of Botany, T.M. Bhagalpur University, Bhagalpur, India

K. K. Bastas et al. (eds.), *Microbial Biocontrol: Molecular Perspective in Plant Disease Management*, Microorganisms for Sustainability 49,
https://doi.org/10.1007/978-981-99-3947-3_16

- Success is attributed to the development of a co-evolutionary arms race between the EF and the host.
- Once mass-produced, the EF is developed into suitable formulations.

16.1 Introduction

In the current decades food security for rising global population is one of the most severe issues. However, the losses caused by the attack of insect or pest during pre- or postharvest storage conditions accelerate this challenge. According to a published report approximately 20% of the grain is lost each year during storage in the developing countries, due to the attack of pest or insects (Phillips and Throne 2010). To combat the pathogenic attack caused by insect pests, various traditional methods such as temperature and humidity regulation (Lord 2010), disinfection of storage site with fumigant (Oppert et al. 2015), application of plant extracts in the storage sites (Athanassiou et al. 2014), etc. have been employed; however, apart from these methods, application of chemical insecticides or pesticides is the most utilized (Arthur and Campbell 2008; Kumar 2022). The detrimental impact of chemical pesticides or insecticides on the consumer food chain, health hazards caused due to their toxic nature, and the emergence of pests resistant to chemicals have shifted the attention toward an alternative that must be nontoxic, economic, and ecofriendly (Rossi et al. 2010; Edde 2012). In this context, biological control using microbial antagonists is an emerging approach to control the pest or insect during storage conditions. In the recent past, entomopathogenic fungi (EF) are finding immense applications to control various insect pests or termites during harvest or postharvest storage (Rath 2000; Sindhu et al. 2011; Batta and Kavallieratos 2018).

EF serve a crucial part in the management of insect pests as they affect non-feeding stages such as eggs and pupae of arthropod hosts and therefore, many of them hold substantial promise as mycoinsecticides that are environmentally compatible and economical. In addition, they also minimize too much dependence on chemical pesticides and can help in safeguarding nature and sustaining food security. EF display prodigal promise for their potential use as bioinsecticides in integrated pest management (IPM) (Sindhu et al. 2011). In the market EF-based formulations are available that find immense applications; for example, the formulations of *Metarhizium anisopliae* can be used to control mosquito vectors like *Aedes aegypti* (Rodrigues et al. 2019) and EF belonging to the class Hyphomycetes can protect ornamental crops and vegetables from the aggressive insect pest *Bemisia tabaci* (Sani et al. 2020).

EF like *Beauveria bassiana* and *Metarhizium anisopliae* find widespread use as biological control agents and they belong to the order Hypocreales of phylum Ascomycota (Wang et al. 2016). EF are pathogenic toward insect hosts and include some aggressively pathogenic forms like *Metarhizium anisopliae* and opportunists like *Mucor hiemalis* (wound pathogen). However, various individual isolates show considerable host specificity; for example, *M. anisopliae var majus* specifically infects Scarabid beetles, while *M. anisopliae var acridum* is specific for grasshoppers and locusts (Driver et al. 2000). In a study, Opisa et al. (2018) reported *M. anisopliae* was most efficient among *Metarhizium anisopliae*, *Beauveria bassiana*, and *Isaria fumosorosea* in controlling pathogen *Spoladea recurvalis* that severely affects Amaranths. The present review article attempts to pursue recent development in entomopathogenic fungi-based biological control including the role nanotechnology can afford in enhancing their effectiveness as biological control agents.

16.2 Action Mechanism of Entomopathogenic Fungi (EF)

The mechanism of action of EF begins with the attachment of EF to the insect cuticle and consummates in the death of the insect with several steps in between as detailed below.

16.2.1 Attachment of EF to the Insect Cuticle

Pest control using EF commences with the attachment on the pest surface or cuticle via propagules through electrostatic or nonspecific hydrophobic interactions. The compounds present in the host cuticle either stimulate or inhibit the conidial germination of EF (Pedrini and Juárez 2008). Besides this, moisture present in the surroundings facilitates the penetration of EF through the folds between the segmented body parts and openings on the insect body (Clarkson and Charnley 1996). A successful attack by EF depends on its ability to avoid the hosts' defense mechanism and it does so by the release of toxins during their passage through a parasitic and a saprotrophic phase in their life cycle (Gillespie et al. 2000; Freimoser et al. 2003a). However, during initial infection by EF, the cuticle is not consumed and it gets finally assimilated during growth and sporulation on the cadaver. The overview of how the EF attack on the pest/insect and their mechanism are presented in Fig. 16.1.

EF surface interaction works to trigger the hosts' innate immune system upon successful attack, in which the cellular immunity leads to phagocytosis, hemocyte aggregation, and pathogen encapsulation, whereas humoral immunity induces the synthesis of antimicrobial peptides, lectins, and prophenoloxidase that imparts resistance against host defense (immunosuppressive agent) and facilitates fungal invasion (Pedrini 2018). Juvenile hormone and dopamine are two well-known insect hormones that elicit an immune response in the insect upon attack by EF. The proteins possessed by the insect display multiple properties of recognition of pathogens, homeostasis, and metabolic portrayal. The cuticle is breached twice: (a) when the EF ingresses into the insect host, and (b) when the fungus exits the dead host.

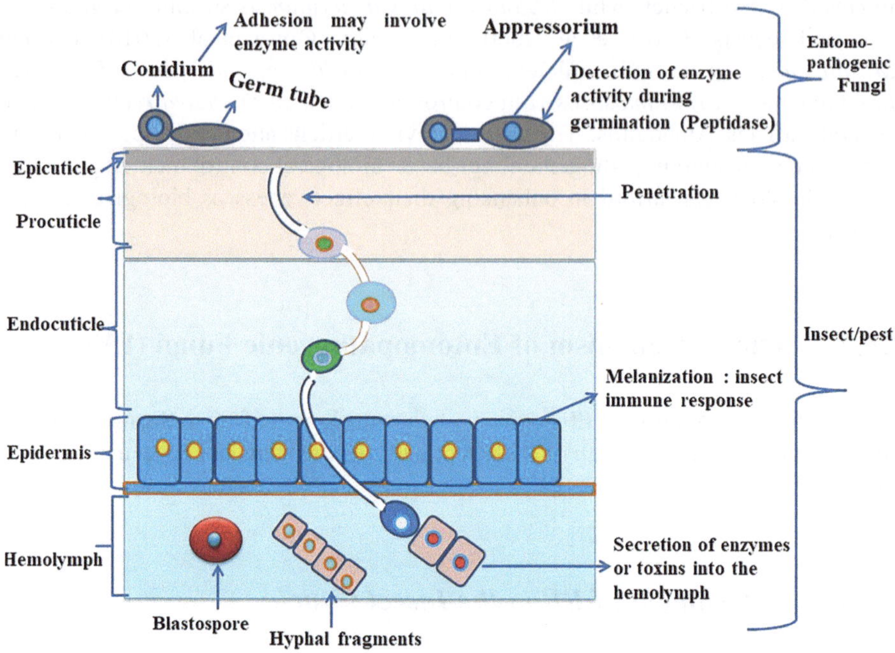

Fig. 16.1 An overview of EF attack on the pest/insect surface and their mechanism of action

It is well known that the insects' immune system works to alleviate stress and improves resistance because of its detoxification and antioxidant systems that work in a synergistic manner (Dubovskiy 2021). Although some EF kill the insect due to the release of toxins, however, especially those belonging to Entomophthorales do not release toxins and kill the insect hosts by the colonization of the tissue (Charnley 2003; Freimoser et al. 2003b). Subsequent events in infection involve melanization of hemolymph and EF form narrow penetration pegs (to prevent dehydration of hosts) and blastopores (for quick assimilation of nutrients of hemolymph) to reduce injury. On the other hand, to disgust, immobilize, or kill EF, the insect produces lectins and phenoloxidase and also recruits hemocytes. The involvement of transferrin and apolipoprotein III is also implicated in the metabolism, homeostasis, and pathogen recognition (Butt et al. 2016).

16.2.2 Penetration of EF Through the Insect Cuticle

The insect cuticle is made up of an outer layer called the epicuticle and lying below it is the procuticle. It is essential to understand the biochemistry involved in the penetration of EF.

16.2.2.1 Penetration Through the Epicuticle

The epicuticle of the insect is composed of hydrocarbons and for the penetration of EF, it needs to be hydrolyzed enzymatically. Hydrocarbons are present in the epicuticle and being hydrophobic allows the attachment of fungal conidia, which is hydrophobic too. The penetrability of the cuticle depends on the type of hydrocarbon and it has been reported that the predominance of saturated hydrocarbons (straight and branched chains) in the epicuticle enhances its penetrability whereas unsaturated hydrocarbons (alkenes or alkadienes) make it less penetrable (Pedrini et al. 2007; Pedrini 2018). The cuticle can be breached by several hydroxylating enzymes secreted by EF (Zhang et al. 2012; Huarte-Bonnet et al. 2018; Keyhani 2018). Evison et al. (2017) have suggested that fungal resistance may be ascribed to the tyrosine content in the cuticle, its thickness, porosity, and biochemistry. The cuticular color varies and, as noted in *Tenebrio molitor*, the color variation is due to limiting levels of the aromatic amino acid tyrosine.

16.2.2.2 Penetration Through the Procuticle

The procuticle is the next layer to be breached and it is composed of protein and chitin. The enzymes involved in its degradation include proteases, peptidases, and chitinases. Genes encoding these enzymes like Pr1 (subtilisin-like protease) and chitinase can be overexpressed either alone or as a fusion protein (protease and chitinase) and lead to the enhancement of virulence of EF. Interestingly, Pr1 is also used as a virulence marker in *Metarhizium* sp. (Pedrini 2018).

The infection caused due to EF attack subsequently kills insects; however, the insect also deploys mechanisms to combat attack and protect itself. Due to multiple interactions taking place on the insect cuticle, co-evolutionary arms race take place as the insect is determined to cause infection (virulence) and the host makes every possible effort in its defense; therefore, a successful infection leads to mycoses, while the victory of the insect announces its defense (Ortiz-Urquiza and Keyhani 2013). The EF secrete secondary metabolites, lytic enzymes, and adhesins that ultimately allow it to enter the hemocoel cavity of the insect (Litwin et al. 2020).

16.2.3 Introgression into the Hemocoel Cavity

The production and release of several secondary metabolites have been implicated in EF, which play a role during penetration into the hemocoel cavity and also act as an immunosuppressant (Rohlfs and Churchill 2011; Trienens and Rohlfs 2012). *Beauveria bassiana* produces secondary metabolites like bassianolide, beauvericin, beauverolides, and oosporein whereas *Metarhizium* is known better to produce destruxins. Fungal pathogenesis and virulence are enhanced due to numerous mycotoxins released by EF (Qasim et al. 2020). *Metarhizium brunneum* secretes destruxins often in contact with the dead tissue and protects the host cadaver from competitive microbial degradation. This observation was more in conformity with Skrobek et al. (2008) who reported that destruxins reach a peak during insect death. It was later found that novel destruxins could be produced on artificial media but not on insects (Xu et al. 2016). In sharp contrast, de Bekker et al. (2013), utilizing the metabolomics approach, showed that *B. bassiana* secreted 80% of the beauverolides in living tissues and therefore played a role in fungal virulence. Süssmuth et al. (2011) found that gene clusters like NRPS (non-ribosomal peptide synthetases) and PKS (polyketides synthetases) and hybrid NRPS-PKS genes were responsible for secondary metabolite production. However, for several gene clusters no secondary metabolite had been ascribed and vice versa (Gibson et al. 2014; Liu et al. 2017). Although genes responsible for the biosynthesis of bassiacridin (a protein toxin released by *B. bassiana*) are not known, they have been isolated, partially purified, and their characterization as virulence factors has been done (Pedrini 2018).

16.2.4 Antifungal Carbonyl Compounds Deposited on the Insect Cuticle

The release of various antifungal carbonyl compounds by the insects deters EF because of their irritating and repulsive properties. At least one compound with a carbonyl functional group is reported in the secretions of some insect pests that are resistant to EF. For instance, the growth of *M. anisopliae* is inhibited by the release of (E)-2-hexenal, (E)-2-octenal, and (E)-2-decenal secreted by rice stalk stink bug (da Silva et al. 2015). In *B. bassiana*, virulence against red flower beetle *Tribolium castaneum* was enhanced due to overexpression of 1,4-benzoquinone reductase but not against other non-quinone secreting beetles (Pedrini et al. 2015). Figure 16.2 highlights the development of adaptive responses in the insect and EF leading to the co-evolutionary arms race.

16.2.5 Lipids in Stress and Virulence of EF

The insect cuticle contains lipids that serve as energy reserves and are important for successful mycosis as these lipids can facilitate or impede EF growth and penetration (Ortiz-Urquiza and Keyhani 2013). Besides these, insect cuticle has been known to display antifungal effects in several insects (Golebiowski et al. 2014a, b, 2015). It has been revealed from studies that melanization of the cuticle is important in defense; for instance, in *B. bassiana*, when TEP-1 and CLIPA-8 (melanization regulators) undergo silencing, they make mosquitoes more susceptible to fungal infection (Yassine et al. 2012). However, other factors like thickened cuticle, the

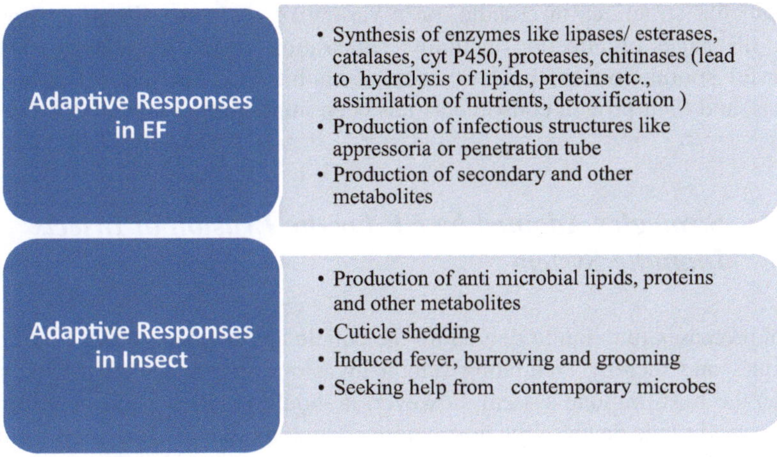

Fig. 16.2 Co-evolutionary arms race: development of adaptive responses in the insect and EF

elevated number of hemocytes, increased phenoloxidase activity, increased ability to encapsulate fungal cells, higher expression of immunity, and stress-related genes were reported in hyper melanized forms ("dark morph") in *Galleria mellonella* strain (Dubovskiy et al. 2013). It has also been reported that the insect molting fluid protein and periodical ecdysis prevented conidial germination and penetration of *B. bassiana* into the hemocoel of pharate instar insects (Zhang et al. 2014). Although studies are challenging due to the existence of diverse lipase enzymes in fungal species, extracellular lipases are known to play role in virulence and/or stress response despite the revelation that expression of several lipases has been confirmed by transcriptomics studies in *B. bassiana* and *M. anisopliae* (Mantilla et al. 2012).

Fungal virulence responses also involve phospholipases that bring about hydrolysis of specific ester bonds in phospholipid substrates, and depending on the type of ester bond they hydrolyze they are classified into A, B, C, and D types. To substantiate this concept, the involvement of phospholipase C during initial infection has been reported in the conidia of *M. anisopliae* (Santi et al. 2010). Various insects have evolved different strategies for overcoming different insect pathogens, e.g., expressed sequence tag (EST) analysis showed the expression of several phospholipases in *M. anisopliae* but not in *Conidiobolus coronatus* growing on insect cuticles (Freimoser et al. 2003a,b). Besides phospholipases, the characterization of perilipins (the multiprotein family that sequesters lipids) and caleosins (calcium-binding proteins associated with lipid bodies) has been carried out in the plant and animal systems. Perilipins are known to restrict access to cytosolic lipases and maintain lipid storage (LD). Loss of MPL-1, a single perilipin ortholog, led to a reduction in LD content, changes in cellular lipid composition, and morphological abnormalities in the germling. Closely embedded in the LD phospholipid monolayer by the proline knot are caleosins, which resemble a hairpin structure. In *B. bassiana*, a single gene ortholog has been characterized and its loss leads to the formation of multilamellar structures in conidia, on a variety of substrates including lipids. In almost all stages of infection including penetration of EF, immune evasion, and successful sporulation on the host cadaver, mobilization of internal lipids, lipid droplets, and their protein constituents has been implicated (Keyhani 2018).

16.2.6 Strategies Adopted by EF for the Evasion of Insects' Immune System

Fungal proteases function in degrading the cuticle and breaking down host defense molecules and thereby facilitating fungal invasion (Vilcinskas 2010). EF should weaken the host immune system; however, it should be neither too harmful to the host nor lead to the death of the host before enough biomass is generated for further infection. EF can bring modifications in its epitope to avoid hemocyte encapsulation or downregulate protease activities to escape recognition by the host immune system. An elevated protease level leads to the enhancement of melanin due to stimulation of the prophenoloxidase cascade and is lethal for both the insect and the EF (Butt et al. 2016; Wang and Wang 2017). Various mechanisms are involved to overcome stressors like high temperature, ultraviolet (UV) rays, visible radiations, oxidative stress, secreted antifungal compounds, etc., and genes related to oxidative stress, secondary metabolite synthesis, and heat shock proteins are known to be activated (Pedrini 2018).

16.3 Epigenetic Regulation

Host response to fungal proteinases that act as virulence factors is epigenetically regulated by histone acetylation and deacetylation; however, the steadiness of response depends on EF's virulence and it synthesizes chymotrypsin-like proteinases and metalloproteinases to degrade defense molecules produced by the host. To neutralize the fungal proteinases, the insect host synthesizes antifungal peptides. It has been suggested by Liquid chromatography-mass spectrometry (LC/MS) and reverse transcription-polymerase chain reaction (RT-PCR) analysis that a correlation exists between host-derived defense molecules and histone acetylation/deacetylation (Mukherjee and Vilcinskas 2018). Pathogenesis and disease development are due to tight conidial attachment to the insect cuticle, and co-evolution of insect and the EF has made possible the selection of appropriate virulence-related proteinases (Hussain 2018).

16.4 Dual RNA-Seq Technique for Further Research on EF and Their Interaction with the Insect Host

For a more thorough understanding of host–pathogen interaction from a molecular standpoint, dual RNA-seq technique is applied (Westermann et al. 2012). This technique can help in simultaneously capturing coding and non-coding transcripts of all kinds of RNA in both the EF and insect, followed by mapping these against reference genomes. The technique was first applied by Chu et al. (2016) to study the interaction between *B. bassiana* and the diamond black moth. Fewer genes were expressed in the infected insects as compared to the non-infected insects after 24 hours of infection. Gene expression doubled at 48 hours after infection and it remained almost the same at 36 hours. Out of the two splicing variants of autophagy-related genes that were expressed, only one was important to mitigate oxidative stress during the growth of EF in hemolymph and thus facilitate virulence of EF. Alternative splicing was cited as the reason for the adaptation of EF to host niches (Pedrini 2018).

16.5 Genetic Engineering of EF

EF can be genetically engineered to enhance its virulence and improve its ability to mitigate abiotic stress and this becomes important to address as reduced virulence and inconsistencies in performance limit the market share of fungal insecticides. Genetic engineering can create novel EF-based bioinsecticides that are environment-friendly economical options with enhanced persistence in contact with the pests and improved infection rates, even with reduced conidial dosage (Zhao et al. 2016). The following genetic engineering strategies may be adopted for enhancing the virulence and its ability to enhance stress.

16.5.1 Genetic Engineering of EF for Virulence Enhancement

Pathogenesis-related genes are tightly regulated and genes from EF that confer pathogenicity and specificity can be identified and transferred from one EF to another for enhancing virulence to create genetically engineered strains (Fang et al. 2009). The following table (Table 16.1) provides a list of genes from EF used to enhance virulence.

Besides this, virulent strains of EF can be generated by utilizing genes responsible for processes such as sterol homeostasis, osmotic balance, food digestion, immunity, and the neural system of the insect (Zhao et al. 2016). When Mr-NPC2a (sterol carrier gene) was transferred from the insects to *B. bassiana*, the engineered EF

Table 16.1 List of genes from EF used to enhance virulence

Pathogenesis-related gene	Source	Target insect	Effect	References
Pr1A (subtilisin-like protease)	*Metarhizium anisopliae*	*Manduca sexta*	Enhanced virulence	St. Leger et al. (1996)
Pr1A(subtilisin-like protease)	*Metarhizium anisopliae*	*Beauveria bassiana*	Enhanced killing rate	Gongora (2004)
CHIT1 (chitinase) overexpression	*B. bassiana*	*B. bassiana*	Enhanced virulence	Fang et al. (2005)
ATM1 (Trehalase)	*Metarhizium acridum*	*M. acridium*	Enhanced growth in hemolymph; requirement of lesser conidial dosage	Zhao et al. (2016), Peng et al. (2015)

possessed enhanced virulence as noted by the disruption of the insects' normal physiology (Zhao et al. 2014, 2016). Several insect proteins like MSDH (*Manduca sexta* diuretic hormone, which regulates insect saltwater balance), when expressed as a transgene, enhanced virulence in various insects such as *M. sexta*, *Galleria mellonella*, and *Anopheles aegypti* (Fan et al. 2012). EF can be engineered so that a lowered conidial dosage may be sufficient to kill the insect pests; for example, when AaIT1, an insect toxin that is a sodium channel blocker, was inserted into *M. anisopliae* strain ARSEF 549, a 22% lower conidial dose was required to kill *M. sexta pests* (Wang and St Leger 2007a, 2007b; Pava-Ripoll et al. 2008). Similarly, it has been reported that in *Lecanicillium lecanii*, expression of toxin BmKit from *Buthus martensii* led to similar mortality rates as the wild type *Aphis gossypii* even at a conidial dose that was 7.1 times lesser compared to the wild type (Xie et al. 2015). Vip3A proteins, produced by entomopathogenic bacteria, are insecticidal and through a series of reactions involving proteolytic activation and binding to the epithelial membrane of the midgut lead to pore formation in the gut of *Lepidopteran* insects. When the Vip3A gene was used to transform *B. bassiana*, the transformed EF were lethal even when ingested (Qin et al. 2010; Zhao et al. 2016; Chakroun et al. 2016). In addition to genetic engineering, Protein engineering can help in constructing fusion proteins; for example, in *B. bassiana*, the expression of fusion protein CDEP1: Bbchit1 (former, coding for protease; and latter, coding for chitinase) accelerated cuticular penetration as compared to the wild type strain or transformants that expressed each gene singly (Fang et al. 2009). The fusion protein was constructed using wild type Bbchit1 and chitin-binding domains (bacteria, insects, or plants), which upon fusion showed high chitin-binding ability; for example, Bbchit1 of *B. bassiana* and chitin-binding domain of silkworm (*Bombyx mori*) after fusion showed enhanced ability (Fan et al. 2007; Zhao et al. 2016). Similarly, when *B. bassiana* containing a hybrid chitinase gene was expressed, a reduction in the death time of insect hosts by 23% was noted as compared to wild type *B. bassiana* (Zhao et al. 2016).

16.5.2 Genetic Engineering of EF to Combat Stress

Various stress responses like osmotic stress, temperature, UV radiation and oxidative stress (UV and heat stress), and production of reactive oxygen species (ROS) also inhibit successful EF invasion, although the EF produce new propagules like microsclerotia (which are understood to be compact, melanized aggregates) and pellets of EF having the capability to withstand desiccation and drying (for further infection) (Huarte-Bonnet et al. 2019). It has been understood that the incorporation of UV protectants and heat stabilizers improves the stability of the fungal spores in the formulation (Behle et al. 2009). Besides these, pigments on the conidia also prevent UV radiation-induced damage and by expressing the tyrosinase gene from *Aspergillus fumigatus*, *B. bassiana* showed enhanced tolerance to UV light (Shang et al. 2012). In *M. anisopliae*, fungal virulence and resistance to UV radiation were improved by the incorporation of melanin synthesis gene cluster polyketide synthases (PKS) from a plant pathogen (Tseng et al. 2011). It has been reported that the overexpression of DNA repair photolyase in *Metarhizium* or *Beauveria* improved resistance to solar radiation (Fang and St. Leger 2012). *Metarhizium robertsii*, genetically engineered with Hsp25 (heat shock protein-encoding gene), exhibited resistance to osmotic and oxidative stress and survival under extreme temperatures (Liao et al. 2014). The involvement of the laccase gene in imparting rigidity to the appressoria wall is hinted and therefore its role in conferring pathogenicity can be believed (Zhao et al. 2016).

16.6 Application of EF Against Phytopathogens

EF can also be present as plant endophytes or in the rhizosphere and can behave as antagonists to phytopathogens (Mantzoukas and Eliopoulos 2020), and it has been proposed that such multitrophic interactions affect the plant in numerous useful ways; for example, *Metarhizium anisopliae* can help in the passage of nitrogen from the larva of *Galleria mellonella* to plants and similarly from *Laccaria bicolor* (*ectomycorrhiza*) to white pine (Behie et al. 2012, Malacrinò 2018) and also act as plant growth promoters (Mantzoukas and Eliopoulos 2020). EF do not harm biodiversity and are environmentally safe; this accounts for their exploitation the world over for managing pests and they can also eliminate phytopathogens or act as phytohormones and thereby stimulate plant growth (Dara 2019; Mantzoukas and Eliopoulos 2020).

Novel strategies of IPM have been described in detail by Mantzoukas and Eliopoulos (2020) utilizing potential biocontrol manifestations of EF, particularly endophytic EF for achieving crop protection. In one report by Russo et al. (2021), corn was protected from attack by *Spodoptera frugiperda* (fall armyworm) by the incorporation of *B. bassiana* as an endophyte. It was observed that reproduction, growth, and food preference of *Spodoptera* were negatively impacted by endophytic colonization. The detrimental effects of herbivore attack are also negated by plant–fungus interaction (Benett et al. 2018). Thus, natural or deliberate inoculation of EF can provide dual protection to plants from the pest and phytopathogens because of the various ecological roles and can therefore be used in multiple ways in integrated pest management (Jaber and Ownley 2018). Additionally, EF can be used together with pesticides of plant origin to obtain synergistic pest mortality; for instance, significantly greater mortality of aphids was observed after the combined application of *Metarhizium anisopliae* and pyrethrum (Fernández-Grandon et al. 2020). Biological control of different insect pests predated by the larva of *Chrysoperla externa* was effectively achieved by EF (Mingotti Dias et al. 2020).

16.7 Production of EF on Mass Scale and Their Utilization for Pest Control

The mass production of EF is an essential need for its large-scale applicability as bioinsecticides in agriculture (Banu and Rajalakshmi 2014). Also, so that inoculum is produced in large amounts, the culture conditions and composition of the culture media should be optimized. Besides, it has been noted that EF vary in their nutritional requirements and in the presence of less favored carbohydrate sources their tolerance to UV light is stimulated due to the enhanced reactive oxygen scavengers (Huarte-Bonnet et al. 2019). Once they are produced on a mass scale, distinct methodologies are available for utilizing EF to control insect pests. One of the methods can be to store cadavers of insects for application in the fields or by biomass production of EF in submerged culture, where the latter can be made into mycelial pellets followed by treatment with desiccation protectants like 10% maltose or sucrose. The dried mycelium can also be microencapsulated in alginate or pregelatinized starch. When applied to agricultural fields, these particles form conidia if adequate moisture is present and infect insect pests. EF can also be produced on agro residues of rice, millet, wheat bran, cracked barley, peat soil, etc. Alternatively, the release of arthropod pests into agricultural fields after deliberately infecting them in the lab has also been practiced in several places. The capacity of *M. anisopliae* conidia to infect *Tenebrio molitor* was increased when the electric field was applied during conidiation in the culture. This strategy is expected to improve field performance; however, how electric field alters cellular mechanism needs further investigation (Rodriguez-Gomez et al. 2020).

16.8 EF Formulations and Their Efficient Use in the Control of Insects and Pests

EF are used as effective biocontrol agents and commercialized as microbial pesticides. As the demands of biopesticides need to be met, standardization of protocols for their mass production, formulation development, and effective delivery systems must be addressed (Álvarez et al. 2017). The insecticidal activities of *Beauveria bassiana*, *Metarhizium anisopliae* (a green muscardine fungus) (Abdel-Raheem 2020), and *Isaria fumosorosea* were studied on various food grains including *Triticum aestivum*, *Oryza sativa*, *Arachis hypogea*, and *Vicia faba*, when attacked during storage by khapra beetle, confused flour beetle, and Mediterranean flour beetle. It was found during the survey that the type of grain and the pests markedly led to variations in the effectiveness of EF (Mantzoukas et al. 2020). Often certain additives including carriers, spreaders, stickers, etc. when added to the conidia or propagules of EF improve potency, virulence, and shelf life. Also, such additives can impart enhancement in the tolerance to high-temperature fluctuations, which improves the adhesiveness or wettability of the propagules of EF. Sometimes EF, when used with plant extract, improve insecticidal capacity; for example extracts of neem and EF are used to control *Spodoptera frugiperda* (Hernandez-Trejo et al. 2019). Biological control of *Nilaparvata lugens* was made possible using *M. anisopliae* conidial suspensions (Atta et al. 2020).

According to Ramanujam et al. (2010), dipping the roots or plant sections or foliar spray on the leaves often improves the delivery and efficacy of EF-based formulations. Besides this, various formulations of EF are available including granular, wettable powder, water disposable powder, liquid (emulsifiable or suspension concentrates), and dust or baits that can be used for soil application. For the formulation, carrier selection is an important factor to consider, and often millet is used as it provides multiple surfaces. For instance, to control wireworms that mostly infect wheat plants, millet-based carriers of *Metarhizium robertsii* DWR 2009, and *Beauveria bassiana* GHA and ERL 836, were reported to be effective (Sharma et al. 2020). Enhanced thermotolerance of *B. bassiana* conidia was noted on whey permeate or by packing millet-based agar, when a comparative study was undertaken to understand thermotolerance of conidia on various substrates (based on millet, whey permeate, or quartered-strength Sabouraud dextrose agar (1/4 SDAY) (Kim et al. 2011). It is important to note that several abiotic (e.g., temperature, humidity, and chemical factors) and biotic (e.g., presence of other antagonistic microbial organisms) factors are important in deciding the effectiveness of granular formulations.

For enhanced stability and viability of conidia during storage conditions, certain absorbents like calcium chloride, silica gel, magnesium sulfate, white carbon, and sodium sulfate are incorporated in the formulations; however, the viability of conidia is best in white carbon. In oil-based formulations of EF, the suitability of methyl oleate (a wetting agent and emulsifier), and corn, cottonseed, and paraffin oils as carriers were studied by Kim et al. (2011). It was found that thermotolerance of

Isaria fumosorosea (strain SFP 198) was best in corn oil when it was subjected to a temperature of 50°C for two hours. During storage, oil-based suspensions exhibited improved conidial thermotolerance (Skinner et al. 2014). In the recent past, various types of EF formulations have been prepared for the effective control of pests. The details of these formulations are discussed in Table 16.2.

16.9 Formulation: Concerns and Challenges

When EF are incorporated into a formulation, they become easier to apply and are ecologically safe to use in the field. To ensure proper blending of the formulation in the spray tank, adjuvants are added. Even after a formulation is out in the market for use, it may be later withdrawn in case of its relatively lesser performance in the field as compared to the studies undertaken in the laboratory. Often this may also emerge when the mycoinsecticide was released without sufficient research and it becomes difficult to handle the EF during fermentation. Such orphaned biopesticides deter their use by agriculturists (Hynes and Boyetchko 2006; Behle and Birthisel 2014). Developing a biopesticide based on a microorganism may at times be challenging because of various reasons; The cost of development of a microorganism based biopesticide is higher as compared to a chemical insecticide, still the microorganism based biofertilizers are gaining prominence. To utilize their potential as a successful formulation, it is essential to improve their persistence (for a lengthened shelf life), virulence, and action spectrum (Behle and Birthisel 2014, Maina et al. 2018). Therefore following a realistic approach, say for example, the performance of formulation should be evaluated under challenging environmental conditions. Enhanced persistence, lengthened shelf life, virulence, and action spectrum would be certain criteria that need careful evaluation to develop a successful formulation (Maina et al. 2018).

16.10 The Recent Innovation in EF

Any innovation, if it complies with the criteria of patentability, can be patented in the name of the inventor at the patent office and upon grant can be commercialized by the inventor to monetize his invention. Patent protection is provided to both products and processes, and novel EF (both strains and species), their isolation methods, enhanced spores/propagules production, and production of secondary metabolites and compounds for potential biotechnological applications including formulation enhancement with improved physical and biochemical properties are protected by patents (Al-Ani 2019; Litwin et al. 2020). Novel species of EF that have been developed include *Isaria javanica*, *I. fumosorosea*, and *Nomuraea rileyi*. An example of a method that can be patented includes sporulation enhancement in *B. bassiana* when chitosan is used as a special media. Recent years have also

Table 16.2 EF formulations and their efficient use in the control of insects and pests

Granular formulations	Fungal baits	Wettable powders	Oil formulations	References
Granular propagules held together by inert clay or grounded residues from plants (grains of rice, barley, and crackled wheat, corn, and millet)	Attract insects in soil (termites, ants, black vine weevil); removes grasshoppers and locusts	Mixed with water, and sprayed onto the plants by hydraulic/ultralow volume applicators; effectiveness can be improved by certain additives	Based on vegetable oils, isoparaffinic hydrocarbon solvents, and mineral oils; methyl oleate (wetting agent and emulsifier); carriers are made up of corn, cottonseed, and paraffin oils	Skinner et al. 2014
Inoculum persistence; inoculum spreads in the desired area; ease of application	For EF, formulations based on granulated bait had satisfactory performance	Certain additives can improve the effectiveness of EF-based wettable powders	In dry conditions, oil formulations help in enhancing spore survival and allow the spores to spread on the leaf surface; also help in stimulating germination and penetrating the insect cuticle by breaching the waxy layer	Nagamoto (2012)
Control of Western flower thrips (WFT), *Frankliniella occidentalis*	Control of *Ixodes scapularis*	Control of the silverleaf whitefly, *Bemisia argentifolii*	Control of greenhouse whitefly (GWF) *Trialeurodes vaporariorum* nymphs	Kim et al. (2013); Williams et al. (2018); Skinner et al. (2012); Wraight et al. (2000)
Control of wireworms	Control of *Microcerotermes diversus* Silv.	Control of spittlebugs and whiteflies (wettable powders and technical unformulated products)	*Cydia pomonella* and *Cydia funebrana*	Sharma et al. (2020); Cheraghi et al. (2013); Mascarin et al. (2019); Herker et al. (2010)
Control of Western flower thrips *Frankliniella occidentalis*	Control of Cuban laurel thrips, *Gynaikothrips uzeli* Zimmerman (Thysanoptera: Phlaeothripidae and the	Control of *Plutella xylostella* by two *Isaria fumosorosea* conidial formulations—	Control of *Rhipicephalus sanguineus*	Zhang et al. (2019); Sánchez-Peña et al. (2011); Nian et al. (2015);

(continued)

Table 16.2 (continued)

Granular formulations	Fungal baits	Wettable powders	Oil formulations	References
	greenhouse whitefly, *Trialeurodes vaporariorum*)	wettable powder and oil-based formulation—combined with *Bacillus thuringiensis*		Barreto et al. (2016)
Control of *Polyphylla fullo* (June Beetle); control of *Riptortus pedestris* using the granular formulation of EF packed in novel screen bags	Control of *Drosophila suzukii* (Matsumura)		Control of tobacco spider mite	Erler and Ates (2015); Yousef et al. (2018); Wekesa (2004); Lee et al. (2016)

witnessed the development of several newer strains of *Beauveria* and *Metarhizium* sp. that have resulted in novel formulations. Besides this, it is possible to detect a strain in a commercial formulation in bulk soil by utilizing simple sequence repeats (SSR) specific to a particular strain (Reineke et al. 2014). However, scientifically validated high-quality research would be required to develop improvised formulations that can be widely used for biological control (Dubovskiy 2021).

16.11 EF-Based Nanoparticles in Biological Control

The popularity and prospects of nanotechnology from the agricultural perspective are tremendous. The fungal members synthesize enzymes either intracellularly, which can be released by mycelial disruption, or extracellularly, which are released in the medium. These extracellular enzymes are easy to handle and offer various advantages including enhanced tolerance to metals and are stable when used for nanoparticle (NP) fabrication (Netala et al. 2016). These attributes are helpful in the utilization of extracellular enzymes for biogenic or green synthesis of nanoparticles; the mycelial disruption step is eliminated (Gade et al. 2008). Besides, the metabolites released during fungal metabolism act as reducing and stabilizing agents during nanoparticle fabrication, and, by carefully tailoring the fungal metabolism, nanoparticles of desired dimensions and morphology can be obtained. However, upon subsequent addition of metal precursors to the fungal metabolites, under optimized conditions of temperature and pH, nanomaterials get synthesized (Gudikandula et al. 2017). The size and charge of nanomaterials are important to consider and various techniques like gel filtration, filtration, ultracentrifugation, and dialysis can sort nanoparticles based on size (Yahyaei and Pourali 2019). For a nanomaterial to be used as a potential antimicrobial in agriculture, it should be biocompatible, possess reduced toxicity, and also have enhanced stability with desirable physicochemical properties (Guilger-Casagrande and de Lima 2019). Despite the existence of the various metal nanoparticles, silver-based nanoparticles display broad-spectrum antimicrobial activity (Loo et al. 2018; Gupta et al. 2017). The evaluation of virulence and pathogenicity of EF-based silver nanoparticles (AgNPs) was carried out by Santos et al. (2021) against the insect model *Tenebrio molitor*; however, upon addition of 1 mM silver nitrate solution, out of 16, only 15 of them yielded silver nanoparticles. Nanoparticle fabrication using extract of EF can be used for the green synthesis of AgNP and applied as biological control. In a study, Yosri et al. (2018) used irradiated, unirradiated, and a combination of irradiated *M. anisopliae* and mycosynthesized titanium nanoparticles (TiNPs) to study its larvicidal effect against the pest *Galleria mellonella*. From their studies, they concluded that a combination of irradiated *M. anisopliae* and mycosynthesized titanium nanoparticles (TiNPs) showed synergistic larvicidal activity. For the management of *Spodoptera litura (Fabricius)*, Iron oxide nanoparticles (FeONPs) were fabricated using *Beauveria brongniartii* and it was found to possess enormous potential for destroying second instar larva (Xu et al. 2020). Silver nanoparticles are toxic and they can also lead to DNA and protein damage (Panpatte and Jhala 2019). In yet another study, it was reported that the effectiveness of *Nomuraea rileyi* (F.) Samson-based nanoparticles against *Spodoptera litura* was enhanced, when the nanoparticles were coated with chitosan as compared to the non-coated ones (Chandra et al. 2013). Vandergheynst et al. (2007) reported enhanced control of mosquito larva when hydrophobic silica nanoparticles of *Lagenidium giganteum* were added. Similarly, Khooshe-Bast et al. (2016) fabricated and utilized zinc oxide nanoparticles (ZnONPs) and *Beauveria bassiana* TS11 to kill adults of *Trialeurodes vaporariorum* pests (Westwood 1856). This brings us to the conclusion that the use of mycogenic nanomaterials can enhance the yield of crops, and help in agriculture that involves nanoformulations for increased crop yield and improvement. Although still in the incipient stage, it would not be wrong to say that these myco-fabricated nano-products hold immense potential that needs to be carefully harnessed (Rao et al. 2017).

16.12 Conclusion

From the foregoing account, it is clear that EF can serve as an effective alternative to chemical pesticides; however, novel and effective strains of EF should be developed and made into formulations. The guidelines for the registration of mycoinsecticides vary in different jurisdictions. The manufacturers of EF-based mycoinsecticidal formulations should use propagules and data from the labs and research stations that comply with Good Laboratory Practices (GLP). Enhancement in the efficacy of EF and lowering the cost of production can help in reducing reliance on chemical pesticides. Molecular biology has been instrumental in understanding the interactions between EF and insects at different stages and dual RNA-seq technology can prove helpful in understanding these interactions. Generalists like *Metarhizium* underwent protein family expansion, extensive gene duplication, and horizontal gene transfer for interaction with larger insect orders, while specialists like *M. acridum* retained existing proteins useful for interaction with the locust. The clarity of the mechanisms underlying the regulation of the cell-dimorphic switch during the colonization of EF in the hemocoel and the functioning of effector-like proteins in EF–arthropod interactions is important for developing pocket-friendly applications. As the acceptance of Genetically Modified Organism (GMOs) are increasing worldwide, while undertaking experiments related to recombinant DNA technology, emphasis should be on developing selectable marker-free transformants, or, even if used, the selectable marker/s should be removed from the transformants after transgenic development. It would also be important to address the problems associated with low virulence and inconsistent performance of EF-based biopesticides together with the biosafety concerns. Nanotechnology can revolutionize agriculture with its many advantages attributable to nano size. Research institutes should devise means to lower the cost incurred during fermentation operations and emphasize technology development to ensure better delivery of EF for the adoption of EF-based biopesticides in biological control.

Acknowledgments The author would like to thank the University Department of Botany, T.M. Bhagalpur University, for providing the necessary infrastructure to carry out the study.

Funding: This study received no grant from any funding agency in the public, commercial, or not-for-profit sectors.

Conflict of Interest No conflict of interest to declare.

References

Abdel-Raheem M (2020) Isolation, mass production and application of Entomopathogenic Fungi for insect pests control. In: Cottage industry of biocontrol agents and their applications. Springer, Cham, pp 231–251

Al-Ani LK (2019) Entomopathogenic Fungi in IP landscape. In: Intellectual property issues in microbiology. Springer, Singapore, pp 223–238

Álvarez SP, Tapia MA, Pérez KI, Guerrero AM (2017) Agriculture applications of entomopathogenic fungi using nanotechnology. In: Fungal Nanotechnology. Springer, Cham, pp 35–53

Arthur FH, Campbell JF (2008) Distribution and efficacy of pyrethrin aerosol to control Tribolium confusum (Coleoptera: Tenebrionidae) in food storage facilities. J Stored Prod Res 44(1):58–64

Athanassiou CG, Rani PU, Kavallieratos NG (2014) The use of plant extracts for stored product protection. *InAdvances in plant biopesticides*. Springer, New Delhi, pp 131–147

Batta YA, Kavallieratos NG (2018) The use of entomopathogenic fungi for the control of stored-grain insects. Int J Pest Manag 64(1):77–87

Behie SW, Zelisko PM, Bidochka MJ (2012) Endophytic insect-parasitic fungi translocate nitrogen directly from insects to plants. Science 336(6088):1576–1577

Behle R, Birthisel T (2014) Formulations of entomopathogens as bioinsecticides. In: Mass production of beneficial organisms. Academic Press, pp 483–517

Behle RW, Compton DL, Laszlo JA, Shapiro-Ilan DI (2009) Evaluation of soyscreen in an oil-based formulation for UV protection of *Beauveria bassiana* conidia. J Econ Entomol 102(5):1759–1766

Butt TM, Coates CJ, Dubovskiy IM, Ratcliffe NA (2016) Entomopathogenic fungi: new insights into host–pathogen interactions. In: Advances in genetics, vol 94. Academic Press, pp 307–364

Chakroun M, Banyuls N, Bel Y, Escriche B, Ferré J (2016) Bacterial vegetative insecticidal proteins (Vip) from entomopathogenic bacteria. Microbiol Mol Biol Rev 80(2):329–350

Chandra JH, Raj LA, Namasivayam SKR, Bharani RA (2013) Improved pesticidal activity of fungal metabolite from Nomureae rileyi with chitosan nanoparticles. In: International conference on Advanced Nanomaterials & Emerging Engineering Technologies, pp 387–390

Charnley AK (2003) Fungal pathogens of insects: cuticle degrading enzymes and toxins. Adv Bot Res 40:241–232

Cheraghi A, Habibpour B, Mossadegh MS (2013) Application of bait treated with the entomopathogenic fungus *Metarhizium anisopliae* (Metsch.) Sorokin for the control of *Microcerotermes diversus* Silv. Psyche 2013:1

Chu ZJ, Wang YJ, Ying SH, Wang XW, Feng MG (2016) Genome-wide host-pathogen interaction unveiled by transcriptomic response of diamondback moth to fungal infection. PLoS One 11(4): e0152908

Clarkson JM, Charnley AK (1996) New insights into the mechanisms of fungal pathogenesis in insects. Trends Microbiol 4(5):197–203

da Silva RA, Quintela ED, Mascarin GM, Pedrini N, Lião LM, Ferri PH (2015) Unveiling chemical defense in the rice stalk stink bug against the entomopathogenic fungus *Metarhizium anisopliae*. J Invertebr Pathol 127:93–100

Dara SK (2019) Interactions of entomopathogens with other pest management options. In: Microbes for sustainable insect pest management. Springer, Cham, pp 299–316

de Bekker C, Smith PB, Patterson AD, Hughes DP (2013) Metabolomics reveals the heterogeneous secretome of two entomopathogenic fungi to ex vivo cultured insect tissues. PLoS One 8(8): e70609

Driver F, Milner RJ, Trueman JW (2000) A taxonomic revision of *Metarhizium* based on a phylogenetic analysis of rDNA sequence data. Mycol Res 104(2):134–150

Dubovskiy IM (2021) Host–pathogen interactions: insects vs. fungi. J Fungi 7:162

Dubovskiy IM, Whitten MA, Kryukov VY, Yaroslavtseva ON, Grizanova EV, Greig C, Mukherjee K, Vilcinskas A, Mitkovets PV, Glupov VV, Butt TM (2013) More than a colour change: insect melanism, disease resistance and fecundity. Proc Royal Soc B Biol Sci 280(1763):20130584

Edde PA (2012) A review of the biology and control of Rhyzopertha dominica (F.) the lesser grain borer. J Stored Prod Res 48:1–18

Erler F, Ates AO (2015) Potential of two entomopathogenic fungi, *Beauveria bassiana* and *Metarhizium anisopliae* (Coleoptera: Scarabaeidae), as biological control agents against the June beetle. J Insect Sci 15(1):44

Evison SE, Gallagher JD, Thompson JJ, Siva-Jothy MT, Armitage SA (2017) Cuticular colour reflects underlying architecture and is affected by a limiting resource. J Insect Physiol 98:7–13

Fan Y, Fang W, Guo S, Pei X, Zhang Y, Xiao Y, Li D, Jin K, Bidochka MJ, Pei Y (2007) Increased insect virulence in *Beauveria bassiana* strains overexpressing an engineered chitinase. Appl Env Microbiol 73(1):295–302

Fan Y, Borovsky D, Hawkings C, Ortiz-Urquiza A, Keyhani NO (2012) Exploiting host molecules to augment mycoinsecticide virulence. Nat Biotechnol 30(1):35

Fang W, St. Leger RJ (2012) Enhanced UV resistance and improved killing of malaria mosquitoes by photolyase transgenic entomopathogenic fungi. PLoS One 7:e43069

Fang W, Leng B, Xiao Y, Jin K, Ma J, Fan Y, Feng J, Yang X, Zhang Y, Pei Y (2005) Cloning of *Beauveria bassiana* chitinase gene Bbchit1 and its application to improve fungal strain virulence. Appl Environ Microbiol 71(1):363–370

Fang W, Feng J, Fan Y, Zhang Y, Bidochka MJ, Leger RJS, Pei Y (2009) Expressing a fusion protein with protease and chitinase activities increases the virulence of the insect pathogen Beauveria bassiana. J Invert Pathol 102(2):155–159

Fernández-Grandon GM, Harte SJ, Ewany J, Bray D, Stevenson PC (2020) Additive effect of botanical insecticide and entomopathogenic fungi on pest mortality and the behavioral response of its natural enemy. Plan Theory 9(2):173

Freimoser FM, Screen S, Bagga S, Hu G, St Leger RJ (2003a) Expressed sequence tag (EST) analysis of two subspecies of *Metarhizium anisopliae* reveals a plethora of secreted proteins with potential activity in insect hosts. Microbiology 149:239–247

Freimoser FM, Screen S, Hu G, Leger RS (2003b) EST analysis of genes expressed by the zygomycete pathogen *Conidiobolus coronatus* during growth on insect cuticle. Microbiology 149:1893–1900

Gade AK, Bonde PP, Ingle AP, Marcato PD, Duran N, Rai MK (2008) Exploitation of *Aspergillus niger* for synthesis of silver nanoparticles. J Biobaased Mater Bioenergy 2(3):243–247

Gibson DM, Donzelli BG, Krasnoff SB, Keyhani NO (2014) Discovering the secondary metabolite potential encoded within entomopathogenic fungi. Nat Prod Rep 31(10):1287–1305

Gillespie JP, Burnett C, Charnley AK (2000) The immune response of the desert locust *Schistocerca gregaria* during mycosis of the entomopathogenic fungus, *Metarhizium anisopliae* var acridum. J Insect Physiol 46(4):429–437

Golebiowski M, Cerkowniak M, Urbanek A, Dawgul M, Kamysz W, Bogus MI, Sosnowska D, Stepnowski P (2014a) Antimicrobial activity of untypical lipid compounds in the cuticular and internal 584 lipids of four fly species. J Appl Microbiol 116:269–287

Golebiowski M, Urbanek A, Oleszczak A, Dawgul M, Kamysz W, Bogus MI, Stepnowski P (2014b) The antifungal activity of fatty acids of all stages of *Sarcophaga carnaria* L. (Diptera: Sarcophagidae). Microbiol Res 169:279–286

Golebiowski M, Cerkowniak M, Urbanek A, Dawgul M, Kamysz W, Bogus MI, Stepnowski P (2015) Identification and antifungal activity of novel organic compounds found in cuticular and internal lipids of medically important flies. Microbiol Res 170:213–222

Gongora B (2004) Transformacion de Beauveria bassiana cepa Bb9112 con los genes de la proteina verde fluorescente y la proteasa pr1A de *Metarhizium anisopliae*. Sociedad Botanica de Entomologia. Bogota (Colombia) 30:1–5

Gudikandula K, Vadapally P, Charya MS (2017) Biogenic synthesis of silver nanoparticles from white rot fungi: their characterization and antibacterial studies. OpenNano 2:64–78

Guilger-Casagrande M, de Lima R (2019) Synthesis of silver nanoparticles mediated by fungi: a review. Front Bioeng Biotechnol 7:287

Gupta RK, Kumar V, Gundampati RK, Malviya M, Hasan SH, Jagannadham MV (2017) Biosynthesis of silver nanoparticles from the novel strain of *Streptomyces* Sp. BHUMBU-80 with highly efficient electroanalytical detection of hydrogen peroxide and antibacterial activity. J Environ Chem Eng 5(6):5624–5635

Herker M, Kleefeldt U, Stephan D (2010) Laboratory experiments with entomopathogenic fungi on artificial hideouts for biocontrol of *Cydia pomonella* and *Cydia funebrana*. In: Ecofruit. 14th international conference on organic fruit-growing. Proceedings for the conference, Hohenheim, Germany. Fördergemeinschaft Ökologischer Obstbau eV (FÖKO), pp 149–155

Hernandez-Trejo A, Estrada-Drouaillet B, López-Santillán JA, Rios-Velasco C, Varela-Fuentes SE, Rodríguez-Herrera R, Osorio-Hernández E (2019) In vitro evaluation of native entomopathogenic fungi and neem (*Azadiractha indica*) extracts on *Spodoptera frugiperda*. Phyton 88(1):47

Huarte-Bonnet C, Kumar S, Saparrat MC, Girotti JR, Santana M, Hallsworth JE, Pedrini N (2018) Insights into hydrocarbon assimilation by eurotialean and hypocrealean fungi: roles for CYP52 and CYP53 clans of cytochrome P450 genes. Appl Biochem Biotechnol 184(3):1047–1060

Huarte-Bonnet C, Mannino MC, Pedrini N (2019) Oxidative stress in Entomopathogenic fungi and its potential role on Mycoinsecticide enhancement. In: Microbes for sustainable insect pest management. Springer, Cham, pp 197–205

Hussain A (2018) Reprogramming the virulence: insect defense molecules navigating the epigenetic landscape of *Metarhizium robertsii*. Virulence 9(1):447–449

Hynes RK, Boyetchko SM (2006) Research initiatives in the art and science of biopesticide formulations. Soil Biol Biochem 38(4):845–849

Jaber LR, Ownley BH (2018) Can we use entomopathogenic fungi as endophytes for dual biological control of insect pests and plant pathogens? Biol Control 116:36–45

Keyhani NO (2018) Lipid biology in fungal stress and virulence: entomopathogenic fungi. Fungal Biol 122(6):420–429

Khooshe-Bast Z, Sahebzadeh N, Ghaffari-Moghaddam M, Mirshekar A (2016) Insecticidal effects of zinc oxide nanoparticles and *Beauveria bassiana* TS11 on *Trialeurodes vaporariorum* (Westwood, 1856)(Hemiptera: Aleyrodidae). Acta Agric Slovenica 107(2):299–309

Kim JS, Je YH, Woo EO, Park JS (2011) Persistence of *Isaria fumosorosea* (Hypocreales: Cordycipitaceae) SFP-198 conidia in corn oil based suspension. Mycopathologia 171:67–75

Kim JS, Je YH, Skinner M, Parker BL (2013) An oil-based formulation of Isaria fumosorosea blastospores for management of greenhouse whitefly *Trialeurodes vaporariorum* (Homoptera: Aleyrodidae). Pest Manag Sci 69(5):576–581

Kumar A (2022) Microbial biocontrol: food security and post harvest management, vol 2. Springer Nature, Switzerland

Lee SJ, Kim S, Skinner M, Parker BL, Kim JS (2016) Screen bag formulation of *Beauveria* and *Metarhizium* granules to manage *Riptortus pedestris* (Hemiptera: Alydidae). J Asia-Pacific Entomol 19(3):887–892

Liao X, Lu HL, Fang W, Leger RJS (2014) Overexpression of a *Metarhizium robertsii* HSP25 gene increases thermotolerance and survival in soil. Appl Microbiol Biotechnol 98(2):777–783

Litwin A, Nowak M, Różalska S (2020) Entomopathogenic fungi: unconventional applications. Rev Environ Sci Biotechnol 19:1–20

Liu H, Xie L, Wang J, Guo Q, Yang S, Liang P, Wang C, Lin M, Xu Y, Zhang L (2017) The stress-responsive and host-oriented role of nonribosomal peptide synthetases in an entomopathogenic fungus, *Beauveria bassiana*. J Microbiol Biotechnol 27(3):439–449

Loo YY, Rukayadi Y, Nor-Khaizura MAR, Kuan CH, Chieng BW, Nishibuchi M, Radu S (2018) In vitro antimicrobial activity of green synthesized silver nanoparticles against selected gram-negative foodborne pathogens. Front Microbiol 9:1555

Lord JC (2010) Dietary stress increases the susceptibility of *Tribolium castaneum* to *Beauveria bassiana*. J Econ Entomol 103:1542–1546

Maina UM, Galadima IB, Gambo FM, Zakaria D (2018) A review on the use of entomopathogenic fungi in the management of insect pests of field crops. J Entomol Zool Stud 6(1):27–32

Mantilla JG, Galeano NF, Gaitan AL, Cristancho MA, Keyhani NO, Gongora CE (2012) Transcriptome analysis of the entomopathogenic fungus *Beauveria bassiana* grown on cuticular extracts of the coffee berry borer (Hypothenemus hampei). Microbiology 158:1826–1842

Mantzoukas S, Eliopoulos PA (2020) Endophytic entomopathogenic fungi: a valuable biological control tool against plant pests. Appl Sci 10(1):360

Mantzoukas S, Lagogiannis I, Karmakolia K, Rodi A, Gazepi M, Eliopoulos PA (2020) The effect of grain type on virulence of Entomopathogenic fungi against stored product pests. Appl Sci 10(8):2970

Mascarin GM, Lopes RB, Delalibera Í Jr, Fernandes ÉKK, Luz C, Faria M (2019) Current status and perspectives of fungal entomopathogens used for microbial control of arthropod pests in Brazil. J Invertebr Pathol 165:46–53

Mingotti Dias P, de Souza LE, Amorim Pessoa LG, Reis Devoz GL, Bárbaro Barbosa Junior G, Macali Werner A, Navarrete AA, Teodoro PE (2020) Selectivity of entomopathogenic fungi to *Chrysoperla externa* (Neuroptera: Chrysopidae). Insects 11(10):716

Mukherjee K, Vilcinskas A (2018) The entomopathogenic fungus *Metarhizium robertsii* communicates with the insect host Galleria mellonella during infection. Virulence 9(1):402–413
Nagamoto NS (2012) Carrying and effect of granulated baits formulated with entomopathogenic fungi among Atta sexdens rubropilosa colonies (Hymenoptera: Formicidae). Sociobiology 59(3):681–689
Netala VR, Bethu MS, Pushpalatha B, Baki VB, Aishwarya S, Rao JV, Tartte V (2016) Biogenesis of silver nanoparticles using endophytic fungus *Pestalotiopsis microspora* and evaluation of their antioxidant and anticancer activities. Int J Nanomedicine 11:5683
Nian XG, He YR, Lu LH, Zhao R (2015) Evaluation of alternative *Plutella xylostella* control by two Isaria fumosorosea conidial formulations–oil-based formulation and wettable powder–combined with *Bacillus thuringiensis*. Pest Manag Sci 71(12):1675–1684
Opisa S, Du Plessis H, Akutse KS, Fiaboe KKM, Ekesi S (2018) Effects of Entomopathogenic fungi and *Bacillus thuringiensis*-based biopesticides on *Spoladea recurvalis* (Lepidoptera: Crambidae). J Appl Entomol 142(6):617–626
Oppert B, Guedes RN, Aikins MJ, Perkin L, Chen Z, Phillips TW, Zhu KY, Opit GP, Hoon K, Sun Y, Meredith G (2015) Genes related to mitochondrial functions are differentially expressed in phosphine-resistant and -susceptible *Tribolium castaneum*. BMC Genomics 16:968
Ortiz-Urquiza A, Keyhani NO (2013) Action on the surface: entomopathogenic fungi versus the insect cuticle. Insects 4:357–374
Panpatte DG, Jhala YK (eds) (2019) Nanotechnology for agriculture: crop production & protection. Springer Nature
Pava-Ripoll M, Posada FJ, Momen B, Wang C, Leger RS (2008) Increased pathogenicity against coffee berry borer, *Hypothenemus hampei* (Coleoptera: Curculionidae) by *Metarhizium anisopliae* expressing the scorpion toxin (AaIT) gene. J Invertebr Pathol 99(2):220–226
Pedrini N (2018) Molecular interactions between entomopathogenic fungi (Hypocreales) and their insect host: perspectives from stressful cuticle and hemolymph battlefields and the potential of dual RNA sequencing for future studies. Fungal Biol 122(6):538–545
Pedrini N, Juárez MP (2008) Entomopathogenic fungi and their host cuticle. In: Encyclopedia of entomology, 2nd ed. Springer-Verlag, Heidelberg, pp 1333–1336
Pedrini N, Crespo R, Juárez MP (2007) Biochemistry of insect epicuticle degradation by entomopathogenic fungi. Comp Biochem Physiol Part C Toxicol Pharmacol 146(1–2):124–137
Pedrini N, Ortiz-Urquiza A, Huarte-Bonnet C, Fan Y, Juárez MP, Keyhani NO (2015) Tenebrionid secretions and a fungal benzoquinone oxidoreductase form competing components of an arms race between a host and pathogen. Proc Nat Acad Sci 112(28):E3651–E3660
Peng G, Jin K, Liu Y, Xia Y (2015) Enhancing the utilization of host trehalose by fungal trehalase improves the virulence of fungal insecticide. Appl Microbiol Biotechnol 99(20):8611–8618
Phillips TW, Throne JE (2010) Biorational approaches to managing stored-product insects. Annu Rev Entomol 55:375–397
Qasim M, Islam SU, Islam W, Noman A, Khan KA, Hafeez M, Hussain D, Dash CK, Bamisile BS, Akutse KS, Rizwan M (2020) Characterization of mycotoxins from entomopathogenic fungi (*Cordyceps fumosorosea*) and their toxic effects to the development of asian citrus psyllid reared on healthy and diseased citrus plants. Toxicon 188:39–47
Qin Y, Ying SH, Chen Y, Shen ZC, Feng MG (2010) Integration of insecticidal protein Vip3Aa1 into *Beauveria bassiana* enhances fungal virulence to *Spodoptera litura* larvae by cuticle and per os infection. Appl Environ Microbiol 76(14):4611–4618
Ramanujam B, Prasad RD, Sriram S, Rangeswaran R (2010) Mass production, formulation, quality control and delivery of *Trichoderma* for plant disease management. J Plant Prot Sci 2(2):1–8
Rao M, Jha B, Jha AK, Prasad K (2017) Fungal nanotechnology: a pandora to agricultural science and engineering. In: Fungal nanotechnology. Springer, Cham, pp 1–33
Rath AC (2000) The use of entomopathogenic fungi for control of termites. Biocontrol Sci Tech 10: 563–581
Reineke A, Bischoff-Schaefer M, Rondot Y, Galidevara S, Hirsch J, Devi KU (2014) Microsatellite markers to monitor a commercialized isolate of the entomopathogenic fungus Beauveria

bassiana in different environments: technical validation and first applications. Biol Control 70: 1–8
Rodrigues J, Borges PR, Fernandes ÉKK, Luz C (2019) Activity of additives and their effect in formulations of *Metarhizium anisopliae* sl IP 46 against *Aedes aegypti* adults and on post mortem conidiogenesis. Acta Trop 193:192–198
Rodriguez-Gomez D, Garate-Velez L, Morales-Piña AL, Gutiérrez-Rojas M, Velasco-Alvarez N (2020) Increase in the mortality of the entomopathogenic fungus *Metarhizium anisopliae* due to the application of an electric field during conidiation. Rev Mex de Ing Quím 19:123–133
Rohlfs M, Churchill AC (2011) Fungal secondary metabolites as modulators of interactions with insects and other arthropods. Fungal Genet Biol 48(1):23–34
Rossi E, Cosimi S, Loni A (2010) Insecticide resistance in Italian populations of Tribolium flour beetles. Bull Insectol 63:251–258
Russo ML, Jaber LR, Scorsetti AC, Vianna F, Cabello MN, Pelizza SA (2021) Effect of entomopathogenic fungi introduced as corn endophytes on the development, reproduction, and food preference of the invasive fall armyworm *Spodoptera frugiperda*. J Pest Sci 94(3): 859–870
Sánchez-Peña SR, Lara JSJ, Medina RF (2011) Occurrence of entomopathogenic fungi from agricultural and natural ecosystems in Saltillo, Mexico, and their virulence towards thrips and whiteflies. J Insect Sci 11(1):1
Sani I, Ismail SI, Abdullah S, Jalinas J, Jamian S, Saad N (2020) A review of the biology and control of whitefly, *Bemisia tabaci* (Hemiptera: Aleyrodidae), with special reference to biological control using entomopathogenic fungi. Insects 11(9):619
Santi L, da Silva WOB, Berger M, Guimaraes JA, Schrank A, Vainstein MH (2010) Conidial surface 744 proteins of *Metarhizium anisopliae*: source of activities related with toxic effects, host penetration and pathogenesis. Toxicon 55:874–880
Santos TS, Passos EMD, Seabra MG, Souto EB, Severino P, Mendonça MDC (2021) Entomopathogenic fungi biomass production and extracellular biosynthesis of silver nanoparticles for bioinsecticide action. Appl Sci 11(6):2465
Shang Y, Duan Z, Huang W, Gao Q, Wang C (2012) Improving UV resistance and virulence of *Beauveria bassiana* by genetic engineering with an exogenous tyrosinase gene. J Invertebr Pathol 109(1):105–109
Sharma A, Jaronski S, Reddy GV (2020) Impact of granular carriers to improve the efficacy of entomopathogenic fungi against wireworms in spring wheat. J Pest Sci 93(1):275–290
Sindhu SS, Rakshiya YS, Verna MK (2011) Biological control of termites by antagonistic soil microorganisms. In: Singh A, Parmar N, Kuhad RC (eds) Bioaugmentation, biostimulation and biocontrol. Springer, Berlin, pp 261–309
Skinner M, Gouli S, Frank CE, Parker BL, Kim JS (2012) Management of *Frankliniella occidentalis* (Thysanoptera: Thripidae) with granular formulations of entomopathogenic fungi. Biol Control 63(3):246–252
Skinner M, Parker BL, Kim JS (2014) Role of entomopathogenic fungi in integrated pest management. In: Integrated pest management. Academic Press, pp 169–191
Skrobek A, Shah FA, Butt TM (2008) Destruxin production by the entomogenous fungus *Metarhizium anisopliae* in insects and factors influencing their degradation. BioControl 53(2): 361–373
St. Leger RJ, Joshi L, Bidochka MJ, Roberts DW (1996) Construction of an improved mycoinsecticide over-expressing a toxic protease. Proc. Natl. Acad. Sci. USA 93:6349–6354
Süssmuth R, Müller J, Von Döhren H, Molnár I (2011) Fungal cyclooligomer depsipeptides: from classical biochemistry to combinatorial biosynthesis. Nat Prod Rep 28(1):99–124
Trienens M, Rohlfs M (2012) Insect–fungus interference competition–the potential role of global secondary metabolite regulation, pathway-specific mycotoxin expression and formation of oxylipins. Fungal Ecol 5(2):191–199

Tseng MN, Chung PC, Tzean SS (2011) Enhancing the stress tolerance and virulence of an entomopathogen by metabolic engineering of dihydroxynaphthalene melanin biosynthesis genes. Appl Env Microbiol 77(13):4508–4519

Vandergheynst J, Scher H, Guo HY, Schultz D (2007) Water-in-oil emulsions that improve the storage and delivery of the biolarvacide *Lagenidium giganteum*. BioControl 52(2):207–229

Vilcinskas A (2010) Coevolution between pathogen-derived proteinases and proteinase inhibitors of host insects. Virulence 1(3):206–214

Wang C, St Leger RJ (2007a) A scorpion neurotoxin increases the potency of a fungal insecticide. Nat Biotechnol 25(12):1455–1456

Wang C, St Leger RJ (2007b) The MAD1 adhesin of *Metarhizium anisopliae* links adhesion with blastospore production and virulence to insects, and the MAD2 adhesin enables attachment to plants. Eukaryot Cell 6(5):808–816

Wang C, Wang S (2017) Insect pathogenic fungi: genomics, molecular interactions, and genetic improvements. Annu Rev Entomol 62:73–90

Wang JB, Leger RS, Wang C (2016) Advances in genomics of entomopathogenic fungi. Adv Gen 94:67–105

Wekesa VW (2004) Evaluation of pathogenic fungi *Beauveria bassiana* and *Metarhizium anisopliae* for the control of tobacco spider mite, *Tetranychus evansi* Baker & Pritchard (Acarina: Tetranychidae) infesting tomatoes. Doctoral dissertation, Jomo Kenyatta University of Agriculture and Technology

Westermann AJ, Gorski SA, Vogel J (2012) Dual RNA-seq of pathogen and host. Nat Rev Microbiol 10(9):618–630

Williams SC, Stafford KC III, Molaei G, Linske MA (2018) Integrated control of nymphal Ixodes scapularis: effectiveness of white-tailed deer reduction, the entomopathogenic fungus *Metarhizium anisopliae*, and fipronil-based rodent bait boxes. Vector-Borne Zoonotic Dis 18(1):55–64

Wraight SP, Carruthers RI, Jaronski ST, Bradley CA, Garza CJ, Galaini-Wraight S (2000) Evaluation of the entomopathogenic fungi *Beauveria bassiana* and *Paecilomyces fumosoroseus* for microbial control of the silverleaf whitefly, *Bemisia argentifolii*. Biol Control 17(3):203–217

Xie M, Zhang YJ, Zhai XM, Zhao JJ, Peng DL, Wu G (2015) Expression of a scorpion toxin gene BmKit enhances the virulence of *Lecanicillium lecanii* against aphids. J Pest Sci 88(3):637–644

Xu YJ, Luo F, Li B, Shang Y, Wang C (2016) Metabolic conservation and diversification of *Metarhizium* species correlate with fungal host-specificity. Front Microbiol 7:2020

Xu J, Zhang K, Cuthbertson AG, Du C, Ali S (2020) Toxicity and biological effects of *Beauveria brongniartii* Fe0 nanoparticles against *Spodoptera litura* (Fabricius). Insects 11(12):895

Yahyaei B, Pourali P (2019) One step conjugation of some chemotherapeutic drugs to the biologically produced gold nanoparticles and assessment of their anticancer effects. Sci Rep 9(1):1–15

Yassine H, Kamareddine L, Osta MA (2012) The mosquito melanization response is implicated in defense against the entomopathogenic fungus *Beauveria bassiana*. PLoS Pathog 8:e1003029

Yosri M, Abdel-Aziz MM, Sayed RM (2018) Larvicidal potential of irradiated myco-insecticide from *Metarhizium anisopliae* and larvicidal synergistic effect with its mycosynthesized titanium nanoparticles (TiNPs). J Radiat Res Appl Sci 11(4):328–334

Yousef M, Aranda-Valera E, Quesada-Moraga E (2018) Lure-and-infect and lure-and-kill devices based on *Metarhizium brunneum* for spotted wing *Drosophila* control. J Pest Sci 91(1):227–235

Zhang S, Widemann E, Bernard G, Lesot A, Pinot F, Pedrini N, Keyhani NO (2012) CYP52X1, representing the new cytochrome P450 subfamily, displays fatty acid hydroxylase activity and contributes to virulence and growth on insect cuticular substrates in the entomopathogenic fungus Beauveria bassiana. J Biol Chem 287(16):13477–13486

Zhang J, Lu AR, Kong LL, Zhang QL, Ling EJ (2014) Functional analysis of insect molting fluid proteins on the protection and regulation of ecdysis. J Biol Chem 289:35891–35906

Zhang X, Lei Z, Reitz SR, Wu S, Gao Y (2019) Laboratory and greenhouse evaluation of a granular formulation of *Beauveria bassiana* for control of western flower thrips, Frankliniella occidentalis. Insects 10(2):58

Zhao H, Xu C, Lu HL, Chen X, Leger RJS, Fang W (2014) Host-to-pathogen gene transfer facilitated infection of insects by a pathogenic fungus. PLoS Pathog 10(4):e1004009

Zhao H, Lovett B, Fang W (2016) Genetically engineering entomopathogenic fungi. In: Advances in genetics, vol 94. Academic Press, pp 137–163